AVIAN ANATOMY

Textbook and Colour Atlas

5m Publishing

AVIAN ANATOMY

Textbook and Colour Atlas

Editors

Horst E. König

Rüdiger Korbel

Hans-Georg Liebich

With contributions from

Hermann Bragulla, Klaus-Dieter Budras, Ana Carretero, Gerhard Forstenpointner, Corinna Klupiec, Johann Maierl, Maren Meiners, Ivan Misek, Christoph Mülling, Marc Navarro, Alexander Probst, Sven Reese, Jesus Ruberte, Ingrid Walter, Gerald Weissengruber and Grammata Zengerling

Translated by

Corinna Klupiec

5m Publishing

First published 2009
This edition published by 5m Publishing 2016

Authorized translation of the second German language edition of König, Horst Erich/ Korbel, Rüdiger/ Liebich, Hans-Georg, Anatomie der Vögel © 2009 by Schattauer GmbH, Stuttgart/Germany

Corinna Klupiec asserts her right to be known as the translator of this work

Published by
5M Publishing Ltd,
Benchmark House,
8 Smithy Wood Drive,
Sheffield, S35 1QN, UK
Tel: +44 (0) 1234 81 81 80
www.5mpublishing.com

A Catalogue record for this book is available from the British Library

ISBN 978-1-910455-60-9

Book layout by Keystroke, Neville Lodge, Tettenhall, Wolverhampton

Printed by Replika Press Pvt Ltd, India

Photos and illustrations by Schattauer

Important note:
Medicine is an ever-changing science, so the contents of this publication, especially recommendations concerning diagnostic and therapeutic procedures, can only give an account of the knowledge at the time of publication. While utmost care has been taken to ensure that all specifications regarding drug selection and dosage and treatment options are accurate, readers are urged to review the product information sheet and any relevant material supplied by the manufacturer, and, in case of doubt, to consult a specialist. From both an editorial and public interest perspective, the publisher welcomes notification of possible inconsistencies. The ultimate responsibility for any diagnostic or therapeutic application lies with the reader.

No special reference is made to registered names, proprietary names, trademarks, etc. in this publication. The appearance of a name without designation as proprietary does not imply that it is exempt from the relevant protective laws and regulations and therefore free for general use.

Editors

O. Univ. Prof. Dr. habil h.c Horst Erich König
Institut für Anatomie, Veterinärmedizinische Universität Wien, Veterinärplatz 1, A-1210 Wien
Horst.Koenig@vu-wien.ac.at

Univ.-Prof. Dr. habil. Rüdiger Korbel, Dipl. ECAMS
Klinik für Vögel, Ludwig-Maximilians-Universität München,
Sonnenstraße 18, D-85764 Oberschleißheim korbel@lmu.de

Univ.-Prof. Dr. h.c. mult. Hans-Georg Liebich
Institut für Tieranatomie, Ludwig-Maximilians-Universität München,
Veterinärstraße 13, D-80539 München
h-g.liebich@anat.vetmed.uni-muenchen.de

Contents

Foreword

The publication of the 2nd German edition of *Avian Anatomy* was met with great enthusiasm and acclaim by students and practitioners alike. The exceptionally vivid anatomical specimens, and their high-quality colour reproduction, were equally as well received as the incorporation of applied clinical anatomy into the instructional material.

It has thus been a long-held desire of the authors, also from a scientific perspective, to make this book available to a wider audience in the form of an English-language version. A further motivation for so doing was to draw attention to the existence of German scientific publications in a field dominated by the English language. Amid a plethora of scientific texts, German editions are largely restricted to parts of Europe, as only relatively few fellow scientists are versed in the German language. This is regrettable, as it means that high-quality work published in German fails to find international resonance, purely due to the language of authorship.

We are therefore all the more grateful to 5m Publishing for including *Avian Anatomy* in their international catalogue, and trust that the resulting worldwide distribution will capture the interest of readers in a wide range of educational and scientific contexts.

The availability of an English version alone, however, does not automatically mean that this objective will be achieved; it was a stroke of good fortune for the publishers and the authors to have found in Dr Corinna Klupiec a capable and technically competent colleague for the translation of this book. Dr Klupiec, who commands both the English and German languages, combines excellent disciplinary knowledge with the ability to integrate anatomical concepts and their clinical application in an instructional context, drawing also on information from the contemporary literature.

Through close cooperation between the authors and the translator, it was possible to supplement the Bibliography with recent publications, resolve discrepancies in the nomenclature and correct errors. This applies also to the clinical components of this preparatory instructional text.

The result is therefore not merely a direct translation, rather an intensively revised and updated version of the 2nd German edition. For this, Dr Klupiec deserves the highest praise and recognition. At the same time, we express our thanks again to 5m Publishing for making, with the publication of this edition, a significant contribution to the dissemination of knowledge in the field of avian veterinary science.

Horst Erich König
Rüdiger Korbel
Hans-Georg Liebich
Vienna and Munich
Spring 2016

Translator's note

The translation of this unique and fascinating book was, in equal parts, a pleasure and a challenge. The avian anatomical literature is less prolific than its mammalian counterpart, with many seminal English-language works currently appearing only in their first or second edition. Thus, the level of agreement and consistency in anatomical nomenclature that exists in the mammalian literature has not yet been reached in the avian realm, even among works published in the same language. Moreover, the German language – like other languages – contains specific terms of common usage for which it is sometimes difficult to find a ready English equivalent.

One of the major undertakings of this translation, therefore, was to achieve alignment and corroboration with English language avian anatomy texts (particularly *Nomina Anatomica Avium* 2nd Edition [1993], King and McLelland [1980–1989] and Smallwood [2014]) while preserving the intellectual, scientific and linguistic integrity of the original work. At the same time, the opportunity arose to incorporate recent advances in selected aspects of avian anatomy and clinical practice, particularly with respect to the avian telencephalon and avian ophthalmology.

On a particular editorial note, it was necessary to determine a specific approach to Anglicisation of anatomical terms. The consistent use of Latin terms in the original work has been carried over into the translation. In the main, equivalent English terms have also been provided for each Latin term. For the muscles, however, only Latin terms have been used, as attempts to convert some names into English was deemed unnecessarily awkward, and a potential detractor from the reader's experience. A small number of exceptions also apply in other body systems, with terms appearing only in Latin or in English, although these have been kept to a minimum.

My sincere thanks go to 5m Publishing for the opportunity to translate this book and, in particular, to Sarah Hulbert and Alessandro Fratta Pasini for their support throughout the process. I am also very grateful to the authors, Professors Horst Erich König, Hans-Georg Liebich and Rüdiger Korbel, for their enthusiasm and cooperation in assisting with the fine-tuning of selected aspects of the work. This collaborative approach was pivotal in ensuring that the book achieves its considerable and significant potential to contribute to the avian scientific and instructional literature in an English language context.

Corinna Klupiec
Autumn, 2016

Authors

Associate Prof. Dr. Hermann Bragulla
Department of Biological Sciences, 202 Life Science Building, Louisiana State University, Baton Rouge, LA 70803-1715, USA

Prof. Dr. Klaus-Dieter Budras
Institut für Veterinär-Anatomie, Fachbereich Veterinärmedizin, Freie Universität Berlin, Koserstraße 20, D-14195 Berlin

Prof. Dr. Ana Carretero
Unidad de Anatomia y Embryologia, Departamento de Patologia y Producciones Animales, Facultad de Veterinaria, Universidad Autonoma de Barcelona, E-08193 Bellaterra, Barcelona

a.o. Univ. Prof. Dr. Gerhard Forstenpointner
Institut für Anatomie, Veterinärmedizinische Universität Wien, Veterinärplatz 1, A-1210 Wien

Dr. Corinna Klupiec
Metamorphomedia NSW, Australia

O. Univ. Prof. Dr. Dr. habil. Dr. h.c. Horst Erich König
Institut für Anatomie, Veterinärmedizinische Universität Wien, Veterinärplatz 1, A-1210 Wien

Univ.-Prof. Dr. Dr. habil. Rüdiger Korbel, Dipl. ECAMS
Klinik für Vögel, Ludwig-Maximilians-Universität München, Sonnenstraße 18, D-85764 Oberschleißheim

Univ.-Prof. Dr. Dr. h.c. mult. Hans-Georg Liebich
Institut für Tieranatomie, Ludwig-Maximilians-Universität München, Veterinärstraße 13, D-80539 München

Priv.-Doz. Dr. Johann Maierl
Institut für Tieranatomie, Ludwig-Maximilians-Universität München, Veterinärstraße 13, D-80539 München

Dr. Maren Meiners
Klinik für Vögel, Ludwig-Maximilians-Universität München, Sonnenstraße 18, D-85764 Oberschleißheim

Prof. Dr. Ivan Misek MVDr., C. Sc.
Institut für Anatomie und Histologie, Veterinär und Pharma- zeutische Universität, Palackeho 1–3, CS-61242 Brno

Prof. Dr. Christoph Mülling
Department of Comparative Biology & Experimental Medicine, Faculty of Veterinary Medicine, University of Calgary, 3330 Hospital Drive NW, Calgary, AB T2N 4N1, Canada

Prof. Dr. Marc Navarro
Unidad de Anatomia y Embryologia, Departamento de Patologia y Producciones Animales, Facultad de Veterinaria, Universidad Autonoma de Barcelona, E-08193 Bellaterra, Barcelona

a.o. Univ. Prof. Dr. Alexander Probst
Institut für Anatomie, Veterinärmedizinische Universität Wien, Veterinärplatz 1, A-1210 Wien

Priv.-Doz. Dr. Sven Reese
Institut für Tieranatomie, Ludwig-Maximilians-Universität München, Veterinärstraße 13, D-80539 München

Prof. Dr. Jesus Ruberte
Departamento de Patologia y Producciones Animales, Facultad de Veterinaria, Universidad Autonoma de Barcelona, E-08193 Bellaterra, Barcelona

a.o. Univ. Prof. Dr. Ingrid Walter
Institut für Histologie und Embryologie, Veterinär- medizinische Universität Wien, Veterinärplatz 1, A-1210 Wien

a.o. Univ. Prof. Dr. Gerald Weissengruber
Institut für Anatomie, Veterinärmedizinische Universität Wien, Veterinärplatz 1, A-1210 Wien

Dr. Grammata Zengerling
Institut für Tieranatomie, Ludwig-Maximilians-Universität München, Veterinärstraße 13, D-80539 München

Abbreviations and directional terms

ant., antt. anterior, anteriores (toward the front)
Art., artt. articulatio, articulationes (joint)
caud., caudd. caudalis, caudales (toward the tail)
cran., crann. cranialis, craniales (toward the head)
dext., dextt. dexter, dextri (right)
dist., distt. distalis, distales (away from the trunk)
dors., dorss. dorsalis, dorsales (toward the back [dorsum])
ext., extt. externus, externi (outer, external)
For., forr. foramen, foramina (opening)
Ggl., ggll. ganglion, ganglia (cluster of nerve cell bodies)
Gl., gll. glandula, glandulae (gland)
inf., inff. inferior, inferiores (below, lower)
int., intt. internus, interni (internal)
lat., latt. lateralis, laterales (toward the side)
Lc. lymphocentrum (lymphocentre)
Lig., ligg. ligamentum, ligamenta (ligament)
Ln., lnn. lymphonodus, lymphonodi (lymph node)
M., mm. musculus, musculi (muscle)
med., medd. medialis, mediales (toward the middle)
N., nn. nervus, nervi (nerve)
post., postt. posterior, posteriores (toward the rear)
Proc., procc. processus (sing. & pl.) (process)
prof., proff. profundus, profundi (deep)
prox., proxx. proximalis, proximales (toward the trunk)
R., rr. ramus, rami (branch)
Rec., recc. recessus (sing. or pl.) (recess, pocket, cleft)
sin., sinn. sinister, sinistri (left)
sup., supp. superior, superiores (above, higher)
supf., supff. superficialis, superficiales (superficial)
ventr., ventrr. ventralis, ventrales (toward the belly [venter])

Introduction

1

H. E. König, J. Maierl, G. Weissengruber and G. Forstenpointner

When the palaeontologist Hermann von Meyer discovered the first evidence of the 'original bird' in 1861, Darwin's seminal work on the origin of the species had only recently been published (1859). At that time, Linnaeus' Systema Naturae was already well over 100 years old. In view of Darwin's and Linnaeus' observations, von Meyer's discovery of **Archaeopteryx lithographica** (= 'old wing of the lithographic limestone'), which had lived around 150 million years ago, was of extraordinary significance. Although recent discoveries suggest that its relationship with modern birds is more complex than first thought, Archaeopteryx arguably remains the most famous fossil discovery in the world of science.

The skeleton of Archaeopteryx exhibits several characteristics that point to an ancestral relationship with small two-legged feathered carnivorous dinosaurs (theropods). Its feet more closely resemble those of dinosaurs than contemporary birds, and the lack of a reversed first toe suggests that Archaeopteryx had limited ability to perch on branches. On the forelimb, the three digits bore claws that may have been used for grasping low scrub. The second toe could be hyperextended backwards, enabling terrestrial locomotion. Its long slender legs provide further evidence that Archaeopteryx was well-suited to running.

1.1 Archaeopteryx bavarica WELLNHOFER, Solnhofer Limestone, Oberjura, Langenaltheimer Haardt, Courtesy of Bavarian State Collection for Palaeontology and Geology, Munich.

As a descendant of feathered dinosaurs, Archaeopteryx represents an intermediate species that retains several reptilian features. These include cranially and caudally

Vogelbůch

Darinn die art/natur vnnd eigenschafft aller vöglen/sampt jrer waaren Contrafactur/angezeigt wirdt: allen Liebhaberen der Künsten/Artzeten/Maleren/Goldschmiden/Bildschnitzeren/Seydensickeren/Weydleüten vnnd Köchen/nit allein lustig zů erfaren/sunder gantz nutzlich vnnd dienstlich zebrauchen. Erstlich durch Doctor Conradt Geßner in Latin beschriben: neüwlich aber durch Růdolff Heüßlin mit fleyß in das Teütsch gebracht/vnnd in ein kurtze ordnung gestelt.

Getruckt zů Zürych bey Christoffel Froschouwer/ im Jar als man zalt M. D. LXXXII.

1.2 Title page: '*Book of Birds*, in which the type/nature and characteristics of all birds/including its true appearance/is displayed: all fanciers of the arts/doctors/painters/goldsmiths/woodcarvers/silk embroiderers/herdsmen and cooks. . .'. "First described by Doctor Conradt Geßner in Latin: recently diligently translated into Teutsch. . . .", Zurich 1582. (Original: Institut für Paläoanatomie und Geschichte der Tiermedizin, Director Prof. Dr Dr habil J. Peters.)

recessed vertebrae, a mobile vertebral column and a cranium containing only a small cavity for the cerebellum. Other distinctive features included a saurian tail and teeth, as well as bird-like plumage on the wings and tail. Of particular significance is the finding that at least one Archaeopteryx fossil specimen has an osseous sternum where breast muscles could have attached, raising questions about the long-held view that the 'original bird' was a poor flyer.

To date, 13 fossil specimens of Archaeopteryx have been found, all within limestone deposits in the Altmühltal in Bavaria, Germany. **Archaeopteryx bavarica** WELLNHOFER, discovered in 1992, is part of the permanent collection of the 'Bayerische Staatssammlung für Paläontologie und historische Geologie' (Bavarian State Collection for Palaeontology and Geology), Munich (Figure 1.1).

History of avian anatomy

While birds differ from other warm-blooded animals in many ways, the ability to fly is undoubtedly their primary and most important distinguishing feature. Accordingly, this fascinating capability was a central theme of the earliest recognised investigations of avian anatomy. These include the observations of Aristotle of Stagira, justifiably acknowledged as the founder of morphology, in the 12th chapter of Book 4 of his treatise on the anatomy of animals (**De partibus animalium**). As well as providing a detailed and comparative description of the internal and external anatomy of birds, Aristotle points out that 'it is of the essence of a bird that it shall be able to fly; and it is by the extension of the wings that this is made possible'. This nevertheless sits somewhat incongruously alongside his views on the aerodynamics of the sternum and pectoral muscles of flying birds, namely that 'The breast in all birds is sharp-edged, and fleshy. The sharp edge is to minister to flight, for broad surfaces move with considerable difficulty, owing to the large quantity of air which they have to displace; while the fleshy character acts as a protection, for the breast, owing to its form, would be weak, were it not amply covered.'

The empirical Aristotelian approach to morphology was soon replaced by dogmatically influenced schools of thought, re-emerging only with the development of humanism at the beginning of the modern era. A notable exception is the altogether independent work of Emperor Frederik II of Hohenstaufen (1194–1250) on the art of falconry ('De arte venani cum avibus [On the art of hunting with birds]'). Its central premise, 'to render things that are, as they are (manifestare... ea, que sunt, sicut sunt)', was quite unrepresentative of its time, embodying a self-assured avowal of experimental science.

While the two surviving volumes of this work lack descriptions of the anatomy of falcons and their prey, it is possible that such information was included in lost portions of the text, along with other aspects of falconry such as medical care.

As one of the earliest protagonists of humanism, **Leonardo da Vinci** was fascinated with the idea of human flight. Da Vinci wrote and compiled a richly illustrated manuscript on the flight of birds (*Sul volo degli Ucelli*) containing sketches documenting his observations on several aspects of bird flight, including flapping of the wings, control of balance, stability, maintaining flight direction and the rigidity of the surface of the wing. He presented detailed descriptions of the function, arrangement and relative resistance and flexibility of the feathers, drawing conclusions about the propulsion and locomotion of birds.

Da Vinci also remarked upon the function of the remiges, attributing to them a role in establishing balance during flight. He noted that they serve to increase the size

Von der Hennen. LXXXIIII

werdend. Man sagt daß die Capaunen auch die fallend sucht ankomme. Vom Capaunenstein ist oben gesagt worden.

Von der Hennen vnd iren eyeren/auch allen eyeren in gemein. Gallina.

Von der gstalt dises vogels.

Hůener werdend die weyblin vnder den haußzuchten genennt / etwan aber wirt diser nam̃ für das gantz geschlächt genommen/als auch bey den Griechen. Denen gibt man etwan namen von der zeyt / als Mertzenhennen / so noch jung vnd nie gelegt habend. Die aber brůtend vnd junge fůrend / werdend Klůckhennen/ Gluggeren/oder Brůthennen genent. Was die Henn für einen vnderscheid von dem Hanen habe/wirst du auß deß selbigen beschreybung leychtlich können abnemen. Dañ die Hennen habend etwan schier keinen kamben/darzů henckt sich etwan der selbig/ vnd stadt nit aufrecht wie deß Hanen. Etliche habẽd auch gantz gefäderte füß/ so mã gehößlete Hůner nennet.

Von natur vnd anmůt der Hennen.

Etliche Hennen legend eins tags zwey mal/also/dz etliche sich ztod gelegt habẽd: dieweyl nit allein die gewächß der erden / sunder auch die vögel außgelärt vnnd vnfruchtbar werdend. Die Hennen in Jllyria legẽd nit nun ein mal eins tags wie ander schwo/sunder zwey oder drey mal/als Aristoteles schreybt. Die Adrianischen Hůner sind auch gantz fruchtbar. Die kleinen Zwerghůnlin sind nit gar fruchtbar/sagt Columella. Die gmeinen Hůner sind auch die fruchtbaristen. Die Hůner legend durch das gantz jar/on in den zweyen keltisten monaten im Winter. Die gůten Leghennen legẽd etwan sechß vnnd fünfzg eyer vor dem sy anfahend brůten / etliche alle tag / an

1.3 'Hens are female domestic chickens/although this word is used for all females/as also by the Greeks. These are named according to their age/Maerzhennen/are so young they have never laid. Those that are brooding and rearing young/are called "Gluggeren"/or brood hens'. From *Book of Birds*, Conrad Gesner, Zurich 1582. (Original: Institut für Paläoanatomie und Geschichte der Tiermedizin, Director Professor Dr Dr Habil J. Peters.)

of the airfoil, allowing the extended wing to be used for braking, particularly when turning.

Also in the sixteenth century, scholars at Italian institutions of learning turned their attention to specific aspects of avian anatomy. Notable individuals included **Ulysse Aldrovandi** (1522–1605) of Bologna and his student **Volcher Coiter** (1534–1576). In Padua, significant contributions were made by **Girolamo Fabricio ab Aquapendente** (1537–1619), who provided the original written account of the bursa of Fabricius or bursa cloacalis, and his pupil **William Harvey** (1578–1657), author of the first accurate description of the circulatory system. Harvey's studies of the developing heart in chicken embryos inspired further research on the egg during incubation and, in subsequent centuries, served as an important source of knowledge for the founders of modern embryology, such as **Caspar Friedrich Wolff** (1734–1794).

1.4 'In Teutsch this bird is called the cock/domestic cock/"Gul" or "Güggel": to which we extensively refer/ generally we speak of cocks or hens/which do not differ from domestic chickens/other than in size/and in that they are noisier'. From *Book of Birds*, Conrad Gesner, Zurich 1582. (Original: Institut für Paläoanatomie und Geschichte der Tiermedizin, Director Professor Dr Dr Habil J. Peters.)

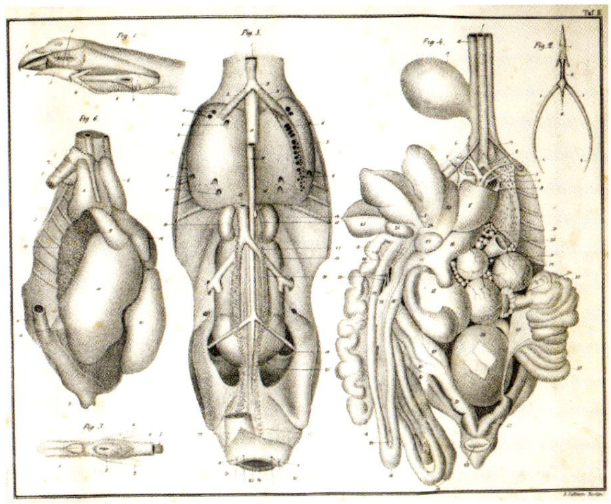

1.5 Drawings of the head, hyobranchial apparatus, viscera and air sacs of a chicken, and the tongue and larynx of a duck. From *Anatomie der Hausvögel* (*Anatomy of Domestic Birds*), Ernst Friedrich Gurlt, Berlin, 1849.

For the great anatomists and natural scientists of the eighteenth and nineteenth centuries, the morphology of the avian body was a frequent source of discussion. **Luigi Galvani** (1737–1798) of Bologna, best known as an electrophysiologist, contributed valuable observations on the urogenital tract and ear, while the renowned palaeontologist **Richard Owen** (1804–1892) produced fundamental works on comparative avian osteology.

In the field of veterinary anatomy, on the other hand, the morphology of birds, particularly domestic birds, tended to play a minor role. By way of example, the earliest recorded representation of the anatomy of birds in a German textbook flowed from the fountain pen of a member of the medical faculty in Dresden (**Carl Gustav Carus**, *Lehrbuch der Zootomie* [*Textbook of Zoology*, 1818]). It would be another 30 years before pertinent work, authored by **Ernst Friedrich Gurlt** in Berlin (Figures 1.5 and 1.6), was produced by a veterinary school. Contributions on the anatomy of domestic birds did not appear in veterinary anatomy textbooks until the second half of the nineteenth century. In Vienna, **Franz Müller** expanded the second edition of his *Lehrbuch der Anatomie der Haussäugethiere* (*Textbook of the Anatomy of Domestic Mammals*, 1871), by including chapters on birds. In the case of Gurlt's *Handbuch der vergleichenden Anatomie der Hausthiere* (*Manual of the Comparative Anatomy of Domestic Animals*) it was not until the eighth edition, published in 1896, that a chapter entitled 'The anatomy of domestic birds' was added, under the editorial supervision of Wilhelm Ellenberger and Carl Müller.

In the twentieth century, particularly in the decades following the Second World War, knowledge of the macroscopic morphology of domestic birds progressed dramatically. A substantial portion of this work was carried out by the groups led by J. McLelland (Edinburgh) and A. S. King (Liverpool), whose intensive research activities

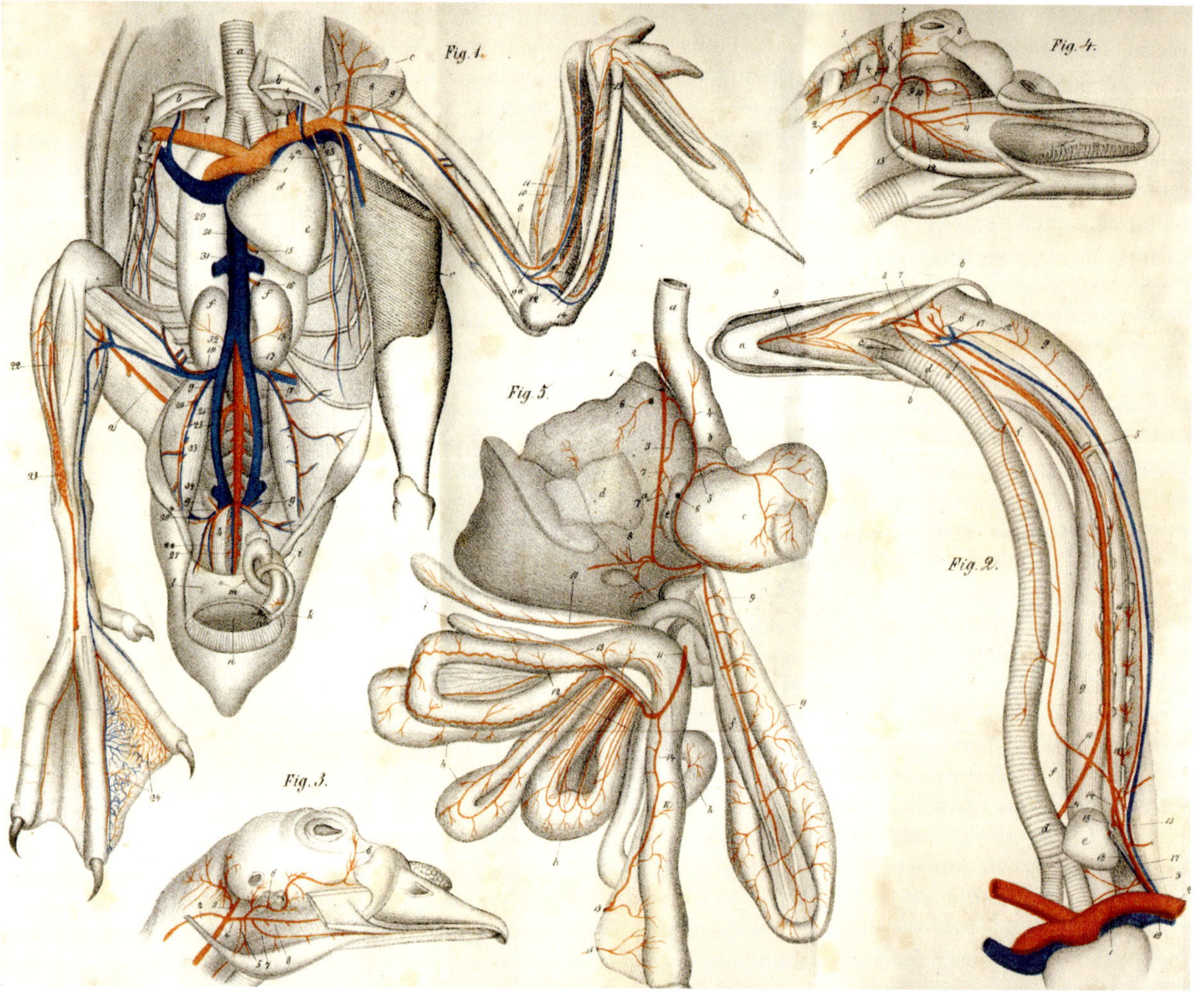

1.6 Drawings of the vessels of the trunk, left wing and right hindlimb of a male Turkish duck. From *Anatomie der Hausvögel* (*Anatomy of Domestic Birds*), Ernst Friedrich Gurlt, Berlin, 1849.

led to the publication of the definitive English-language text *Form and Function in Birds* (1979–1989). A series of important and fundamental contributions on the topographical anatomy of the domestic chicken was also developed at the Department of Veterinary Anatomy at the University of Nagoya in Japan, under the leadership of Mikio Yasuda.

In the German literature, a description of the systematic anatomy of domestic birds first appeared in 1973 with the first edition of the fifth volume of Richard Nickel, August Schummer and Eugen Seiferle's *Lehrbuch der Anatomie der Haustiere* (*The Anatomy of Domestic Animals*).

The publication by J. Baumel of the *Nomina Anatomica Avium* (NAA), first in 1979 then in second edition in 1993, contributed significantly to the consistent use of standardised anatomical terms, an essential requirement for methodical documentation of anatomical information.

Overview of the locomotion and anatomy of birds

Birds are **feathered**, **warm blooded**, **egg-laying vertebrates** that, in the main, are **capable of flight**. In particular, this capacity for flight accounts for the uniformity of the basic anatomical structure of the roughly 10,000 extant bird species, and more than 20,000 subspecies. Significant variation in body size and shape is encountered primarily in the few flightless birds, such as penguins and ratites (Figures 1.7 and 1.20).

Birds evolved considerably later than mammals from their common reptile-like ancestors (birds and reptiles being known collectively as sauropsids) and therefore possess greater similarities with reptiles than with existing mammals.

One of the significant **evolutionary adaptations** of birds is a **progressive reduction in body weight**. Concomitantly, the heavy parts of the body became positioned close to the centre of gravity. Compared with the relatively elongated, mobile torso of reptiles, birds thus developed a relatively **short and compact trunk**. The flight muscles that generate the thrust required for flying were exempted from this reduction in body weight, and constitute approximately 15–20 per cent of the total body weight of modern birds.

Feathers

Feathers are birds' most important 'evolutionary invention'. This lightweight body covering contributes to **temperature regulation** (**homeothermy**) and, in most avian species, also enables flight. Exceptions include the ratites, whose soft, loose feathers provide good insulation, but are ill-suited to flying. In contrast, the fastest flyers, such as falcons, possess the most rigid feather coat.

Feathers are the most significant determinant of a bird's colour. Some species may develop abnormal colouring when kept in captivity, often due to the lack of a particular nutrient. Flamingos kept in zoos develop white rather than red feathers unless their diet is supplemented with carotene. A further example is the European goldfinch, in which the consumption of large quantities of hemp causes the naturally brown plumage to turn black.

1.7 Common ostriches (Struthio camelus) are the largest living birds. Mature males weigh up to 150kg and grow up to 3m in height. Ostriches are flightless but are very good runners and can reach maximum speeds of 50 to 70 km/h. Courtesy of T. Angermayer, Tierpark Hellabrunn, Munich.

Skeletal adaptations for locomotion

The bird skeleton exhibits **numerous adaptations** that serve to facilitate flight (Figures 1.33 to 1.40). In contrast to the dense, marrow-filled long bones of mammals, the large avian long bones (humerus, femur) are pneumatised (although there are also some good flyers, such as seagulls, in which the humerus is devoid of air cavities and is filled instead with marrow).

Other adaptations include a reduction in the **number of bones** and an overall shortening of the body. The distal coccygeal vertebrae are fused into a single small bone. Several of the digits of the manus and pes are absent or vestigial. The result of all of these modifications is that the weight of the skeleton of a pigeon, for example, is only **4.5 per cent of its total body weight**. In a mammal of the same weight, the equivalent figure is approximately 6 per cent.

Apart from weight considerations, the high degree of **stability** required for flight plays an important role in the structure of the skeleton. Segments of the **vertebral column are fused**, reducing the need for ligaments and muscles. Caudally directed rib projections (processus uncinati) form struts that overlap the subsequent rib. In some diving birds, these even traverse two ribs to protect the torso from the high pressures encountered in deep water.

The relative size of the **sternum**, on the other hand, cannot be reduced as this is where the flight muscles arise. Indeed, its broad trough-like structure and prominent

1.8 Scarlet ibis (Eudocimus ruber) with bright red coloured plumage. Courtesy of PD Dr S. Reese, Munich.

1.9 The peregrine falcon (Falco peregrinus) is ideally adapted to long-distance aerial hunting. To catch its prey – consisting mostly of smaller birds – it stoops at speeds of up to 300km/h. Courtesy of H.-K. Hussong, Fürth.

keel provide a **large surface area** for muscular attachment (Figures 2.29 to 2.31). In addition, the **coracoid bones** between the shoulder joints and the sternum act as a substantial brace to withstand the strain placed on the shoulder during the beating of the wings.

Types of locomotion

Some bird species are capable of remarkable feats of flying. Falcons, for example, can reach speeds of 200km/h, considerably more in a stoop (Figure 1.9). Condors fly at altitudes of up to 4,000m, while a goose has been observed at 8,800m. It is estimated that a common swift with two nestlings flies 1,000km per day. Migratory birds cover much greater distances. Plovers, which migrate from Nova Scotia to Argentina, fly 3,300–4,000km without stopping. The white stork covers around 20,000km per year, the Arctic tern up to 35,000–40,000km per annum.

As well as conferring upon birds the ability to fly, the specialised avian anatomy facilitates (and/or does not preclude) quick and agile movement on **land** and in **water**.

During **flight**, the body is suspended from the shoulder joints, such that the centre of gravity is positioned, very efficiently, directly under the shoulder and well in front of the hips. To facilitate balance when on **land**, the femur is directed cranially, allowing the knee to be positioned close to the centre of gravity (Figures 1.37 to 1.40).

For many birds, **climbing** is the preferred form of locomotion. Woodpeckers and treecreepers climb trees by hopping along the bark. A similar technique is used by wallcreepers in order to scale steep cliffs. The nuthatch is the only species of bird that moves head-first **down a tree**. All of these climbing species are **short-legged**, enabling them to carry their centre of gravity close to the tree trunk while using their tail for support. Woodpeckers and nuthatches also splay their fourth digit laterally to cling on to the trunk.

Large wings, while useful for flying, are a hindrance in the water. Several species of duck, including the long-tailed duck, common scoter and the velvet duck, have solved this problem by **beating their folded wings** while diving. Plunge-divers such as gannets and boobies, several pelicans and petrels, as well as terns and kingfishers plummet directly from the air into the water, sparing themselves the effort of 'underwater flying'.

Those water birds that are poor divers obtain their food from the water while swimming. As masters of both swimming and diving, penguins have **dispensed with flight altogether**, their wings transformed into **flippers** (Figure 1.20). Most penguins can reach depths of 30m, rockhopper penguins up to 100m. Emperor penguins are capable of diving to 500m, remaining underwater for 20 minutes. This is particularly remarkable given that most birds rarely remain submerged for more than a minute at a time, and usually at depths of less than 10m.

Flight

Two main theories are put forward regarding the origin of flight. One proposes that the ability to fly began with small animals that moved on two legs. These animals developed **feathers** in order to thermoregulate, to present displays to other animals and to achieve better balance. When the feathers of the forelimbs became longer, these animals were able to make progressively longer leaps, and eventually developed the capacity for flight. The alternative hypothesis assumes that flight originated among animals that were good climbers, possessed feathers and hopped from branch to branch in trees. When their feathers eventually lengthened, and other adaptations had taken place, the animals were able initially to glide and ultimately to fly.

The **ability to fly** is of considerable importance in the conservation of avian species. Flight provided the evolutionary advantage of being able to access food sources that were not available on the ground. Moreover, the manoeuvrability achieved through flight allowed birds to search quickly over a large area for both food and shelter, and to better evade land-based predators. Aerial locomotion also facilitated territorial expansion across hostile environmental boundaries. Yet another benefit of flight is the ability to undertake seasonal migration in order to access favourable feeding and nesting sites.

The anatomical structure of the wings is presented only briefly here, with cross references to the corresponding chapters in which more detail is provided.

The **bones of the wings** are comprised of the humerus of the brachium, the radius and ulna of the antebrachium and the markedly reduced bones of the manus (see Chapter 3). The elbow and carpus function as hinge joints. They are

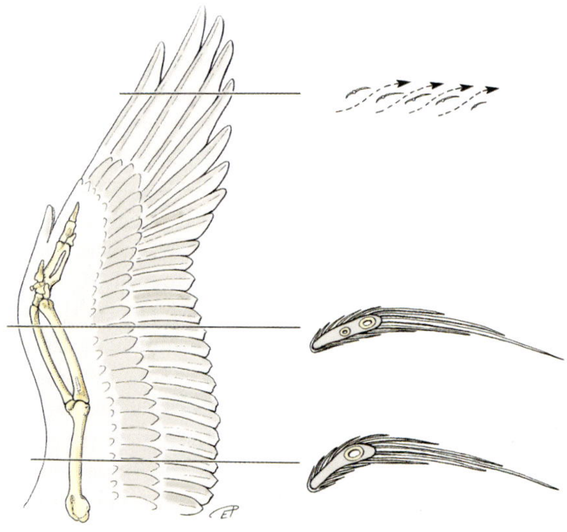

1.10 Cross sectional profile (schematic) of a wing at the level of the brachium, antebrachium and manus; the dashed arrows between the cross sections of the primary remiges represent the flow of air at the tip of the wing.

mutually dependent and can therefore, in a practical sense, only be extended or flexed in tandem. On the manus, only the alular digit has a 'thumb-like' range of movement.

The key **muscles** of the wings (see Chapter 3) are those that raise and lower the limb, particularly the pectoral muscle (lowers the wing) and the supracoracoideus muscle (raises the wing).

The **skin** and its specialised appendages, the **feathers**, are arranged on the wing in a highly specific fashion (see Chapter 17). Folds of skin (**patagia**) span the bones of the forelimb, creating a uniform surface over the angular skeletal architecture of the wing. The cranially facing angle between the shoulder and the bones of the forearm is enclosed by the **propatagium**. The **metapatagium** performs the equivalent function between the humerus and the trunk, while the **postpatagium** spans the angle between the bones of the forearm and the manus.

Feathers known as **tectrices** (coverts) combine with the basic structural framework of the wing to give it its streamlined shape and characteristic cross sectional appearance. Along the caudal contour of the wing, the tectrices are replaced by the longer **remiges**. In several bird species, particularly those that glide at low speeds, the tips of the primary remiges are separated (Figure 1.10; see also below).

Gliding represents the 'simplest' form of flight. Young birds glide from nests or cliffs as a means of testing their flight muscles. Noteworthy in this regard are the 'flying schools' of the blue-footed boobies of the Galapagos Islands. Flocks of young birds move their wings as they walk along towering seaside cliffs. In a favourable wind, they push off from the cliff's edge and glide over the open ocean, using movements of their wings and tail to steer their maiden flight. In soaring flight, birds make use of thermals and ridge lift to maintain altitude without needing to beat their wings.

The ability to glide, and achieve other forms of flight, is related to the airfoil shape of the wing. At the level of the secondary remiges and humeral (tertial) feathers (see Chapter 17), the cross-sectional profile of the wing is distinctly rounded, with a bulging leading edge (Figure 1.10). Caudally the cross section tapers off, ending with the remiges. When air passes over the outstretched wing, the air stream is parted. Due to the shape of the wing, the air passing over the upper surface moves faster than the air flowing beneath the wing. In accordance with the Bernoulli principle, the air pressure associated with the faster-flowing air above the wing is lower than the air pressure below the wing. This results in powerful **lift** at the cranial edge of the wing, decreasing towards its caudal border. Lift can be increased by elevating the leading edge, such that the wing faces the oncoming air at an angle (angle of attack, or α; Figure 1.11). On the underside of the wing, the downward deflection of airflow creates an increase in pressure that adds to the lift. The resulting pressure differential above and below the wing is in the vicinity of 3:1.

The degree of lift is related to two main factors: the square of the velocity of the air passing over the wing, and the projected surface area (equivalent to the shadow cast by the bird). In simplified terms, this means that birds that glide at low speeds, where the velocity of the air passing over the wings is relatively low, have broad (i.e., wide and deep) wings. This results in a large surface area, as seen in eagles and vultures (Figure 1.12).

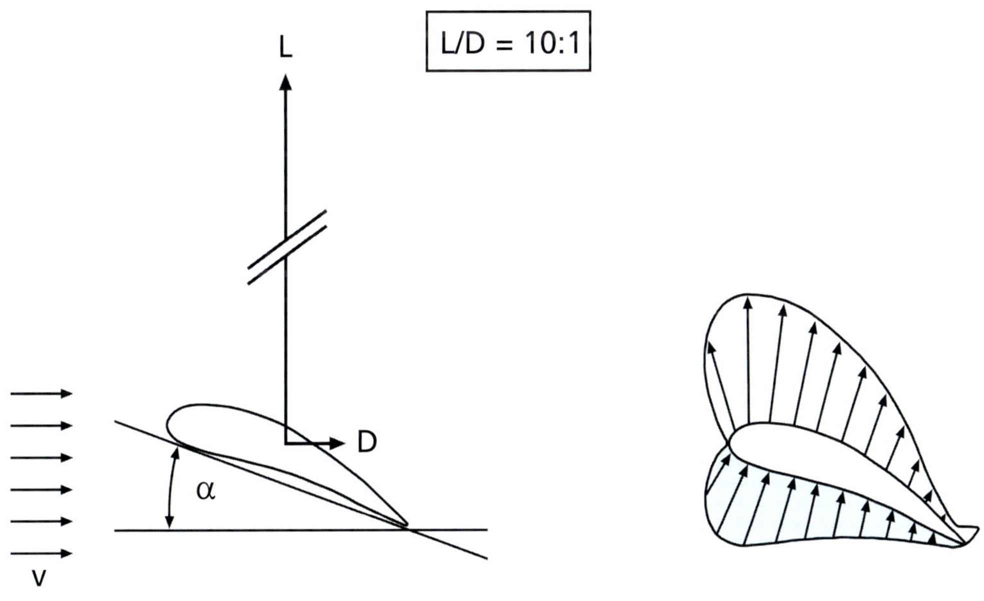

1.11 Schematic representation of the relationship between lift (L) and drag (D) at an angle of attack α and air velocity v (left); schematic representation of the cross section of a wing indicating distribution of pressure (right).

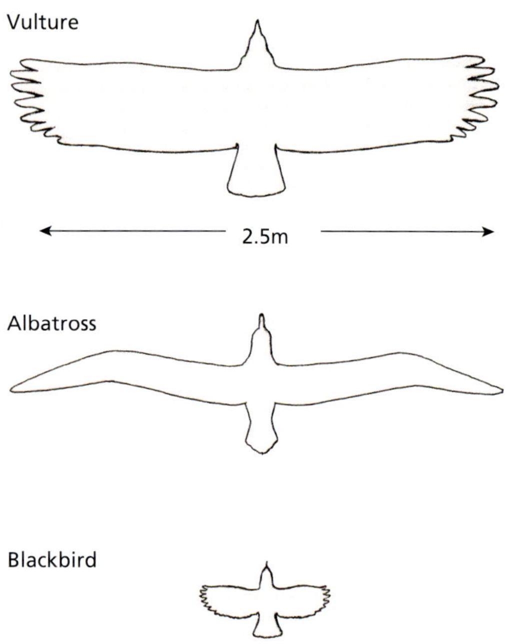

1.12 Schematic representation of different wing shapes based on the flight silhouette. Vulture (Gyps rueppellii) (1:20); albatross (Diomedea exulans) (1:20); common blackbird (Turdus merula) (1:10).

In these birds, the separation of the tips of the **primary remiges** at the end of the wings provides an additional source of lift and propulsion. Due to the difference in pressure above and below the wing, air streams upwards at the wing tips (Figure 1.10). As the vanes of the remiges of the manus are asymmetrical, the splayed remiges act as additional small airfoils.

In birds exposed to constant and sometimes strong winds, for example when flying over the ocean, the situation is quite different. These birds are **fast flyers**, with correspondingly narrow wings. They are not able to glide at slow speeds, even in low wind conditions, because the smaller surface area of their wings would generate insufficient lift. Birds that fall into this category include the albatross (Figure 1.12) and various types of gull.

In many smaller birds, such as the blackbird (Figure 1.12), the shape of the wing renders them poor gliders, irrespective of wind speed. These birds rely on beating their wings for generating lift and propulsion, and thus primarily utilise **flapping flight** (see below).

Birds in flight are also exposed to the slowing effect of drag. The optimal ratio of lift to drag is in the order of 10:1 (Figure 1.11).

All of these mechanical principles are based on the assumption that air flow over the wing is laminar, with the air passing close to the wing. **Turbulent** air movement above the wing significantly reduces lift. A larger angle of attack not only increases lift and drag, but also raises the likelihood of turbulence, such that the maximum attack angle is around 15°. Above this limit, the air stream becomes separated from the wing with rapid loss of lift.

Birds have various methods at their disposal to shift turbulent air towards the caudal edge of the wing. One of these is the **alula**, which redirects the stream of air back towards the surface of the wing. This increases the **air flow velocity** in the critical region at the leading edge. In this way, turbulence at the surface of the wing is reduced, even at somewhat larger angles of attack (Figure 1.13). The elastic compliance of the tectrices provides an additional means of shifting the point of air stream separation as far caudally as possible. In this situation, the surface of the wing conforms to some extent to the altered direction of air flow (Figure 1.13).

The **differences in pressure** above and below the wings give rise to a current of air that moves outwards from the bird's body (Figure 1.14). When a bird is gliding smoothly, this results in a vortex at the tips of the wings, setting a certain volume of air in motion. This phenomenon has a braking effect with a corresponding loss of kinetic energy. Different types of vortex are formed, depending on the type of flight (see below, 'Flapping flight' and Figure 1.15). In all cases, they create drag that opposes the propulsive force.

Flapping flight is produced by the continuous **upwards** and **downwards** movement of the wings. At positive

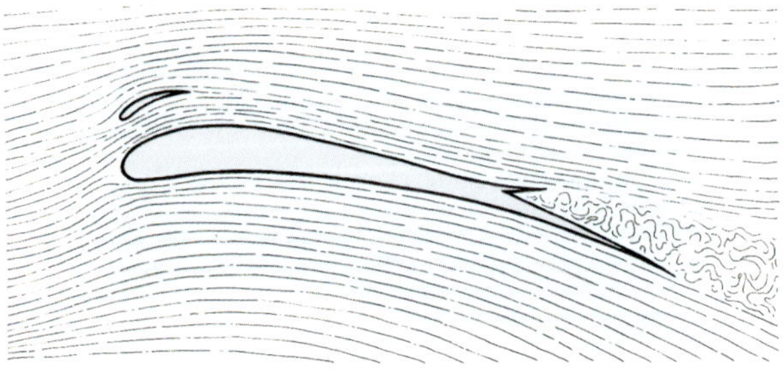

1.13 Schematic representation of the aerodynamic effect of the alula and the elastically compliant tectrices.

1.14 Formation of vortices at the tips of the wings in a gliding bird.

1.15 Schematic representation of the phases of flapping flight in the pigeon. The closed ring vortices resulting from each downstroke are depicted on the right.

angles of attack, a high velocity of air flow over the wing is generated during the downward stroke. As with gliding, this results in both lift and thrust. In the simplest scenario, the upstroke is required purely to return the wing to a position from which it can perform another downstroke. The wing is partly folded during the upstroke to reduce its surface area.

This is typical of the so-called 'bounding flight' of small birds. Rapid beating of the wings, at frequencies of around 20Hz, propels the bird upwards. The wings are then folded, the flight muscles are rested and, losing some of its altitude, the bird advances at high speed. After a time, the flapping phase begins again and altitude is regained.

In **medium-sized birds**, such as pigeons, gulls and hawks, the downstroke also produces considerable lift, with moderate propulsion. Due to their body size, and the corresponding moment of inertia, these birds beat their wings at frequencies of only 3–10Hz. To account for their body mass, they must therefore also utilise the upstroke for achieving lift. This applies especially during **climbing flight**, when the primary remiges (in particular) are drawn closer to the body by the action of the wings, and are rotated by the air pressure. Due to the asymmetry of the vanes, spaces open up between the remiges and air flows between the feathers. In this way, each individual feather acts as a small airfoil, contributing to lift and thrust. During **level flight** at higher speeds, air passes over the entire outstretched wing, with the secondary remiges also contributing to propulsion during each flapping phase.

As with gliding, a **pressure differential** is generated between the upper and lower surfaces of the wing during flapping flight. Thus, the same principles apply with respect to generation of vortices. The nature of the vortex is related to the shape and movement of the wing, as well as the speed of flight. The wings of gulls and pigeons flying at high speed, for example, produce continuous vortices that follow a zig-zagging path. This type of flight is associated with the generation of lift during both the upstroke and downstroke. In small songbirds and in pigeons flying at relatively low speeds, the vortex assumes a closed **ring shape** (Figure 1.15). Here, lift is only produced during the downstroke.

Hovering is a specialised form of flapping flight in which the bird pauses in the air, either by utilising a headwind or by generating lift by flapping (true hovering). Kestrels, for example, use the former method to watch over their prey, waiting for an opportune moment to tuck in their wings and swoop onto their target.

The most extreme form of true hovering is seen in the **hummingbird**, during which these birds typically flap their wings 30–50 times per second (Figure 1.16). Their body is held close to vertical, with the wings beating forwards and backwards. On the downstroke, the wing is brought forward at a positive angle of attack, generating lift. At the end of the downstroke, the wing is inverted such that the upper surface faces downwards. In this way, a positive angle of attack is maintained on the upstroke, helping to maintain altitude. During each flapping cycle, the tips of the wings describe a horizontal figure eight.

1.16 A hummingbird hovering (Juan Fernández firecrown; Sephanoides fernandensis). Courtesy of Professor Dr Daniel Gonzalez-Acuna, Chillan, Chile.

Land- and water-based locomotion

While most birds are masters of flight, a form of locomotion seen only rarely in other vertebrates, some birds are restricted to terrestrial, and in some cases aquatic environments. Particularly good runners are found among the Struthioniformes, while excellent swimmers are represented in several orders.

The flightless Struthioniformes are comprised of the ostrich, rhea, emu, cassowary and kiwi. Together with extinct species (elephant birds, moas) these birds are referred to as ratites. This name is derived from the Latin word 'ratis' (raft), a reference to the absence of a keel (carina) on the sternum, thus distinguishing them from flying (carinate) birds. Closely related to the Struthioniformes are the tinamous (Tinamiformes), which are ground-dwelling but capable of flight.

Flightlessness is not limited to the Struthioniformes. Penguins, Galapagos cormorants, the extinct great auk as well as the kakapo (owl parrot; Strigops habroptilus) and takahe (Porphyrio hochstetteri and extinct P. mantelli) of New Zealand are (were) also denied access to the sky.

While many types of bird may be deemed 'good walkers', only a few species – such as the greater roadrunner, the emu and the African ostrich – are worthy of the title 'good runners'. As bipedal digitigrade animals, walking and running generally involves bringing the left and right caudally positioned hindlimb forward in an alternating fashion, with one leg in the swing phase while the other is in the stance phase. Another form of terrestrial locomotion involves jumping or hopping, in which both legs are extended simultaneously. Some species, including the native blackbird, exclusively use a hopping gait when on the ground while others, such as the house sparrow, include both hopping and walking in their repertoire.

The ostrich can reach maximum speeds of 80km/h (emu: c. 50km/h) and can maintain high speeds for periods of up to half an hour. This comprehensively surpasses the performance of racehorses and makes the ostrich the fastest (self-powered) bipedal animal.

All of the body systems, particularly the **hindlimbs**, play a role in supporting this high level of performance. Among the large ratites (emu, cassowary, rhea, ostrich) the structure and proportions of the hindlimbs, and the position of the joints, exhibit many similarities. The **centre of**

1.17 Positioning of the limbs with respect to the centre of gravity during two-legged standing and walking (chicken, viewed from the front).

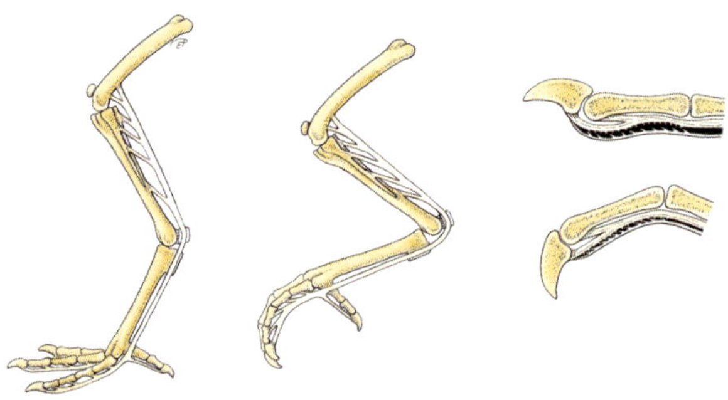

1.19 Automatic digital flexor mechanism (left and middle; combined action of the intertarsal and interdigital joints) and digital tendon locking mechanism (right; processes on the flexor tendons interlock with folds in the tendon sheath).

1.18 Corresponding phases of the walking (chicken, above) and running (common ostrich, below) gaits.

gravity in these species lies **medial to the distal end of the femur**. The muscles of the hindlimb alone account for upwards of one quarter of the total body weight. Much of this muscle mass lies close to the trunk at the proximal end of the limbs and thus close to the pivot point (= hip joint) of the corresponding limb. This is energetically advantageous for the pendulum-like movements of the hindlimbs during running, and assists with acceleration. The joints of the distal limb are primarily responsible for achieving efficient forward progress.

1.20 Despite growing to a metre in height, Emperor penguins (Aptenodytes forsteri) are skilful swimmers and divers. Courtesy of T. Angermayer, Tierpark Hellabrunn, Munich.

1.21 Mallard duck (Anas platyrhynchos) with ducklings. The young birds are adept swimmers from birth.

In these birds, **high speeds** are achieved by increasing the frequency of the gait cycle, reducing the time that the foot contacts the ground and increasing the depth of stride. The femur is short and, as in quadruped athletes, the distal components of the limb are relatively long (noting, however, that the distal segments of the limbs of wading birds are longer than those in fast-running birds). Distally, from around the middle of the shank (crus), the muscular system is represented only by tendons, in which elastic energy can be stored.

The **fusion** of the tarsal and metatarsal bones to form the **tarsometatarsus** is generally regarded as an adaptation to land-based locomotion (cf. metacarpus and metatarsus of even-toed ungulates). The ostrich is unique among birds in that the number of toes is reduced to two (third and fourth digits), and the proximal phalanx is not in contact with the ground. As in hoofed animals, this reduced footprint is beneficial for **acceleration** and **maintaining high speeds**. The fact that the ostrich uses the smaller fourth toe for balance probably explains why the number of toes has not undergone further evolutionary reduction. The knee, intertarsal and metatarsophalangeal joints are connected by muscles and their tendons in an obligate arrangement, such that their movement is interdependent. As in quadrupeds, this likely contributes to the precision of the gait and represents a means of conserving energy during locomotion.

In many bird species, **perching** is facilitated by the **automatic digital flexor mechanism** (in which flexion of the intertarsal joint places tension on the digital flexor tendons, causing the toes to grip the perch/branch) and the so-called **digital tendon locking mechanism** (in which the flexor tendons are held in place by processes that interlock with folds in the surrounding tendon sheath) (Figure 1.19). This allows the bird to conserve energy that would otherwise be needed to actively maintain the perching posture.

In birds that have become adapted for **climbing**, locomotion is also assisted by the beak (parrots), the tail feathers (woodpeckers) or the wings (hoatzin).

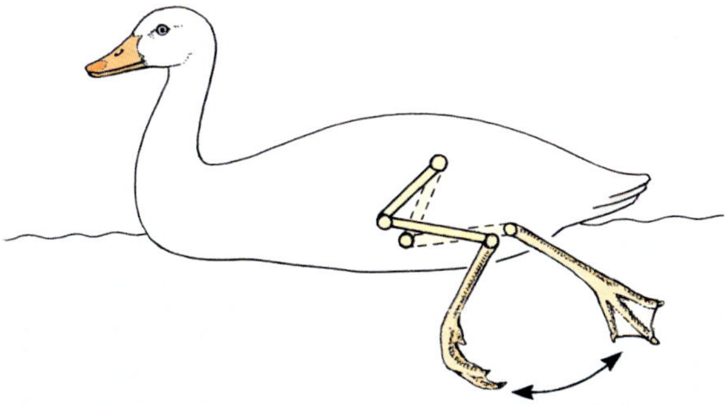

1.22 During surface swimming, movement principally occurs at the intertarsal joint. During forward motion, the toes are spread as the intertarsal joint is extended, tensing the interdigital web.

During **surface swimming**, the movement of the hindlimbs mainly involves the intertarsal joint (Figure 1.22), with many species moving their limbs alternately. As the hindlimb is brought forward, the toes are flexed and drawn together, such that the interdigital webbing is folded. The digital tendon locking mechanism is also engaged by many birds when swimming, suggesting that it aids this form of locomotion. Depending on species, diving birds use movements of the limbs and/or wings to propel themselves through the water.

Digestive system

The anatomy of the organs of the digestive system conforms to the principle of concentrating mass near the body's centre of gravity. A **lightweight beak** takes the place of heavy, teeth-bearing jaws. The **oversized beaks** of several bird species (Figure 1.26), including toucans and hornbills, are deceptive in this regard, as they are light and based on a framework of pneumatised spongy bone.

The attachment between the upper jaw and the cranium permits a variable degree of movement, while the **lower jaw** is freely moveable relative to the skull. Through their opposing movements, the upper and lower jaw are able to produce a secure, tweezer-like grip.

The **shape of the beak** is adapted to the bird's **dietary habits** (Figures 1.24 to 1.27). A straight, **pointed beak** is well suited to **catching insects** as well as gathering **seeds** and **berries**. In **woodpeckers**, the union between the beak and the cranium has shock-absorbing properties. These soften the impact of the repeated striking used to chisel deep holes in hardwood to reach insects within the tree. The most **powerful beak among European grain eating species** is found in the hawfinch. Its strong jaw muscles allow it to crack the pits of cherries and olives with ease.

Powerful jaw musculature is also found in raptors, which use their hooked beaks to tear their often sizeable prey into pieces that are small enough to swallow. Most **ducks** and **geese** have **broad beaks** that are well adapted for picking off pieces of plants. The beaks of mergansers (sawbills) are long and slender, with **serrated edges** that are used when fishing to grip their slippery prey. Avocets feed by 'scything' (slicing from side to side) their upturned, slightly opened bills through shallow water. When the bill makes contact with prey, it rapidly snaps shut. In the spoonbill, this reaction is lightning-fast, indeed it is one of the fastest reflex movements in the animal kingdom.

Flamingos and pelicans have **highly specialised beaks**. The former use their beak as a filter, placing it upside-down in the water. Comb-like keratinised bristles at the edges of the beak serve to trap fine food particles, crabs and algae. Pelicans plunge their open beaks into the water, curving the sides of the lower beak to create a wide-mouthed scoop (Figure 1.23). When a fish enters this trap, the mandibles spring back and the upper beak clamps down. The food is then usually moved caudally (towards the centre of gravity) by swallowing it into the crop.

1.23 Pelicans, such as the Dalmatian pelican (Pelecanus crispus), use the highly distensible throat pouch located between the mandibles as a scoop during feeding.

1.24 The grey heron (Ardea cinerea) hunts by striking at high speed with its long pointed beak, literally 'stabbing' their prey. Courtesy of Dr Petra Kölle, Munich.

1.25 The marabou stork (Leptoptilos crumeniferus) is a scavenger that pecks open the abdominal wall of animal cadavers with its powerful wedge shaped beak. Courtesy of Dr S. Reese, Munich.

1.26 The large colourful beak of toucans, such as the white throated toucan (Ramphastos tucanus), can be longer than the body.

1.27 As in most true macaws, the sides of the face of the red-and-green macaw (Ara chloroptera) are largely bare. Macaws are very social and intelligent birds. The weight of its brain relative to its body weight is among the highest of all birds. Courtesy of Professor Dr Sabine Kölle, Giessen.

1.28 Egg size of the common ostrich (left) versus the chicken (right). The unique number and distribution of pores in the egg shell can be used (e.g., in raptors) to associate an egg with an individual bird.

Particularly in **seed-** and **grain-eating** species, the **crop serves as a storage container**. Several seabirds and also swifts travel large distances to find food for their young, carrying it to them in their crops. Pigeons and flamingos feed their nestlings nutrient-rich liquid for the first few days after hatching. This substance is produced in the crop in pigeons (crop milk) and by glandular regions in the oesophagus in flamingos.

Birds have a **glandular** and a **muscular** stomach. In the glandular stomach, pepsin and hydrochloric acid initiate the digestion of protein. Ingesta then passes into the muscular stomach where hard leaves, grasses, grain and insects are crushed through muscular contraction, or ground by 'grinding plates' (formed by the cuticle). Many birds also consume **grit** to assist the mechanical function of the stomach. Fish- and meat-eating birds do not require the same degree of muscular development. In frugivores, the muscular stomach is often similarly reduced. Some birds consume large quantities of food that then passes very quickly through the digestive tract.

The **small intestine** is relatively long in **grain-** and **cellulose-eating** birds, and shorter in **carnivores** and birds that consume soft-bodied insects. The two caeca join the digestive tract at the distal end of the small intestine. They are usually short, although only rarely rudimentary. Ostriches, chickens and ducks have particularly long caeca, containing large numbers of cellulolytic bacteria.

The **rectum** opens into the **cloaca**. The first cloacal compartment, the **coprodeum**, is an important site of water absorption. This can be so efficient that some larks, for example, do not need to drink.

Respiratory system

The respiratory organs of birds are conspicuously light and very efficient. Compared with other vertebrates, the **avian lung** makes better use of inspired air. Instead of flowing into blind-ended structures, the air passes through a continuous system of **delicate tubules**.

The avian respiratory apparatus also includes air sacs that are connected to the lungs by the bronchi. **Gas exchange** takes place in various regions of the lung both during **inspiration**, as **air travels to the air sacs**, and **expiration**.

The structures responsible for **vocalisation** are incorporated into the breathing apparatus. While hissing noises are generated by the larynx, the sounds typically associated with birds arise from the **syrinx**, located at or near the **bifurcation of the trachea** (Figures 7.8 to 7.11 in Chapter 7). The production of **deep tones** is also related to the length of the trachea. In swans and cranes, the trachea is arranged in numerous coils, which are sometimes housed within the sternum (e.g., whooper swan). In the magpie goose, the tracheal coils are located external to the sternum, between the skin and the muscles of the breast. **Gular pouches**, areas of inflatable skin, are found on the neck of some species including the **prairie chicken** and the male frigate bird. In the latter species, inflation of the pouch is used to produce a spectacular display during courtship.

Urogenital apparatus

Birds are endowed with a very **efficient excretory system**. The kidneys produce a watery **uric acid solution**, from which much of the water is reabsorbed in the rectum. The resulting semi-solid urine constitutes the white layer that typifies bird faeces. This 'water recycling' mechanism helps birds to fly long distances without needing to drink.

The **absence of a bladder saves on body weight**, as does the development of only **one ovary** and **oviduct** in the female of most bird species. In males and females, the gonads are positioned at the body's gravitational centre. Both the testes and the ovary become considerably enlarged during the breeding season, the latter due to the development of the substantial avian follicles.

Avian egg and incubation period

In most species, fertilised eggs are laid by the female within days to weeks of mating. The **size of the clutch** varies widely from just a single egg to more than 20, with songbirds typically laying between five and eight eggs.

The **incubation period** also exhibits considerable genus- and species-specific variation, ranging from approximately ten days to three months.

The **chicken** lays eggs at intervals of 1–2 days. Incubation only begins after about 3–4 weeks when the nest is full. The hen then only leaves the nest once or twice a day to eat and drink. Consistent and sustained warming of the eggs by the hen ensures that the development of the chicks within the eggs is essentially synchronous. Consequently, all chicks hatch at around the same time after 21 days of brooding. The chicks are raised exclusively by the hen, who shelters them from the elements and cares for them until they are self-sufficient.

Brooding activity is innate and commercial laying hens also exhibit this behaviour if they are given an opportunity to nest and tend to their eggs.

Female budgerigars lay eggs at an interval of approximately two days and commence incubation from the first

1.29 Eggs of the southern lapwing (Vanellus chilensis) (left), yellow winged blackbird (Agelaius thilius) (middle) and the wren-like rushbird (Phleocryptes melanops) (right). Courtesy of Professor Dr Daniel Gonzalez-Acuna, Chillan, Chile.

day of laying. The young birds hatch three weeks later at corresponding two-day intervals. Predominantly protein-rich food is provided to the hatchlings by the female. The young birds usually spend the first four weeks in the nest, with the male also participating in their care in the latter stages. Fledging occurs at approximately five weeks of age.

Brooding activity varies markedly among other species. In most European birds, the incubation period is 2–4 weeks, typically two weeks in songbirds. In other species, however, the incubation period can range from as little as ten days to as many as 80 days. It is poorly correlated with the size of the birds, and unrelated to the size of the clutch. Generally, both males and females contribute to incubation, although the involvement of the male is variable.

Brooding in hens is influenced by breed, day length and ambient temperature. The urge to brood, induced by elevated circulating prolactin levels, is driven by photoperiod and also by the pressure of the developing eggs (up to 30) on the underbelly of the hen. Brooding behaviour ceases after 21 days or when the first hatchlings begin to chirp.

Markedly different incubation behaviour is observed in several of the large ratites. Some allow their eggs to be incubated by the sun, while in others incubation is performed by the male. The male rhea, for example, is exclusively responsible for both incubating and rearing the chicks. The female wanders away and mates with additional males.

The avian egg is covered in a **calcified shell** and has a characteristic species-dependent **colour** and **shape**.

In chickens, egg colour varies with the **genetic strain** of the bird. White-shelled eggs are predominantly produced by smaller breeds arising from the Mediterranean and North-West European regions, by the so-called Polish chicken and related species, and by the bantam. Pure breed chickens with white earlobes also usually lay white eggs.

In contrast, brown, yellowish or yellow eggs are mainly produced by the medium to large Asiatic breeds, which are often characterised by having red earlobes. The South American araucana is known for its blue-green eggs.

Egg colour is the product of **three pigments**: **biliverdin** produces blue tones**, protoporphyrin** is responsible for red/yellow/brown shades and a **zinc-biliverdin chelate** produces a green colour. The final colour of the shell is determined by the relative proportions in which these pigments are secreted by the uterine epithelium.

It has generally been thought that the colour of the egg shell aids in **camouflage**, the pigmentation rendering the eggs less visible to predators. However, recent findings have expanded this theory. For example, concealment may not be the predominant role of the red-brown spots and patches found on the eggs of many birds. Rather, these may also play an important role in the **structural integrity** of the egg by **increasing the elasticity** of the shell. In the red-brown areas sometimes seen on the typically white eggs of Passeriformes, the shell is thinner and more flexible, providing a degree of shock absorption. Moreover, it has been established that egg shells containing protoporphyrin are better able to reflect incident infrared light than unpigmented calcified shell. These eggs therefore absorb less heat from the sun and lose less water through evaporation, reducing the likelihood of the developing embryo succumbing to dehydration.

Cardiovascular system

With some minor exceptions, the circulatory system of birds is similar to that of mammals. However, differences are observed in erythrocyte morphology, avian red blood cells being nucleated and relatively more numerous. Blood pressure is higher in birds and the **renal portal system** of their reptilian ancestors has been preserved.

Brain and sense organs

Relative to body weight, the brain of birds is five to 20 times larger than that of reptiles. Within the class Aves, relative brain weight is lowest in **ostriches, chickens** and **pigeons** and largest in **parrots**. The occipital lobes and the cerebellum are relatively large in birds.

The **sense organs**, especially the eyes, are **particularly well developed** in birds (with the exception of the kiwi). In relation to overall body size, avian eyes are considerably larger than those of domestic mammals. The muscles that control the movement of the eye are poorly developed. This is compensated for, however, by the mobility of the head. The **eyes** are usually **positioned laterally**, resulting in a **very wide field of vision**. **Stereoscopic vision**, on the other hand, is limited to an arc of around 6–10 degrees directly in front of the bird. The African penguin is at the extreme end of the spectrum, its eyes being directed so far laterally that the left and right visual fields do not overlap. A wide visual field is of particular significance for

1.30 Eye of a Magellanic horned owl (Bubo magellanicus). Courtesy of Professor Dr Daniel Gonzalez-Acuna, Chillan, Chile.

1.31 Eye position in a cocoi heron (Ardea cocoi). Courtesy of Professor Dr Daniel Gonzalez-Acuna, Chillan, Chile.

1.32 Head of a short-eared owl (Asio flammeus). Courtesy of Professor Dr Daniel Gonzalez-Acuna, Chillan, Chile.

ground-dwelling birds that are vulnerable to attack from any angle. In contrast, the eyes of owls, raptors and some other species are positioned further forward, giving rise to a binocular visual field of 60–70 degrees (Figure 1.32).

Birds, including domestic varieties, are able to perceive **colour**. Indeed, colouring is used as a prominent form of signalling and display, particularly by males. The exact manner in which colours are recognised and interpreted is difficult to establish and can only be speculated upon. Nevertheless, the high concentration of photoreceptors in the avian retina is consistent with well-developed colour perception.

Poultry and pigeons are capable of discerning red, yellow and green, but are less able to distinguish blue. Additional studies suggest that blue tones also play a limited role in many other, yet not all bird species. Nevertheless, birds appear to have excellent colour vision, particularly with respect to yellow and 'leaf green' shades. It is unclear whether a relationship exists between colour recognition by birds and their ability to perceive light in the UV spectrum. Noteworthy in this regard is the observation that UV light is reflected by the feathers of some, but not all, avian species. There is certainly considerable evidence that birds perceive the colours in their environment differently to humans, and probably with greater discrimination. This applies particularly to UV light perception, which plays an important role in enabling birds to recognise their own eggs, based on differences in the reflection of UV light by matt versus glossy shells.

Bird **migration** counts as one of the most formidable journeys undertaken by members of the animal kingdom. The mechanisms by which birds navigate and orient themselves during migration are incompletely understood. Preparation for migration is triggered by **changing day length**, which induces hypothalamic and hypophyseal secretions that act on the endocrine glands such as the thyroid gland. This usually corresponds with the end of the moult, and its associated changes in metabolism, and with the laying down of significant fat stores.

According to contemporary thinking, programming of the **migratory route** is genetically based. Many migratory species appear to utilise a 'sun compass' (polarised light). Having taken off in the right direction, they are able to identify the path ahead, probably using **physical landmarks** such as islands.

Vision is believed to be the **most important of the senses** during migratory flight. Birds that migrate at night make use of **fixed stars** to assist with navigation. Further evidence suggests that **wind direction** and **olfactory cues** are also relevant, and that some birds make use of the earth's **magnetic field**.

As well as being endowed with excellent vision, birds have a highly developed **auditory sense**, comparable to that of humans. Owls are the most accomplished avian species in terms of localising sound.

Olfaction is of limited significance for the majority of birds. Exceptions include several ducks and geese, the kiwi, various species of vulture and the Procellariidae (e.g., petrels and shearwaters), in which the sense of smell, and thus the olfactory bulb, are well developed. **Gustation (taste)** appears to be less developed in birds than in mammals.

Large numbers of **touch receptors** are found in the skin, at sites of feather attachment, in the oral cavity and beak and in the forearms and the muscles of the shank. These regions are also associated with perception of **pain** and **temperature**.

Locomotor apparatus

Skeleton (systema skeletale)

Osteology (osteologia)

Despite considerable species variation, the basic design of the skeleton is consistent across the class Aves (Figures 1.33 to 1.40). As in mammals, many of the bones of the avian skeleton develop from a cartilaginous template that is gradually replaced by bone.

1.33 Skeleton of a common buzzard (Buteo buteo).

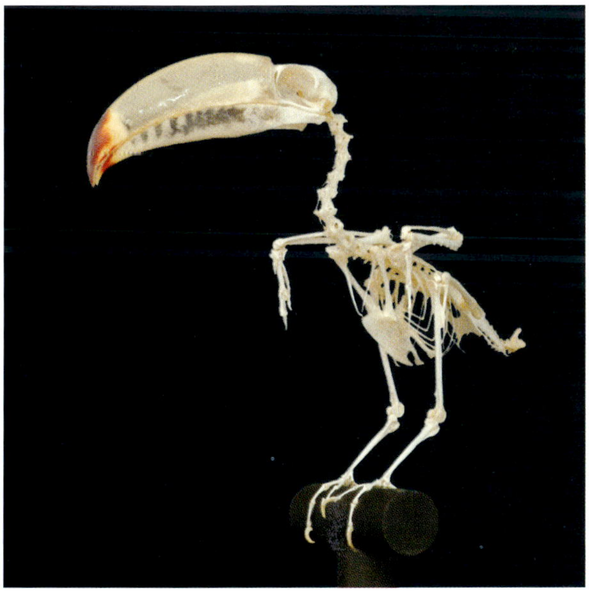

1.34 Skeleton of a toucan (Ramphastos).

In the long bones, this process begins with the conversion of connective tissue around the mid-section of the cartilage template into a thin bony collar, through a process known as **perichondral ossification** (strictly, **perichondral intramembranous ossification**, as the bone forms directly

1.35 Skeleton of a white stork (Ciconia ciconia).

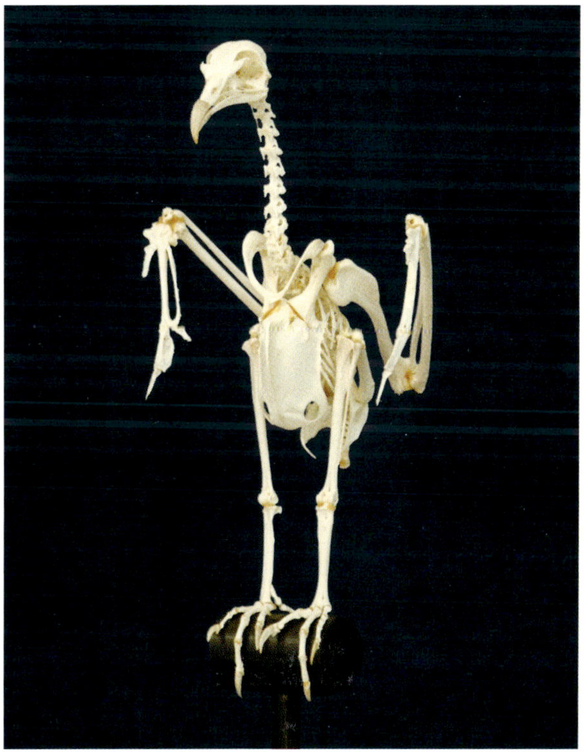

1.36 Skeleton of an Egyptian vulture (Neophron percnopterus).

1.37 Skeleton of a domestic pigeon (Columba livia dom.). The expanded sternal carina provides a large surface for attachment of the flight muscles.

1.40 Skeleton of a goose (Anser anser).

1.38 Skeleton of a chicken (Gallus gallus dom.).

1.39 Skeleton of Mallard duck (Anas platyrhynchos).

from connective tissue). The bone formed in this way consists of immature, woven bone. This rigid shell restricts the widening of the cartilage model, resulting in **lengthwise growth**. Thus, at the ends of the cartilage model chondrocytes become arranged in columns. With ongoing ossification, the bony collar extends in both directions towards the extremities.

In the next stage, the cartilage of the **diaphysis** is replaced by bone matrix (**endochondral ossification**). Blood vessels invade the interior of the cartilage model, along with various cell types, including osteoblasts and osteoclasts.

Resorption of cartilage results in the formation of marrow spaces and bony trabeculae. During ongoing development, primitive bone is resorbed and remodelled, resulting in the formation of lamellar bone. This also occurs from the middle of the shaft, or diaphysis, towards the extremities. In contrast to mammals, there are only **a few** bones in which **secondary centres of ossification develop** (Figure 1.42). There are also no true **epiphyseal (growth) plates**, although zones of proliferation are evident. Notable in birds, compared with mammals, is that blood vessels are found in these regions. In the **tibiotarsus**, the proximal row of tarsal bones is fused with the tibia. This is recognisable in young birds as a separate centre of ossification (Figure 1.42). The same applies to the distal row of tarsal bones, which becomes completely fused with the metatarsal bones in the adult to form the **tarsometatarsus**. The distal row of carpal bones is similarly joined with the metacarpals, forming the carpometacarpus.

Skeletal maturity occurs at a **relatively early age** in birds. Of the original cartilaginous model, only the hyaline cartilage at the articular surfaces remains. Cartilage is also found in the intervertebral discs and menisci of adult birds.

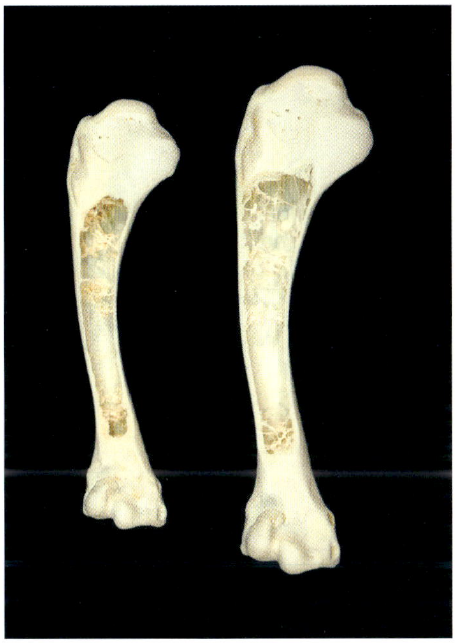

1.41 Right humerus of a hen (Gallus gallus dom.) (left) and of a cock (Gallus gallus dom.) (right). Both bones are pneumatised (medullary cavity has been opened).

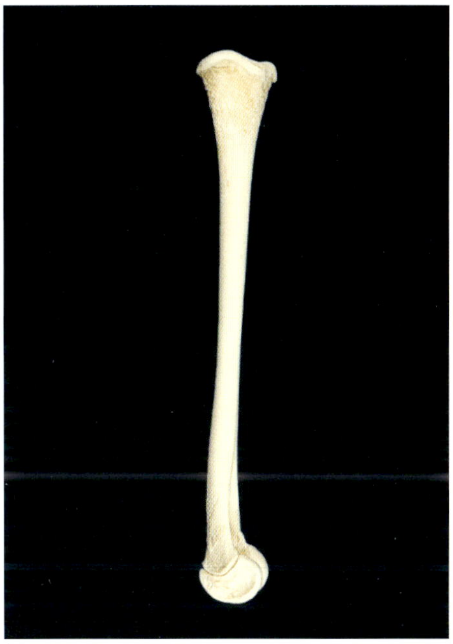

1.42 Left tibiotarsus of a young chicken (Gallus gallus dom.) with distal centre of ossification.

STRUCTURE OF MATURE BONE

Relative to body weight, the skeleton of birds is **lighter** than that of mammals. Cell types found in bone include numerous osteoblasts and osteocytes, as well as osteoclasts. These are responsible for the maintenance and remodelling of bone.

Around half of the extracellular bone matrix is made up of **mineral** compounds, predominantly in the form of **hydroxyapatite crystals** (calcium phosphate [85 per cent], calcium carbonate [10 per cent]). The remaining components include sodium, magnesium, nitrate, fluorine and trace elements. Approximately a quarter of the matrix is composed of **organic macromolecules**, primarily **collagen** (90–95 per cent). Others include various glycosaminoglycans (chondroitin-4-sulphate, chondroitin-6-sulphate, keratan sulphate) and proteoglycans. The water content of **extracellular bone matrix** is approximately 25 per cent.

Due to the high proportion of inorganic substances, the bones of birds are relatively brittle and have a tendency to splinter. These characteristics are an important consideration in the management of fractures.

For the purpose of **weight reduction**, several bones (e.g., humerus, coracoid, sternum) are **pneumatised** (air-filled; Figure 1.41). In the head, the paranasal sinuses communicate with the nasal cavity or the auditory tube. In the torso, **air sacs** send out diverticulae (see Chapter 7 'Respiratory system') that pass through openings in the bone (foramina pneumatica or pori pneumatici) into the medullary cavity, particularly in long bones, and displace the bone marrow (Figure 1.41). This system is developed more extensively in birds that are good flyers.

The **bone marrow** of young animals is red. With increasing age, this transforms into yellow marrow.

TYPES OF BONE

In addition to compact bone and spongy bone, a third specialised form, known as medullary bone, is found in birds. **Compact bone** (Figures 1.43 and 1.44), as seen in the diaphysis of long bones, exhibits essentially the same lamellar structure as seen in mammals. **Spongy bone** (Figure 1.43) is found in the epiphyses of long bones and in the vertebrae.

Medullary bone (Figures 1.43 and 1.44) is a form of **calcium reservoir** found in female birds during the breeding period. In the hen, for example, the high demand for calcium required for egg shell formation cannot be met exclusively through intestinal absorption. Consequently, osteoclasts are stimulated to liberate calcium through bone resorption.

Formation of medullary bone occurs in well vascularised long bones approximately **two weeks before the start of laying**. In the hen, medullary bone production continues throughout the entire laying period. Medullary bone is formed by the ingrowth of bony spicules from the endosteum into the medullary cavity. In the chicken, much of the medullary cavity comes to be occupied by medullary bone. Osteons are not formed in medullary bone, with collagen fibres and hydroxyapatite crystals exhibiting no particular orientation. This lack of organisation reflects the non-structural nature of this specialised type of bone.

Medullary bone formation is initiated by the hormonal changes that precede the onset of egg laying. Under the influence of oestrogens and androgens, absorption of

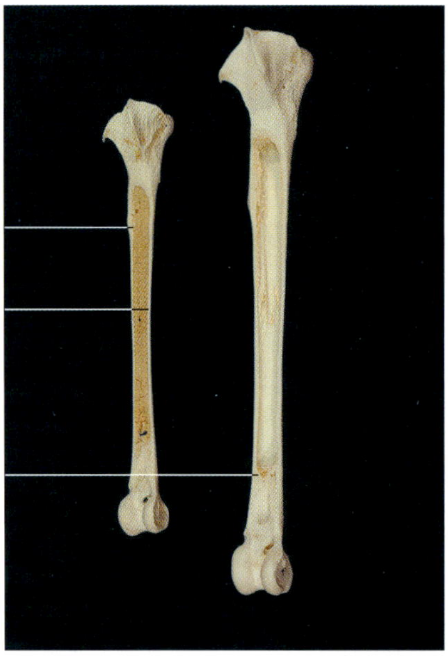

1.43 Left tibiotarsus of a hen (Gallus gallus dom.) with medullary bone (left) and of a cock (Gallus gallus dom.) with bone marrow removed (right).

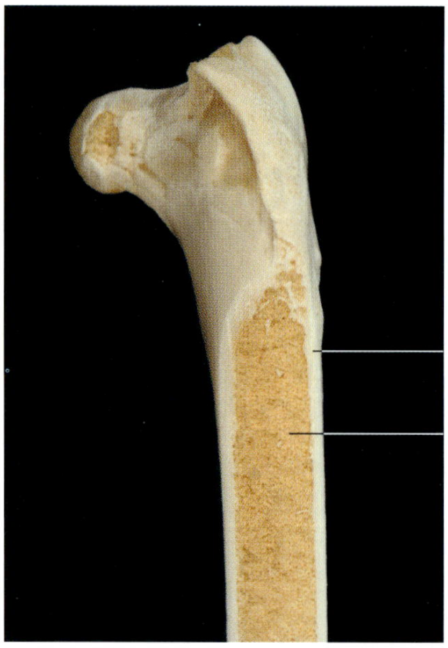

1.44 Left femur of a hen (Gallus gallus dom.) with medullary bone (medullary cavity opened).

calcium and phosphate from the gut is increased, and the medullary bone becomes well mineralised. As a result, the weight of the skeleton is increased by approximately 20 per cent. **During the ovulation–oviposition cycle** the medullary bone undergoes alternating phases of growth and regression. When the uptake of calcium and phosphate is insufficient for normal calcification of the egg shell, spicules of medullary bone are resorbed, becoming thinner and shorter. Thus, the medullary bone functions as a depot that can be used to balance out fluctuations in the absorption of dietary minerals.

The presence of medullary bone must be taken into consideration when interpreting bone density using diagnostic imaging (radiography, computed tomography).

In the event of **hormonal disturbances,** medullary bone can also be produced by male birds. Oestrogen-secreting Sertoli cell tumours in budgerigars can result in extensive osteogenesis, such that even pneumatised long bones (e.g., humerus) may become completely filled with bone.

Arthrology (syndesmologia)

The joints of birds are essentially comparable with those of mammals. Joint mobility is variable, depending on whether the bones are joined by fibrous connective tissue (**syndesmoses**), cartilage (**synchondroses),** bone (**synostoses**) or a joint capsule (**diarthroses**). Highly tensile collagenous connections are exemplified by the fibrous bands joining the bones of the shoulder girdle.

Cartilaginous joints include those between the individual vertebrae of the tail. Bony unions are generally more common in birds than in mammals. The bones of the adult skull are completely fused, such that it is difficult to identify lines of demarcation. Extensive fusion of the thoracic and lumbar vertebrae (**notarium, synsacrum**) assists in stabilising the skeleton.

The **freely movable joints** (diarthroses; synovial joints) exhibit the same typical features seen in mammals. These include:

- extremities lined with articular cartilage (cartilago articularis),
- joint capsule (capsula articularis) with
- ligaments (ligamenta articularia) and
- an enclosed joint cavity (cavum articulare) containing
- synovial fluid (synovia).

Synovial joints are found in the limbs and between the bodies and articular processes of the cervical vertebrae. The classification of synovial joints with respect to shape is similar to that used in mammals. Articular incongruities, as found in the knee and intertarsal joints, are evened out by fibrocartilaginous menisci or articular discs.

Myology (myologia)

The skeletal musculature of birds contains a lower proportion of connective and adipose tissue than that of mammals. **Muscle fibre density** is generally higher in birds. The functions of the connective tissue are to:

- bind muscle fibres together,
- provide channels for vessels and nerves,
- allow movement of individual muscle bundles,
- enclose the muscle belly while allowing it to move with respect to its surroundings.

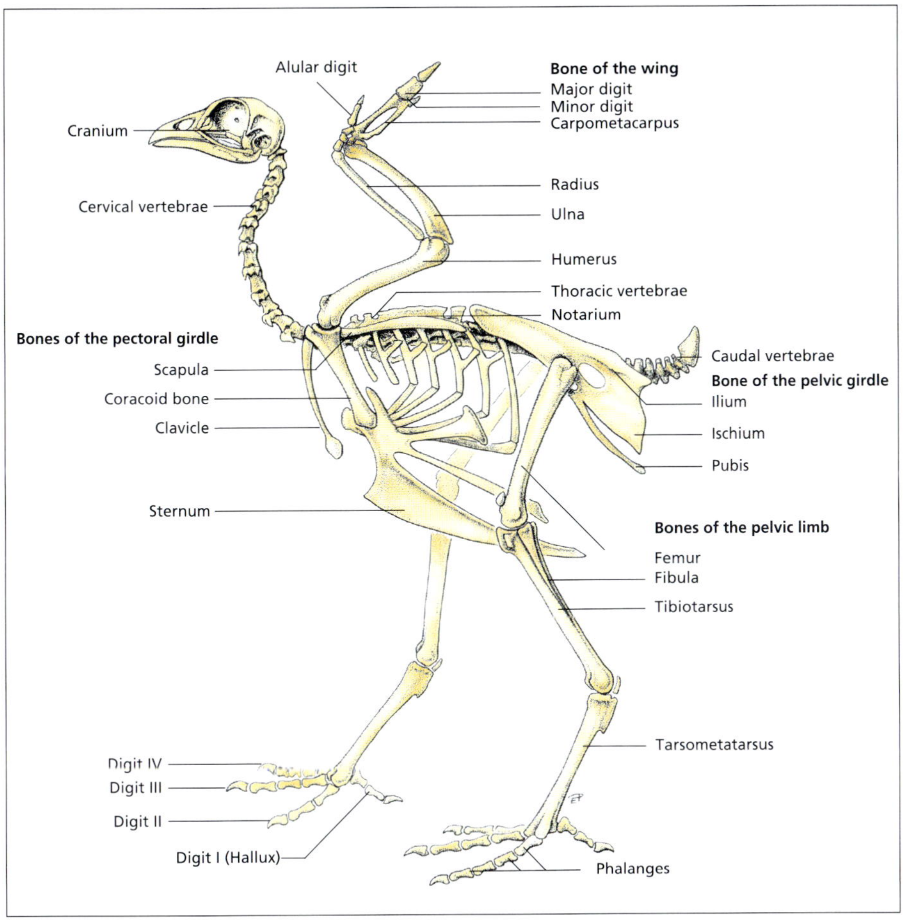

1.45 Skeleton of the chicken (Gallus gallus dom.) (schematic; left lateral view).

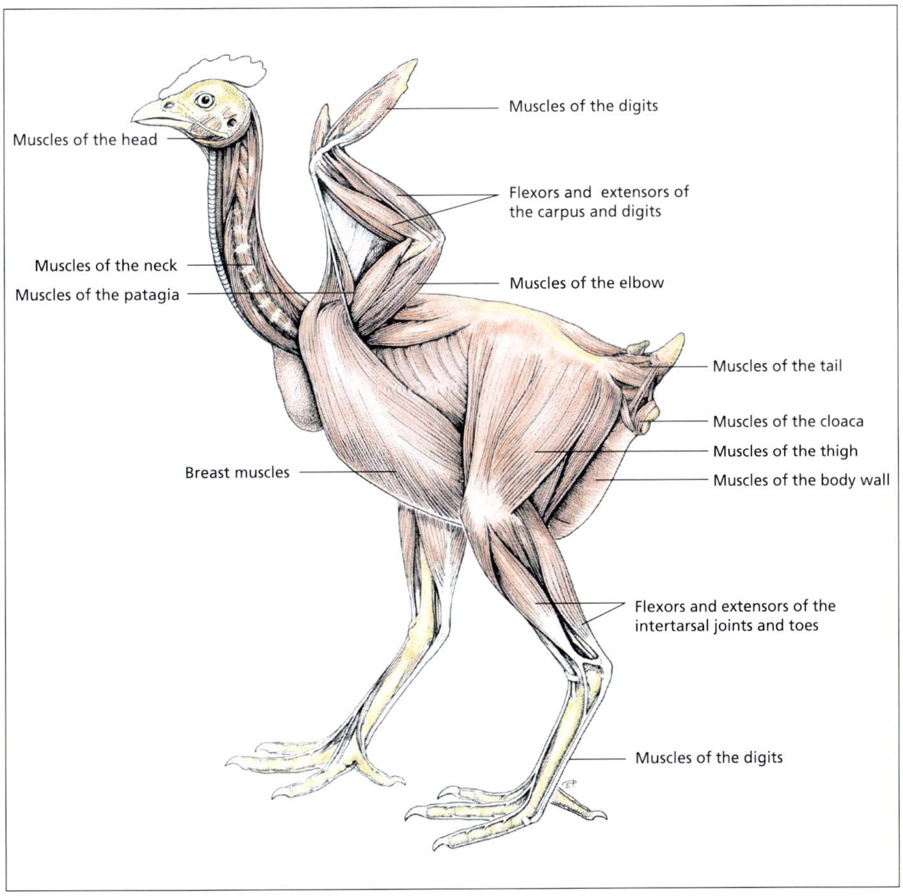

1.46 Superficial and middle muscle layers of the chicken (Gallus gallus dom.) (schematic; left lateral view).

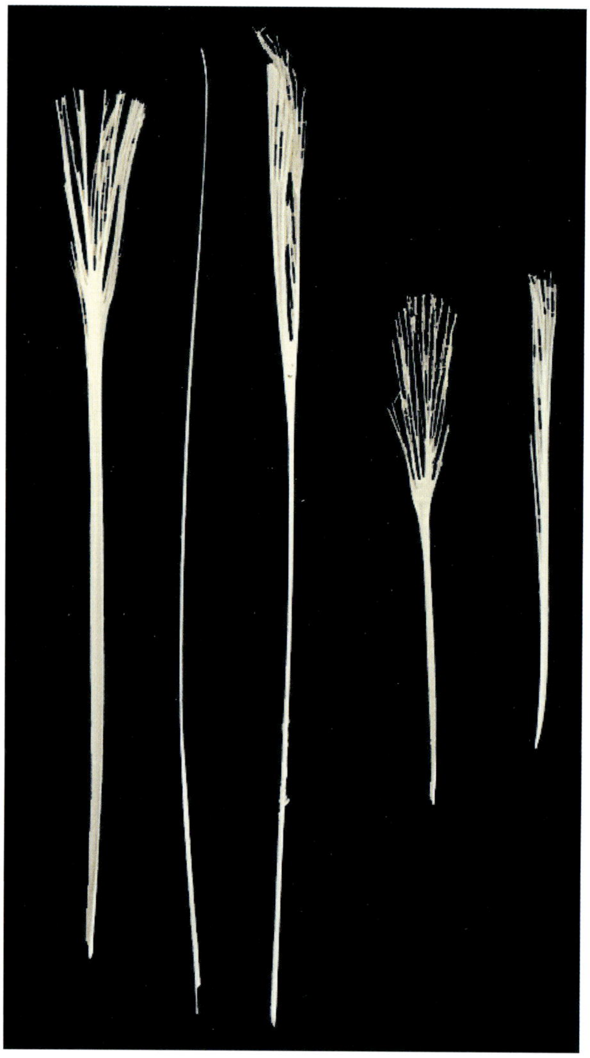

1.47 Ossified tendons of pelvic limb muscles of various species of crane (Grus spp.).

Connective tissue elements also serve to subdivide the muscle tissue: individual muscle fibres are surrounded by **endomysium**. Groups of muscle fibres are enclosed within **perimysium** to form fasciculi. Bundles of fasciculi make up the muscle belly, which is encased in the **epimysium**.

The basic **histological structure** of muscle is similar in birds and mammals. As such, a detailed description is not included here.

In the chicken, light (white) and dark (red) muscle can be distinguished **macroscopically**. The difference in colour reflects the varying proportions of different muscle fibre types and their corresponding myoglobin content. Light muscle fibres have a proportionally higher concentration of muscle fibrils, but contain less myoglobin and cytochrome. These fibres rely mainly on anaerobic glycolysis for energy production and therefore contain plentiful stores of glycogen. In terms of function, these cells are characterised by **fast contraction** ('fast twitch fibres') and rapid fatigue.

In contrast, the cytoplasm of dark muscle fibres contains fewer myofibrils. These cells also contain much greater quantities of myoglobin and enzymes of the electron transport chain, the latter stemming from their large numbers of mitochondria. These features enable the cells to utilise fat as an energy source. Since these cells are slower to contract ('slow twitch fibres') and fatigue, they are well suited to sustained effort. **Intermediate muscle fibres** have functional characteristics of both light and dark fibres.

In general, muscles contain a mixture of **all three fibre types**. Species differences exist, however, with the composition of individual muscles varying according to requirements. The flight musculature of good flyers, such as the pigeon, consists predominantly of dark muscle. In the hummingbird, the pectoral muscle is composed exclusively of dark fibres. Diving birds have deep red muscles. Oxygen stored in the muscle cells is released during diving and utilised for aerobic metabolism.

The **musculature** of birds is adapted for the particular requirements of flight or terrestrial movement, and is therefore distributed differently compared with mammals (Figure 1.46). The muscles of the breast and forelimb are well developed, as are those of the pelvis and thigh. In addition, the well-differentiated neck muscles contribute to balance during flight, while the powerful tail musculature controls the steering feathers (rectrices).

In contrast, the muscles of the thoracic and lumbar vertebrae are notably less developed, there being less need for muscular support in these relatively fixed and inflexible segments of the vertebral column.

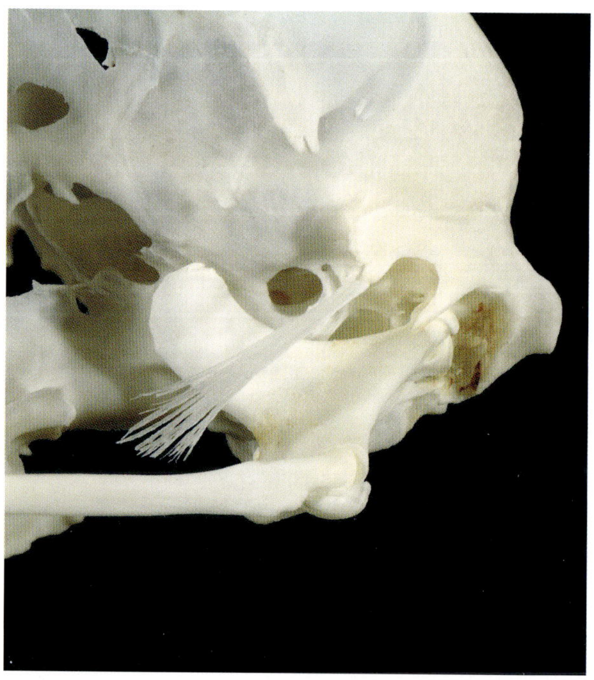

1.48 Ossified tendon at the origin of the elevator of the lower jaw (m. adductor mandibulae externus rostralis) in a crane. Courtesy of Dipl.-Biol. Martin Kobienia, Munich.

The bellies of the limb muscles are located proximally, as close to the body as possible, to facilitate the positioning of the **centre of gravity** close to the middle of the body. Consequently many of the tendons of the limb muscles are very long. These tendons are also particularly likely to exhibit evidence of ossification (Figure 1.47).

Ossification of tendons also occurs in the long muscles of the back. This provides additional stabilisation and is especially important in birds such as ducks and geese in which the thoracic vertebrae are not fused.

The **diaphragm is absent in birds**. Thus, the muscles of the thoracic and abdominal walls are utilised for respiration. The driving force for breathing is the expansion or contraction of the rib cage. During **inspiration**, this is achieved on the one hand by movement of the ribs, and on the other by elevation of the sternum at the sternocoracoid joints (such that the body of the sternum is displaced cranioventrally). These movements are reversed for **expiration**. It is crucial to take this into account when restraining a bird during physical examination to prevent asphyxiation (see also Chapter 20).

Head and trunk

2

J. Maierl, H.-G. Liebich, H. E. König and R. Korbel

Skeleton of the head

Several features of the avian skull reflect an **ancestral relationship with reptiles**. The unpaired occipital condyle, for example, permits a greater range of movement than in mammals. Furthermore, the quadrate and pterygoid bones are mobile, and the quadrate bone forms a joint with the articular bone of the mandible. The mandible is usually composed of five or six smaller bones.

The bones of the head consist of thin plates that are formed either from connective tissue or from cartilaginous templates. To assist with flight, the bones of the head are extensively pneumatised. This is facilitated by the fusion of these bones in the relatively early stages of growth. Remnants of sutures are rarely seen in adult birds. The skull is divided into two components, the **cranium** (neurocranium) and the **skeleton of the face** (viscerocranium).

Bones of the head

Cranium (ossa cranii)

The external shape of the cranium is defined principally by the large **orbits** (orbitae) and the ample **cranial cavity** (cavum cranii) (Figures 2.2 and 2.3). These provide protection for the substantial eyeballs and the brain. The orbits form a rostral indentation in the bulbous skeleton of the cranium (Figure 2.2). Since the volume of the brain increases relatively little with respect to body size, smaller birds have a comparatively larger head than bigger species.

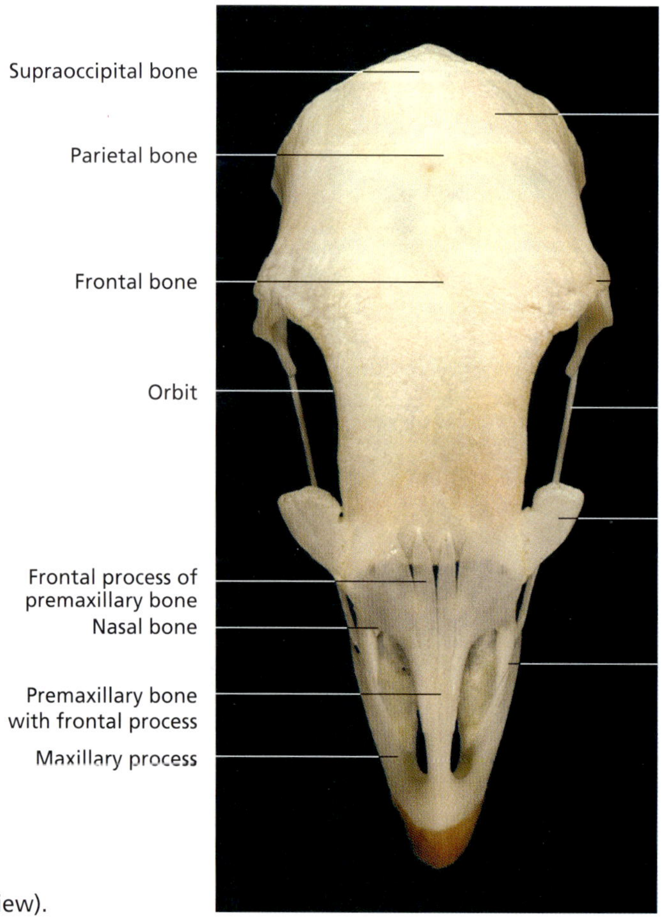

2.1 Skull of a chicken (dorsal view).

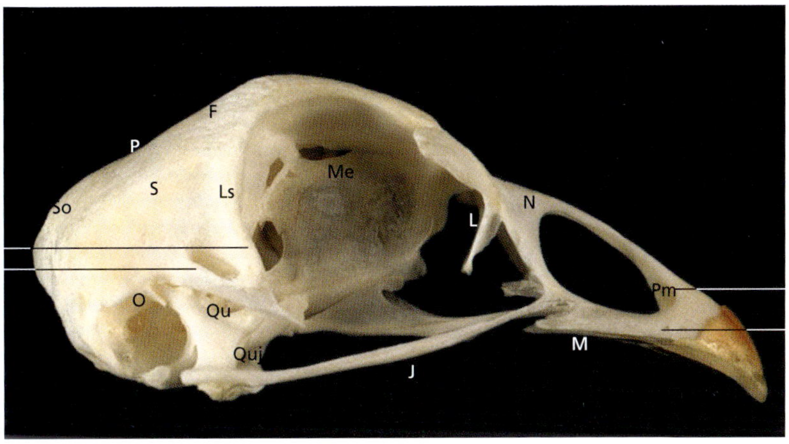

2.2 Skull of a chicken (right lateral view).

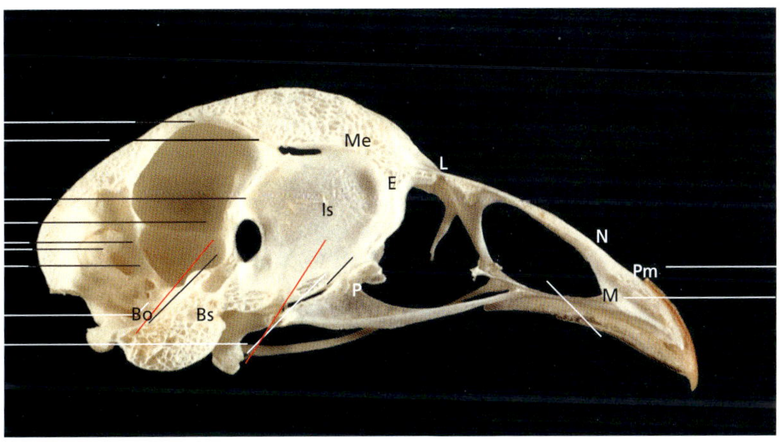

2.3 Skull of a chicken (paramedian section, left side of skull viewed from the right).

The bones of the cranium (Figures 2.1ff.) are the:

- **occipital bone (os occipitale)**, comprising the:
  - unpaired basioccipital bone (os basioccipitale),
  - unpaired supraoccipital bone (os supraoccipitale),
  - exoccipital bones (ossa exoccipitalia);
- **sphenoid bone (os sphenoidale)**, comprising the:
  - unpaired basisphenoid bone (os basisphenoidale),
  - laterosphenoid bone (os laterosphenoidale),
  - unpaired parasphenoid bone (os parasphenoidale);
- **squamosal bone (os squamosum)**,
- **parietal bone (os parietale)**,
- **frontal bone (os frontale)**,
- **otic bones (ossa otica)**, comprising the:
  - epiotic bone (os epioticum),
  - opisthotic bone (opisthoticum),
  - pro-otic bone (os pro-oticum),
  - metotic bone (os metoticum);
- **unpaired mesethmoid bone** (os mesethmoidale),
- **unpaired ectethmoid bone** (os ectethmoidale) and
- **lacrimal bone (os lacrimale)**.

The **occipital bone** (os occipitale) (Figure 2.4) forms the caudal wall of the brain case and surrounds the **foramen magnum**, through which the spinal cord exits the skull. It consists of the basioccipital bone (os basioccipitale), supraoccipital bone (os supraoccipitale) and the paired exoccipital bones (ossa exoccipitalia). A significant feature of this region is the **parabasal fossa** (fossa parabasalis) that provides passage for blood vessels and nerves.

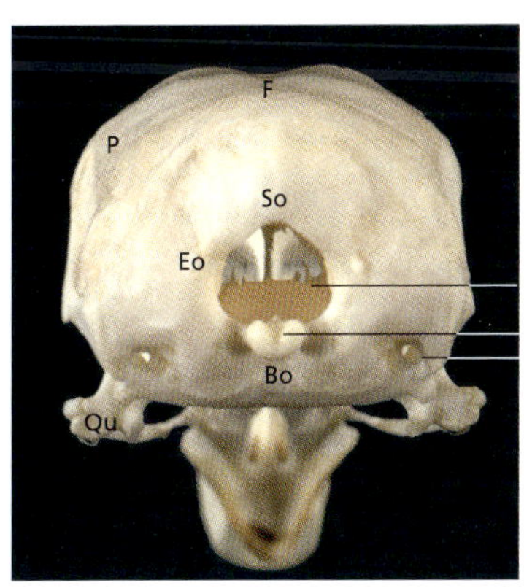

2.4 Skull of a chicken (caudal view).

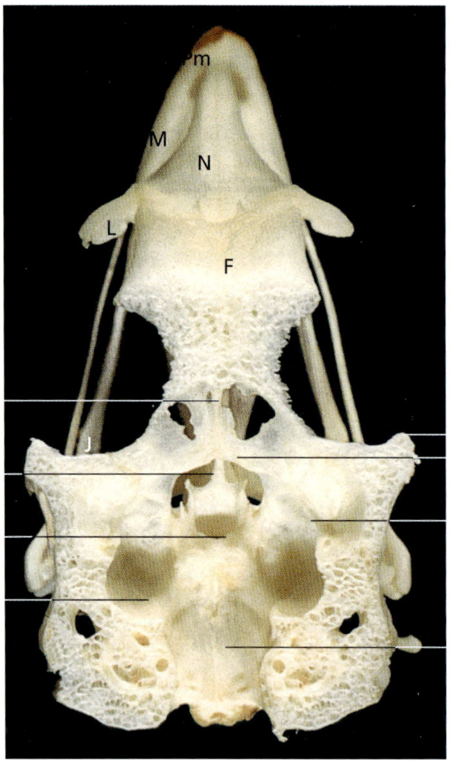

2.5 Skull of a chicken (caudodorsal view, cranial cavity opened).

The **basisphenoid bone** (os basisphenoidale) lies rostral to the basioccipital bone (Figure 2.3), forming part of the base of the skull. Further rostrally, the **parasphenoid bone** (os parasphenoidale) completes the floor of the cranial cavity. The parasphenoid lamina (lamina parasphenoidalis) constitutes the major part of the base of the cranium. From there, the parasphenoid rostrum (rostrum parasphenoidale) extends rostrally, connecting the parasphenoid bone with the palate and forming the ventral component of the **interorbital septum** (septum interorbitale) (Figure 2.3). The **exoccipital bones** are positioned caudolaterally, on either side of the base of the skull (Figure 2.4).

The **laterosphenoid bone** (os laterosphenoidale) (Figure 2.2) forms part of the caudoventral wall of the orbit. Several nerves penetrate this bony plate to reach the orbit and the facial component of the skull. The **squamosal bone** (os squamosum) (Figure 2.2) lies further dorsal. Together with the **post-orbital process** (processus postorbitalis) of the laterosphenoid bone, the squamosal bone forms much of the lateral wall of the cranium. The **temporal fossa** (fossa temporalis) lies caudal to the post-orbital process.

The fused **otic bones** (ossa otica) are interposed within the ventrocaudal aspect of the cranium (Figure 2.2). This bony complex houses the sense organ responsible for balance and hearing (see Chapter 16 'The ear').

The dorsal wall of the cranium is formed by the **frontal bone** (os frontale) and the **parietal bone** (os parietale) (Figure 2.1). Rostral processes of the frontal bone create a roof-like projection over the interorbital septum. The **mesethmoid bone** (os mesethmoidale) makes up much of the septum, creating an almost complete bony division between the orbits (Figure 2.3) (see also Chapter 15 'The eye', Figure 15.2). Rostrally, the orbit is separated from the nasal cavity by the **ectethmoid bone** (os ectethmoidale). The **lacrimal bone** (os lacrimale) forms the rostral and ventral continuation of the lateral orbital margin (Figure 2.3).

The base of the cranial cavity is divided into the **rostral cranial fossa** (fossa cranii rostralis), **the middle cranial fossa** (fossa cranii media) and the **caudal cranial fossa** (fossa cranii caudalis) (Figures 2.3 and 2.5). The **rostral compartment** houses the olfactory bulb. Accommodated within the **middle cavity** are the diencephalon (ventral, superficial components), the optic nerve and optic chiasm, and the hypophysis within the hypophyseal fossa (Figures 2.3 and 2.5).

The **fossa for the mesencephalic tectum** (fossa tecti mesencephali) (Figure 2.5) lies lateral and caudal to the middle cranial fossa. This is joined caudally by the caudal cranial fossa, the surface of which contains an impression formed by the **medulla oblongata**.

The curved roof of the cranium contains two large cavitations (Figure 2.3): the rostral **cerebral fossa** (fossa cerebri) for the cerebral hemispheres and the caudal **cerebellar fossa** (fossa cerebelli) that encloses the cerebellum. A lateral excavation, the **internal acoustic fossa** (fossa acoustica interna), is traversed by the vestibulocochlear and facial nerves as they exit the cranial cavity (Figure 2.3).

The **tympanic cavity** (cavum tympanicum) and the orbits, including their species-specific features, are addressed in further detail in Chapter 16 'The ear' and Chapter 15 'The eye'.

Skeleton of the face (ossa faciei)

The bones of the facial portion of the skull are the:

- premaxillary bone (os premaxillare),
- nasal bone (os nasale),
- palatine bone (os palatinum),
- maxillary bone (os maxillare),
- jugal/quadratojugal bone (os jugale/os quadratojugale),
- vomer (usually unpaired),
- pterygoid bone (os pterygoideum),
- quadrate bone (os quadratum),
- mandible (mandibula) and
- hyobranchial apparatus (apparatus hyobranchialis, unpaired).

The boundary between the cranium and the skeleton of the face is located in front of the eyes at the level of the zona flexoria between the forehead and the nasal bone.

The conformation of the **facial skeleton** (facies, viscerocranium) is influenced considerably by the shape and mobility of the beak. In the chicken, the beak resembles a pyramid with the base directed towards the eyes. The facial bones take the form of **thin plates** or **delicate bony rods**.

They enclose the nasal cavity and provide the osseous foundation for the **upper** and **lower beak** (Figures 2.6 to 2.12).

The rostral tip of the face is formed by the **premaxillary bone** (Figures 2.1ff.), which varies in shape according to species and is richly supplied by nerves (ophthalmic nerve, V3) and blood vessels. The premaxillary bone has three projections (Figures 2.1 and 2.2), comprising the:

- **palatine process (processus palatinus)**, forming the rostral border of the hard palate,
- **frontal process (processus frontalis)**, which forms a flexible attachment to the frontal bone, and the
- **maxillary process (processus maxillaris)**, of which the free border forms the edge of the upper beak.

The bony nostril is bounded by the frontal and maxillary processes of the premaxillary bone, and the delicate nasal bone (Figures 2.1 and 2.2) The **premaxillary** and **maxillary** processes of the nasal bone form the caudodorsal angle of the nostril.

The **palatine bone** forms the continuation of the incomplete hard palate. It extends caudally as far as the parasphenoid rostrum (rostrum parasphenoidale) with which it articulates (Figures 2.3 and 2.16).

The relatively small **maxillary bone** (Figure 2.3) joins the premaxillary bone to form the short, caudal terminal portion of the upper beak. Its **palatine process** also forms part of the hard palate. The **jugal process** (Figure 2.3) establishes a connection between the maxillary bone and the fused (rostral) **jugal** and (caudal) **quadratojugal**

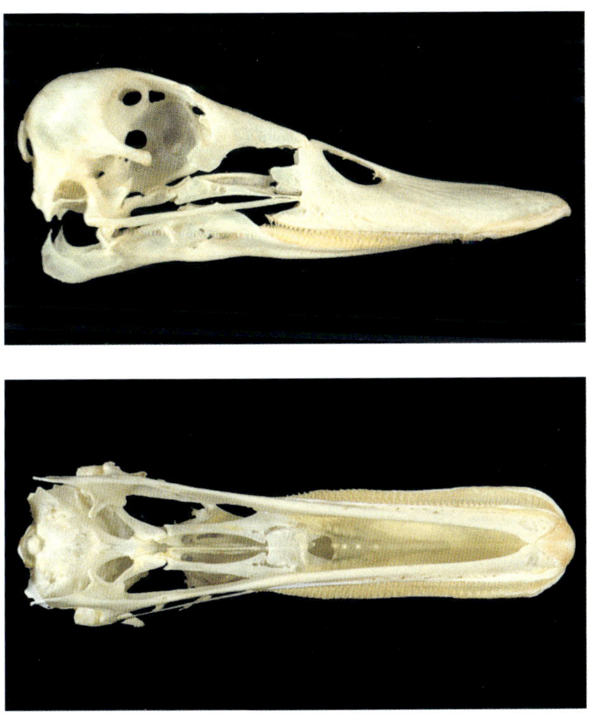

2.6 Skull of a mallard duck (lateral and ventral view).

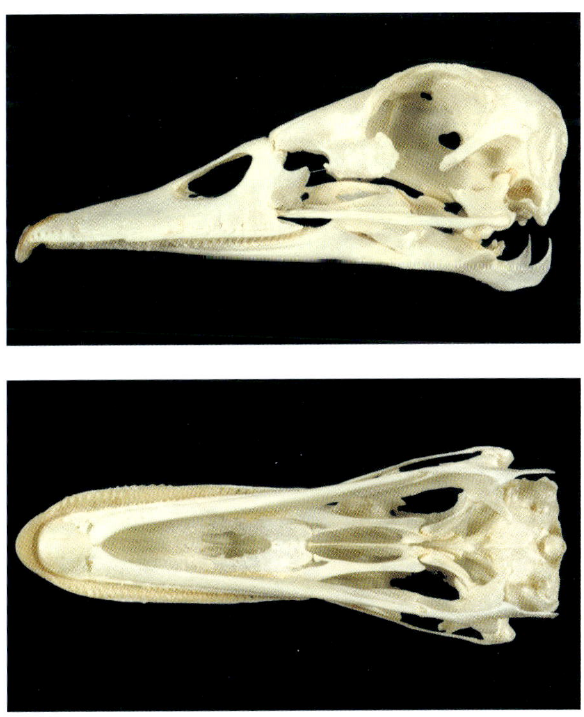

2.7 Skull of a domestic goose (lateral and ventral view).

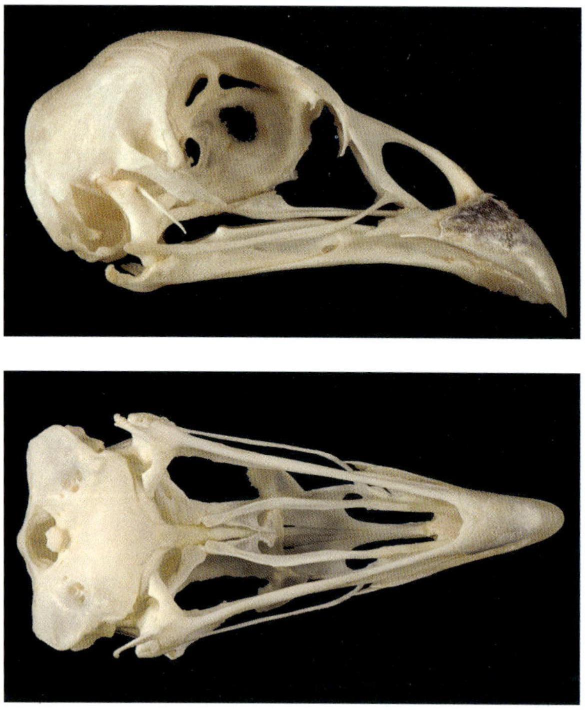

2.8 Skull of a common pheasant (Phasianus colchicus subspec.) (lateral and ventral view).

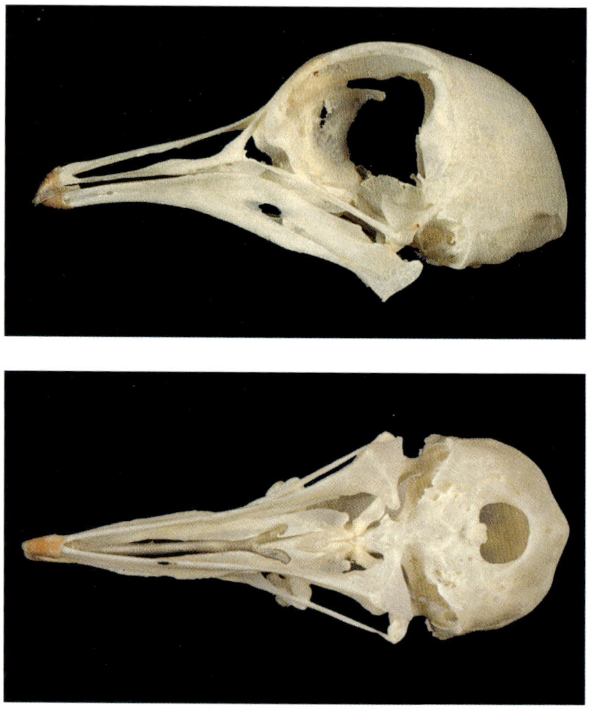

2.9 Skull of a domestic pigeon (Columba livia domestica) (lateral and ventral view).

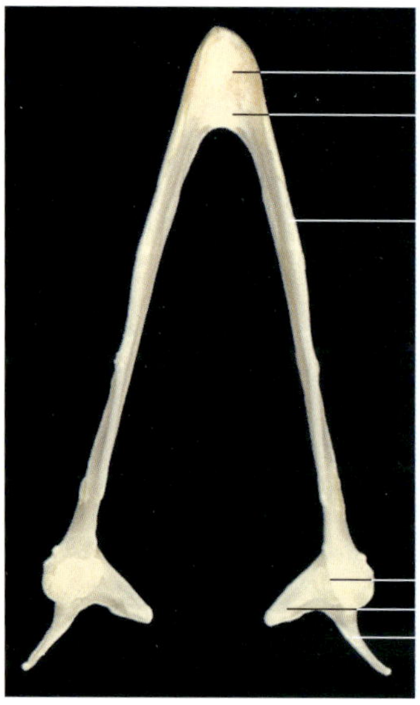

2.10 Mandible of a chicken (dorsal view).

bones. These latter bones form a long thin bar, the **jugal arch** (arcus jugalis or zygomaticus), that joins the **quadrate bone** (Figure 2.2).

The **vomer** is rudimentary in chickens, while in ducks and geese it completes the nasal septum.

The **pterygoid bone** (Figure 2.12) is bar-shaped. At its rostral end, it forms a gliding joint with the parasphenoid rostrum and also articulates with the palatine bone. Together, the pterygoid and palatine bones form the palatoquadrate bridge. The caudal end of the pterygoid bone articulates with the **quadrate bone** (Figures 2.15 and 2.16).

The quadrate bone plays a key role in the movement of the maxillopalatine apparatus (see below). It articulates with the **mandible** to form the principal **joint of the lower jaw**. The temporomandibular joint of mammals is equivalent to the secondary joint of the mandible in birds.

The quadrate bone has three processes. A caudodorsally directed **otic process** extends from the body of the quadrate bone to articulate with the squamosal and prootic/episthotic bones. The **mandibular process** articulates with the mandible. It also forms movable joints with the quadratojugal bone (laterally) and the pterygoid bone (medially). The **orbital process** (Figure 2.15) serves as a site of muscular attachment.

The **mandible** (Figures 2.10 and 2.12) is a laterally flattened bone that makes only a limited contribution to the vertical dimensions of the head. Its **six pairs of fused bones** form an acutely angled, caudally open structure (Figure 2.10).

From rostral to caudal, the bones of the mandible consist of the:

- dental (dentary) bone (os dentale),
- splenial bone (os spleniale),
- angular bone (os angulare),
- supra-angular bone (os supra-angulare),
- prearticular bone (os praearticulare) and
- articular bone (os articulare).

The apical region of the mandible (**rostrum mandibulae**) is formed by the two dental bones, joined at the **mandibu-**

lar symphysis. A channel within the mandible allows for the passage of nerves and vessels (canalis neurovascularis mandibulae). The sharp dorsal edge of the mandible, the **tomial crest** (crista tomialis) (Figure 2.12), lies opposite its maxillary equivalent in the upper jaw.

The **articular bone** bears an articular surface, the **quadrate articular fossa** (fossa articularis quadratica), for the joint between the mandible and the quadrate bone. Lateral and medial processes of the articular bone act as sites of muscle attachment.

The **hyobranchial apparatus** (apparatus hyobranchialis) (Figures 2.13 and 2.14) consists of the:

- paraglossum (or entoglossum),
- basihyale (basibranchiale rostrale),
- urohyale (basibranchiale caudale) and the
- bilateral cornu branchiale, consisting of the ceratobranchiale and the epibranchiale.

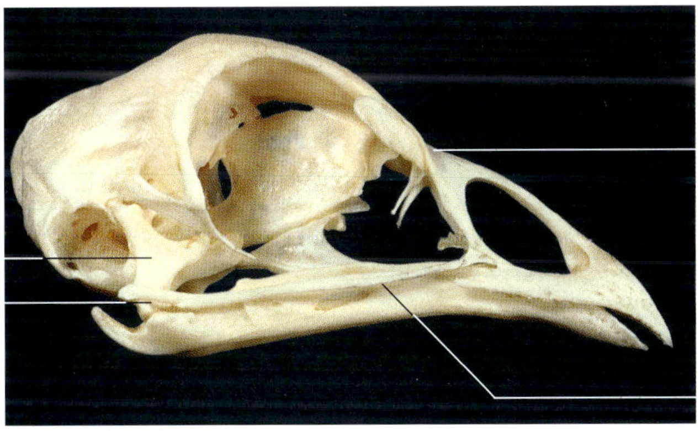

2.11 Skull of a chicken with raised mandible (right lateral view).

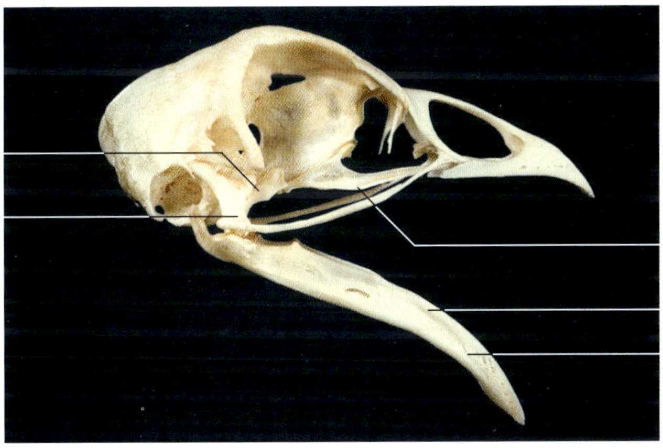

2.12 Skull of a chicken with lowered mandible (right lateral view).

2.13 Hyobranchial apparatus of a chicken (lateral view).

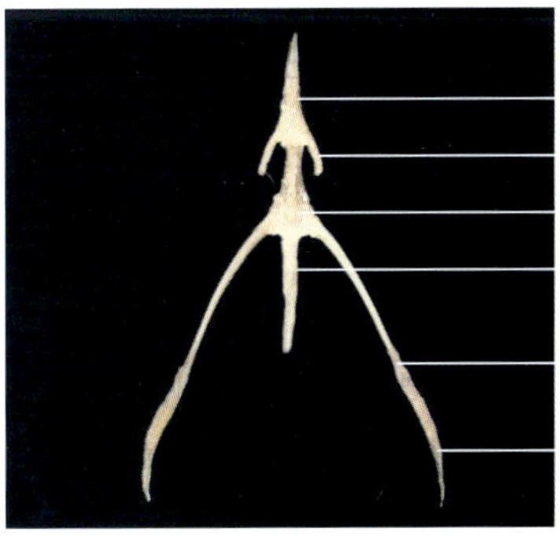

2.14 Hyobranchial apparatus of a chicken (dorsal view).

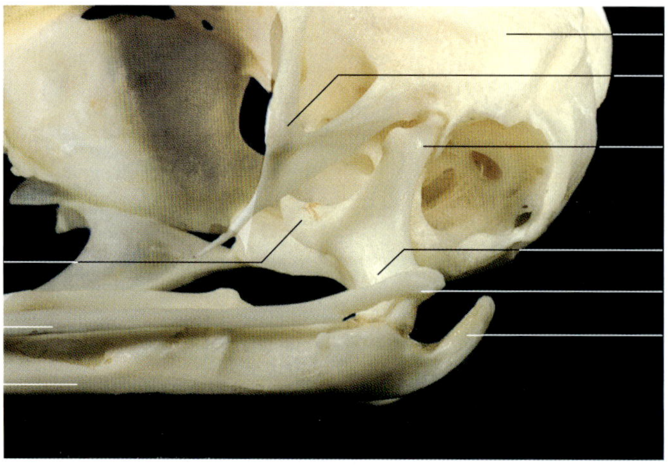

2.15 Closed mandibular joint of a chicken (left lateral view).

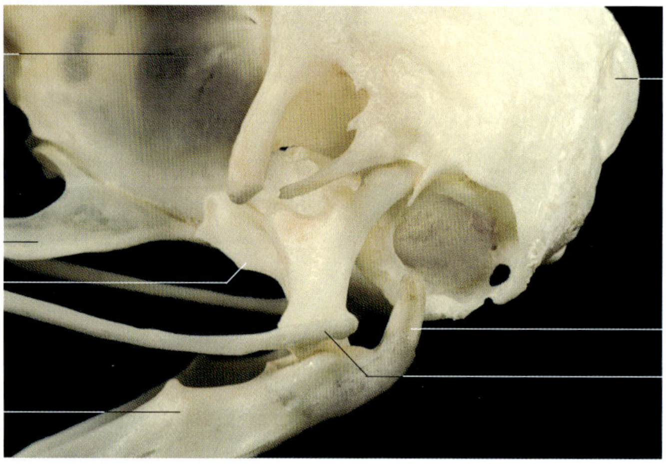

2.16 Open mandibular joint of a chicken (left lateral view).

Together, the paraglossum, basi- and urohyale form an unpaired rod that lies in the middle of the tongue. The paired **cornua branchialia,** comprising the **ceratobranchiale** and the **epibranchiale**, are attached to the basihyale. In contrast to mammals, the cornua do not end at the base of the skull. Instead, they form a sling that extends to the caudal surface of the cranium (Figure 2.17). This arrangement allows for considerable freedom of movement, as exemplified by the highly protrusible tongue of the woodpecker.

Joints of the head

The sutures that are initially evident between the individual bones of the cranium become ossified in early development. **Diarthroses** of the head serve primarily to **open the beak**. They are divided into **two** groups, the joints of the upper jaw (sometimes referred to simply as the maxilla) and the joints of the mandible (Table 2.1).

The first group exclusively serves the **maxillopalatine apparatus** (Figures 2.11 and 2.12), the function of which is centred around the **quadrate bone** and its articular connections with the base of the skull. During opening of the beak, the quadrate bone (Figures 2.15 and 2.16) moves dorsally and rostrally (via the quadrato-squamoso-otic and quadrato-pterygoid joints), causing the bones of the palatoquadrate bridge (pterygoid and palatine) and the jugal arch to slide rostrally.

The pterygoid bone articulates with the medial aspect of the quadrate bone (caudally) and with the parasphenoid rostrum (pterygorostral joint) (Figure 2.8). In the latter region, there is an additional articulation (variably movable depending on species) with the caudal end of the palatine bone, completing the palatoquadrate bridge. Thus, rostral movement of the pterygoid bone also drives the palate forward.

The **jugal arch** (**jugal** and **quadratojugal bones**) articulates with the lateral aspect of the quadrate bone (quadrato-quadratojugal joint). Rostrally, the jugal arch ends in a 'flexion zone' (**zona flexoria**).

The **zonae flexoriae** are specialised regions of the bones of the head that permit a degree of flexion and extension. Together with the synovial joints, they play an important

Table 2.1 Overview of the key joints of the head.

Joint	Bones		Function
Joints of the upper jaw			
Quadrato-squamoso-otic joint (artc. quadrato-squamoso-otica)	Quadrate bone, otic process	Squamosal bone, otic bones	Caudal, movable attachment of the quadrate bone
Quadrato-pterygoid joint (artc. quadratopterygoidea)	Quadrate bone, medial articular surface	Pterygoid bone, caudal end	Movement of the palatoquadrate bridge
Pterygorostral joint (artc. pterygorostralis)	Pterygoideum, rostral articular surface	Rostrum of parasphenoid bone	Movement of the palatoquadrate bridge
Quadrato-quadratojugal joint (artc. quadrato-quadratojugalis)	Quadrate bone, lateral articular surface	Quadratojugal bone, quadratic condyle	Movement of the jugal arch
Joints of the lower jaw			
Quadrato-mandibular joint (artc. quadratomandibularis)	Quadrate bone, mandibular process	Mandible	Raising and lowering of the lower jaw
Mandibulo-sphenoid joint (artc. mandibulosphenoidalis)	Parasphenoid bone	Mandible, medial process	Raising and lowering of the lower jaw

role in the movement of the skull. Of particular significance is the **craniofacial hinge** (zona flexoria ['elastica'] craniofacialis; Figures 2.6 and 2.11), located between the frontal and the fused nasal and premaxillary bones. This flexible zone at the boundary between the cranial and facial skeletons forms the fulcrum around which movement of the maxillopalatine apparatus is converted into **upward movement of the upper jaw** during opening of the beak.

The **joints of the mandible** of birds (Figures 2.15 and 2.16, Table 2.1) are comprised of the articulation between the quadrate bone and the mandible (quadrato-mandibular joint) and a joint between the parasphenoid bone and the medial process of the mandible (mandibulo-sphenoid joint).

The joints of the head are stabilised by various ligaments. Species variation precludes a general description of their arrangement. In summary, however, the principal function of the ligamentous apparatus is to support and guide the movements of the bones during opening of the beak. In functional terms, the arrangement of the bones, joints and ligaments of the head results in mechanical coupling of the opposing movements of the upper and lower jaw, allowing for a wide gape (opening of the mouth).

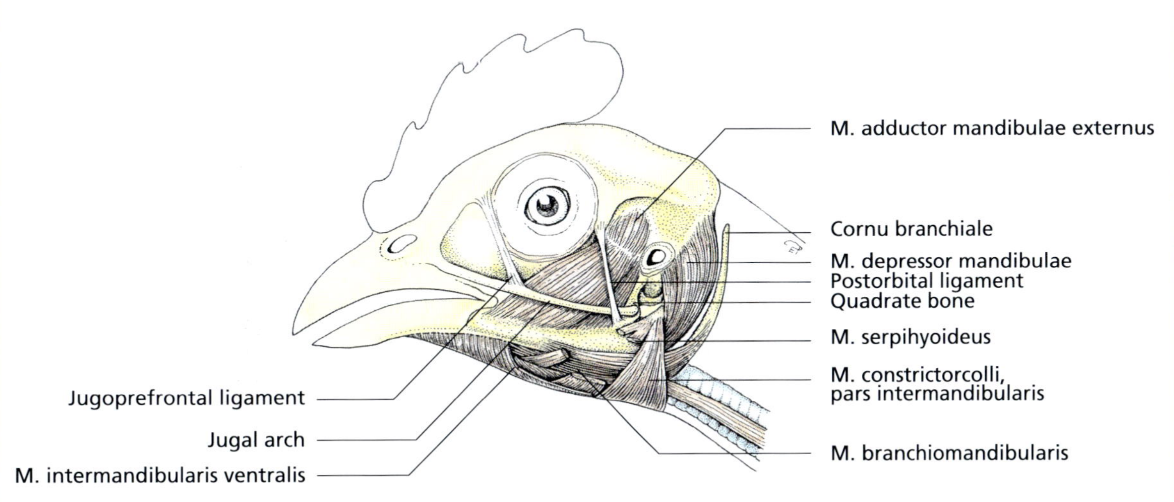

2.17 Muscles of the head of the chicken (schematic; lateral view).

Table 2.2 Muscles of the jaw.

Name Innervation	Origin	Insertion	Action
M. adductor mandibulae externus Mandibular nerve	Postorbital process, temporal fossa	Lateral aspect of mandible, from quadrato-mandibular joint to angle of jaw	Elevate the lower jaw
M. pseudotemporalis superficialis Mandibular nerve	Temporal fossa	Lateral aspect of mandible	Elevate the lower jaw
M. pseudotemporalis profundus Mandibular nerve	Orbital process of the quadrate bone	Mandible	Close the beak
M. adductor mandibulae ossis quadrati Mandibular nerve	Body and otic process of the quadrate bone	Lateral aspect of mandible, rostral to quadrato-mandibular joint	Close the beak
M. pterygoideus Mandibular nerve	Palatine bone, pterygoid bone	Medial process of mandible	Close the beak
M. protractor pterygoidei et quadrati Mandibular nerve	Interorbital septum	In two parts: pterygoid bone and quadrate bone (body, orbital process)	Elevate the upper jaw
M. depressor mandibulae Facial nerve	Exoccipital bone, squamosal bone, basioccipital bone	Caudoventral aspect of mandible; medial process of mandible	Depress the lower jaw

Muscles of the head

Many of the muscles of the head are associated functionally with organs, including the eye, middle ear, tongue, larynx and trachea. These are discussed in the context of the relevant organ systems. Only the **muscles of the jaw** are described here (Figure 2.17, Table 2.2).

The muscles of the jaw are grouped according to their action. Those that pull the mandible towards the skull serve to **elevate the lower jaw**. The muscles that draw the upper and lower beak together, by rotating the quadrate bone, act to **close the beak**. Elevation of the upper jaw is the sole function of the m. protractor pterygoidei et quadrati, while **depression of the lower jaw** is brought about by the m. depressor mandibulae.

Skeleton of the trunk

Vertebral column (columna vertebralis)

Compared with mammals, the segments of the vertebral column of birds are more difficult to differentiate. The transition from the cervical to the thoracic vertebrae in particular is the subject of debate. Consequently there is variation in the literature reporting the number of vertebrae in each vertebral segment.

Cervical vertebrae (vertebrae cervicales)

The typically S-shaped avian **cervical vertebral column** (Figure 2.18) is generally considerably more mobile, and contains more vertebra, than that of mammals. There are 14 cervical vertebrae in the chicken, 17 in the duck and 12 in the pigeon.

The **first cervical vertebra** (atlas) (Figure 2.19) is a ring-shaped bone with a dorsal **arch** (arcus) and a ventrally located **body** (corpus). A recess on the cranial surface of the body, the **condyloid fossa** (fossa condyloidea) forms the articular surface for the occipital condyle of the occipital bone. The dorsal surface of the body bears the articular surface for the dens of the axis. A further articular surface for the axis (facies articularis axialis) is located on the caudal surface of the body of the atlas. Processes on the caudal aspect of the arch of the atlas articulate with cranial processes on the axis.

The **second cervical vertebra, or axis,** (Figure 2.20) is notably larger than the atlas. Its elongated **body** (corpus axis) articulates caudally with the third cervical vertebra. Its **cranial surface** (facies articularis atlantica) forms a joint with the atlas that also incorporates the **dens**. The **arch of the axis** (arcus axis) features an unpaired dorsal **spinous process** (processus spinosus) as well as articular processes. Two **cranial processes** (processus articulares craniales)

2.18 Skull and cervical vertebrae of a chicken with bones of the pectoral limbs (right lateral view).

articulate with the corresponding caudal processes on the atlas. The caudal processes of the axis form **plane (gliding) joints** with the cranial processes of the third vertebra. As in mammals, the **caudal incisure** (incisura caudalis arcus) of the axis and the **cranial incisure** (incisura cranialis arcus) of the next vertebra give rise to the **intervertebral foramen** (foramen intervertebrale).

On **subsequent vertebral bodies** (Figures 2.21 and 2.22), the articular surface is **saddle-shaped**. The **cranial articular surface** is dorsoventrally convex, and concave in the transverse plane. The caudal articular surface is shaped in an opposite, **complementary** fashion. **Carotid processes** are evident ventrally from the mid-cervical region. The **groove** formed by these paired structures (sulcus caroticus) carries the **internal carotid arteries** (aa. caroticae internae). A median ventral crest (crista ventralis) is also present. The cranial and caudal articular processes, known as **zygapophyses**, lie in close proximity to a low spinous process. Bilateral **transverse processes** (processus transversi) are perforated near their origin by a **transverse foramen** (foramen transversi). Together, the transverse foramina give rise to a **transverse canal**, equivalent to the canalis transversarius of mammals. A pointed, caudally

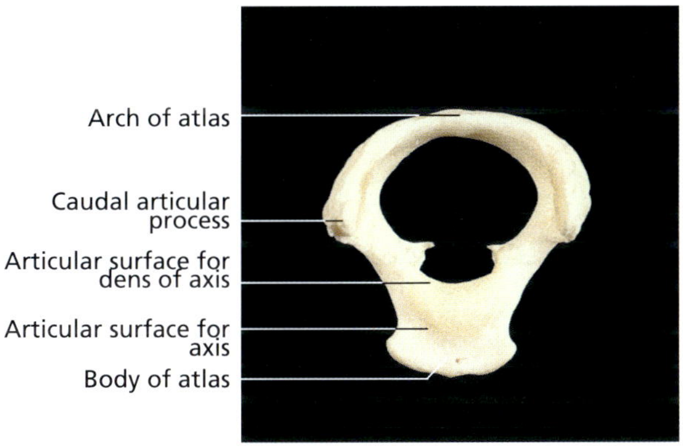

2.19 First cervical vertebra of a chicken (caudal view).

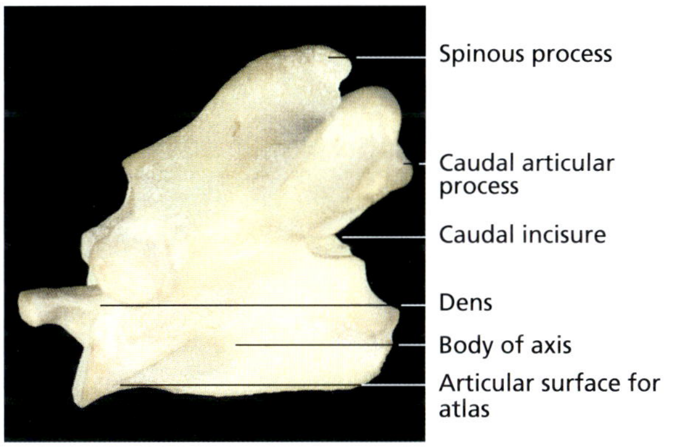

2.20 Second cervical vertebra of a chicken (left lateral view).

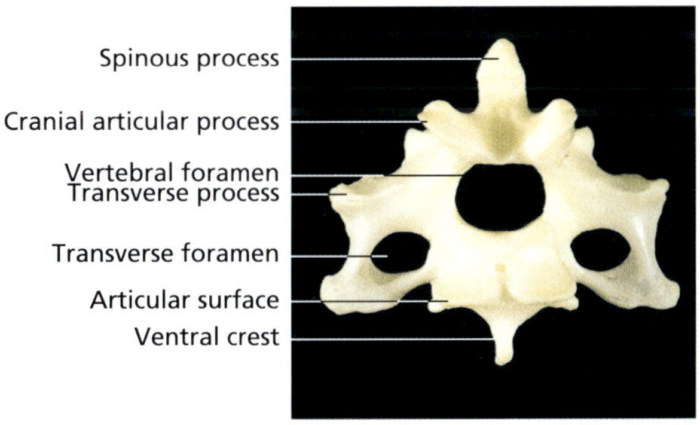

2.21 Tenth cervical vertebra of a chicken (cranial view).

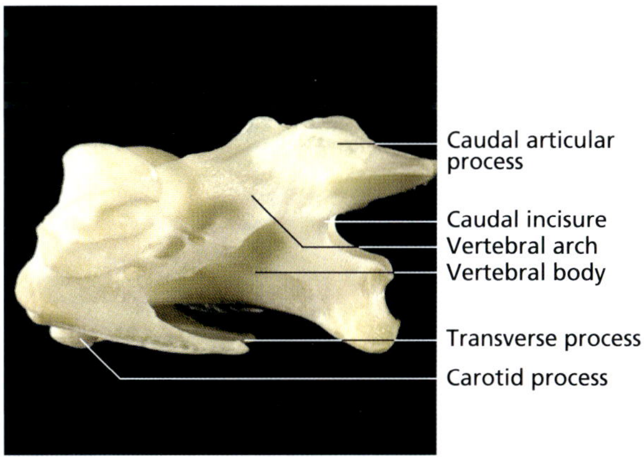

2.22 Tenth cervical vertebra of a chicken (left lateral view).

directed **costal process** (processus costalis), the rudiment of a cervical rib, extends from the caudal aspect of the transverse processes. Deep cranial and caudal incisures give rise to broad intervertebral foramina.

Thoracic vertebrae (vertebrae thoracicae)

In terms of function (breathing), **the boundary between the cervical and thoracic vertebrae** can be considered as the point at which the first free ribs appear. Other definitions, not adopted here, define the first thoracic vertebra as that with the first complete rib. In the adult chicken, only the first and sixth thoracic vertebrae occur as separate bones. The second to the fifth vertebrae are fused to form the **notarium** (os dorsale) (Figures 2.23 and 2.24), while the last vertebra (the seventh) is fused with the **synsacrum**.

In addition to the typical vertebral processes (spinous, articular), the vertebrae of the thoracic segment possess characteristic **surfaces for articulation with the ribs**. The head of the rib articulates with the **costal fovea** (fovea costalis) on the **costolateral eminence** (eminentia costolateralis), located on the lateral aspect of the vertebra of the same number. Another costal fovea on the transverse process receives the **tubercle of the rib**.

The **notarium is a rigid osseous unit** that, among domestic birds, is present only in **chickens** and **pigeons**. In other species (e.g., ducks and geese), stabilisation of the thoracic vertebral column is achieved by ossification of tendons and ligaments.

On the notarium, the ventral processes are partly ankylosed, giving rise to the incomplete plate-like **ventral crest** (crista ventralis). This serves as the origin of the **horizontal septum** (septum horizontale) and the **oblique septum** (septum obliquum) (see Chapter 5 'Body cavities'). The **spinous processes** (processus spinosi) form a dorsal **spinous crest** (crista spinosa or dorsalis). Fusion of the transverse processes gives rise to a continuous plate, the **transverse lamina** (lamina transversa). **Openings in the lamina** (fenestrae intertransversariae) are penetrated by dorsal branches of the spinal nerves.

Synsacrum

The **synsacrum** (Figures 2.25ff.) is formed by the last thoracic vertebra, the lumbar vertebrae, sacral vertebrae and the first caudal vertebrae.

In the chicken, a **total** of 15–16 **vertebrae** contribute to the synsacrum, while in other species the number varies from nine to 22. Approximately midway along the synsacrum, the vertebral canal is widened for the **lumbosacral intumescence** (intumescentia lumbosacralis) of the spinal cord. The spinal nerves emerge through the **intervertebral foramina** and their dorsal branches pass through the **fenestrae intertransversariae**.

The **spinous processes** (Figure 2.25) coalesce to form a continuous ridge, the **spinous crest** (crista spinosa or dorsalis). This is ankylosed with the dorsomedial edges of the two iliac bones to form the **iliosynsacral crest** (crista iliosynsacralis). The latter serves as the origin for muscles of the back. In the adult animal, a strong osseous connection is established between the ilium and the **transverse** and **costal processes** of the synsacrum. These features form a **stable framework for the transmission of forces** during walking. A ventral crest on the cranial ventral surface of the synsacrum gives way caudally to a ventral groove (Figure 2.26).

Caudal vertebrae (vertebrae caudales)

Some of the caudal vertebrae (Figures 2.25 to 2.28) are incorporated into the **synsacrum**, while others exist as individual vertebrae (5–6 in number). The **vertebral canal is relatively wide** in this region as, in contrast to mammals, the spinal cord does not exhibit an ascensus medullae spinalis (shortening of the spinal cord relative to the vertebral canal) and thus extends to the last segments of the vertebral column.

The caudal vertebrae have distinct transverse and well-developed spinous processes. At the end of the

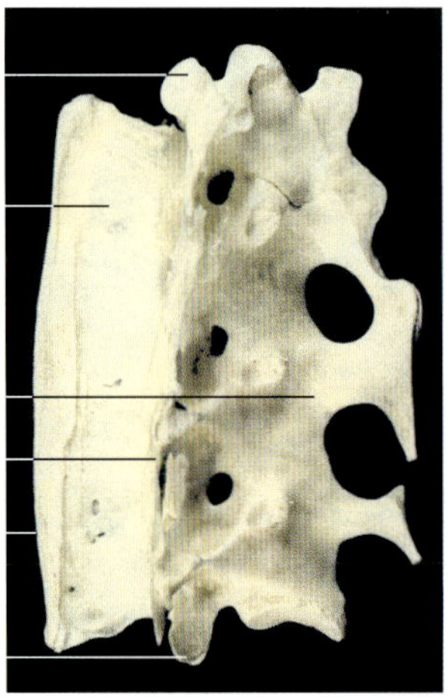

2.23 Notarium of a chicken (left lateral view).

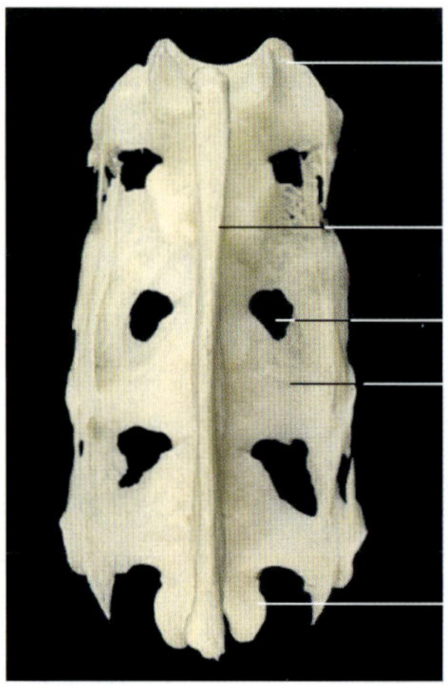

2.24 Notarium of a chicken (dorsal view).

vertebral column, the **terminal caudal vertebrae are fused** to form the **pygostyle** (Figure 2.28). The appearance of this plate-like bone varies with species from almost triangular to parallelogram-shaped. Its relatively broad **base** articulates with the last free caudal vertebra while the caudodorsally directed **apex** helps to support the **rectrices** and, where present, the **ornamental tail** feathers.

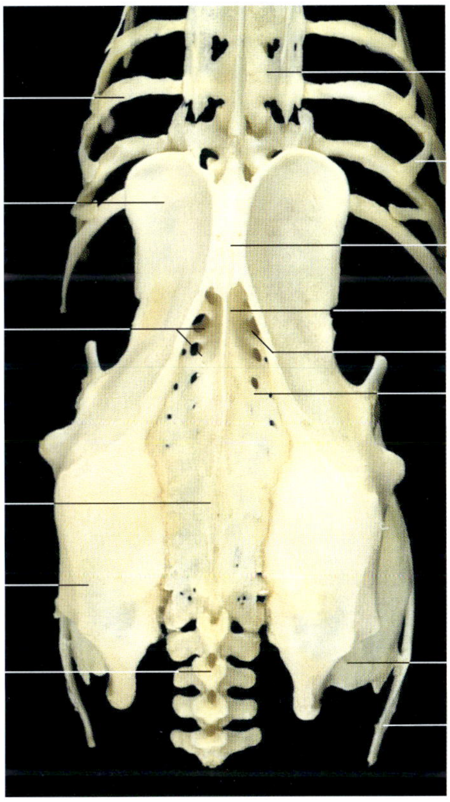

2.25 Caudal thoracic vertebrae, synsacrum, caudal vertebrae and pelvic bones of a chicken (dorsal view).

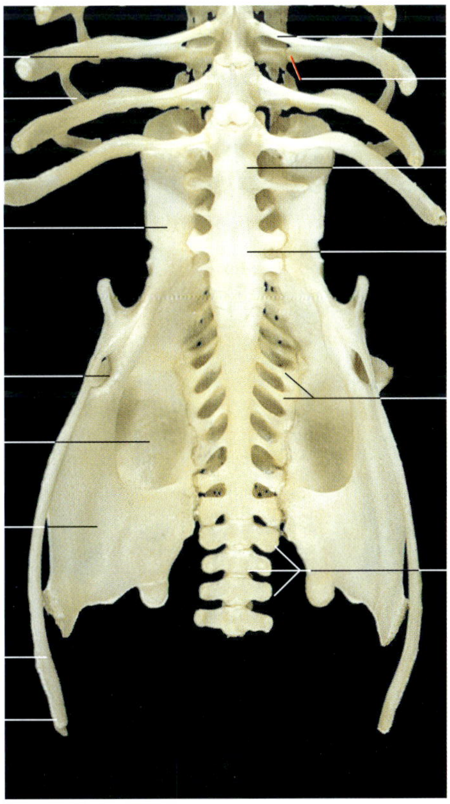

2.26 Caudal thoracic vertebrae, synsacrum, caudal vertebrae and pelvic bones of a chicken (ventral view).

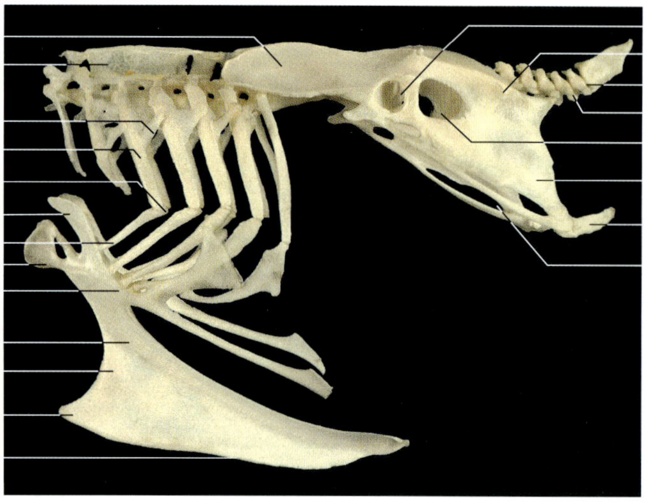

2.27 Skeleton of the trunk and caudal vertebrae of a chicken (left lateral view).

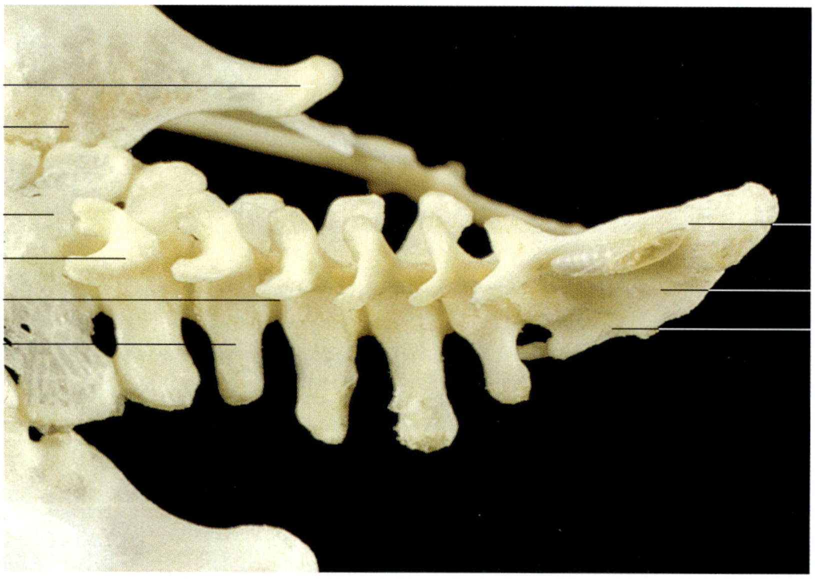

2.28 Free caudal vertebrae and pygostyle of a chicken (left dorsolateral view).

The lateral surface of the pygostyle is referred to as the **lamina.** Several muscles are involved in controlling the movement of the tail feathers (see p. 40, 'Muscles of the tail').

Ribs (costae)

The ribs (Figure 2.27) are classified as **sternal** (complete), or **asternal** (incomplete). There are seven pairs of ribs in the chicken, turkey, quail and pigeon. Ducks and geese have more, with nine pairs each.

The development of the ribs varies with species, although the first two (maximum three) and the last ribs are asternal. The free ends of these 'floating' ribs are embedded in the muscle of the body wall.

The **sternal ribs** consist of two components, a **vertebral portion** (costa vertebralis) and a **sternal portion** (costa sternalis), the latter being equivalent to the mammalian costal cartilage. As seen in mammals, the vertebral component has a **head** (capitulum costae) (Figure 2.26) separated by a tapered **neck** (collum costae) from the **costal tubercle** (tuberculum costae).

At the **costal angle** (angulus costae) the rib turns ventrally to become the **body** (corpus costae). This is flattened and has an internal and external surface, as well as a cranial and caudal margin. The **uncinate process** extends caudally from the caudal margin (**margo caudalis**) (Figure 2.27), usually reaching as far as the lateral surface of the subsequent rib. In diving birds, this arrangement **strengthens the lateral body wall** and thus **assists with breathing**. A **cartilaginous intercostal joint** (synchondrosis intercostalis) (Figure 2.27) connects the vertebral and sternal ribs. The bar-shaped **sternal rib** articulates distally with the costal margin (margo costalis or lateralis) of the sternum.

Sternum

The sternum (Figures 2.29 to 2.31) is the major **supportive element** within the ventral body wall. Its shape varies with species.

The **body of the sternum** (corpus sterni) features a dorsally concave plate with a cranial median **bony projection**, the **sternal rostrum** (rostrum sterni). Lateral to the rostrum, the **coracoid pillar** (pila coracoidea) forms a sturdy buttress that bears the channel-like **articular surface for the coracoid bone** (sulcus articularis coracoideus).

A substantial **craniolaterally oriented process** (processus craniolateralis) represents the limit of the cranial opening of the ribcage. **Costal incisures** (incisurae costales) are present on the costal (or lateral) margin of the sternum for articulation with the distal ends of the sternal ribs. The greatest species variation is observed at the caudal sternal margin. In the chicken, the **lateral trabecula** (trabecula lateralis) juts out laterally while the **intermediate trabecula** (trabecula intermedia) extends caudolaterally. These projections form the boundary of the **lateral incisure** (incisura lateralis). The **median trabecula** (trabecula mediana) lies in the midline and, together with the intermediate trabecula, delineates the **medial incisure** (incisura medialis). Both incisures are closed by **connective tissue membranes**. In waterfowl, the medial incisure is considerably smaller, since the body of the sternum extends caudally as a wide bony sheet (Figure 2.30).

On the inner, **visceral surface of the sternum** (facies visceralis) (Figure 2.30) a distinction is made between **hepatic** (pars hepatica) and **cardiac** (pars cardiaca) parts. The visceral surface also bears the laterally situated **pneumatic pores** (pori pneumatici) and a cranial **pneumatic foramen** (foramen pneumaticum). Pneumatisation is derived from **diverticulae** originating from the clavicular air sac.

The **outer, muscular surface of the sternum** (facies muscularis) is the site of origin of the major **flight muscles** (m. pectoralis, m. supracoracoideus). This surface is

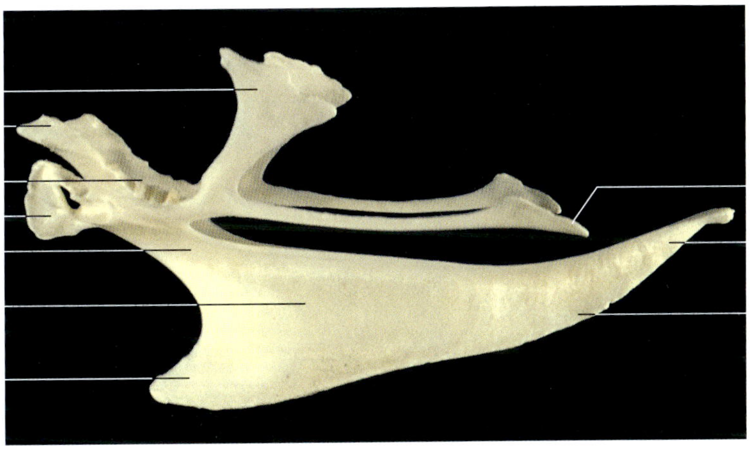

2.29 Sternum of a chicken (left lateral view).

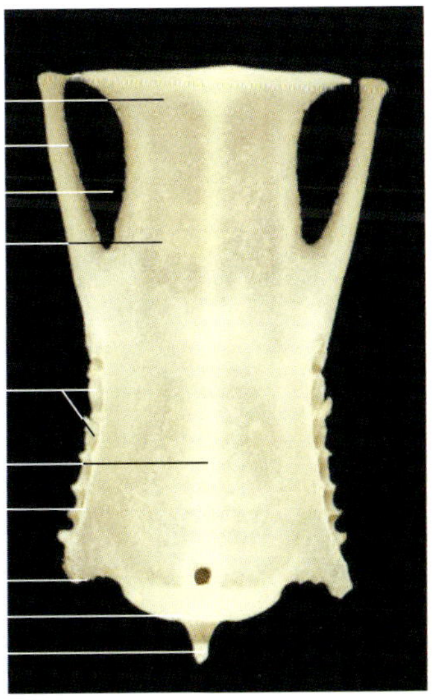

2.30 Sternum of a mallard duck (dorsal view).

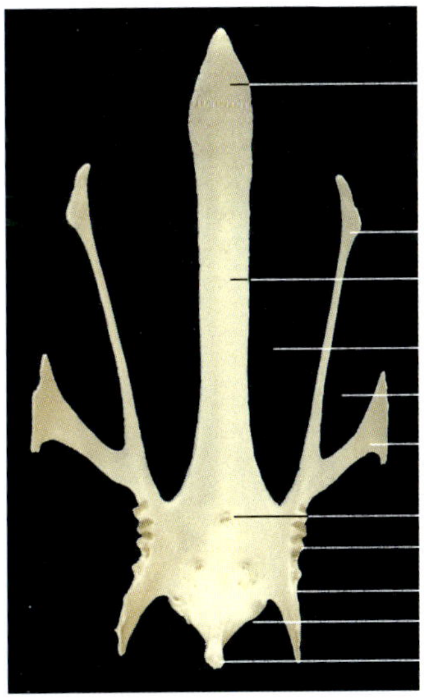

2.31 Sternum of a chicken (dorsal view).

enlarged by an elongated median projection, the **carina** (carina sterni) (Figure 2.29), particularly in species that are good flyers. The carina lies between the pectoral muscles like the keel of a ship. Its ventral margin (**margo ventralis**) is easily palpable through the skin (see Chapter 18 'Clinical examination'). At the **apex** (apex carinae), the ventral margin meets the cranial margin, which extends towards the body of the sternum.

Based on the degree of development of the carina sterni, birds are categorised as **ratites** (poorly developed carina; e.g., ostrich) or **carinate birds** (well-developed carina).

Joints of the trunk

Based on its mobility, the vertebral column can be divided into two sections: the highly movable cervical spine and the extensively fused, and thus extremely stable, thoracic, lumbar, sacral and caudal spine. The characteristics of both sections confer advantages for flying. Cervical mobility is useful for balance, while the rigidity of the trunk supports the action of the muscles.

Joints of the vertebral column (juncturae columnae vertebralis)

The first articulation of the vertebral column is the **atlanto-occipital joint** (articulatio atlanto-occipitalis). Based on the shape of the occipital condyle and the condyloid fossa of the atlas, this is classified as a **rotary joint**. Compared with the first two vertebral joints in mammals, which act as a functional unit, the atlanto-occipital joint is more important in birds as it is associated with a **greater range of movement**. The **atlantoaxial articulation** (articulatio atlantoaxialis) consists of three joints: an articulation exists between the bodies of the atlas and axis, the dens slides within a corresponding recess in the atlas and the caudal articular processes of the atlas articulate with their cranial equivalents on the axis. This imposes considerable limitations on the movement of the second vertebral joint.

The **subsequent cervical vertebral articulations** are typically composed of **two synovial joints**. Saddle-shaped articulations (articulationes intercorporales) are present between the **vertebral bodies**, while plane (gliding) joints (articulationes zygapophysiales) unite the **articular processes** of adjacent vertebrae. These impart stability without excessively restricting mobility. Depending on species, **menisci** may also be present in the joints between vertebral bodies. Dorsal, lateral and ventral ligaments, many of which are elastic, also serve to stabilise the vertebral column.

The aforementioned characteristics also apply to the free joints of the thoracic vertebrae. The remaining thoracic vertebrae are incorporated into the **notarium** via **synostoses**. Vertebrae contributing to the **synsacrum** are similarly ankylosed. The last caudal vertebrae, proximal to the pygostyle, are connected by movable joints (synchondroses).

Joints of the ribs (juncturae costarum)

A cartilaginous joint (synchondrosis capitis costae) exists between the head of the vertebral component of the rib and the vertebral column. This is strengthened ventrally by fibrous tissue. The **costal tubercle** forms a synovial joint with the **transverse process** (articulatio costotransversaria). Another cartilaginous union (synchondrosis intercostalis) is found between the vertebral and sternal parts of the rib (Figure 2.27). The synovial **sternocostal articulation** (articulatio sternocostalis) is formed by the distal end of the sternal portion of the rib and the lateral margin of the sternum (Figure 2.27).

The joint between the **uncinate process** and the **vertebral component of the subsequent rib** takes the form of either a **fibrous joint** (sutura costouncinata) or an osseous union (synostosis).

Joints of the sternum (juncturae sterni)

The sternum completes the ribcage ventrally via the previously described **sternocostal articulation** (Figure 2.27) between the sternum and the sternal component of the ribs. Another synovial articulation, the **sternocoracoid joint** (articulatio sternocoracoidea), connects the sternum with the coracoid bone. This is described further in the context of the joints of the pectoral girdle (see Chapter 3 'Thoracic limb'). In addition to these synovial joints, numerous ligamentous connections exist between the sternum and the bones of the pectoral girdle. These can be considered largely as localised reinforcement for the **sternocoracoclavicular membrane** (membrana sternocoracoclavicularis).

Muscles of the trunk (musculi trunci)

Muscles of the vertebral column (musculi vertebrales)

Particularly in the cervical region, the muscles of the vertebral column are extensively divided and arranged in multiple segments. They can be grouped into the:

- mm. craniocervicales,
- mm. cervicales dorsales,
- mm. cervicales ventrales and
- mm. cervicales laterales.

The **mm. craniocervicales** (Table 2.3) arise predominantly in the cranial section of the cervical spine and insert on the head. As such, they move the head with respect to the cranial end of the neck. The **mm. cervicales dorsales**, **ventrales** and **laterales** are responsible for moving the head relative to the trunk (Table 2.3).

Table 2.3 Muscles of the vertebral column (excluding the tail).

Name Innervation	Origin	Insertion	Action
M. biventer cervicis Dorsal branches of cervical nerves	Spinous process of 2nd thoracic vertebra	Cranial and caudal belly (tendinous intersection); supraoccipital bone	Elevate the head; Extend (cranial third) and dorsiflex (middle and caudal thirds) the neck
Mm. craniocervicales			
M. complexus Dorsal branches of cervical nerves 2 & 3	Articular processes of cervical vertebrae 3–5	Supraoccipital bone	Elevate the head and extend the cervical spine; unilaterally: turn the head
M. splenius capitis Dorsal branches of first cervical nerves	Arch and spinous process of the axis	Supraoccipital bone	Elevate the head; unilaterally: turn the head
M. rectus capitis dorsalis Dorsal branches of cervical nerve 1	Via several muscle slips from lateral aspect of cervical vertebrae 5 to 3	Parasphenoid lamina	Bow the head (flex the occipital joint)
M. rectus capitis lateralis Ventral branches of first cervical nerves	Via several muscle slips from ventral aspect of cervical vertebrae 5 to 2	Exoccipital bone	Turn and elevate the head; extend cranial part of the neck
M. rectus capitis ventralis Ventral branches of first cervical nerves	Ventral aspect of cervical vertebrae 6 to 1 (may be bisected by common carotid artery)	Both parts: parasphenoid lamina	Bow and turn the head; flex cranial part of the neck
Mm. cervicales dorsales			
M. longus colli dorsalis Dorsal branches of cervical and thoracic nerves	**Pars cranialis**: spinous processes of cervical vertebrae 3–9 **Pars caudalis**: spinous processes of cervical vertebrae 2–4 **Pars profunda**: spinous processes of cervical vertebrae 8–11 **Pars thoracica**: iliosynsacral crest and notarium	**Partes cranialis** and **caudalis** caudal aspect of axis; **Pars caudalis** also cervical vertebrae 7–14 **Pars profunda**: cervical vertebra - two vertebrae cranial to its respective origin (i.e. 6–9) **Pars thoracica**: cranioventrally on the transverse processes up to thoracic vertebra 2	Elevate and extend the neck; unilaterally: flex the neck laterally
Mm. ascendentes: Dorsal branches of cervical and thoracic nerves M. ascendens thoracicus M. ascendens cervicalis	Lateral aspect of thoracic vertebrae 1–3 Articular processes of cervical vertebrae 7–15	Craniomedially oriented on cervical vertebra 11 to thoracic vertebra 2 Craniomedially oriented on articular processes up to cervical vertebra 3	Elevate and turn base of the neck Extend and straighten the neck; unilaterally: turn the head
Mm. iliocostalis et longissimus dorsi Dorsal branches of thoracic, lumbar and sacral nerves	Cranial margin of preacetabular wing of ilium	Transverse processes of thoracic vertebrae; proximal ribs	Fix the thoracic vertebral column; continued cranially by the mm. ascendentes
Mm. intercristales Mm. interspinales Dorsal branches of cervical nerves	Connect the spinous processes of adjacent vertebrae from the arch of the atlas to the first thoracic vertebra; mm. interspinales are less clearly defined than the mm. intercristales		Straighten the cervical vertebral column

Table 2.3 continued.

Name Innervation	Origin	Insertion	Action
Mm. cervicales laterales			
Mm. intertransversarii Dorsal branches of cervical nerves	Connect the transverse processes of adjacent cervical vertebrae		Flex the vertebral column laterally
Mm. inclusi: Dorsal branches of cervical nerves Mm. inclusi dorsales Mm. inclusi ventrales	Dorsal and ventral surfaces of the transverse processes to the arch (dorsal) and body (ventral) of the preceding vertebra		Turn (dorsal) and flex (ventral) the vertebral column
Mm. cervicales ventrales			
M. longus colli ventralis Ventral branches of cervical and thoracic nerves	Ventral crests of thoracic vertebrae 3–5, cranial portions directly from the vertebrae	Ventral surface of the cervical vertebrae to the third thoracic vertebra	Flex the vertebral column
M. flexor colli lateralis **M. flexor colli medialis** Ventral branches of the cervical nerves	Lateralis: ventrolaterally on cervical vertebrae 2–5 Medialis: laterally on cervical vertebrae 3–8	Lateralis: ventral surface of atlas Medialis: ventral crest of axis; costal processes of cervical vertebrae 3 and 4	Flex the cranial section of the cervical vertebral column

Functionally, the cervical vertebral column (Figure 2.18) can be divided into three segments that vary in their degree of mobility. Using the resting S-shaped curve of the neck as a reference, the first cervical segment (from the head to about the fifth cervical vertebra) is characterised by its capacity for **marked ventroflexion**. This part of the neck can be straightened at best, but not extended any further. The following section (c. cervical vertebrae 6–10) is capable of pronounced **dorsal flexion**. From approximately cervical vertebra 11 to the first thoracic vertebra, the cervical spine can again be **flexed ventrally**, but is also capable of considerable **lateral movement** and some **dorsiflexion**.

Muscles of the thoracic and abdominal walls (musculi thoracis et abdominis)

The muscles of the trunk (Tables 2.4 and 2.5) complete the **thoracic and abdominal walls**. Since there **is no diaphragm** in birds, and the **position** and **volume of the lung** is fixed by its encasement in the rib cage, the air sacs (see Chapter 7 'Respiratory system') and the muscles of the thoracic wall are of particular importance with respect to breathing.

During **inspiration**, the cranially directed angle between the vertebral and sternal components of the ribs is enlarged by the action of specific muscles of the thoracic wall. This serves to increase the **distance between the vertebral column and the sternum**. The resulting reduction in pressure within the body cavity is equalised by an influx of air into the air sacs. In this process, the sternum does not move ventrally along its whole length. Rather, it **pivots about an axis** through the shoulder joints, to which the sternum is **tightly connected** by the **coracoid bone** (see Chapter 20, Figure 20.2).

During the **expiratory phase**, the sternum returns to its original position and the angle between the sternal and vertebral components of the ribs decreases. This raises the pressure in the body cavity, leading to the **expulsion of air from the air sacs**.

The **abdominal muscles** are thin. Fibre direction within these muscles is comparable with their equivalents in mammals. However, there is no rectus sheath or inguinal canal.

Muscles of the tail (musculi caudae)

When classified **on the basis of function** (steering during flight, copulation, defaecation and communication) the muscles of the tail can be categorised as (Table 2.6):

- elevators of the tail and rectrices and
- depressors of the tail and rectrices, that also produce lateral movement when acting unilaterally.

Alternatively, the tail muscles (Figure 2.33) can be classified according to their position and attachments (Table 2.6):

- muscles that arise on the vertebrae and extend to other vertebrae and the rectrices,
- muscles that connect the pelvis with the rectrices, and
- muscles that exclusively extend between the rectrices.

Table 2.4 Muscles of the thoracic wall.

Name Innervation	Origin	Insertion	Action
M. scalenus Ventral branches of cervical nerves	**Pars cranialis**: transverse process, cervical vertebra 13 **Pars caudalis**: transverse process, thoracic vertebra 1	**Pars cranialis**: first asternal rib **Pars caudalis**: uncinate process of second rib	Inspiration: draw first rib cranially
Mm. levatores costarum Dorsal branches, thoracic nerves	Transverse processes of thoracic vertebrae 2–5 (7)	Caudovetrally oriented on proximal ends of ribs 3–6 (9)	Inspiration: draw the ribs cranially
Mm. intercostales externi Intercostal nerves	Span the intercostal spaces of the vertebral components of the ribs, fibres run caudoventrally		Inspiration: draw the ribs cranially
Mm. intercostales interni Intercostal nerves	Span the intercostal spaces of the vertebral components of the ribs, fibres run cranioventrally		Expiration: draw the ribs caudally
M. costosternalis: Intercostal nerves Pars major Pars minor	**Pars major/minor**: craniolateral process of sternum	**Pars major**: sternal ribs 2-6 **Pars minor**: first two ribs	**Pars major**: inspiration **Pars minor**: expiration
M. costoseptalis Intercostal nerves	Medial aspect of intercostal joints of ribs 3–5	Radiates into horizontal septum	Expiration: tensing of the horizontal septum
M. sternocoracoideus Intercostal nerves	Medial aspect of craniolateral process of sternum	Coracoid bone, toward the shoulder	Fix the sternocoracoid joint

Table 2.5 Muscles of the abdominal wall.

Name Innervation	Origin	Insertion	Action
M. rectus abdominis Intercostal nerves, ventral branches of lumbar nerves	Intermediate trabecula of sternum, caudally on last rib (sternal component)	Via an aponeurosis on the caudal third of the pubis	Expiration; compress the abdomen
M. obliquus externus abdominis Intercostal nerves, ventral branches of lumbar nerves	Uncinate processes; ventral borders of the ilium and pubis, fibres run caudoventrally	Via an aponeurosis on the median trabecula; blends with tendon of m. rectus abdominis	Expiration; compress the abdomen
M. obliquus internus abdominis Intercostal nerve VI, ventral branches of lumbar nerves	Ventral border of ilium; cranial half of pubis	Fibres run cranioventrally to attach to the last rib	Expiration; compress the abdomen
M. transversus abdominis Intercostal nerves V/VI, ventral branches of lumbar nerves	Ribs 5–7; ilium and pubis	Transverse fibre direction; median aponeuroses attach to the sternum	Expiration; compress the abdomen

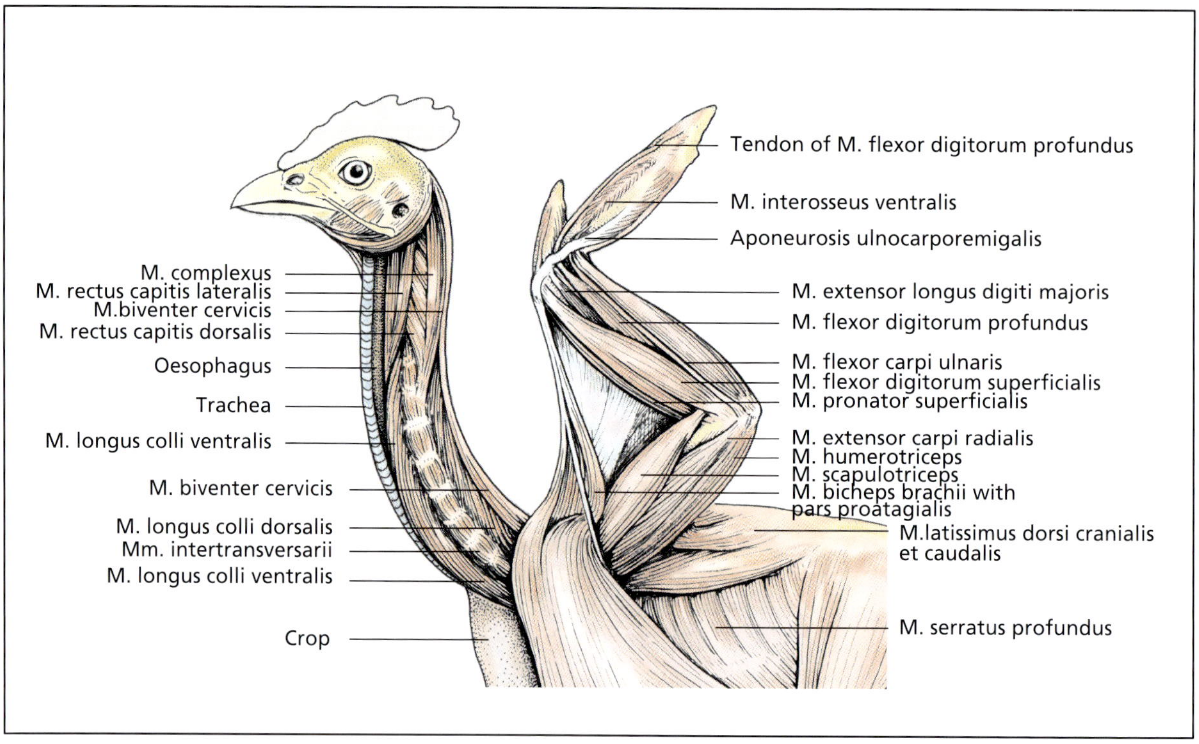

2.32 Muscles of the neck and pectoral limb of the chicken (schematic; left lateral view).

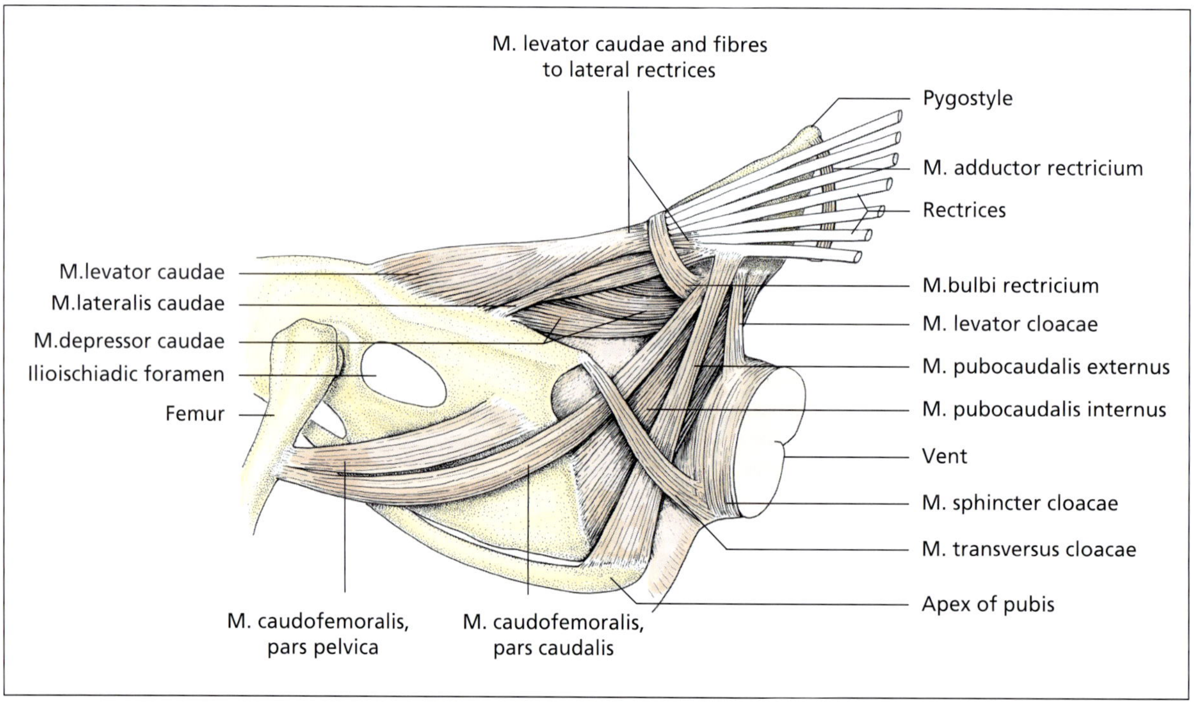

2.33 Muscles of the tail of the chicken (schematic; lateral view).

Table 2.6 Muscles of the tail.

Name Innervation	Origin`	Insertion	Action
M. levator caudae: Dorsal branches of caudal nerves – **Pars rectricalis**	**Pars rectricalis**: post-acetabular wing of the ilium	**Pars rectricalis**: caudal vertebrae, lateral rectrices	Elevate the tail and lateral rectrices
– **Pars vertebralis**	**Pars vertebralis**: caudal border of ilium	**Pars vertebralis**: dorsal border of pygostyle	
M. lateralis caudae Pudendal plexus	Caudal border of ilium, caudal vertebrae 1/2	Lateral rectrices	Draw rectrices laterally
M. depressor caudae Pudendal plexus and caudal plexus	Caudal end of synsacrum and ventrally on caudal vertebrae	Ventral aspect of base of pygostyle	Depress the tail
M. caudofemoralis: Caudal coxal nerve – **Pars caudalis**	Ventral pygostyle	Mid femur	Depress the tail and draw tail laterally (see also muscles of the thigh)
M. pubocaudalis externus Pudendal plexus	Distal pubis	Follicles of lateral rectrices	Depress rectrices and draw rectrices laterally
M. pubocaudalis internus Pudendal plexus	Caudal ischium, medial aspect of apex of pubis	Ventral aspect of base of pygostyle	Depress the tail and draw tail laterally
M. bulbi rectricium Caudal plexus	Connects the lateral surfaces of the pygostyle with the last caudal vertebrae, encompassing the feather follicles dorsally and ventrally		Adjust the position of the rectrices
M. adductor rectricium Pudendal plexus	Apex of pygostyle	Ventrally on the calamus of all rectrices	Draw rectrices together

On a systematic and topographical basis, it would be appropriate to describe the **cloacal muscles** here. Instead, taking a functional approach, these are addressed in Chapter 6 'Digestive system' and Chapter 8 'Urinary system', and in Chapters 9 and 10 (organs of the genital system).

Clinical aspects

The premaxillary and maxillary region is a common site of **fractures** and luxations of the joints of the upper beak. These may result from events such as head trauma, bite wounds and attacks (e.g., following unsuccessful mating attempts in cockatoos). Depending on the circumstances, euthanasia may be indicated as the patient's bite strength can make it difficult to achieve satisfactory surgical fixation using cerclage wire or prostheses.

The proximity of the orbits in most diurnal birds and the presence of the sclerotic ring (anulus ossicularis sclerae; see Chapter 15 'The eye') are predisposing factors for **contre-coup injuries**. These render the avian eye more susceptible to intra-ocular haemorrhage, typically occurring on the side contralateral to the trauma site. It is essential therefore to perform ophthalmoscopy in trauma cases (the incidence of haemorrhage, usually originating from the pecten oculi, is more than 30 per cent).

In conducting an ophthalmic examination, it is important to note that, due to the often thin or soft tissue nature of the interorbital septum, the illumination of one pupil with a point source of light can result in **retroillumination** of the other eye. This effect may produce a consensual pupillary light reflex. The latter is not usually expected in birds due to the almost complete decussation of optic nerve fibres in most avian species (60–70 per cent decussation in owls), as opposed to the partial decussation observed in domestic mammals and humans.

The cranial and caudal margins of the sternum serve as **important landmarks** for identifying the correct site for injection into the breast muscle (refer to Chapter 21 'Medication and blood collection techniques'). This is located in the **cranial third** of a line passing between the two aforementioned margins, paramedian to the carina of the sternum.

If the injection is given too far caudally, there is a risk that the needle may pass through the medial incisure of

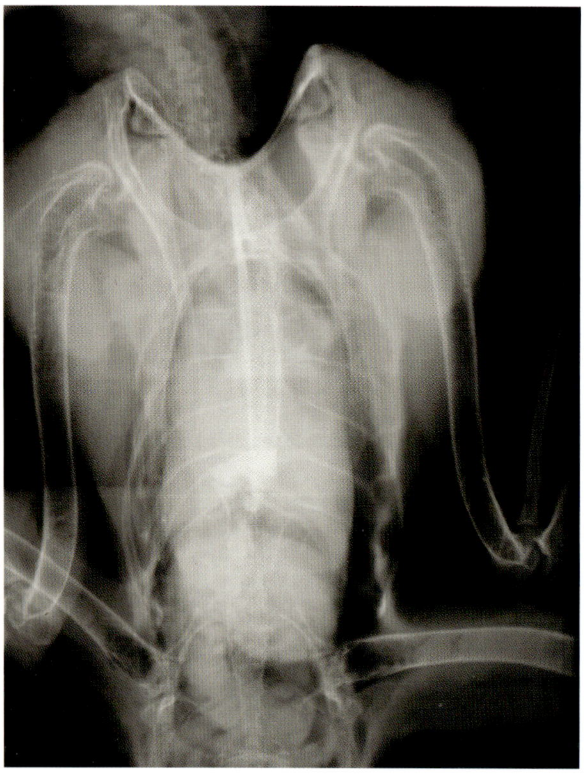

2.34 Radiograph (ventrodorsal view) of a vertebral luxation and fracture at the 'locus minoris resistentiae' between the notarium and the synsacrum in a gyr-saker hybrid falcon (Falco rusticolus x Falco cherrug).

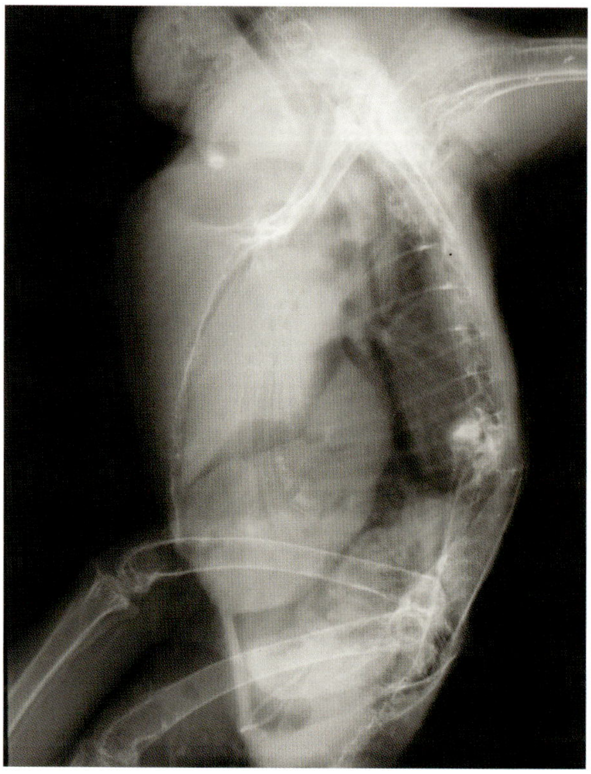

2.35 Radiograph (lateral view) of a vertebral luxation and fracture at the 'locus minoris resistentiae' of the vertebral column (same patient as in Figure 2.34).

the sternum in chickens (or its counterpart, the foramen ovale, in parrots) resulting in accidental intra-abdominal injection, usually with penetration of the liver.

The joint between the notarium and synsacrum represents a 'locus minoris resistentiae' (site of lesser resistance) in this otherwise rigid section of the vertebral column. Fractures and luxations at this site, as well as head trauma, are typical injuries in birds that have been involved in accidents (e.g., collisions with windows or car windscreens, hunting accidents in raptors) (Figures 2.34 and 2.35).

Clinical manifestations of this type of injury include loss of superficial and deep sensation. Mild cases may present with malpositioning or flaccid paralysis of the 'rump' (tail). In severe cases, signs may include paralysis of the cloaca, through to paraplegia with flaccid paralysis of the hindlimbs. The diagnosis is established using radiography.

Treatment includes resolution of the traumatically induced oedema of the spinal cord. If normal function cannot be re-established within ten days of treatment, the prognosis is generally poor.

Thoracic limb (membrum thoracicum)

J. Maierl, H. E. König, H.-G. Liebich and R. Korbel

The skeleton of the thoracic limb comprises the **bones of the pectoral girdle** (ossa cinguli membri thoracici) and the **bones of the wing** (ossa alae) (Figure 3.1). In birds, the pectoral girdle reaches its full expression, consisting of three bones. Its structure reflects its crucial role in **supporting flight**. The bones of the pectoral girdle and the associated flight muscles are incorporated into the ovoid shape of the trunk, thus reducing wind resistance.

Directional terms relating to the wings are applied based on a posture in which the animal is standing with its forelimbs extended laterally. The upper side (dorsal surface) and underside (ventral surface) of the wing are continuous at the cranial and caudal borders. The same designations are utilised when the animal's wings are folded.

3.1 Skeleton of the pectoral girdle and wing (schematic) of a peregrine falcon (Falco peregrinus).

Skeleton of the pectoral girdle and wing

Skeleton of the pectoral girdle (ossa cinguli membri thoracici)

The fully developed avian pectoral girdle (Figures 3.2 to 3.5) consists of the:

- coracoid bone (os coracoideum),
- scapula,
- clavicle (clavicula).

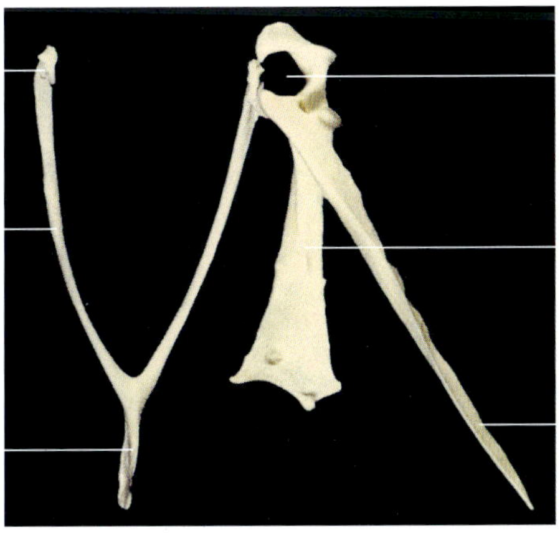

3.2 Bones of the right pectoral girdle of a chicken (caudodorsal view).

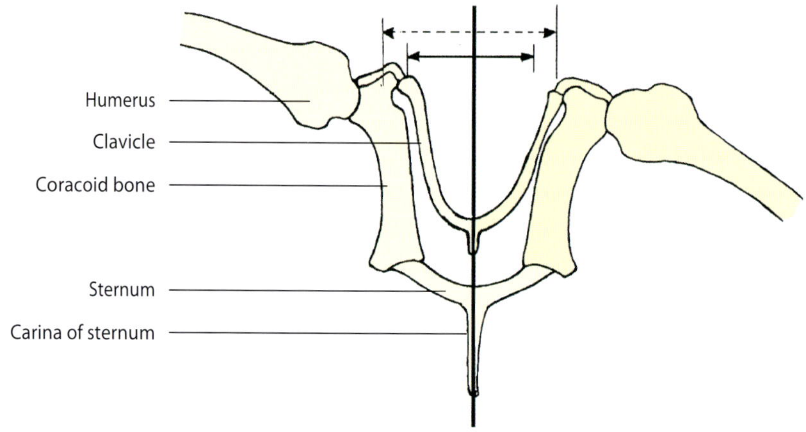

3.3 Dynamics of the skeleton of the pectoral girdle during flight (schematic). The right side represents the downstroke, the left side the upstroke. The spring-like effect of the furcula (fused clavicles) draws the shoulder joints closer together during the upstroke.

These bones are joined to the trunk and limbs in an arrangement that is stable under tension and pressure.

Coracoid bone (os coracoideum)

The coracoid bone (Figures 3.2 and 3.4) is a rod-shaped bone that connects the cranial border of the sternum with the shoulder joint. It is the **strongest bone** in the pectoral girdle. At its **distal (sternal) extremity** (extremitas sternalis) an approximately transversely positioned condylar surface forms a joint with the **articular groove** (sulcus articularis coracoideus) of the **coracoid pillar** (pila coracoidea) of the sternum (Figures 2.30 and 2.31). Oval in cross-section, the body of the coracoid presents a **ventral** and **dorsal** surface (facies ventralis and facies dorsalis) that merge at the **lateral** and **medial margins** (margo lateralis and margo medialis).

The **dorsal surface** faces the trunk. Depending on species, it features a **pneumatic foramen** (foramen pneumaticum) for an outpouching of the **clavicular air sac** (Figure 3.4). The ventral surface merges smoothly with an intermuscular line. Proximally the shaft becomes rounder in cross-section.

The **omal extremity** (extremitas omalis) (Figure 3.4) is divided into several protuberances. Projecting caudolaterally, the **glenoid process** (processus glenoidalis) forms the major component of the **articular surface for the head of the humerus** (facies articularis humeri). A separate **articular surface for the scapula** (facies articularis scapularis) is also present.

The most prominent projection, the hook-like **acrocoracoid process** (processus acrocoracoideus) (Figure 3.4), extends medially forming the **supracoracoid groove** (sulcus supracoracoideus) for the tendon of the supracoracoideus muscle. Distally this groove is flanked by the **procoracoid process** (processus procoracoideus) (Figure 3.4), thus completing the lateral component of the **tri-**

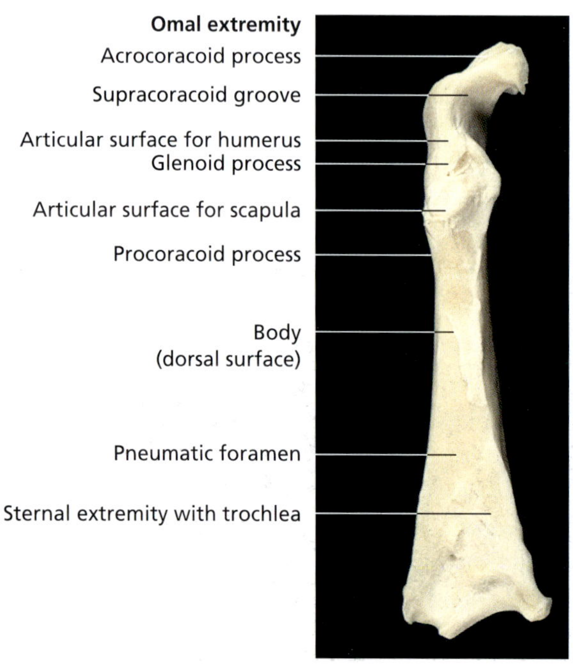

Fig 3.4 Left coracoid bone of a chicken (dorsal view).

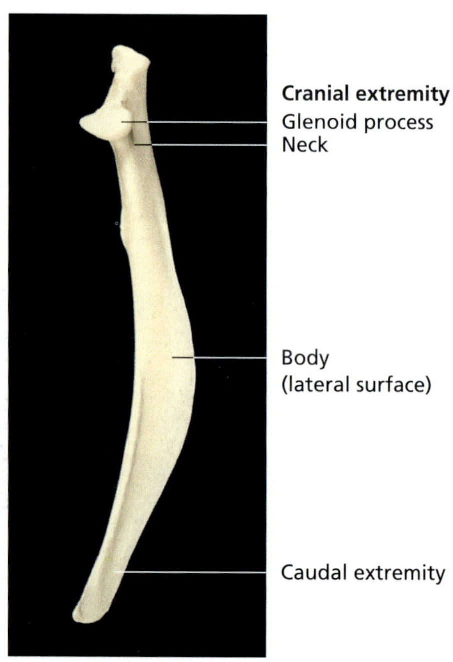

3.5 Left scapula of a chicken (lateral view).

osseal canal (canalis triosseus) (Figure 3.2). Medially the canal is bounded by a fibrous connection between the **acrocoracoid process** and the clavicle.

Scapula

The scapula is sabre-shaped (Figure 3.5). At its **cranial extremity** (extremitas cranialis) it is firmly bound with the coracoid bone. The laterally directed **glenoid process** of the scapula completes the articular surface for the head of the humerus. Craniomedially the scapula is joined to the clavicle (Figure 3.2). The **supracoracoid groove** lies immediately lateral to this union. A **neck** (collum scapulae) separates the cranial extremity from the **body** of the scapula. The long **caudal extremity** (extremitas caudalis) is slightly curved and lies approximately parallel to the vertebral column, extending almost to the **ilium**.

Clavicle (clavicula)

The **clavicle** (Figure 3.2) is a thin, curved rod. At its proximal (omal) end, it is connected to the **omal extremity** of the coracoid and the **scapula**. Following a cranially convex curve (Figure 3.1), it passes ventromedially to meet with its opposite number.

The two clavicles are joined by an osseous union to form the **'wishbone'** (furcula). A bony appendage**,** the **apophysis furculae** (hypocleidum), extends from the furcula (Figures 3.2 and 3.11). This is blade-shaped in the chicken. Depending on species, the furcula joins the sternal rostrum by either a direct or fibrous connection. The furcula functions as a **spring-like brace** between the shoulder joints (Figure 3.2). In parrots, the clavicles are reduced and are not ankylosed to form a furcula, while in some flightless birds the clavicles are absent.

Skeleton of the wing (ossa alae)

The bones of the wing (Figures 3.6 to 3.8) consists of the:

- humerus,
- ulna and radius,
- carpus (ossa carpi),
- metacarpus (ossa metacarpalia) and
- digits (ossa digitorum manus).

The skeleton of the wing is characterised by reductions and simplifications in the form of ankyloses, particularly at the tip of the limb.

Humerus

Based on the standard anatomical position (wing laterally extended), the body of the humerus (Figures 3.6 and 3.7) has two flattened surfaces, termed cranial and caudal, that merge at the dorsal and ventral margins. The ellipsoid **head of the humerus** (caput humeri) forms part of the shoulder joint. On the cranial surface, just distal to the head, there is a **transverse groove** (sulcus transversus) for the **acrocoracohumeral ligament** (lig. acrocoracohumerale) (Figure 3.6). Both dorsally and ventrally, a tuberosity merges with a more distally located ridge: the smaller **dorsal tubercle** (tuberculum dorsale) continues as the **deltopectoral crest** (crista deltopectoralis or pectoralis) and the distinctly larger **ventral tubercle** (tuberculum ventrale) merges with the **bicipital crest** (crista bicipitalis). The area between the

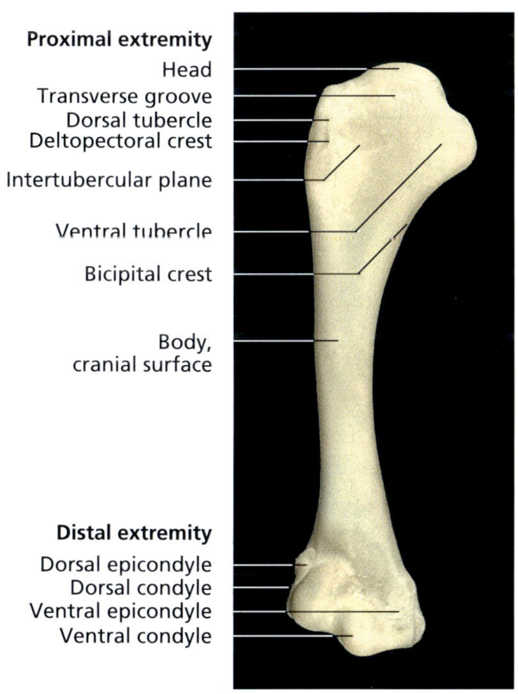

3.6 Right humerus of a chicken (cranial view).

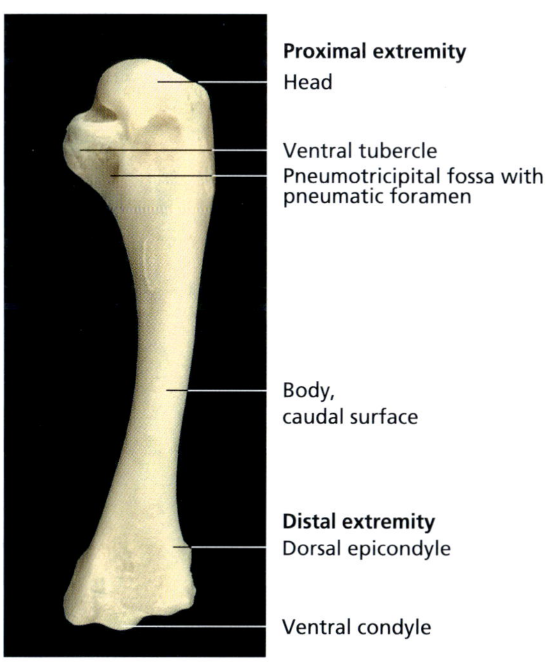

3.7 Right humerus of a chicken (caudal view).

tubercles, cranially, is the **intertubercular plane** (planum intertuberculare). A deep recess, the **pneumotricipital fossa** (fossa pneumotricipitalis) (Figure 3.7) lies immediately adjacent to the ventral tubercle on the caudal surface. It houses a pneumatic foramen.

Continuing at a slight angle from the proximal extremity, the shaft (body) of the humerus is oval in cross section. The **distal extremity** (extremitas distalis), with its prominent transversely positioned barrel-like **dorsal** and **ventral condyles** (condyli dorsalis and ventralis), forms part of the elbow joint (Figure 3.6). Just proximal to the condyles, the **ventral** and **dorsal epicondyles** (epicondylus ventralis and dorsalis) serve as sites of muscular attachment. When the wings are folded, the distal end of the humerus extends to the cranial border of the ilium (e.g., chicken) or as far as the hip joint (e.g., goose and duck).

Ulna and radius

As in mammals, the ulna and radius form the skeleton of the antebrachium (Figure 3.8). With some variation among species, these bones are generally of approximately equal length and lie parallel to one another when the wing is folded. In this posture, the smaller radius lies dorsal to the stouter ulna.

The **proximal extremity** of the ulna features a relatively poorly developed **olecranon** as well as two surfaces, the **dorsal** and **ventral cotyla** (cotyla dorsalis and ventralis), for articulation with the condyles of the humerus. An articular facet for the radius, the **radial incisure** (incisura radialis), is etched into the bone, distal to the cotyla dorsalis. The caudodorsal surface of the **body** (corpus ulnae) is devoid of muscle and features small rounded **projections** (papillae remigales) and transversely oriented troughs. These are associated with the attachments of the follicles of the remiges.

The **distal extremity** (extremitas distalis) (Figures 3.8, 3.14 and 3.15) is characterised by the **carpal trochlea** (trochlea carpalis), which consists of a pair of condyles. Components of the trochlea form articulations with each of the two carpal bones. A small recess, the **depressio radialis,** serves as the site of articulation with the distal radius.

At the **proximal extremity** of the **radius** (Figures 3.8, 3.12 and 3.13) there is an articular surface for the humerus (cotyla humeralis) as well as an articular facet for the ulna (facies articularis ulnaris). The **body** (corpus radii) follows a relatively straight course to its slightly thickened **distal extremity** (extremitas distalis) (Figures 3.8, 3.14 and 3.15), which bears surfaces for articulation with the radial

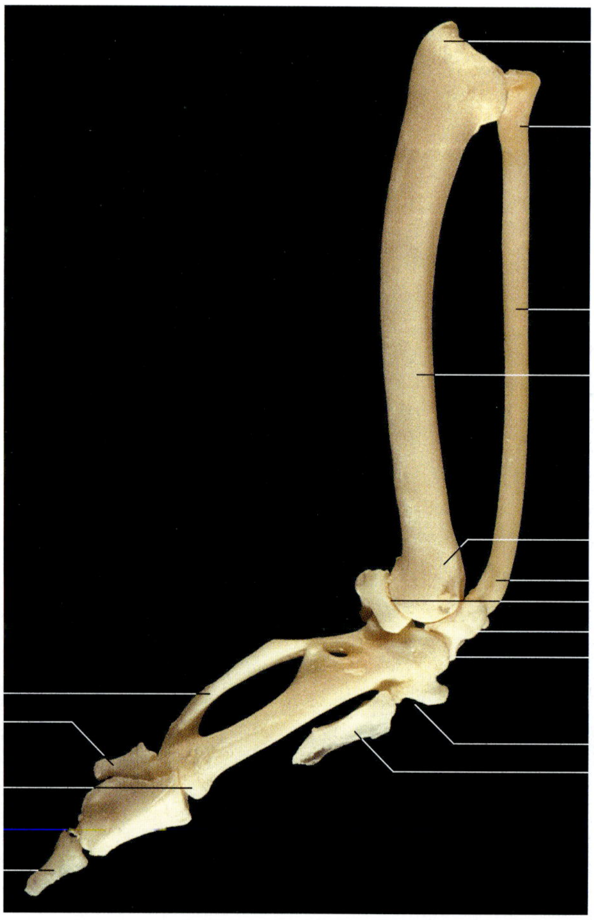

3.8 Bones of the right antebrachium and manus of a chicken (dorsal view).

carpal bone (**facies articularis radiocarpalis**), and with the **depressio radialis** of the ulna (**facies articularis ulnaris**).

Carpal bones (ossa carpi)

Of the **five embryonic carpal bones** (ossa carpi), only the **ulnar carpal bone** (os carpi ulnare) and the **radial carpal bone** (os carpi radiale) remain in the adult bird (Figure 3.8). These originate from the proximal row of carpal bones, while the central carpal bone and the distal row of bones are incorporated into the carpometacarpus (see below).

The compact radial carpal bone bears articular surfaces for the ulna, radius and carpometacarpus. In contrast, the ulnar carpal bone is larger and distinctly angular in shape, with a long and a short limb separated by the **metacarpal incisure** (incisura metacarpalis). Articular surfaces for the ulna and carpometacarpus are present.

Metacarpal bones (ossa metacarpalia)

The metacarpus (Figure 3.8, 3.14 and 3.15) is reduced to **three elements**. Embryonically, the metacarpus consists of an **alular**, **major** and **minor metacarpal bone** (os metacarpale alulare, ossa metacarpalia majus and minus). In the adult bird, these are fused with the distal row of carpal bones, giving rise to the **carpometacarpus**. At the proximal end of the carpometacarpus, the **carpal trochlea** (trochlea carpalis) articulates with the carpal bones. The alular metacarpal bone, a stub-like projection from the dorsal surface of the carpometacarpus, bears the articular surface for the alular digit (digitus alularis) (Figure 3.8).

Two fused bridges of bone, the major metacarpal bone and the smaller minor metacarpal bone, extend distally towards the digits. At its distal extremity, each metacarpal bone bears a surface (facies articularis digitalis major and minor) for articulation with the phalanges.

Bones of the digits (ossa digitorum manus)

The bones of the digits (Figure 3.8) are considerably reduced. Generally the alular and minor digits possess only one cone-shaped phalanx, while the major digit has two phalanges. The proximal phalanx consists of a thickened dorsal border that merges with an expanded ventral bony plate.

Joints of the pectoral girdle and wing

Joints of the pectoral girdle (juncturae cinguli membri thoracici)

The **saddle-shaped** synovial **sternocoracoid joint** (articulatio sternocoracoidea), between the sternum and the coracoid bone (Figure 3.11), permits flexion and extension in the sagittal plane (hinge-like movement) and gliding movement in the transverse plane. It is stabilised by the **collateral sternocoracoid ligaments** (ligg. collateralia sternocoracoidea). **Syndesmoses** unite the proximal end of the coracoid bone with the **clavicle** and the **scapula**. The clavicle is also connected to the sternum by a syndesmosis. These fibrous joints are referred to as the:

- coracoscapular joint (syndesmosis coracoscapularis),
- acrocoracoclavicular joint (syndesmosis acrocoracoclavicularis),
- sternoclavicular joint (syndesmosis sternoclavicularis).

The **coracoscapular joint** (synd. coracoscapularis) is stabilised principally by the **coracoscapular interosseous ligament** (lig. coracoscapulare interosseum). This structure not only connects the coracoid bone and the scapula, but also forms a **cartilaginous lip** that invests the glenoid

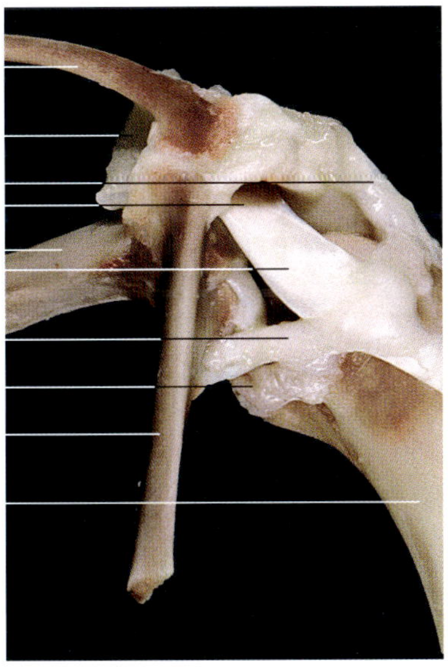

3.9 Right shoulder joint of a chicken (dorsomedial view), Courtesy of Dr R. Macher, Vienna.

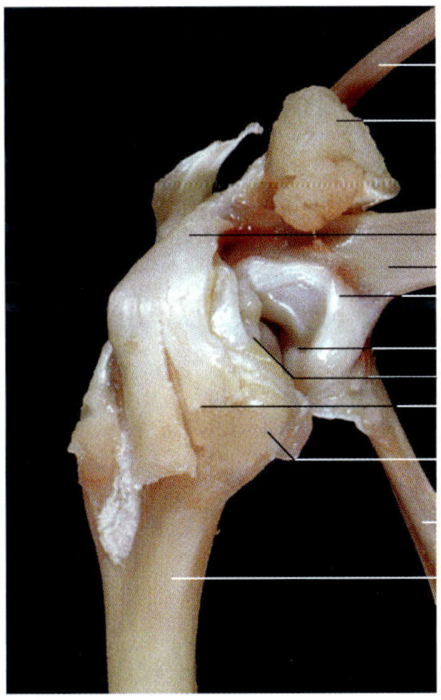

3.10 Right shoulder joint of a chicken (dorsolateral view), Courtesy of Dr R. Macher, Vienna.

processes of both bones. It thus constitutes the basis of the **glenoid cavity** (cavitas glenoidalis), which conforms to the shape of the humeral head during movement of the joint.

The **acrocoracoclavicular joint** is a fibrous union between the acrocoracoid process and the proximal end of the clavicle. It completes the medial wall of the **triosseal canal** (Figure 3.9). The **acrocoracoacromial ligament** (lig. acrocoracoacromiale) closes the remaining gap between the coracoid bone and the scapula.

In the **sternoclavicular joint** (synd. sternoclavicularis), the hypocleidum is connected with the cranial margin of the sternal carina by the sternoclavicular ligament (ligamentum sternoclaviculare). The **sternocoracoclavicular membrane** (membrana sternocoracoclavicularis) spans the space between the clavicle, the cranial edge of the sternum and the coracoid. This contains reinforcing tracts that can be considered as discrete ligaments. The membrane is covered by the m. pectoralis and the m. supracoracoideus.

Joints of the wing (juncturae alae)

The joints of the wing (Figures 3.9ff.) are comprised of the:

- **shoulder joint (articulatio humeri** or **coracoscapulohumeralis)**,
- **elbow joint (juncturae cubiti)** comprising the:
  - humeroradial joint (articulatio humeroradialis),
  - humeroulnar joint (articulatio humeroulnaris),
  - proximal radioulnar joint (articulatio radioulnaris proximalis); and
- **joints of the carpus and manus (juncturae carpi et manus)**.

Shoulder joint (articulatio humeri)

The shoulder joint (Figures 3.9 to 3.11) is formed by the **glenoid cavity** and the ellipsoid **head of the humerus**. The composition of the glenoid cavity is described above (**coracoscapular joint**). As an ellipsoid joint, the shoulder has a wide range of movement and an extensive joint capsule. The **fibrocartilago humeroscapularis** is incorporated into the dorsal section of the capsule, acting as the functional equivalent of a sesamoid bone for the m. deltoideus major.

The **acrocoracohumeral ligament** (Figures 3.9 to 3.11) plays a particularly important role in the shoulder joint. It connects the **acrocoracoid process** with the **transverse groove** on the proximal cranial aspect of the humerus. Additional stabilisation is provided by several ligaments extending between the scapula and the humerus (**ligg. scapulohumeralia**) (Figures 3.9 and 3.11) and by two intracapsular ligaments, covered in synovial folds. The collective function of these ligaments, together with the surrounding musculature, is to prevent **excessive rotation** of the humerus.

In the **resting position** (wings adducted) the humerus **lies along the trunk**. During gliding flight, the humerus is abducted by up to 90 degrees. Rotation of the shoulder joint is used to determine the so-called **angle of attack** of the wing (Figure 3.11). After the wing is raised from the outstretched position, the pectoral muscles (see below) produce the powerful downstroke (Figures 3.3 and 3.23). Rather than moving in the vertical plane, however, the wings move diagonally from a position above and in front, to below and behind. In this process, the rib cage serves to prevent excessive movement of the wings. Backward rota-

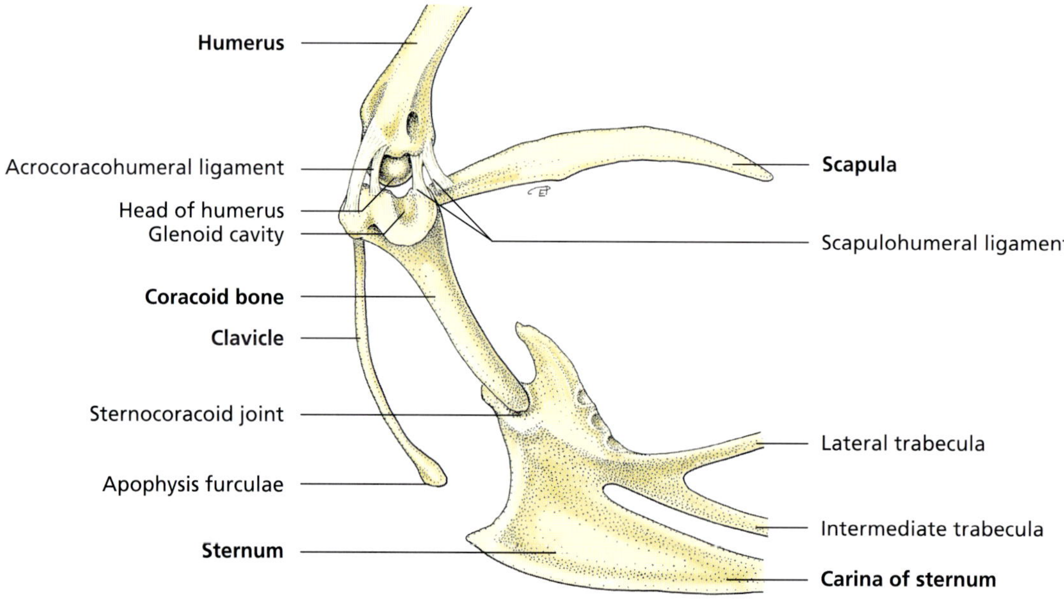

3.11 Left shoulder joint of the chicken (schematic; lateral view).

tion of the humerus, and thus the entire wing, is referred to as **supination,** the opposite movement as **pronation**.

Elbow joint (juncturae cubiti)

As in mammals, the elbow (Figures 3.12 and 3.13) consists of three distinct joints:

- humeroulnar joint (articulatio humeroulnaris),
- humeroradial joint (articulatio humeroradialis),
- proximal radioulnar joint (articulatio radioulnaris proximalis).

The composition of these joints is summarised in Table 3.1.

The **joint capsule** encloses **all three articulations**. The **collateral ligaments** (ligg. collateralia ventrale and dorsale) attach proximally at the **ventral** and **dorsal epicondyles** of the humerus (Figures 3.12 and 3.13). The ventral ligament ends on the ulna near the ventral cotyla. Exhibiting greater species-related variation in its thickness and course, the dorsal ligament ends on the dorsal radius.

An intracapsular ligament, the **radioulnar meniscus** (meniscus radioulnaris), is interposed between the dorsal humeral condyle and the dorsal cotyla of the ulna. It is attached by short ligaments to the radius and ulna, uniting these bones.

Table 3.1 Joints of the elbow.

Name	Proximal articular surface	Distal articular surface
Humeroradial joint	Dorsal condyle of the humerus	Humeral cotyla of the radius
Humeroulnar joint	Ventral and dorsal condyle of the humerus	Ventral and dorsal cotyla of the ulna
Proximal radioulnar joint	Radial incisure of the ulna	Ulnar articular surface of the radius

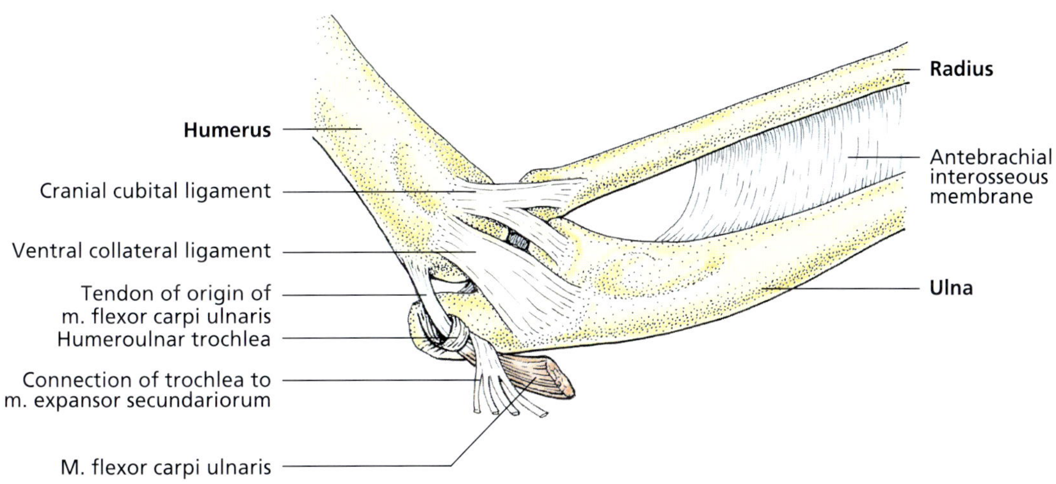

3.12 Left elbow of the chicken (schematic; ventral view).

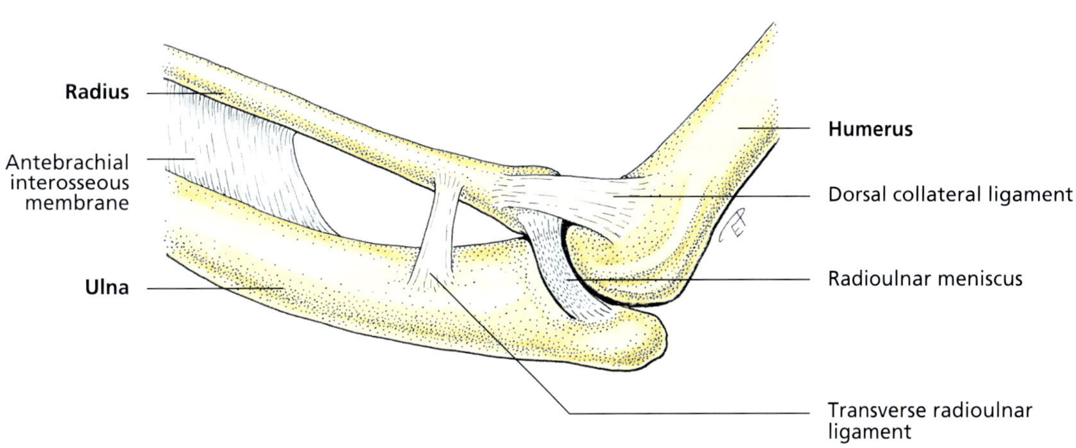

3.13 Left elbow of the chicken (schematic; dorsal view).

The cranial section of the joint capsule is invested with elastic fibres. This region, referred to as the **cranial cubital ligament** (lig. craniale cubiti), is thought to assist with initial flexion of the elbow joint from the fully extended position. The **humeroulnar trochlea** (trochlea humeroulnaris) arises from the ulna and loops around the tendon of origin of the m. flexor carpi ulnaris (Figure 3.12). Thus, the trochlea alters the direction of pull of the ulnar carpal flexor and establishes a connection between the ventral epicondyle of the humerus and the ventral surface of the ulna.

The humeroulnar trochlea is attached to the **distal tendon** (tendo distalis) of the m. expansor secundariorum. This connection assists in coordinating the **arrangement of the secondary remiges** when the muscle contracts.

Movement of the elbow joint consists mainly of flexion and extension. A functional coupling exists between the elbow and the wrist, whereby extension of the elbow results in **concordant extension of the carpus.** This results from longitudinal displacement of the radius with respect to the ulna ('drawing parallels' action) during movement of the elbow joint (Figure 3.16). The shape of the humeral joint surfaces and the arrangement of the ligaments of the elbow contribute to this phenomenon.

Full extension of the elbow also results in **rotation of the bones of the antebrachium** about the longitudinal

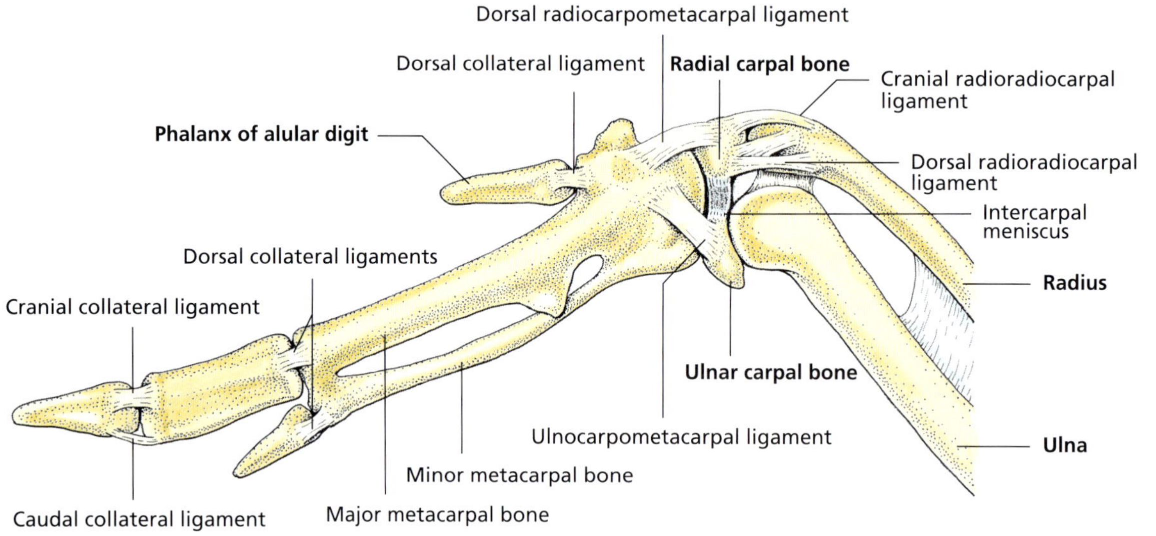

3.14 Attachments of the joints of the left manus of the chicken (schematic; dorsal view), adapted from Ghetie, 1976.

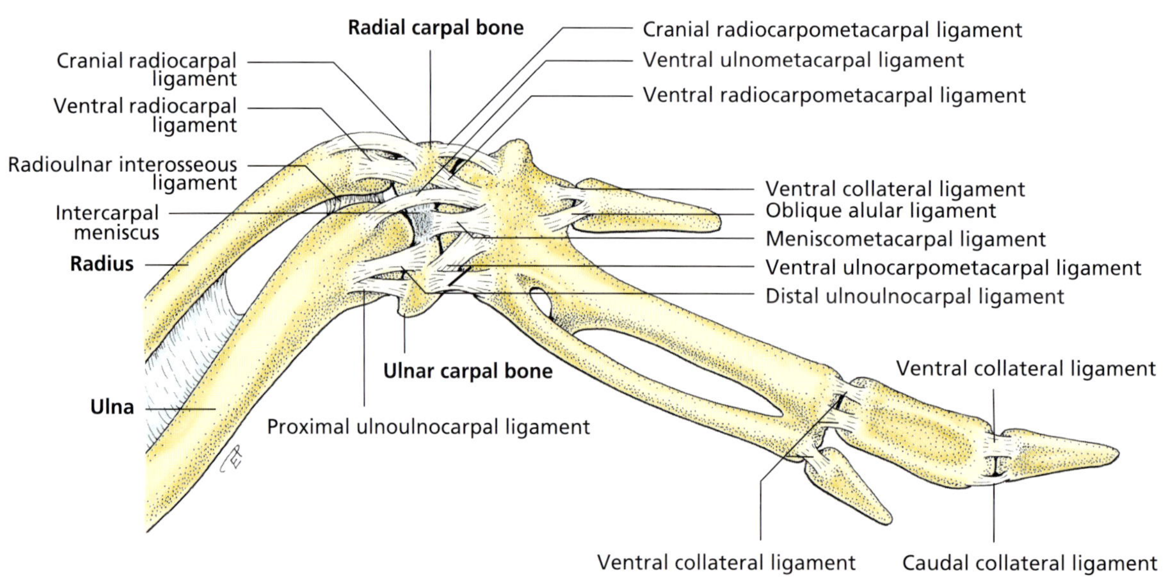

3.15 Attachments of the joints of the left manus of the chicken (schematic; ventral view), adapted from Ghetie,1976.

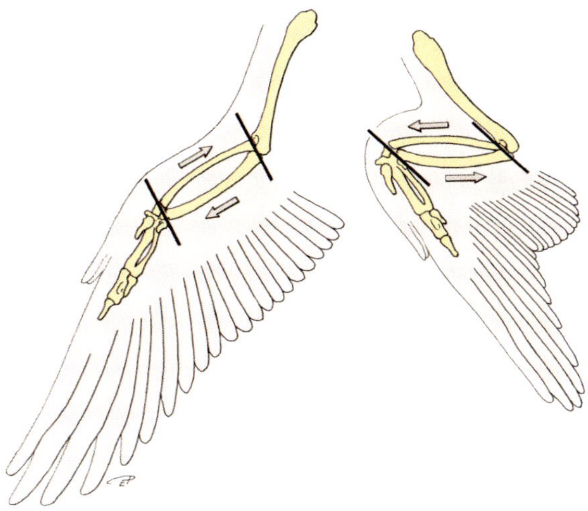

3.16 Synchronous movement of the elbow and carpus according to the 'drawing parallels' action of the radius and ulna.

axis. This occurs because the olecranon is deflected by the ventral humeral epicondyle, causing the ulna to rotate and the radius to turn with it.

The bones of the antebrachium are connected proximally by the **transverse radioulnar ligament** (lig. radioulnare transversum) (Figure 3.13). Further distally, the radius and ulna are bound together by the antebrachial interosseous membrane (membrana interossea antebrachii) (Figure 3.13).

The movements described above are made possible by the radioulnar joints. Proximally, the radius and ulna are united by a synovial joint (art. radioulnaris proximalis), which is stabilised by ligaments. Distally, the **radioulnar interosseous ligament** (lig. radioulnare interosseum) creates a fibrous union (syndesmosis radioulnaris distalis) between the two bones.

Joints of the carpus and manus (juncturae carpi et manus)

The joints of the carpus and manus (Figures 3.14 and 3.15) are comprised of the carpal and digital joints. Bones contributing to the carpal joints include the radius and ulna, the radial and ulnar carpal bones and the carpometacarpus.

The **carpus** essentially functions as a hinge joint that, as a consequence of the sliding movements of the radius and ulna (see above), is extended and flexed in synchrony with the elbow. Numerous ligaments interconnect the **bones of the carpal joint** (Figures 3.14 and 3.15). The **intercarpal meniscus** (meniscus intercarpalis) is interposed between the radial and ulnar carpal bones (Figure 3.14 and 3.15). On the ventral aspect of the joint, a **flexor retinaculum** (retinaculum flexorum) for the tendons of the flexor muscles extends from the distal radius to the ulnar carpal bone. This fibrous band forms the proximal part of the **ventral aponeurosis** (aponeurosis ventralis) whose distal component, the **aponeurosis ulnocarporemigalis**, passes to the ulnar carpal bone and the metacarpal bone of the minor digit. From there it radiates towards the follicles of the primary remiges. All of the **digits** (alular, major, minor) articulate with the carpometacarpus by means of synovial metacarpophalangeal joints.

An interphalangeal joint is also present in the major digit (Figure 3.14). The ligamentous attachments of the digits allow the alular digit to move in **various directions**, while the articulations of the major and minor digits function primarily as **hinge** joints. Due to its connection with the major digit by the **interosseous ligament** (lig. interosseum), the minor digit follows the movements of its larger counterpart.

Muscles of the pectoral girdle and wing

A distinction can be made between the flight muscles and the muscles of the pectoral girdle. The latter are less developed, as the pectoral girdle is stabilised primarily by its sturdy and tightly interconnected bony framework (particularly the coracoid), with the muscles serving to provide additional support.

The **flight musculature** incorporates some of the **most powerful muscles in the body**. These include the pectoral muscle (m. pectoralis), which is responsible for the downstroke of the wing**.** The elevators of the wing (e.g., m. supracoracoideus) are notably weaker than the pectorals, since their action is assisted by gravity.

As well as flexing and extending the shoulder joint, the **muscles of the shoulder** are able to abduct and adduct the humerus. Some rotation of the humerus about the longitudinal axis of the outstretched wing is also possible, although this is limited by the ellipsoid shape of the humeral head and by various ligaments. Pronation of the wing – whereby its leading edge is tilted ventrally – reduces the angle of attack, resulting in a decrease in lift (see Chapter 1 'Introduction'). Supination has the opposite effect, increasing the angle of attack and associated lift.

The remaining, intrinsic muscles of the wing are comparatively poorly developed. Individual muscles are categorised according to their location and function. On the whole, the flexors are less developed than the extensors. The latter perform the crucial role of stabilising the wing during flight while also tensing the patagia.

The **muscles acting on the elbow** are divided mainly into flexors and extensors. Rotation is also possible, by virtue of the radioulnar joints, and is brought about by the muscles of supination and pronation.

As the bones of the antebrachium slide past one another during flexion and extension ('drawing parallels', see above), the carpus – with assistance from specialised carpal muscles – moves in synchrony with the elbow. This mechanism is advantageous both for extension of the wings during flight and for coordinated folding of the wings when the forelimbs are retracted against the trunk (Figure 3.16).

Numerous individual muscles at the tip of the wing act to adjust the position of the digits, and thus the alular and primary remiges, as required during flight.

The **major digit** is capable of limited movement in various planes. This is particularly important for fine adjustments of the primary remiges. Correct placement of the alula is of similar aerodynamic significance. This is controlled by movement of the alular digit, which can be flexed and extended, and also abducted and adducted. In this way the alula makes an important contribution to maintaining laminar air flow over the upper surface of the wing throughout a range of angles of attack and air flow velocities.

The **nomenclature**, innervation and attachments of individual muscles of the pectoral girdle and wing are summarised in Tables 3.2 to 3.4 and in various schematic diagrams (Figures 3.17 to 3.22).

Table 3.2 Muscles of the pectoral girdle and wing.

Name Innervation	Origin	Insertion	Action
Tensors of the patagium			
In addition to the muscle described here, various components of other muscles participate in tensing the patagium. This function is specified in the 'action' column for each relevant muscle.			
M. expansor secundariorum Anconeal nerve	M. subcoracoideus and m. scapulohumeralis	Follicles of the proximal secondary remiges near the elbow	Spread the secondary remiges in the outstretched wing by exerting proximal tension
Muscles of the pectoral girdle and wing			
Mm. rhomboidei superficialis/profundus Accessory brachial plexus	Spinous process of the caudal cervical and thoracic vertebrae	Dorsal border and medial surface of the scapula	Fix the scapula dorsally
M. serratus superficialis Accessory brachial plexus	Vertebral components of the ribs	Ventral border of the scapula	Fix the scapula ventrally **Pars metapatagialis**: tense the metapatagium
M. serratus profundus Accessory brachial plexus	Vertebral components of first ribs	Ventral border of the scapula	Fix the scapula ventrally
M. scapulohumeralis cranialis/caudalis Subscapular nerve	Distally on the scapula	Humerus, near the pneumotricipital fossa	Retract the humerus
M. subcoracoideus Subcoracoscapular nerve	By two heads proximally on the coracoid bone and cranially on the scapula	Ventral tubercle of the humerus	Adduct and supinate the humerus
M. subscapularis Subscapular nerve	Ventral border of the scapula	Ventral tubercle of the humerus	Adduct and pronate the humerus
M. coracobrachialis cranialis Medianoulnar nerve	Proximally on the coracoid bone	Cranioventrally and proximally on the humerus	Protract humerus
M. coracobrachialis caudalis Pectoral nerve	Lateral border of the coracoid bone	Ventral tubercle of the humerus	Retract and pronate the humerus
M. latissimus dorsi Nerve of m. latissimi dorsi	Thoracic vertebrae and the preacetabular wing of the ilium	Caudodorsally and proximally on the humerus	Adduct and supinate the wing **Pars metapatagialis**: tense the metapatagium
M. pectoralis Pectoral nerves Pars sternobrachialis Pars costobrachialis	**Pars sternobrachialis:** carina of sternum; sternoclavicular ligament **Pars costobrachialis:** clavicle; coracoid bone	Deltopectoral crest of the humerus	Depress and pronate the humerus (generate lift and thrust) **Pars propatagialis:** tense the propatagium

Name Innervation	Origin	Insertion	Action
M. supracoracoideus Supracoracoid nerve	Sternum; coracoid bone, sternocoracoclavicular membrane	Via the triosseal canal, on the dorsal tubercle of the humerus	Elevate and supinate the humerus
M. deltoideus Axillary nerve Pars major Pars minor Pars propatagialis	**Pars major:** acromion; neck of scapula **Pars minor:** coracoid bone; acromion	Together on the deltopectoral crest of the humerus	**Pars major:** elevate and supinate the wing **Pars minor:** elevate the wing **Pars propatagialis:** tense the propatagium

Table 3.3 Muscles of the pectoral girdle and wing: muscles of the elbow and carpus.

Name Innervation	Origin	Insertion	Action
Muscles of the elbow			
M. triceps brachii:		Together on the olecranon	Extend the elbow
Radial nerve M. scapulotriceps M. humerotriceps M. coracotriceps	**M. scapulotriceps:** scapula **M. humerotriceps:** humerus, near the pneumotricipital fossa **M. coracotriceps:** coracoid bone (proximally)		
M. biceps brachii: Bicipital nerve Caput coracoideum Caput humerale	**Caput coracoideum:** coracoid bone **Caput humerale:** humerus	By a common tendon on the radius and ulna	Flex the elbow **Pars propatagialis**: tense the propatagium
M. brachialis Median nerve	Humerus	Proximally on the ulna	Flex the elbow
M. pronator superficialis/ profundus Median nerve	Both heads: ventral epicondyle of the humerus	**M. pronator superficialis:** distally and ventrally on the radius **M. pronator profundus:** distally and caudally on the radius	Flex the elbow, pronate the wing
M. entepicondyloulnaris Ulnar nerve	Ventral epicondyle of the humerus	Proximally on the ventral surface of the ulna	Flex the elbow
M. ectepicondyloulnaris Radial nerve (superficial branch)	Dorsal epicondyle of the humerus	Dorsocranially on the ulna	Flex the elbow and supinate the forearm
M. supinator Radial nerve	Dorsal epicondyle of the humerus	Proximally on the cranial radius	Flex the elbow and supinate the forearm
Muscles of the carpus			
M. extensor carpi radialis Radial nerve	Dorsal epicondyle of the humerus	Carpometacarpus	Extend the carpus
M. extensor carpi ulnaris Radial nerve	Dorsal epicondyle of the humerus	Carpometacarpus	Extend the carpus
M. flexor carpi ulnaris Ulnar nerve	Ventral epicondyle of the humerus	Ulnar carpal bone	Flex the elbow and carpus **Pars remigalis**: rotate the remiges

Table 3.3 continued.

Name Innervation	Origin	Insertion	Action
M. ulnometacarpalis dorsalis Radial nerve (deep branch)	Distal extremity of the ulna	Minor metacarpal bone	Flex the carpus
M. ulnometacarpalis ventralis Median nerve (deep branch)	Proximally on the ventral surface of the ulna	Dorsally on the metacarpus	Pronate the manus

Table 3.4 Muscles of the pectoral girdle and wing: muscles of the digits.

Name Innervation	Origin	Insertion	Action
Muscles of the digits			
M. extensor digitorum communis Radial nerve (superficial branch)	Dorsal epicondyle of the humerus	In two parts: 1. caudally on the phalanx of the alular digit; 2. cranially on the proximal phalanx of the major digit	Adduct the alular digit; Extend the major digit; also: extend the carpus
M. extensor longus digiti majoris Radial nerve (deep branch)	By two heads, from the radius and ulnar carpal bone	Cranial edge of the distal phalanx of the major digit	Extend the major digit
M. interosseus dorsalis Radial nerve (deep branch)	Dorsally on the carpometacarpus	Distal phalanx of the major digit	Extend the major digit
M. interosseus ventralis Ulnar nerve (cranial branch)	Ventrally on the carpometacarpus, interosseous space	Dorsally on the distal phalanx of the major digit	Flex the major digit
M. flexor digitorum superficialis Median nerve	Via the humerocarpal ligament, on the ventral epicondyle of the humerus	Cranially on the proximal phalanx of the major digit	Flex the major digit
M. flexor digitorum profundus Median nerve	Proximal third of the ulna	Base of the distal phalanx of the major digit	Flex the major digit
M. abductor digiti majoris Median nerve (deep branch)	By two heads, dorsally and ventrally on the carpometacarpus	Ventrally on the proximal phalanx of the major digit	Abduct the major digit
M. extensor longus alulae Radial nerve (deep branch)	By two heads, on the radius and ulna	Cranial edge of the phalanx of the alular digit	Extend the carpus and alular digit
M. extensor brevis alulae Radial nerve (deep branch)	Alular metacarpal bone	Phalanx of the alular digit	Extend the alular digit
M. flexor alulae Median nerve (deep branch)	Ventrally on the alular metacarpal bone	Ventrally on the phalanx of the alular digit	Flex the alular digit
M. abductor alulae Median nerve (deep branch)	Radial carpal bone; Alular metacarpal bone	Ventrally on the phalanx of the alular digit	Abduct the alular digit

Name Innervation	Origin	Insertion	Action
M. adductor alulae Radial nerve (deep branch)	Cranially on the major metacarpal bone	Caudally on the phalanx of the alular digit	Adduct the alular digit
M. flexor digiti minoris Ulnar nerve	Caudal edge of the minor metacarpal bone	Caudally on the phalanx of the minor digit	Flex and abduct the minor digit

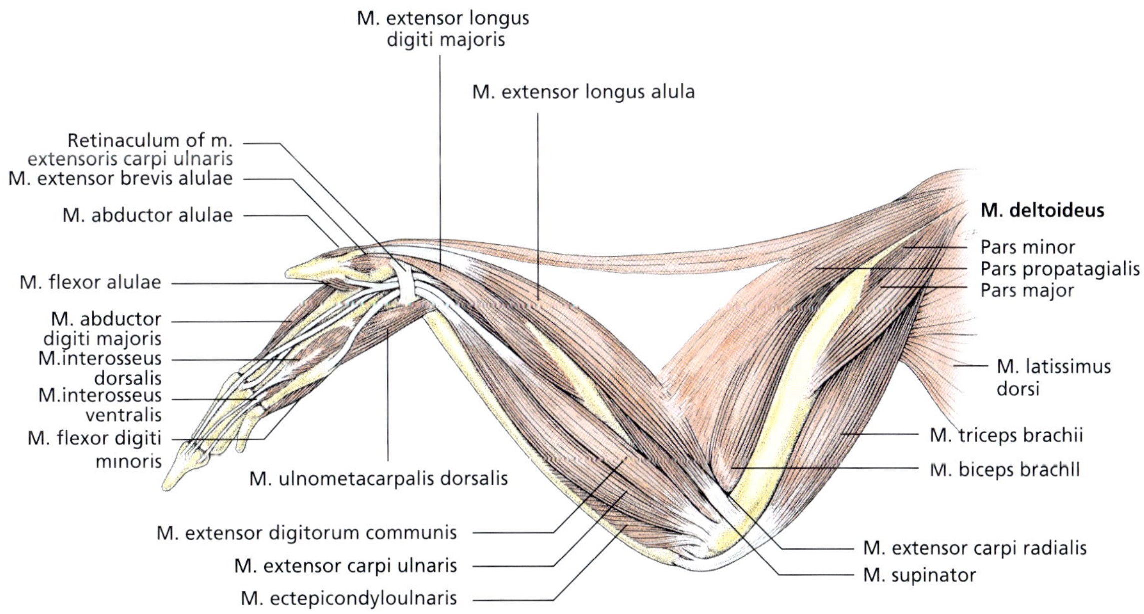

3.17 Muscles of the left wing in the chicken (schematic; dorsal view).

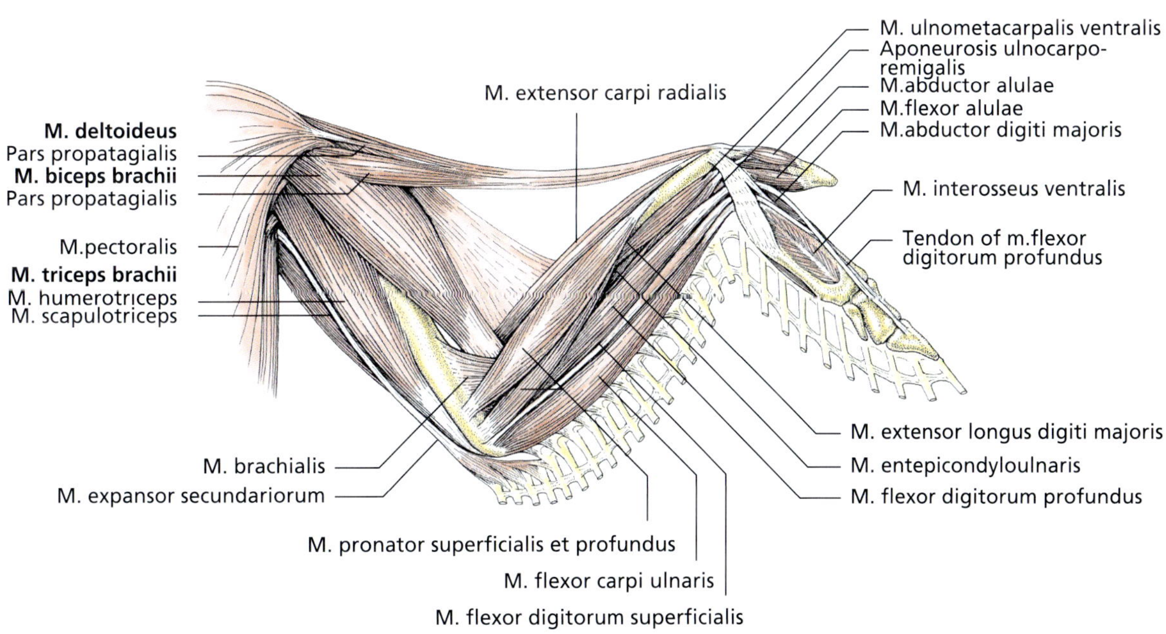

3.18 Muscles of left wing of the chicken (schematic; ventral view).

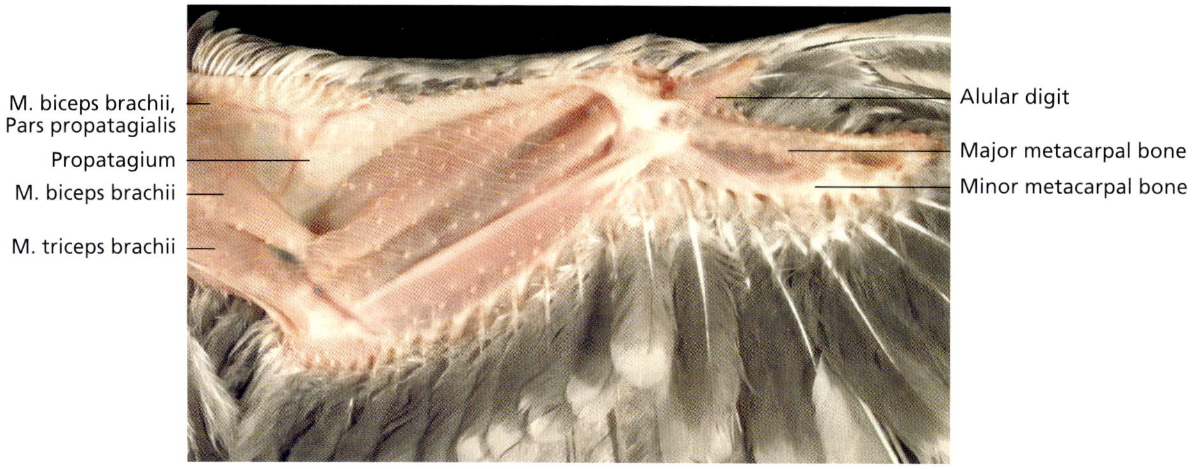

3.19 Overview of the ventral surface of the left wing of a chicken, feathers removed. Courtesy of Dr Annette Kaiser, Munich.

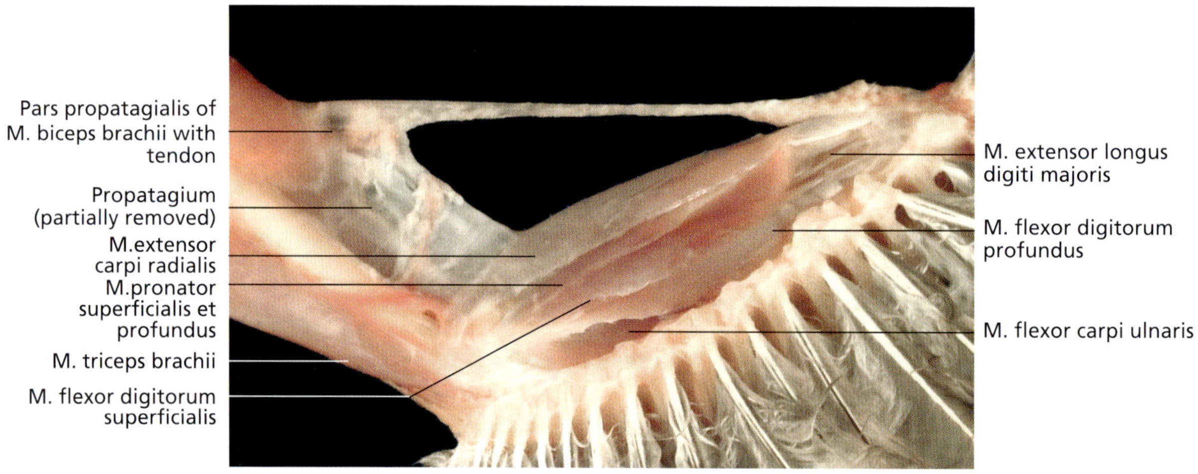

3.20 Muscles of the left wing of a chicken (ventral view). Courtesy of Dr Annette Kaiser, Munich.

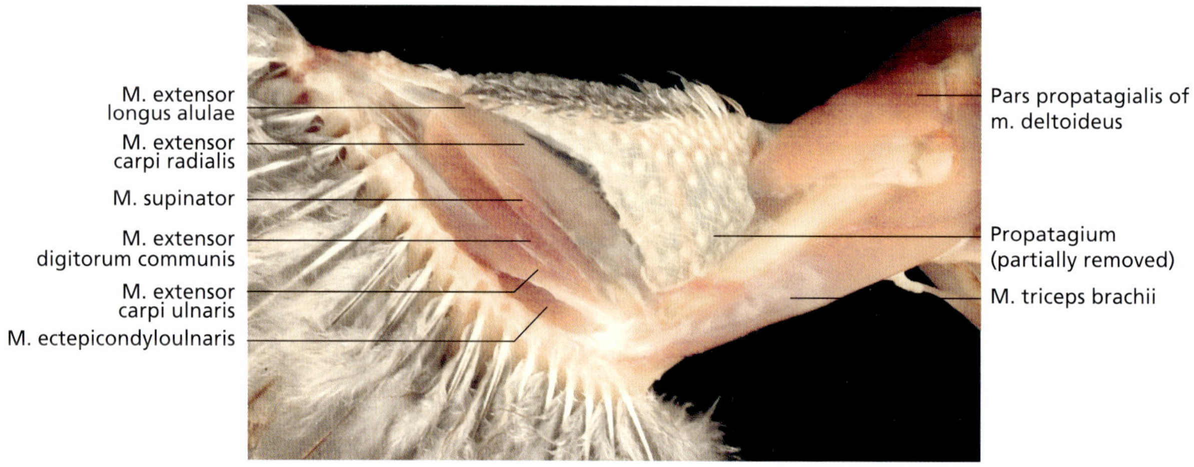

3.21 Muscles of the left wing of a chicken (dorsal view). Courtesy of Dr Annette Kaiser, Munich.

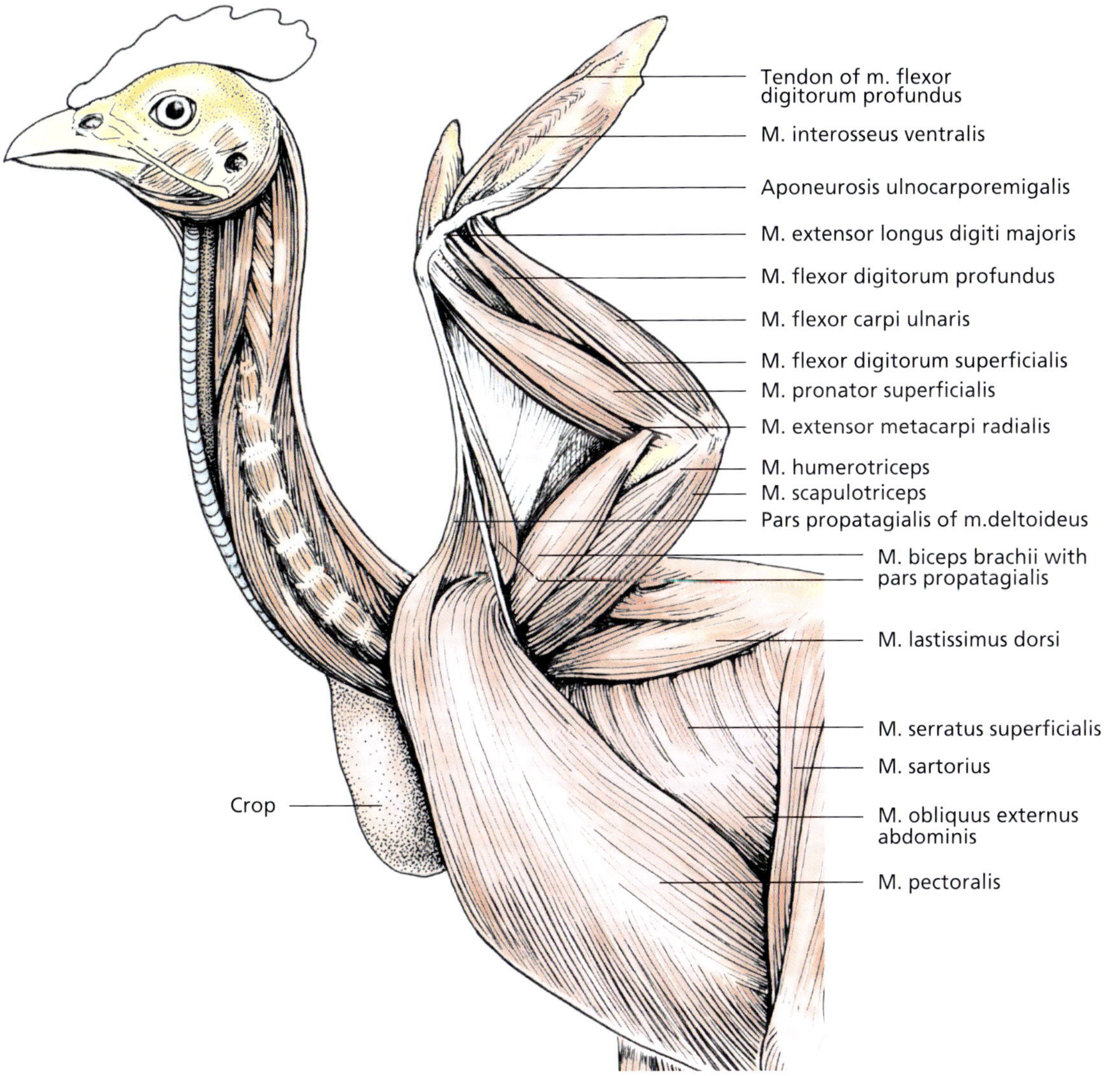

3.22 Superficial muscles of the left pectoral girdle and wing in the chicken (schematic; lateral view), adapted from Ghetie, 1976.

Clinical aspects

Due to the complex anatomical relationships in the shoulder region, **diagnosis of luxations and fractures of the coracoid bone, scapula and clavicle** is relatively challenging (Figures 3.24 and 3.25). When using diagnostic imaging, it is essential that the patient is placed in **standardised** and **symmetrical positions** for both the lateral and ventrodorsal views. Due to the many anatomical differences that exist between avian species, appropriate positioning is essential to facilitate identification of relevant landmarks for comparing the left and right sides of the body.

In a conscious unrestrained patient, luxations and fractures do not always manifest as asymmetrical positioning of the wings. Thus, in addition to routine manual examination, it is important to examine the wing under **general anaesthesia** with the patient placed in **dorsal recumbency**. When the patient is viewed at the level of the table surface, from the head towards the tail, the relaxed injured wing lies flat on the table surface. In contrast, a normal wing becomes increasingly elevated from the table towards its extremity, forming a ventrally oriented hollow.

When using a figure-of-eight **bandage in the management of thoracic limb injuries** (refer to Chapter 23 'Surgical fracture management'), it is crucial to prevent pressure and tension on the delicate and relatively poorly vascularised patagia (pro- and metapatagium) by avoiding excessively tight bandaging. In the presence of inappropriate bandaging, and in the absence of adequate regular physiotherapy, the obligatory immobilisation of the wing during fracture healing can lead to irreversible contracture of the patagia, leaving the bird unable to fly. To reduce the likelihood of these complications, physiotherapy should commence on day three to five postoperatively. Therapy includes rhythmic stretching of the patagium, by means of maximal passive extension of the wing, for 10–15 minutes per day.

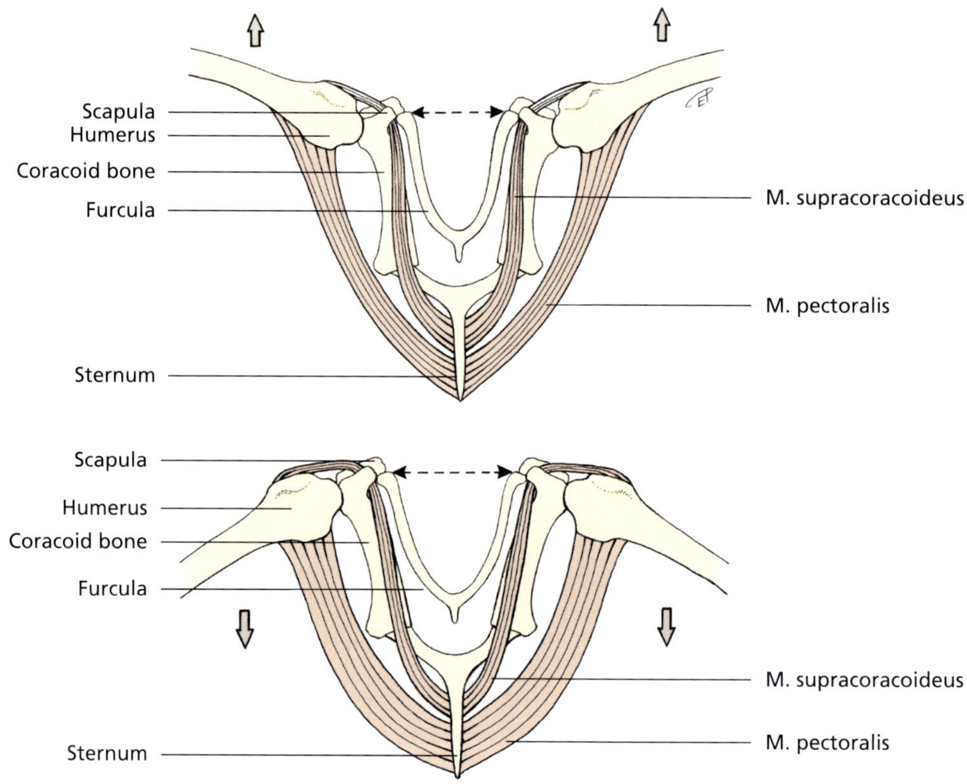

3.23 Actions of the m. pectoralis (down arrows) and the m. supracoracoideus (up arrows) during flapping of the wings (schematic). On the downstroke (mm. pectorales), the distance between the shoulder joints increases. During the upstroke (mm. supracoracoidei), the pectoral girdle springs back into its original position.

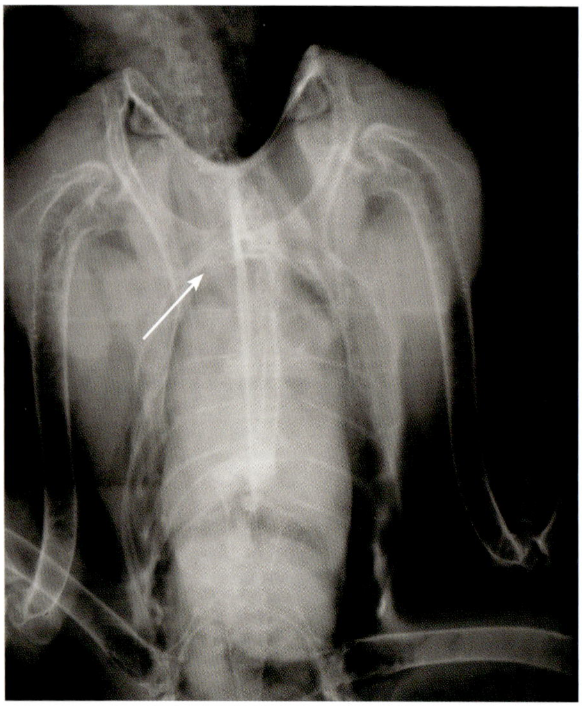

3.24 Radiograph (partial ventrodorsal view) of a coracoid fracture (arrow) in an Indian runner duck (Anas platyrhynchos).

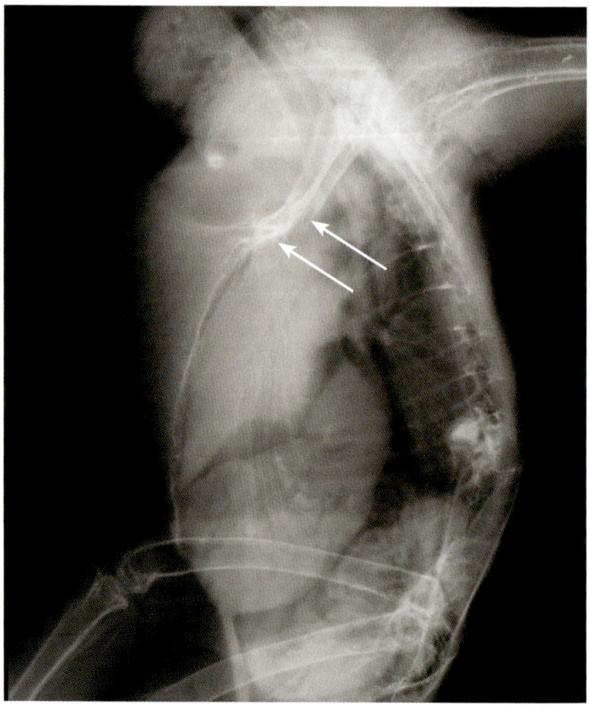

3.25 Radiograph (partial ventrodorsal view) of a coracoid luxation (arrows) in a great crested grebe (Podiceps cristatus).

Contracture of the patagium due to inadequate supportive therapy can be an indication for euthanasia on animal welfare grounds, even if fracture healing is successful. Thus, appropriate attention to the post-operative health of the patagium warrants particular attention to avoid compromising the outcome of fracture management.

Open fractures of the humerus are sometimes associated with foamy haemorrhages (synchronised with respiration) at the fracture site. This results from **damage to the diverticulum of the clavicular air sac**, such that the patient breathes through the fractured bone. These patients must be treated with antibiotics with gram positive coverage, and possibly also with antifungals, to prevent microbial contamination of the respiratory tract. Possible aspiration of blood into the respiratory system via this route must also be considered.

The development of synostoses can **interfere with the normal longitudinal displacement of the radius and ulna** ('drawing parallels' action) (Figure 3.16). This mechanism is part of the normal movement of the wing and is therefore indispensable for flight. Synostoses can result from inappropriate fracture management, excessive callus formation following infection, and various other causes. They invariably compromise flying ability and may render the bird completely flightless. Depending on the bird's captive environment, or its living conditions in the wild, this can represent grounds for euthanasia.

Amputation of the wing above the carpometacarpal joint is questionable in terms of animal welfare, depending on the bird's husbandry or natural environment, since amputation both precludes flight and affects the bird's thermal insulation. This, in turn, can lead to other undesirable consequences such as infection (e.g., aspergillosis).

Surgical intervention ('pinioning') for preventing flight (e.g., to prevent the escape of zoo birds or pet birds kept outdoors) was once carried out routinely. This involves unilateral amputation of the wing at the level of the carpometacarpus. By amputating on one side only, birds can no longer achieve balanced flight. The alular digit is preserved to protect the amputation site and prevent the development of excessive granulation tissue through repeated trauma to the exposed stump. This procedure is now prohibited or restricted in some jurisdictions unless it can be justified for medical reasons.

4

Pelvic limb (membrum pelvinum)

J. Maierl, H.-G. Liebich, H. E. König and R. Korbel

As well as serving to carry the weight of the body during land-based locomotion, the pelvic limbs of birds are used for scratching, climbing, grasping and swimming. Due to the bipedal avian gait and the physiological requirements of egg laying, the bones of the pelvic limb exhibit several specialised features.

Skeleton of the pelvic girdle and pelvic limb

Skeleton of the pelvic girdle (ossa cinguli membri pelvici)

As in mammals, the pelvic girdle is made up of three bones (Figures 4.1ff.):

- ilium (os ilium),
- ischium (os ischii),
- pubis (os pubis).

In the adult bird, these bones are fused, forming the **os coxae**. Furthermore, the latter is ankylosed with the **synsacrum**. The rigid connection thus formed between the pelvic girdle and the vertebral column is well suited to supporting the animal's **body weight**. In most birds, including the chicken, the pelvis is open ventrally. **To facilitate the passage of eggs, there is no pubic symphysis.**

The **acetabulum** (Figure 4.1) is formed, as in mammals, by the ilium, ischium and pubis. An **acetabular foramen** (foramen acetabuli) lies deep within the articular socket. It is closed by the **acetabular membrane** (membrana acetabuli). Caudodorsally, the articular surface of the acetabulum merges with that of the **antitrochanter**, which forms a joint with the **neck** of the **femur** (collum femoris) and the **major trochanter** (trochanter major) (Figure 4.5). While this arrangement reduces mechanical stresses on the femur, it also limits the range of motion of the hip joint.

The **obturator foramen** (foramen obturatum) (Figure 4.1) is located caudoventral to the acetabulum between the pubis and the ischium. It allows for passage of the m. obturatorius medialis and egress of the obturator nerve. The **ilioischiadic foramen** (foramen ilioischiadicum), lying directly caudal to the acetabulum, is traversed by the ischiadic nerve. A further opening, the **ischiopubic window** (fenestra ischiopubica) is located caudally between the ischium and pubis.

Ilium (os ilium)

The ilium (Figure 4.1ff.) is the largest of the three pelvic bones. It is divided into the:

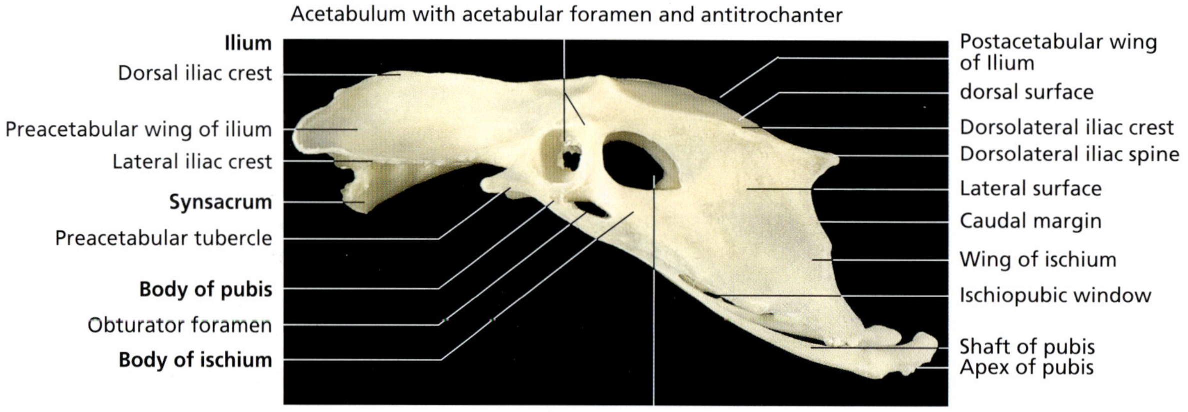

4.1 Left pelvic bones of a chicken (lateral view).

- preacetabular wing (ala praeacetabularis),
- body (corpus ossis ilii),
- postacetabular wing (ala postacetabularis).

The **preacetabular wing** (ala praeacetabularis) is bounded dorsally by the **dorsal iliac crest** (crista iliaca dorsalis). In the chicken, this ridge joins with the fused spinous processes of the synsacrum, forming the iliosynsacral crest,

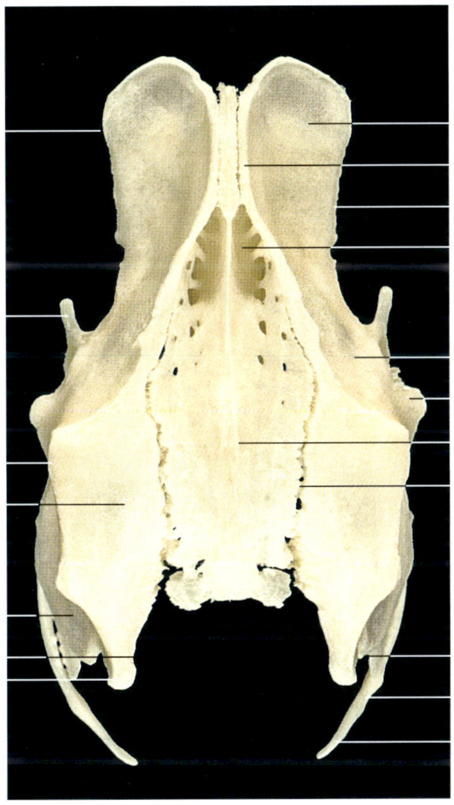

4.2 Bones of the pelvic girdle and synsacrum of a chicken (dorsal view).

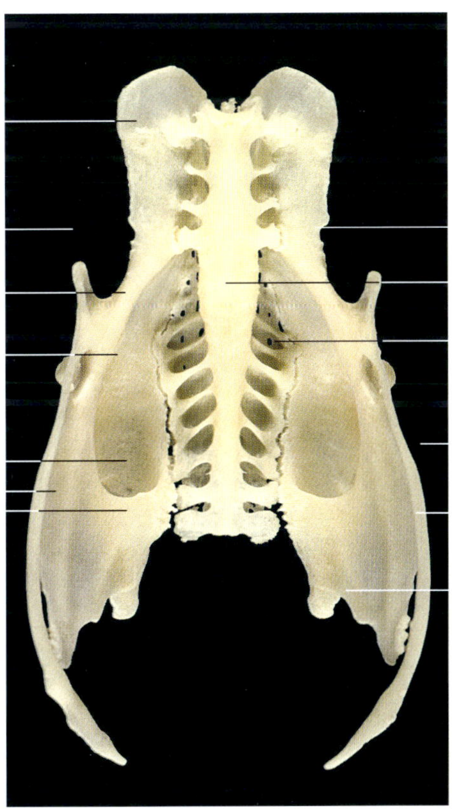

4.3 Bones of the pelvic girdle and synsacrum of a chicken (ventral view).

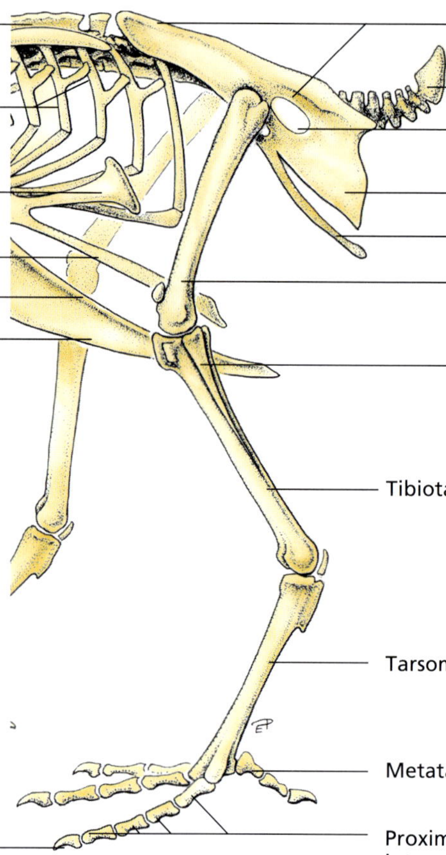

4.4 Skeleton of the pelvic limb of the chicken (schematic; lateral view).

which further contributes to the stability of this region. This union also gives rise to the **iliosynsacral canal** (canalis iliosynsacralis) (Figure 4.2), which is enclosed by the medial surface of the preacetabular wing and the spinous and transverse processes of the synsacrum. The canal is occupied by muscles of the vertebral column.

The **lateral iliac crest** (crista iliaca lateralis) marks the cranial and lateral margin of the preacetabular wing. This provides an expanded surface for attachment of muscles of the limb (Figures 4.1 and 4.3).

The **body of the ilium** (Figures 4.1 and 4.2) contributes dorsally and cranially to the acetabulum and accommodates the **antitrochanter**. The **preacetabular tubercle** (tuberculum praeacetabulare) projects cranioventrally. In some species it may arise from the pubis.

The **postacetabular wing** is divided into two parts, presenting a **dorsal** and **lateral surface** (facies dorsalis and lateralis) (Figures 4.1 and 4.2). Its dorsal surface abuts the synsacrum medially and is separated from the lateral surface by the **dorsolateral iliac crest** (crista dorsolateralis ilii). This tapers caudally as the **dorsolateral iliac spine** (spina dorsolateralis ilii).

The lateral surface of the ilium merges with the **wing of the ischium** without a distinct boundary. Caudally the termination of the **postacetabular wing** is delineated by the **caudal margin** (margo caudalis). The main feature of the **ventral surface** (facies ventralis) (Figure 4.3) is the **renal surface** (facies renalis) for the caudal division of the kidney. Caudally this joins the **caudal recess of the renal fossa** (recessus caudalis fossae) from which the m. obturatorius medialis arises.

Ischium (os ischii)

The ischium (Figure 4.2) consists of the:

- body (corpus ischii),
- wing of the ischium (ala ischii).

The **body of the ischium** contributes to the acetabulum caudally and forms part of the ventral antitrochanter.

Caudally, the **wing of the ischium** (ala ischii) forms the ventral continuation of the lateral surface of the ilium. It lacks distinctive features and serves as the origin of muscles of the pectoral limb.

Pubis (os pubis)

The pubis (Figures 4.1 and 4.4) is the most ventral of the three pelvic bones. Its **body** (corpus pubis) completes the acetabulum ventrally. The body continues caudally as the **shaft** (scapus pubis). This is incompletely fused with the ischium, the resulting gaps forming the obturator foramen (cranially) and the ischiopubic window (caudally).

The caudal tip of the shaft bears the slightly thickened, palpable **apex of the pubis** (apex pubis).

Skeleton of the pelvic limb (ossa membri pelvici)

The bones of the pelvic limb (Figure 4.4) consist of the:

- femur (os femoris),
- the tibiotarsus and the fibula,
- tarsometatarsus and
- phalanges.

Femur (os femoris)

The femur (Figures 4.5 and 4.6) is cylindrical with a slight cranial curvature. At its **proximal extremity** (extremitas proximalis), the round **femoral head** (caput femoris) bears a small **cavity for the ligament of the femoral head** (fovea ligamenti capitis).

The **neck of the femur** (collum femoris) (Figures 4.5 and 4.6) connects the head with the shaft. This transitional region includes the articular surface (facies articularis antitrochanterica) (Figure 4.6) for the joint with the antitrochanter. The laterally positioned **femoral trochanter** (trochanter femoris) projects proximally. A bony ridge, the **trochanteric crest** (crista trochanteris), extends distally from the trochanter. Muscular lines are present on the cranial and caudal surfaces of the **body** of the femur.

As in mammals, the **distal extremity** (extremitas distalis) features **lateral** and **medial condyles** (condylus lateralis and medialis) separated by an **intercondylar groove** (sulcus intercondylaris). The lateral surface of the lateral condyle presents a circular recess, the **fibular trochlea** (trochlea fibularis) (Figure 4.6), for articulation with the fibula. Lateral and medial epicondyles are situated proximal to their respective condyles. Recesses for the collateral and cruciate ligaments are evident.

Impressions left by various muscles, including the m. tibialis cranialis, the m. gastrocnemius lateralis and medialis and the m. iliofibularis are also discernible. On the cranial surface (Figure 4.5), **the patellar groove** (sulcus patellaris) provides a gliding articular surface for the patella. It is bounded on both sides by bony crests.

The **patella**, the sesamoid bone of the mm. femorotibiales, is lined with cartilage on its caudally directed articular

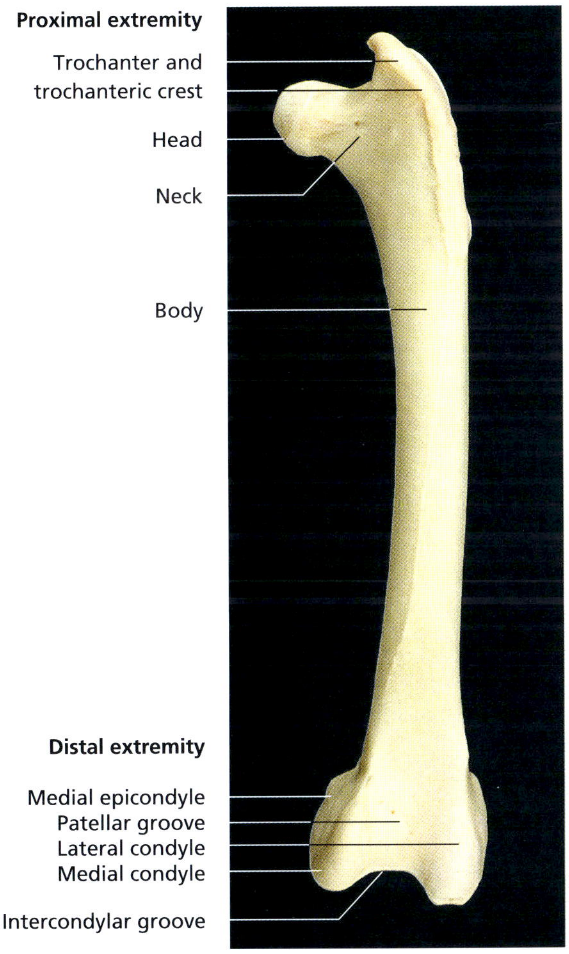

4.5 Left femur of a chicken (cranial view).

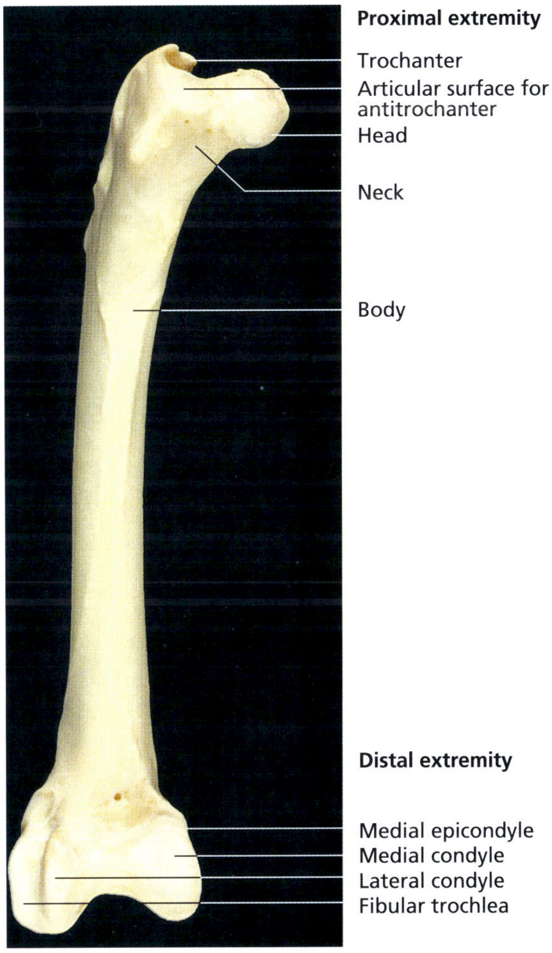

4.6 Left femur of a chicken (caudal view).

surface. Its cranial surface may be grooved by the tendon of the m. ambiens.

Tibiotarsus

The term tibiotarsus (Figures 4.7 and 4.8) applies to the structure formed by the **fusion** of the **tibia** with the **proximal row of tarsal bones.** At its expanded **proximal extremity**, or **tibial head** (caput tibiae), the smaller **lateral** and larger **medial articular surface** (facies articularis lateralis and medialis) are separated by the **interarticular area** (area interarticularis). The articular surfaces form **incongruent joints** with the respective condyles of the femur. Lateral and medial **menisci** enhance the conformity of the joint.

The **cranial cnemial crest** is a prominent feature of the proximal tibiotarsus (Figure 4.7). Its proximal end is connected by the transversely oriented **patellar crest** (crista patellaris) with the **lateral cnemial crest** (crista cnemialis lateralis) (Figure 4.7). The patellar crest also serves as the site of attachment of the patellar ligament. Lying between the cnemial crests, the **intercnemial groove** (sulcus intercnemialis) is the origin of the m. extensor digitorum longus. On the caudal aspect of the tibiotarsus, the m. flexor digitorum longus arises from the **flexor fossa** (fossa flexoria) (Figure 4.8). Further distally, the approximately triangular cross section of the tibiotarsus gives way to an ovoid profile.

The **extensor groove** (sulcus extensorius) is located cranially, just proximal to the distal extremity (Figure 4.7). It is spanned by the delicate **supratendinal bridge** (pons supratendineus), which guides the distal tendons of the extensors of the toes. Immediately distal to the extensor groove are the **medial** and **lateral condyles** (condylus medialis and lateralis), separated by the **intercondylar incisure** (incisura intercondylaris). The articular surfaces of the condyles coalesce caudally, forming the **trochlea for the tibial cartilage** (trochlea cartilaginis tibialis) (Figure 4.8). This articulates with the **tibial cartilage** (cartilago tibialis), a protective structure over which the tendons of the m. gastrocnemius and digital flexors glide.

The **tibial cartilage** (Figures 4.17, 4.18, 4.24 and 4.25) is composed of **fibrocartilage** and may ossify in adults into an **intertarsal sesamoid bone** (os sesamoideum intertarsale). **Medial** and **lateral epicondyles** (epicondylus medialis and lateralis) are located proximal to the condyles. These are extensively sculpted by the attachment sites of ligaments. The tibiotarsus has sole responsibility for weight bearing in the crus.

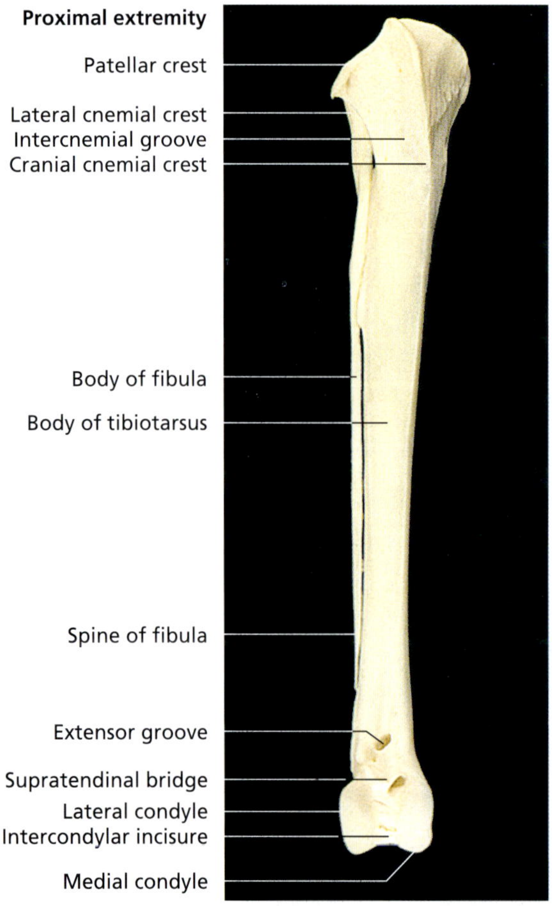

4.7 Right tibiotarsus and fibula of a chicken (cranial view).

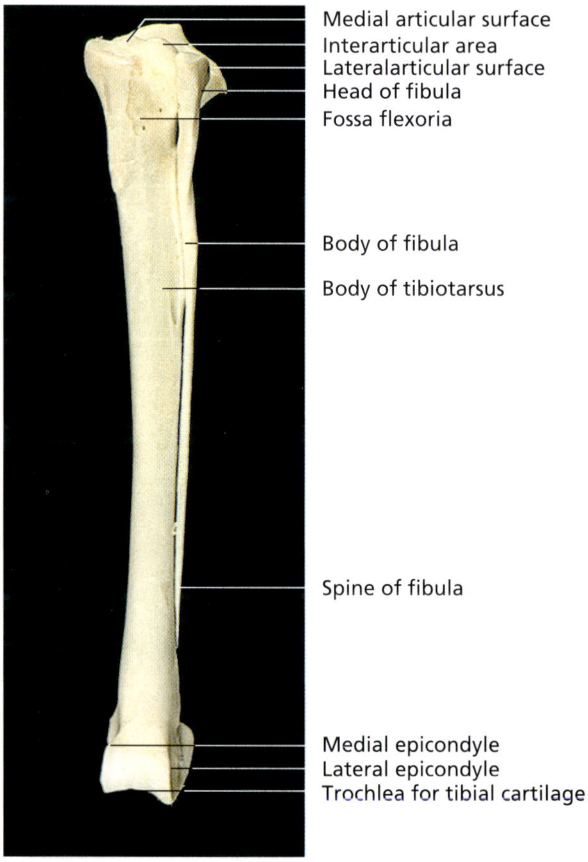

4.8 Right tibiotarsus and fibula of a chicken (caudal view).

Fibula

The fibula (Figures 4.7 and 4.8) has a **head** and a slender **body**. Its head features two articular surfaces, the **facies articularis tibialis** for the tibiotarsus and the **facies articularis femoralis** for the lateral condyle of the femur. The thin rod-like body tapers to a pointed **spine** (spina fibulae).

Tarsometatarsus

The **tarsometatarsus** (Figures 4.9 and 4.10) is formed by fusion of the **central** and **distal tarsal bones** and **metatarsal bones II–IV**. Thus, **a discrete tarsus** is not present in the avian skeleton. Only the first metatarsal (os metatarsale I) is separate, the fifth metacarpal has been lost.

At the proximal extremity, the recessed **medial** and **lateral cotylae** (cotyla medialis and lateralis) form the articular surfaces for the condyles of the tibiotarsus. The cotylae are separated by the **intercotylar eminence** (eminentia intercotylaris). Prominent markings for the attachment of ligaments are evident laterally. The **hypotarsus**, arising from components of the distal tarsal bones, is located on the plantar surface. It consists of longitudinally oriented **crests** (cristae hypotarsi) and **channels** (sulci hypotarsi). Depending on species, the channels may become enclosed to form **canals** (canales hypotarsi). The structures of the hyptotarsus serve to guide the tendons of the flexor muscles.

Elliptical in cross-section, the **body** of the tarsometatarsus has a **dorsal** and a **plantar surface** (facies dorsalis and plantaris). In male chickens, turkeys and pheasants, a **calcaris process** (processus calcaris) projects in a medioplantar direction from the plantar surface (Figure 4.10). This is the osseous foundation of the **spur** (calcar).

In domestic poultry, a small **fossa** for articulation with the **first metatarsal bone** (fossa metatarsi I) (Figure 4.9) lies just distal to the location of the calcaris process.

The derivation of the tarsometatarsus from multiple metatarsal bones becomes readily apparent at its distal end in the form of articular trochleae for **pelvic digits II–IV** (trochleae metatarsi II–IV).

The middle trochlea extends most distally (Figure 4.10). It is separated from the others by the **medial** and **lateral intertrochlear incisure** (incisura intertrochlearis medialis and lateralis). Small pits on either side of each trochlea signify the attachment site of collateral ligaments.

Metatarsal bone I projects distomedially from its articulation with the tarsometatarsus, ending in a **trochlea** (trochlea metatarsi I or hallucis) (Figure 4.9).

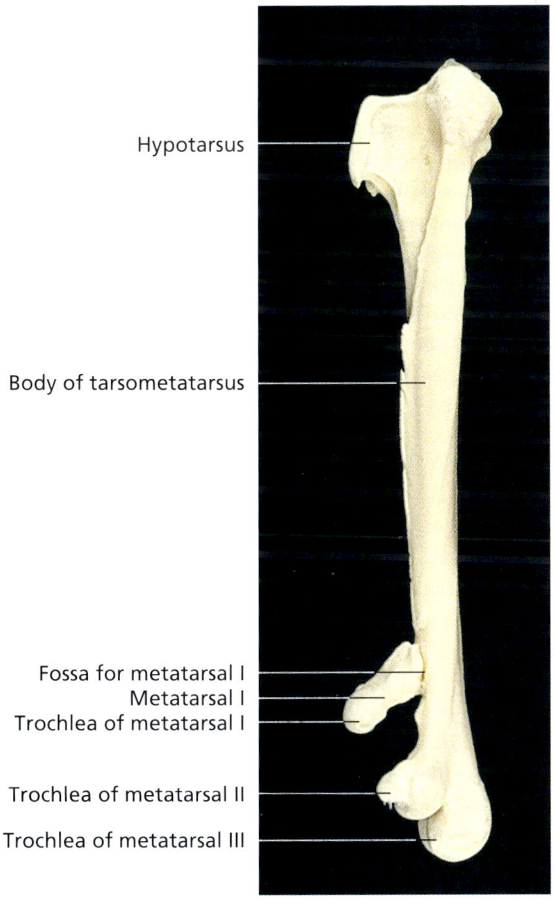

4.9 Left tarsometatarsus and metatarsal I of a chicken (medial view).

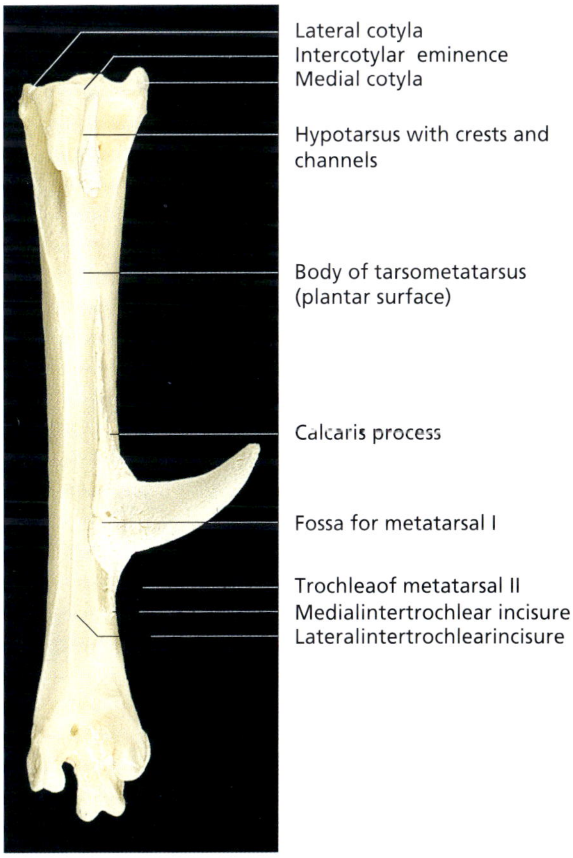

4.10 Left tarsometatarsus of a male chicken (plantar view).

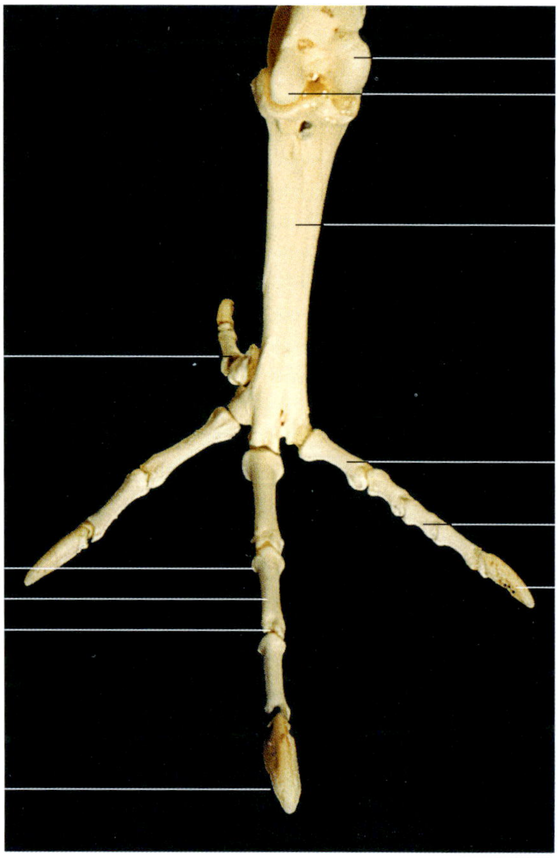

4.11 Skeleton of the left foot of a chicken (dorsal view).

Notwithstanding particular specialisations, the digits of the pelvic limb essentially conform to one of **two basic arrangements**: **anisodactyly**, where only the first digit extends in a plantar direction, and **zygodactyly** where both the first and fourth digits assume a plantar orientation.

Digits

In birds, the pelvic digits (Figure 4.11) contain different numbers of phalanges. While the first digit has only two, this number increases by one with each digit, such that the fourth digit has five phalanges. The structure of the **proximal** and **intermediate phalanges** (phalanges proximales et intermediae) is relatively consistent (Figures 4.4 and 4.11). Their features include a **base** (basis) with a concave articular surface, a **body**, and a **capitulum** incorporating a trochlear articular surface. Recesses for the attachment of collateral ligaments are located on the sides of the trochlea (Figure 4.11). Attachment sites of the tendons of the digital extensors and flexors are recognisable on the dorsal and ventral surfaces as proximal bone thickenings.

The **distal phalanx** (phalanx ungualis or terminalis) consists of a **base** and an **apex**, the base articulating with the penultimate phalanx. As suggested by its shape, the tip of the distal phalanx forms the bony substructure of the claw.

Joints of the pelvic girdle and pelvic limb

Joints of the pelvic girdle (juncturae cinguli membri pelvici)

A rigid connection exists between the os coxae and the **synsacrum** (Figures 4.2 and 4.3). This bony union includes the **iliosynsacral synostosis** (synostosis iliosynsacralis) between the transverse processes and the ilium, and the **interiliospinal synostosis** (synostosis interiliospinalis) between the spinous processes and the dorsal edge of the ilium.

The three components of the ox coxae are themselves extensively **ankylosed** with one another in the adult bird. All of the openings seen in anatomical pelvic specimens are closed by soft tissue in the living animal.

Joints of the pelvic limb (juncturae membri pelvici)

In addition to synostotic and fibrous joints, several substantial synovial joints are found in the pelvic limb. These are the:

- hip joint (junctura coxae),
- knee joint (juncturae genus),
- intertarsal joint (articulatio intertarsalis) and
- joints of the foot (juncturae tarsi et pedis).

The hip joint is composed of two articulations (Figure 4.12), the coxocapital joint (articulatio coxocapitalis) and the coxotrochanteric joint (articulatio coxotrochanterica).

In the **coxocapital joint,** a fibrocartilaginous **acetabular labrum** (labrum acetabulare) articulates with the head of the femur, which is anchored to the acetabular socket by the ligament of the femoral head. The joint is further supported by ligaments passing from the three pelvic bones to the femur. These are the:

- iliofemoral ligament (lig. iliofemorale),
- ischiofemoral ligament (lig. ischiofemorale),
- pubofemoral ligament (lig. pubofemorale).

The **coxotrochanteric joint** connects the antitrochanter with the articular surface of the femoral neck and the trochanter. This significantly restricts abduction of the limb, but also considerably reduces bending stress on the femur.

Knee joint (juncturae genus)

The knee joint (Figures 4.13 to 4.16) consists of **four individual joints**, enclosed by an extensive intercommunicating joint cavity. These are the:

- femorotibial joint (articulatio femorotibialis),
- femoropatellar joint (articulatio femoropatellaris),
- femorofibular joint (articulatio femorofibularis) and
- tibiofibular joint (articulatio tibiofibularis).

Two menisci serve to augment the congruence of the **femorotibial joint**. The **medial meniscus** (meniscus medialis) tapers cranially and caudally, permitting direct contact centrally between the femur and the tibiotarsus. Several ligaments connect the medial meniscus with the femur and the tibiotarsus, namely the:

- caudal meniscotibial ligament (lig. meniscotibiale caudale),
- meniscofemoral ligament (lig. meniscofemorale),
- transverse ligament (lig. transversum genus).

Table 4.1 Openings of the pelvis.

Opening	Associated structure
Acetabular foramen	Acetabular membrane
Ilioischiadic foramen	Ilioischiadic membrane: openings present cranially for passage of the ischiadic nerve and blood vessels
Ischiopubic window	Ischiopubic membrane
Obturator foramen	M. obturatorius medialis; passage for obturator nerve

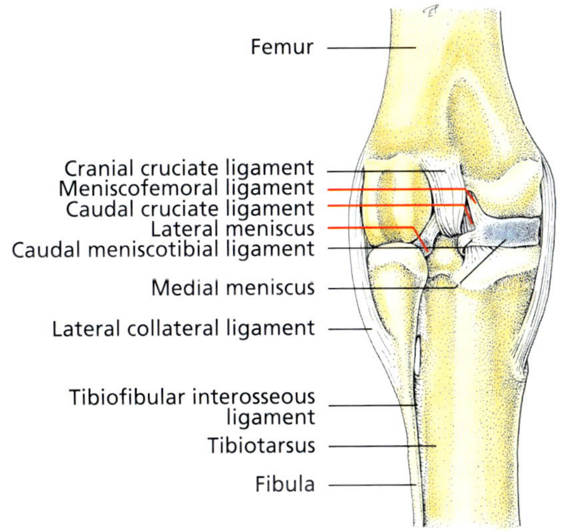

4.13 Left knee joint of the chicken (schematic; caudal view).

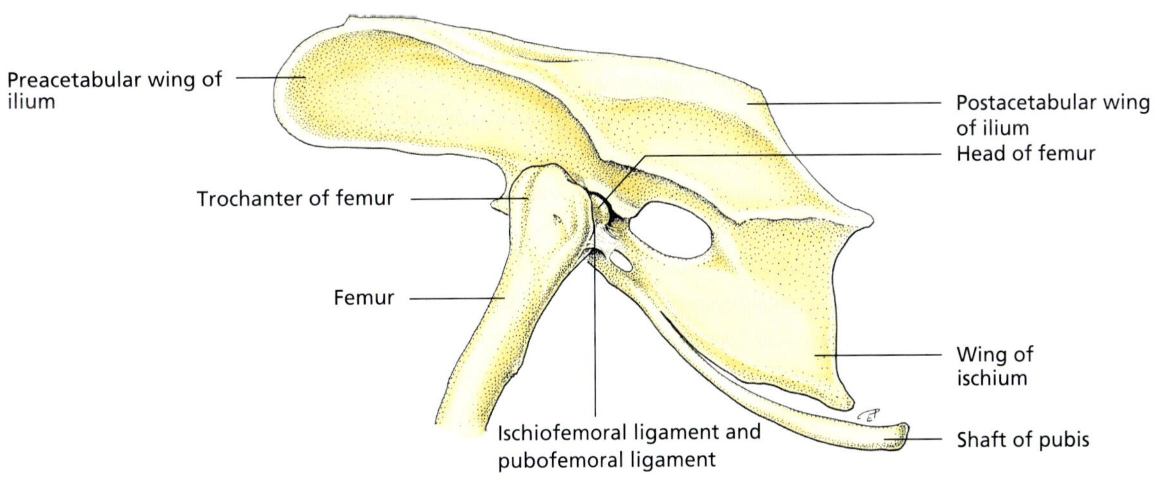

4.12 Left hip joint of the chicken (schematic; lateral view).

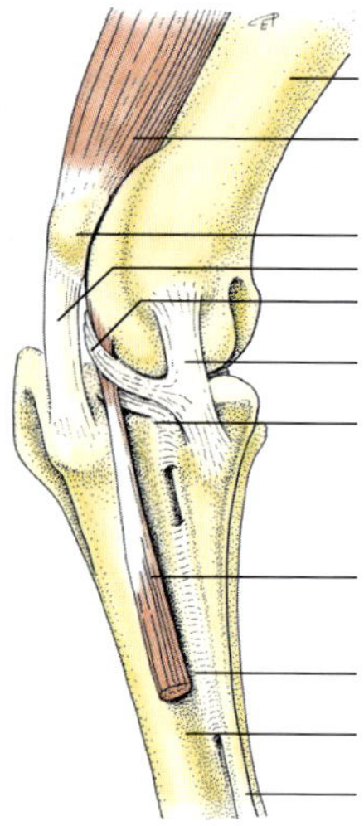

4.14 Left knee joint of a chicken (schematic; lateral view).

The **lateral meniscus** (meniscus lateralis) is an ovoid cartilaginous disc situated between the lateral condyle of the femur and the lateral articular surface of the tibiotarsus. Laterally, it is in contact with the medial border of the head of the fibula. Ligaments pass from all but the medial border of the lateral meniscus to the tibiotarsus, fibula and the femur (**cranial meniscotibial ligament,** caudal meniscofibular ligament**,** meniscocollateral ligament**,** meniscofemoral ligament)**.**

Medial and **lateral collateral ligaments** (lig. collaterale mediale and laterale) (Figure 4.13) guide the movement of the joint, together with the **cruciate ligaments** (lig. cruciatum craniale and caudale) (Figure 4.13).

As in mammals, the **femoropatellar** joint is formed by the patella and the patellar groove of the femur. The so-called **patellar ligament** (lig. patellae) (Figure 4.14) is the tendon of the mm. femorotibiales. It inserts on the **patellar crest** (Figure 4.7), forming a major portion of the cranial wall of the femorotibial joint cavity.

Due to the relatively proximal position of the fibula with respect to the tibiotarsus (Figures 4.13 and 4.15), the head of the fibula articulates with the fibular trochlea of the femur, giving rise to the femorofibular joint.

The proximal joint between the **tibiotarsus** and **fibula** is a synovial joint, while the bones are joined distally by a **syndesmosis**. A well-developed array of ligaments connects these bones along their entire length:

- **proximally:**
  - cranial tibiofibular ligament (lig. tibiofibulare craniale),
  - oblique tibiofibular ligament (lig. tibiofibulare obliquum);
- **distal:**
  - tibiofibular interosseous ligament (lig. tibiofibulare interosseum) and
  - crural interosseous membrane (membrana interossea cruris) (Figures 4.13, 4.14 and 4.15).

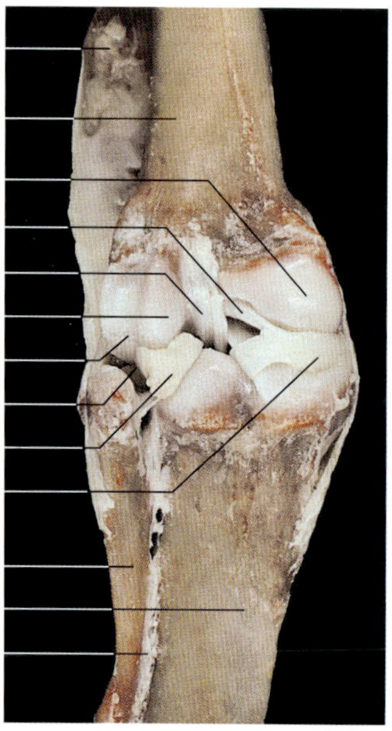

4.15 Left knee joint of a chicken (caudal view). Courtesy of Dr R. Macher, Vienna.

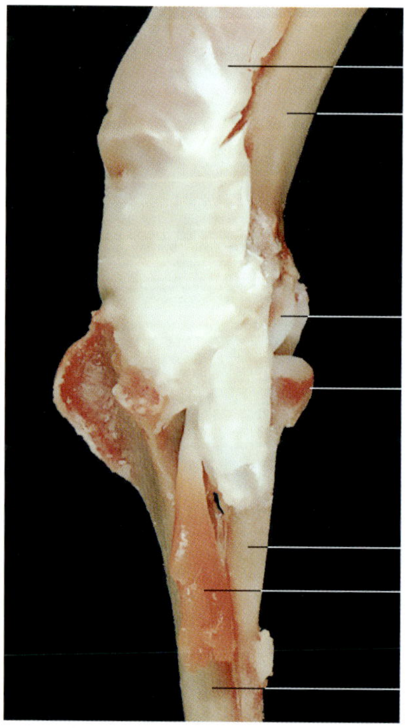

4.16 Left knee joint of a chicken (lateral view). Courtesy of Dr R. Macher, Vienna.

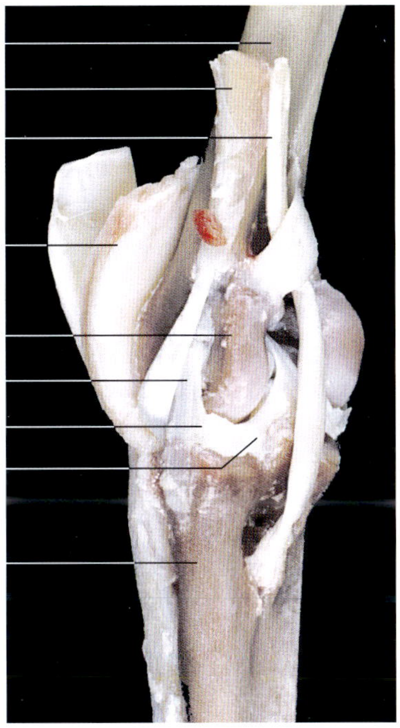

4.17 Right intertarsal joint of a chicken (dorsolateral view)

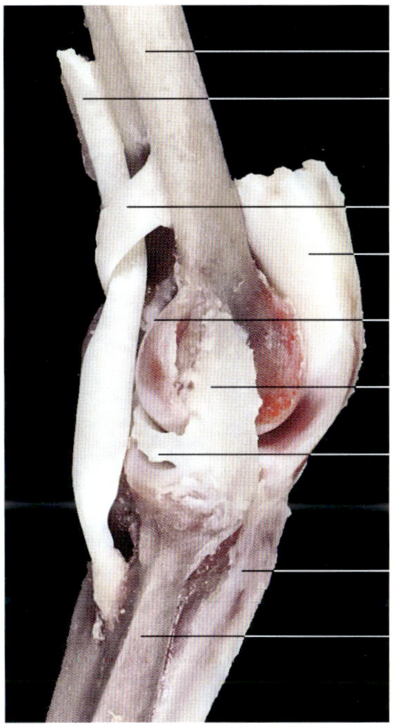

4.18 Right intertarsal joint of a chicken (dorsomedial view).

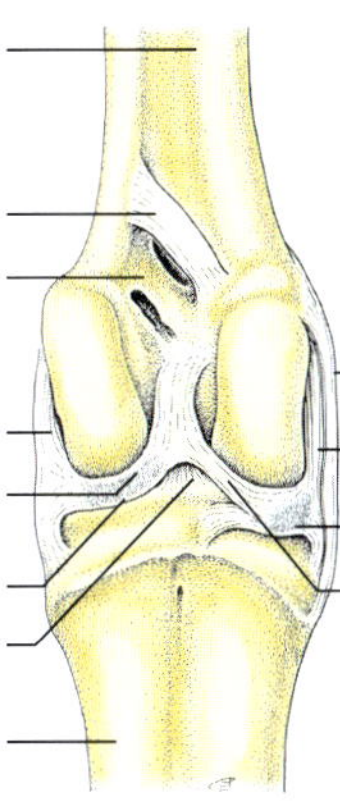

4.19 Left intertarsal joint of a chicken (schematic; dorsal view).

Intertarsal joint (articulatio intertarsalis)

The intertarsal joint, between the **tibiotarsus** and the **tarsometatarsus** (Figures 4.17ff.), is located at the site of the embryonic tarsal bones. As it is **incongruent**, a lateral and usually also a medial C-shaped meniscus (exhibiting species variation in shape and development) are interposed between the articulating bones (Figures 4.17 to 4.19). These are attached by ligaments to the tarsometatarsus and tibiotarsus.

The two **collateral ligaments** (lig. collaterale mediale and laterale) have an important mechanical role, restricting the movement of the joint to flexion and extension. In addition, the **intercondylar tibiometatarsal ligament** (lig. tibiometatarsale intercondylare) connects the intercondylar eminence of the tarsometatarsus with the intercondylar incisure of the tibiotarsus. This limits sliding movement, in a similar manner to the cruciate ligaments of the knee.

The **fibrocartilaginous tibial cartilage** (Figures 4.17ff.) is attached to the caudal aspect of the intertarsal joint. This **pressure-relieving structure** facilitates the passage of the m. gastrocnemius and the superficial digital flexors, which run through a groove on its external surface, as well as the deep flexors, which pass through canals within the cartilage. The tibial cartilage is connected on both sides with the **medial** and **lateral retinaculum** (retinaculum mediale and laterale) of the tibiotarsus (Figures 4.24ff.). A **flexor retinaculum** (retinaculum flexorium) fixes the superficial tendons to the caudal surface of the tibial cartilage. The tendons course further distad and are directed through the grooves or canals of the **hypotarsus** (Figure 4.10).

Joints of the metatarsal bones

The individual metatarsal bones and the distal row of tarsal bones are fused. Only **metatarsal I** remains **separate**, joined by a **syndesmosis** with **metatarsal II**. Various partly elastic ligaments extend between metatarsal I and the tarsometatarsus, permitting a slight splaying of the former bone (Figure 4.9).

The **transverse metatarsal ligament** (lig. transversum metatarsale) (Figure 4.20) connects the first and fourth digits, forming a **plantar band** that binds the flexors within the thus formed **plantar flexor canal** (canalis flexorius plantae).

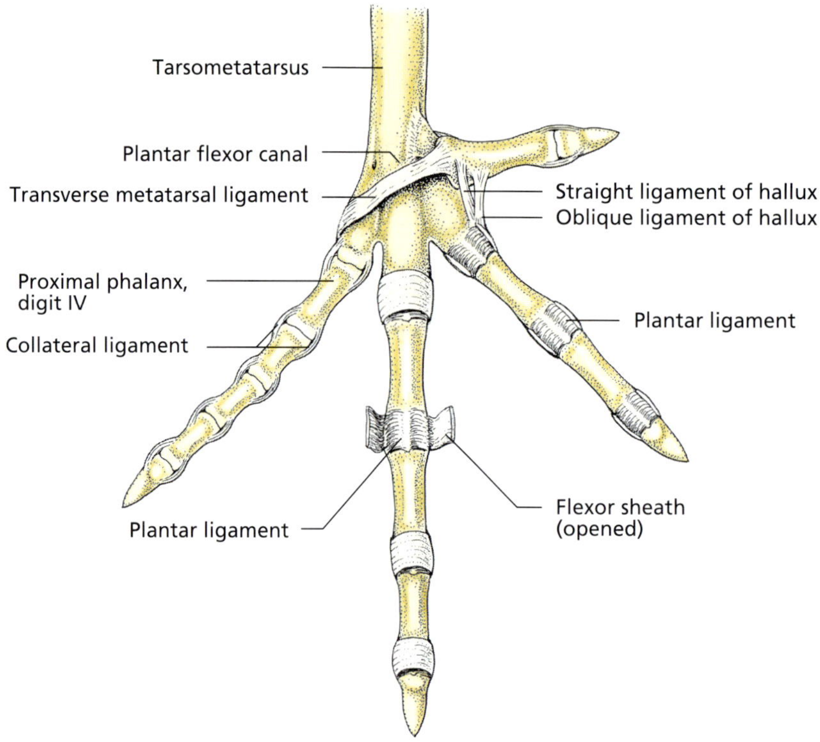

4.20 Joints of the pedal digits of the left foot of the chicken (schematic; plantar view).

Metatarsophalangeal joints (articulationes metatarsophalangeales)

The metatarsophalangeal joints (Figure 4.20) are typically associated with paired collateral ligaments. On the plantar aspect, the joint capsule is reinforced by **fibrocartilage** to form a **plantar ligament** (lig. plantare) that **reduces pressure** on the flexor tendons (Figure 4.20). **Straight** and **oblique ligaments** (lig. rectum hallucis and lig. obliquum hallucis) prevent overextension of the caudally directed first digit (**hallux**) (Figure 4.20). Following slightly different courses, both of these ligaments connect the base of the proximal phalanges of pedal digits I and II.

A **deep fat pad** (corpus adiposum plantare profundum) cushions the joints while a superficial fat body (corpus adiposum plantare superficiale) provides additional protection as part of the metatarsal pad. The pedal digits can be flexed and extended, and also abducted and adducted, allowing the digits to be spread for placing the foot on the ground or tensing the interdigital web.

Interphalangeal joints (articulationes interphalangeales)

The interphalangeal joints (Figure 4.20) are generally consistent in structure. They feature paired collateral ligaments that stabilise the joint, as well as a **plantar ligament**. Manifesting as fibrocartilaginous reinforcement of the joint capsule (Figure 4.20), the latter guards against hyperextension. A shallow groove formed by the plantar ligament is spanned by a **flexor sheath** (vagina fibrosa) (Figure 4.20) that holds the flexor tendons firmly against the bones of the digits.

Muscles of the pelvic limb

The bellies of the pelvic limb muscles (Figures 4.21 to 4.33) are positioned towards the **body's centre of gravity**. Consequently the distal muscles, such as those that move the digits, are **purely tendinous** from approximately the level of the distal crus. Further distally, only small isolated muscles are present. The tendons are very long and pass in close proximity to the bones of the crus and pes.

This is facilitated by several retaining structures, the **simplest** being the fibrous **retinaculae** including the fibular muscular retinaculum, tibiotarsal extensor retinaculum and tarsometatarsal extensor retinaculum (retinaculum musculi fibularis, retinaculum extensorium tibiotarsi, retinaculum extensorium tarsometatarsi). A particularly **rigid form** of retinaculum is formed by the ossification of connective tissue in the **supratendinal bridge**, on the craniodistal tibiotarsus (Figures 4.7 and 4.19).

The fibrocartilaginous **tibial cartilage** is attached to the caudal tibiotarsus (Figures 4.24ff.). Tendons of the extensors of the intertarsal joint and flexors of the digits glide through its superficial grooves and internal canals. The superficially located tendons, particularly the tendon of the m. gastrocnemius, are held in place by a flexor retinaculum. Due to the arrangement of the musculature, and the angles between the bones, several muscles of the pelvic limb have multiple functions. These include:

4.21 Superficial musculature of the pelvis, tail and proximal pelvic limb of the chicken (schematic; lateral view).

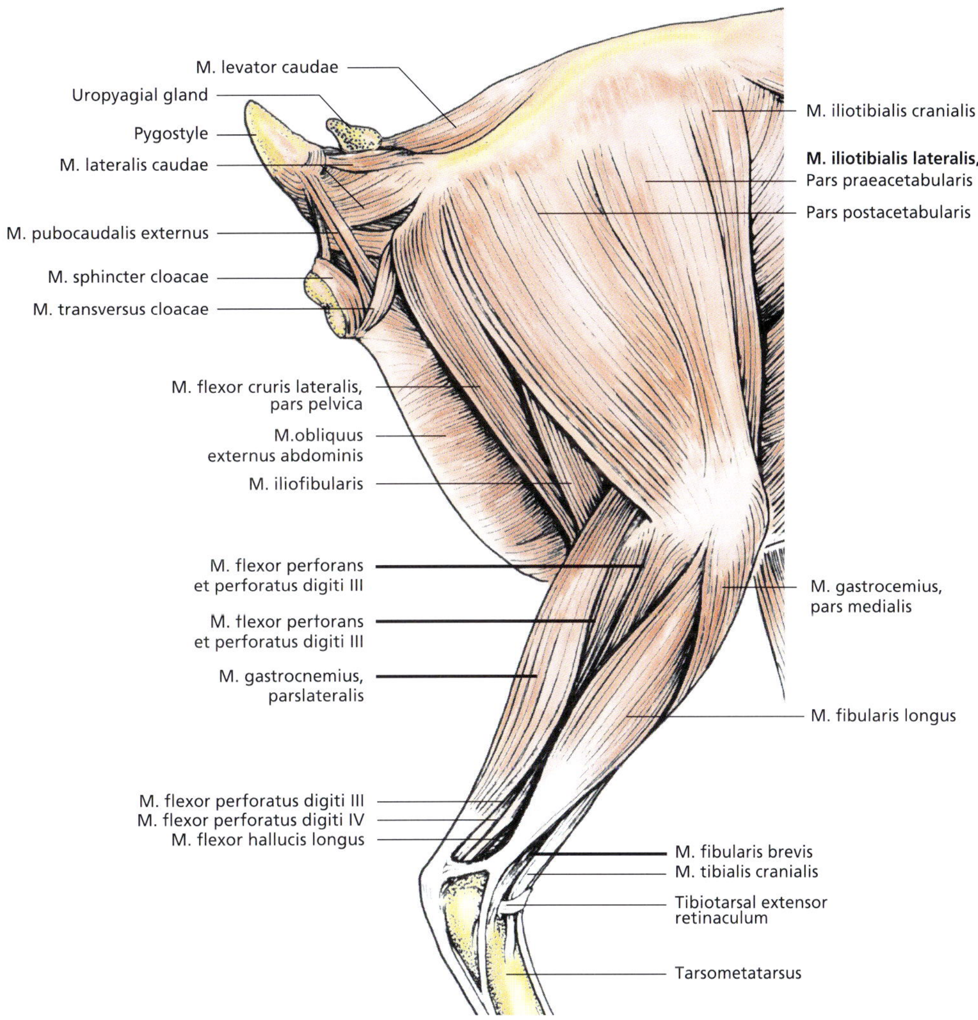

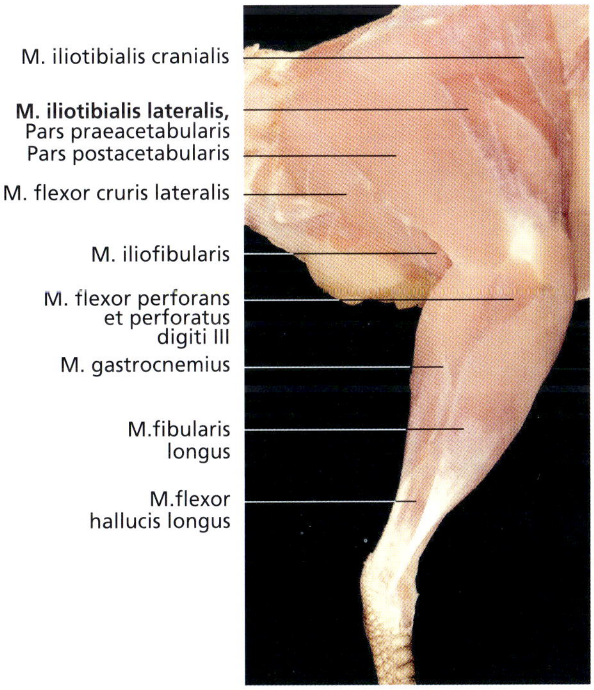

4.22 Superficial muscles of the pelvis and muscles of the proximal pelvic limb of a chicken (lateral view). Courtesy of Dr Annette Kaiser, Munich.

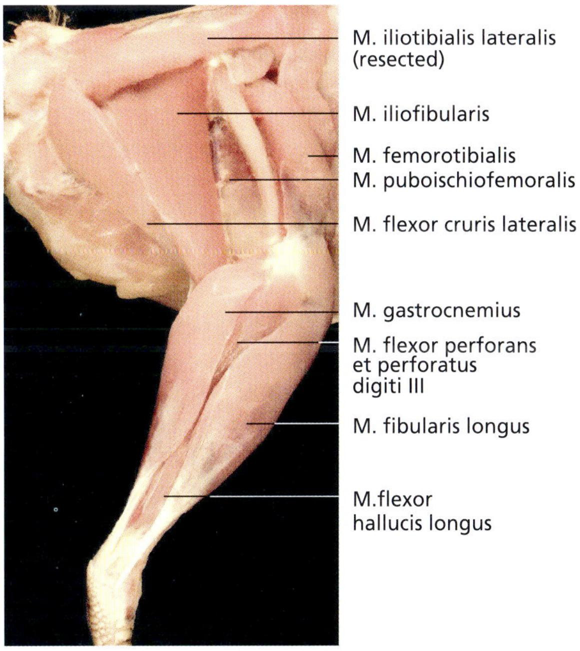

4.23 Deep muscles of the pelvis and muscles of the proximal pelvic limb of a chicken (lateral view). Courtesy of Dr Annette Kaiser, Munich.

- extensors of the hip and flexors of the knee,
- flexors of the hip and extensors of the knee,
- extensors of the intertarsal joint and flexors of the digits,
- flexors of the intertarsal joint and extensors of the digits.

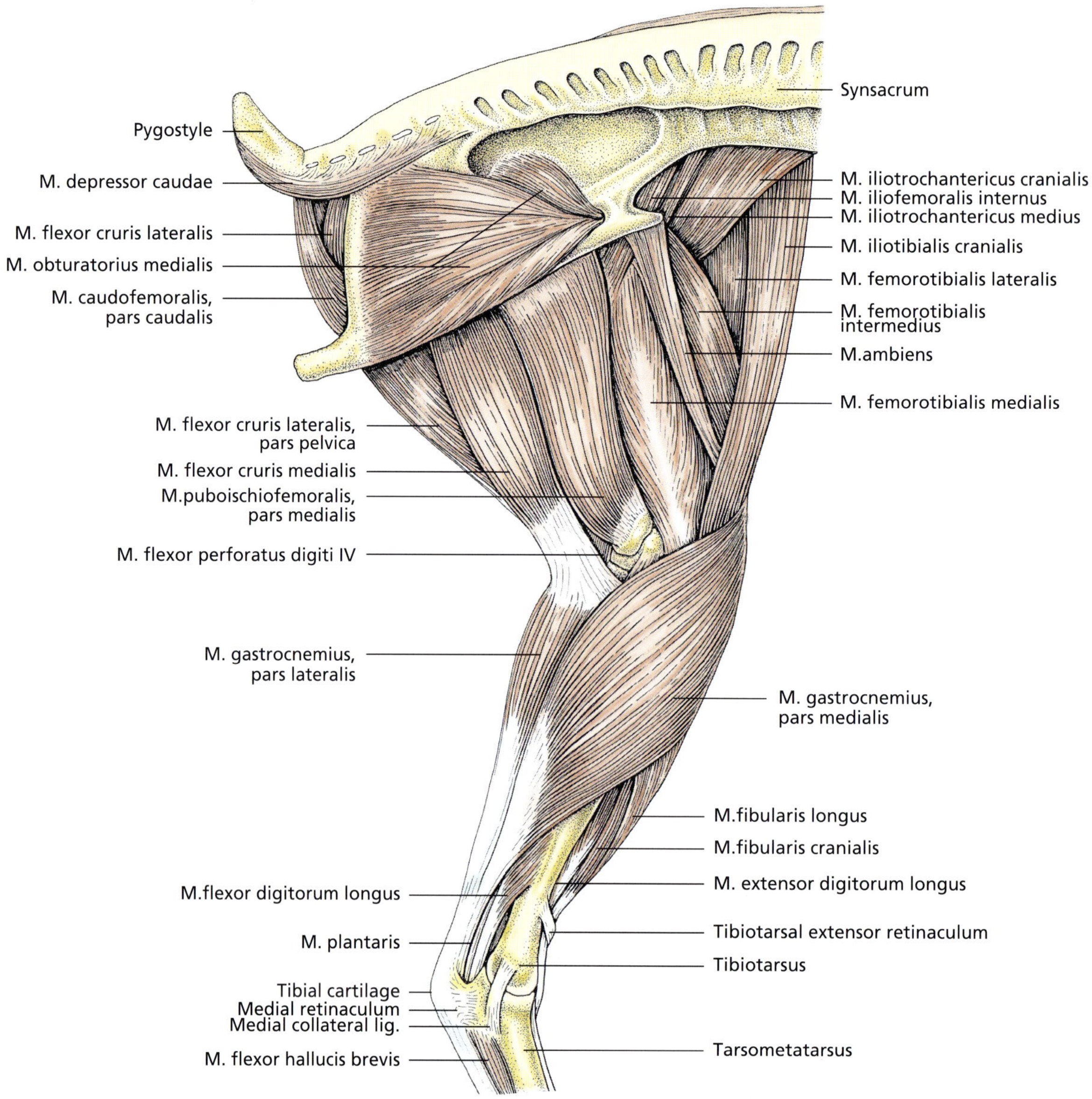

4.24 Muscles of the pelvis, thigh and crus of the chicken (schematic; medial view), adapted from Ghetie, 1976.

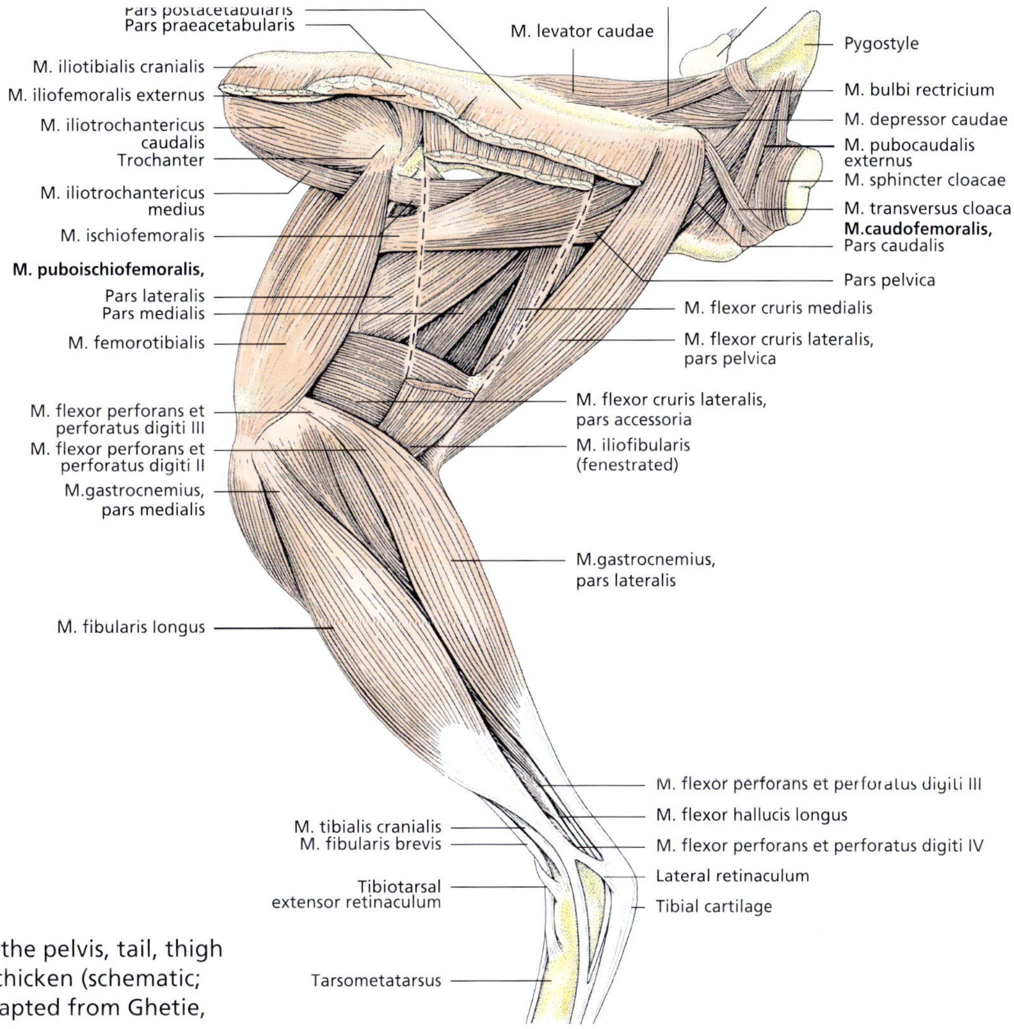

4.25 Muscles of the pelvis, tail, thigh and crus of the chicken (schematic; lateral view), adapted from Ghetie, 1976.

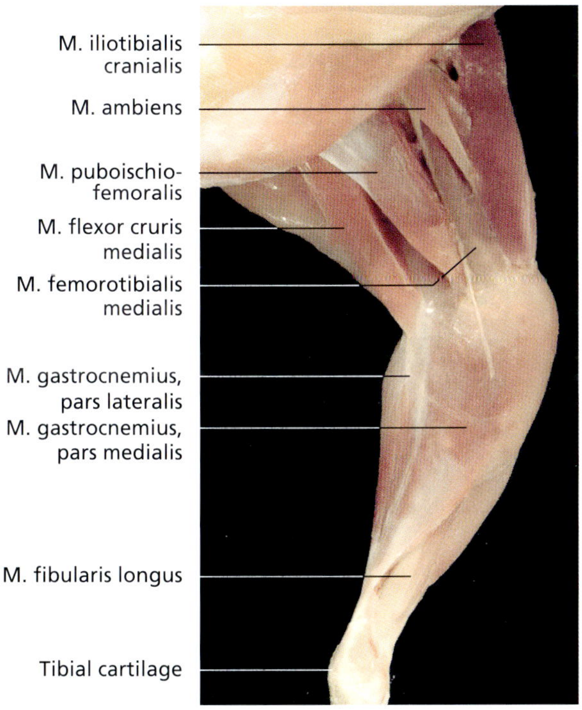

4.26 Muscles of the thigh and crus of a chicken (medial view). Courtesy of Dr Annette Kaiser, Munich.

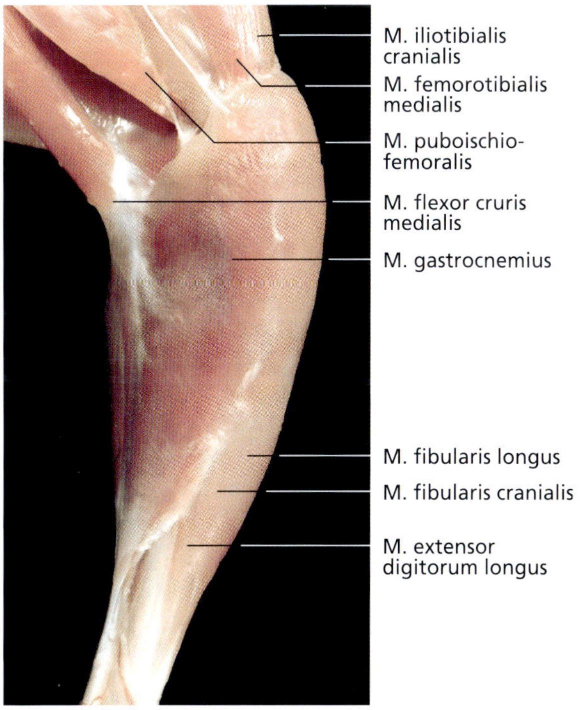

4.27 Muscles of the thigh and crus of a chicken (detailed medial view). Courtesy of Dr Annette Kaiser, Munich.

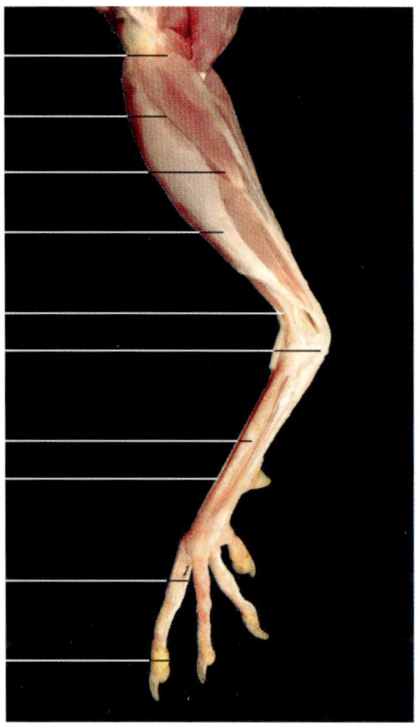

4.28 Muscles and tendons of the crus and foot of a male chicken (lateral view; pars lateralis of m. gastrocnemius removed). Courtesy of Dr Annette Kaiser, Munich.

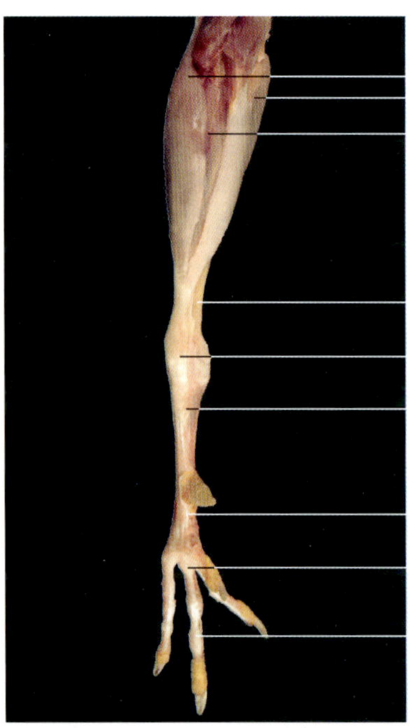

4.29 Muscles and tendons of the crus and foot of a male chicken (caudal view). Courtesy of Dr Annette Kaiser, Munich.

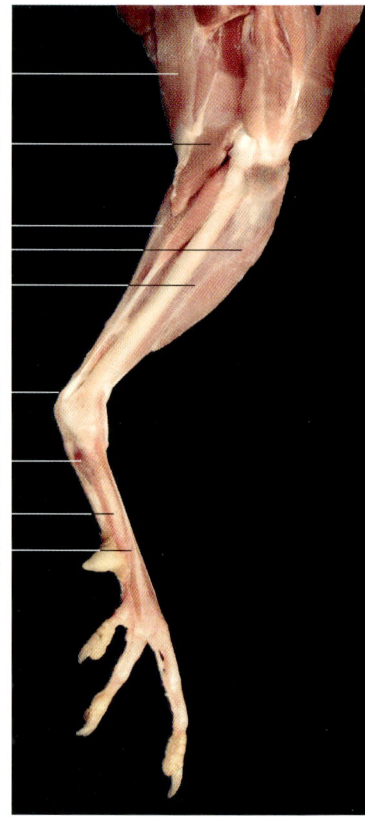

4.30 Deep muscles of the crus of a chicken (medial view). Courtesy of Dr Annette Kaiser, Munich.

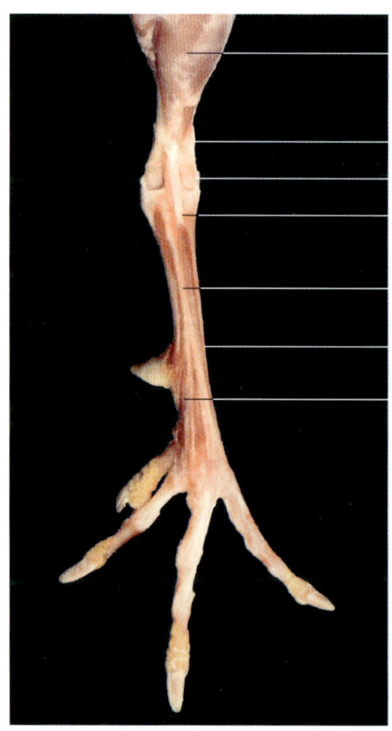

4.31 Muscles and tendons of the crus and pedal digits of a chicken (dorsal view). Courtesy of Dr Annette Kaiser, Munich.

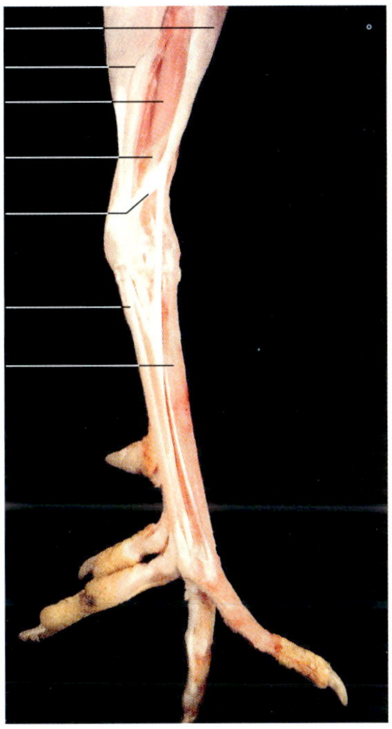

4.32 Muscles and tendons of the crus and pedal digits of a chicken (lateral view). Courtesy of Dr Annette Kaiser, Munich.

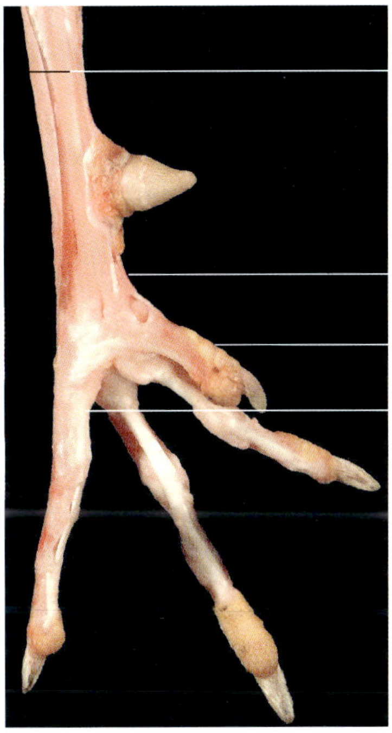

4.33 Muscles and tendons of the pedal digits of a chicken (plantar view). Courtesy of Dr Annette Kaiser, Munich.

The nomenclature of the individual muscles of the pelvic limb is summarised, together with their innervation, origin, insertion and action, in Table 4.2.

Table 4.2 Muscles of the pelvic limb.

Name Innervation	Origin	Insertion	Action
Muscles of the hip and knee joints			
M. iliofibularis Ischiadic nerve (muscular branch)	Dorsolateral iliac crest	Fibula (after passing through a fibrous loop, the ansa musculi iliofibularis)	Extend the hip joint; flex the knee joint
M. iliotibialis cranialis Lateral cutaneous femoral nerve (muscular branch)	Craniodorsal border of preacetabular wing of the ilium	Medially and proximally on the tibiotarsus	Flex the hip joint; extend the knee joint; advance the limb
M. iliotibialis lateralis Sacral plexus	In two parts (preacetabular and postacetabular) on the dorsal border of the ilium	In conjunction with the patellar ligament on the proximal tibiotarsus	**Pars praeacetabularis**: Flex the hip joint; extend the knee joint **Pars postacetabularis**: Extend the hip joint; flex the knee joint; abduct the limb
Mm. iliotrochanterici cranialis/medius/caudalis Cranial coxal nerve	Preacetabular wing of the ilium and cranioventral border of the wing	Distal to the major trochanter, craniolaterally on the shaft of the femur	Inwardly rotate the thigh; flex the hip joint
M. iliofemoralis internus Medial cutaneous femoral nerve (muscular branch)	Preacetabular region of the ilium; near m. iliotrochantericus medialis	Caudomedially at the proximal end of the femur	Outwardly rotate the thigh; flex the hip (weak action)

Table 4.2 continued.

Name Innervation	Origin	Insertion	Action
M. iliofemoralis externus Ischiadic nerve (muscular branches)	Dorsal to the acetabulum	Outer side of the major trochanter	Abduct the thigh; extend or flex the hip joint (depending on posture)
Mm. femorotibiales lateralis/intermedius/ medialis Femoral nerve (muscular branches)	Laterally, cranially and medially on the shaft of the femur, major trochanter	Lateral and cranial cnemial crests, patellar crest	Extend the knee joint; inwardly rotate the limb (medialis)
M. flexor cruris lateralis: Caudal coxal nerve – **Pars pelvica** Tibial nerve – **Pars accessoria**	Pars pelvica: Dorsolateral iliac spine; caudal vertebrae **Pars accessoria:** caudally and laterally on the femur, near the lateral condyle	Together proximally and medially on the tibiotarsus (together with m. flexor cruris medialis)	Extend the hip joint; flex the knee joint
M. flexor cruris medialis Caudal coxal nerve	Laterally and caudally on the ischium	Proximally and medially on the tibiotarsus (together with m. flexor cruris lateralis)	Flex the knee joint; extend the hip joint
M. caudofemoralis: Caudal coxal nerve – **Pars pelvica** – **Pars caudalis**	**Pars pelvica**: Lateral ilium **Pars caudalis**: Ventral pygostyle	Together on the caudal proximal third of the femur	**Pars pelvica**: Extend the hip joint **Pars caudalis**: Depress the tail
M. ischiofemoralis Sacral plexus	Laterally on the ischium	Caudolaterally on the femur, distal to the major trochanter	Outwardly rotate the thigh; extend the hip joint
M. obturatorius lateralis: Lateral obturator nerve – **Pars dorsalis** – **Pars ventralis**	Caudoventral to the acetabulum on the edge of the obturator foramen	Major trochanter and caudolaterally on the proximal extremity of the femur	Outwardly rotate the thigh
M. obturatorius medialis Medial obturator nerve	Internal surface of the pelvis, covers the ischiopubic window and part of the ilioischiadic foramen	Tendon traverses the obturator foramen, inserts on major trochanter	Outwardly rotate the thigh
Muscles of the hip and knee joints			
M. puboischiofemoralis: Lateral obturator nerve – **Pars lateralis** – **Pars medialis**	Ventrolateral ischium and pubis	Caudally on the distal two thirds of the femur	Extend the hip joint; adduct the thigh
M. ambiens Femoral nerve (muscular branches)	Preacetabular tubercle of the ilium, extends distally along a medial course; enters groove in patella to become more lateral	On the aponeuroses of the mm. flexores perforati II–IV	Extend the knee joint; support flexion of the digits (via its insertion)
M. popliteus Medial sural nerve (tibial nerve)	Flexor fossa of tibiotarsus	Caudally on the head of the fibula	Fix the head of the fibula

Name Innervation	Origin	Insertion	Action
Muscles of the intertarsal joint			
M. tibialis cranialis: Fibular nerve Caput femorale Caput tibiale	**Caput femorale**: Lateral femoral condyle **Caput tibiale**: Cranial cnemial crest of the tibiotarsus	By a common tendon passing under the tibiotarsal extensor retinaculum onto the tarsometatarsus	Flex the intertarsal joint
M. fibularis longus Fibular nerve	Lateral cnemial crest of the tibiotarsus	By a tendon on the tibial cartilage; merges with m. flexor perforatus digiti III	Extend the intertarsal joint; support flexion of digits
M. fibularis brevis Fibular nerve (superficial branch)	By two heads on the craniolateral tibia and fibula	Proximal and lateral tarsometatarsus	Inwardly rotate the tarsometatarsus
M. gastrocnemius: Lateral sural nerve – **Pars lateralis** Medial sural nerve (tibial nerve) – **Pars intermedia** – **Pars medialis**	**Pars lateralis**: near the lateral femoral condyle **Pars intermedia**: near the medial femoral condyle **Pars medialis**: Patella, patellar ligament	All three components combine within the Achilles tendon and pass over the tibial cartilage to reach the medial and lateral hypotarsal crests and the plantar surface of the body of the tarsometatarsus	Extend the intertarsal joint (strong action)
M. plantaris Medial sural nerve (tibial nerve)	Proximally on the caudomedial surface of the tibiotarsus	Laterally on the proximal tarsometatarsus	Extend the intertarsal joint
Long muscles of the pedal digits			
M. extensor digitorum longus Fibular nerve	Proximal tibiotarsus (intercnemial groove)	Via the extensor groove of tibiotarsus, on phalanx II to distal (ungual) phalanx of digits 2 to 4	Extend digits 2 to 4; flex the intertarsal joint
M. flexor perforans et perforatus digiti II Lateral sural nerve (tibial nerve)	Near the fibular trochlea of the femur, passes through canals in the tibial cartilage	Via a bony groove on the hypotarsus; plantar aspect of base of phalanx II of digit 2	Flex digit 2; extend the intertarsal joint
M. flexor perforans et perforatus digiti III Lateral sural nerve (tibial nerve)	Lateral cnemial crest of tibiotarsus; laterally on patellar ligament; passes over surface of tibial cartilage	Via bony groove on hypotarsus; plantar aspect of base of phalanx III of digit 3	Flex digit 3, extend the intertarsal joint
M. flexor perforatus digiti II Lateral sural nerve (tibial nerve)	By four heads: femur; tibiotarsus	Via the tibial cartilage and hypotarsus; plantar aspect of base of phalanx I of digit 2	Flex digit 2, extend the intertarsal joint
Long muscles of the pedal digits			
M. flexor perforatus digiti III Lateral sural nerve (tibial nerve)	By two heads: femur, distal tibiotarsus	Via the tibial cartilage and hypotarsus; plantar aspect of base of phalanx II of digit 3	Flex digit 3; extend the intertarsal joint; tendon of insertion receives tendon of m. fibularis longus

Table 4.2 continued.

Name Innervation	Origin	Insertion	Action
M. flexor perforatus digiti IV Lateral sural nerve (tibial nerve)	By four heads: femur, fibula; receives tendon of m. ambiens	Via the tibial cartilage and hypotarsus; plantar aspect of base of phalanx II/III/IV of digit 4	Flex digit 4; extend the intertarsal joint
M. flexor digitorum longus Medial sural nerve (tibial nerve)	Caudally on the shaft of the fibula and the proximal end of the tibiotarsus	Via the tibial cartilage and hypotarsus; plantar aspect of distal phalanx and other phalanges of digits 2 to 4	Flex digits 2 to 4; extend the intertarsal joint (also origin of m. lumbricalis)
M. flexor hallucis longus Medial sural nerve (tibial nerve)	By two heads: distally on the caudal surface of the femur	Via the tibial cartilage and hypotarsus; plantar aspect of distal phalanx of digit 1	Flex digit 1; extend the intertarsal joint
Short muscles of the pedal digits			
M. extensor hallucis longus Fibular nerve (deep branch)	Dorsomedial tarsometatarsus	Extensor tubercle of distal phalanx of digit 1	Extend digit 1
M. flexor hallucis brevis Parafibular nerve (tibial nerve)	Hypotarsus and caudomedial tarsometatarsus	Plantar aspect of base of proximal phalanx of digit 1	Flex digit 1
M. abductor digiti II Fibular nerve (deep branch)	Medial and distal plantar surface of the tarsometatarsus	Medial aspect of base of proximal phalanx of digit 2	Abduct digit 2
M. adductor digiti II Parafibular nerve (tibial nerve)	Distal plantar surface of the tarsometatarsus	Lateral aspect of base of proximal phalanx of digit 2	Adduct digit 2
M. extensor brevis digiti III Fibular nerve (superficial branch)	Distal dorsal surface of the tarsometatarsus	Extensor tubercle of proximal phalanx of digit 3	Extend digit 3
M. extensor brevis digiti IV Fibular nerve (superficial branch)	Laterally on the dorsal surface of the tarsometatarsus	Medial aspect of proximal phalanx of digit 4	Adduct and extend digit 4
M. abductor digiti IV Parafibular nerve (tibial nerve)	Laterally on the plantar surface of the tarsometatarsus	Lateral aspect of base of proximal phalanx of digit 4	Abduct digit 4
M. lumbricalis Parafibular nerve (tibial nerve)	Tendon of m. flexor digitorum longus	Plantar ligament of metatarsophalangeal joint of digits 2 and 3	Draw the plantar ligament proximally (protect flexor tendons against compression)

Clinical aspects

Due to its rigid anatomical configuration, the pelvis is a relatively uncommon site of fractures in birds. **Hindlimb lameness** is of greater clinical significance. Causes include space occupying lesions in the body cavity (e.g., kidney tumours, egg binding) that press the lumbosacral nerve plexus against the bony synsacrum. Such aetiologies need to be distinguished from other differential diagnoses, including fractures and luxations of the hindlimb.

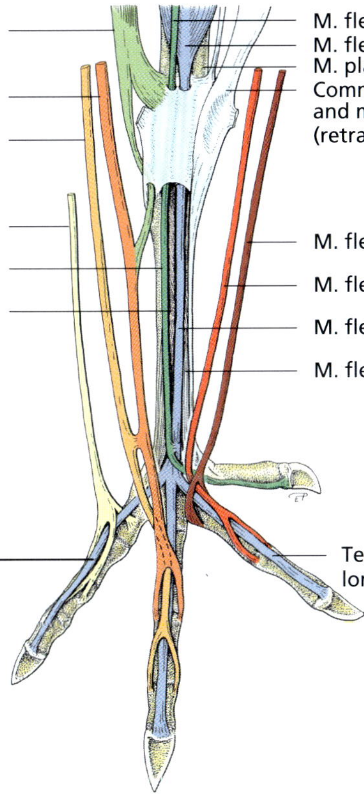

4.34 Flexor tendons of the pedal digits of the left foot of the chicken (schematic; tendons separated, plantar view), adapted from Ghetie, 1976.

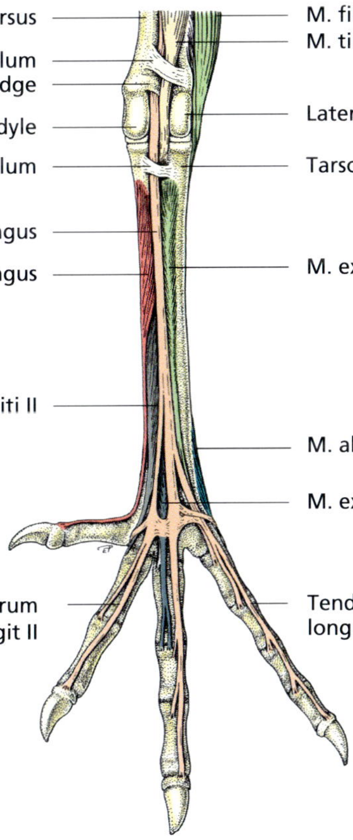

4.35 Extensor tendons of the pedal digits of the left foot of the chicken (schematic; dorsal view), adapted from Ghetie, 1976.

The great crested grebe (Podiceps cristatus) and common swift (Apus apus) are commonly found recumbent in the wild and are presented for veterinary care with suspected paraplegia or hindlimb fractures. However, these assumptions are often incorrect, the actual cause being related to **species-specific anatomical characteristics**. These birds have often been brought down by the drag of a truck or by a storm. The relatively caudal positioning of the pelvic joints in the grebe, and the very short pelvic limbs of the swift, hinder their ability to take off from the ground. Assistance with take-off can be rendered by placing them on water or launching them into the air over a grassy surface.

Fractures of one hindlimb are accompanied by non-weight-bearing on the affected side with additional load placed on the contralateral limb. This increases the risk of ulcerative pododermatitis (bumblefoot) developing in the uninjured limb. Seen more in certain species, bumblefoot results from pressure-related disturbance of the microvasculature in the metatarsal pad and limited (compared with mammals) fibrinolytic activity during the breakdown of inflammatory fibrin clots.

In managing hindlimb fractures it is therefore very important to care also for the unaffected limb by applying specialised **bandages** to the plantar aspect of the foot (ball or shoe bandage; see Chapter 23 'Surgical fracture management'). This facilitates even distribution of pressure and **reduces the risk of pododermatitis**.

Due to the anatomical relationships of the thigh, the use of external fixation techniques in **orthopaedic management of femoral fractures** (see Chapter 23 'Surgical fracture management') is challenging, with considerable muscle mass encountered on the lateral surgical approach. Treatment of femoral fractures is therefore relatively laborious.

The femur, like the humerus, is pneumatised in many bird species by a diverticulum of the abdominal air sac (saccus abdominalis). Both of these bones are therefore unsuitable for **intraosseous** replacement of fluid and electrolytes in patients that are dehydrated or have experienced significant blood loss, as this could allow fluid to enter the respiratory system causing death from asphyxiation.

The prognosis for injuries involving the digits depends largely upon which of the digits exhibits **functional deficiencies** or postural abnormalities, or requires amputation on medical grounds. In raptors, the **caudally directed first digit**, the so-called hallux, is used to grasp prey in the air or on the ground. Functional deficiency or loss of this digit means that the bird is no longer able to hunt.

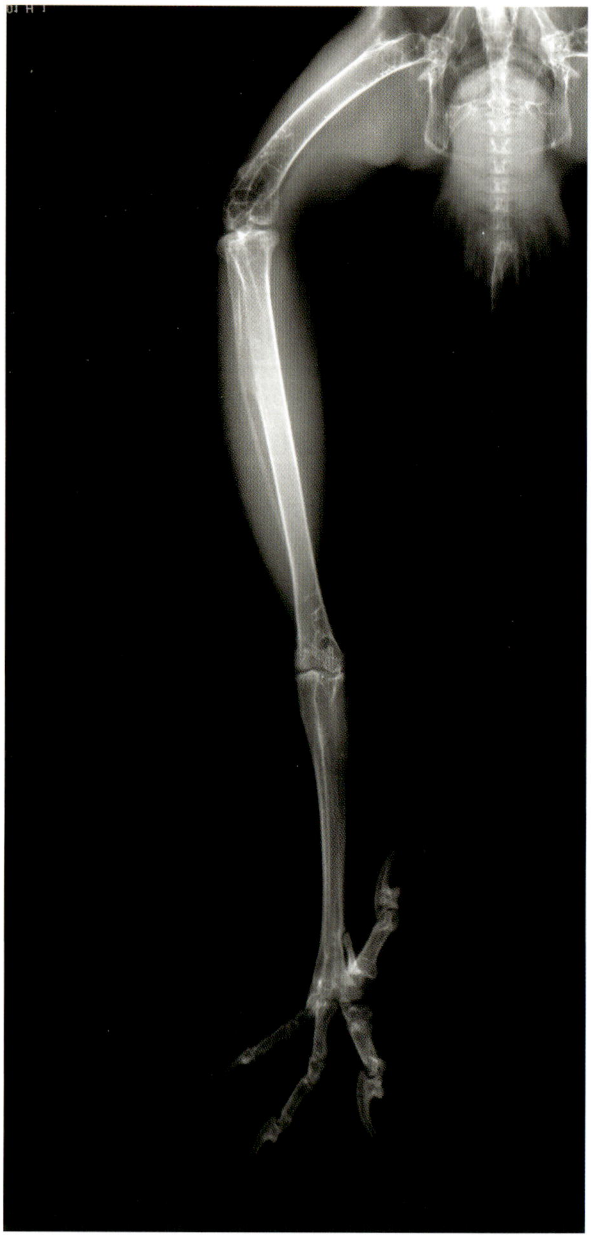

4..36 Radiograph of the pelvic limb (ventrodorsal view) of a common buzzard (Buteo buteo).

The **second, medially directed digit** is used by birds of prey to immobilise their quarry or to hold their food. Thus the absence of this digit can lead to problems with prehension and feeding. Abnormal function of the **third and/or fourth digits** is of limited significance in raptors.

Abnormalities or loss of highly specialised digits in certain species, such as the first digit of woodpeckers or the webbed feet of ducks, can prohibit successful rehabilitation for release into the wild, and may represent an indication for euthanasia due to welfare considerations.

5

Body cavities

H. E. König, A. Probst, H.-G. Liebich and R. Korbel

The diaphragm is absent in birds. Therefore, the use of terms such as 'thoracic' and 'abdominal' with respect to divisions of the avian body cavity relates to equivalent anatomical regions in mammals, as delineated by homologous bones and muscles (Figure 5.1). Birds, like reptiles, have a horizontal septum (septum horizontale) that forms the ventral boundary of the lungs. A second, oblique septum (septum obliquum), located more ventrally, is formed by ingrowth of the cervical air sacs (see below; Figure 5.5).

The horizontal septum passes from the ventral crest (crista ventralis) of the thoracic vertebrae to the ribs. At the level of its costal attachment, five small skeletal muscles (mm. costoseptales) arise from the ribs and fan out into the septum. The oblique septum also begins at the ventral crest of the thoracic vertebrae, extending further ventral as far as the lateral edge of the sternum. In the chicken, the body cavity contains a total of 16 discrete spaces. These are divided into air sacs and serosa-lined subdivisions of the coelomic cavity. The eight air sacs (sacci pneumatici) of the chicken (Figures 5.1ff.) are described in more detail in Chapter 7 'Respiratory system'. These are comprised of:

- an unpaired cervical air sac (saccus cervicalis),
- an unpaired clavicular air sac (saccus clavicularis),
- paired cranial thoracic air sacs (sacci thoracici craniales),
- paired caudal cervical air sacs (sacci thoracici caudales) and
- paired abdominal air sacs (sacci abdominales).

The subdivisions of the coelomic cavity consist of:

- two pleural cavities (cava pleurae),
- four hepatic peritoneal cavities (cava hepatica peritonei),
- an intestinal peritoneal cavity (cavum intestinale peritonei) and
- a pericardial cavity (cavum pericardii).

The lining of the pleural cavities (cava pleurae) (Figure 5.5) is only complete during embryonic development. By the time of hatching, the pleura are reduced to mere remnants and much or all of the cavity is lost. The lung is attached to

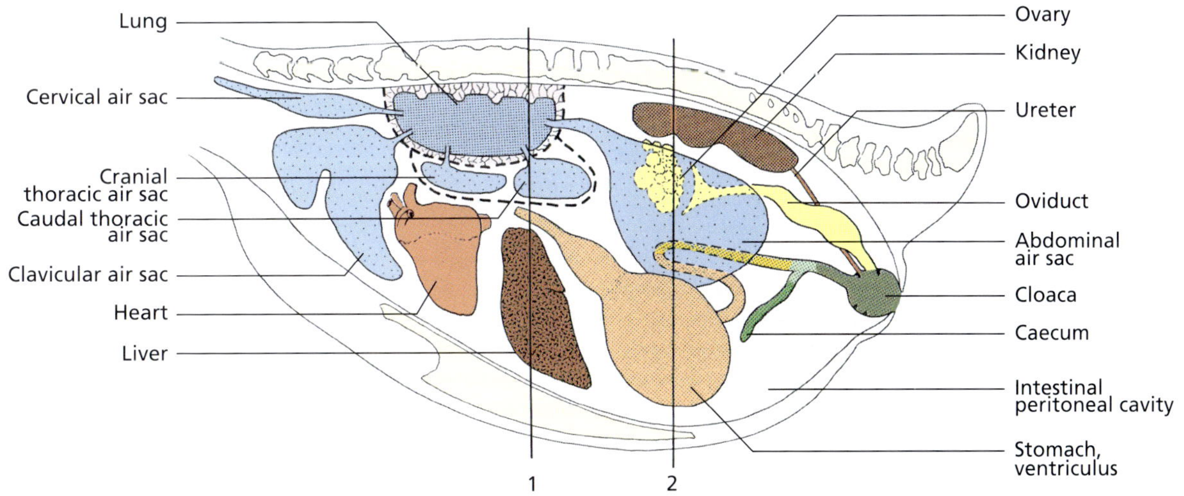

5.1 The coelomic cavity of the chicken (schematic), 1 and 2 indicate the level of the sections shown in Figures 5.5 and 5.6, adapted from Vollmerhaus, 2004.

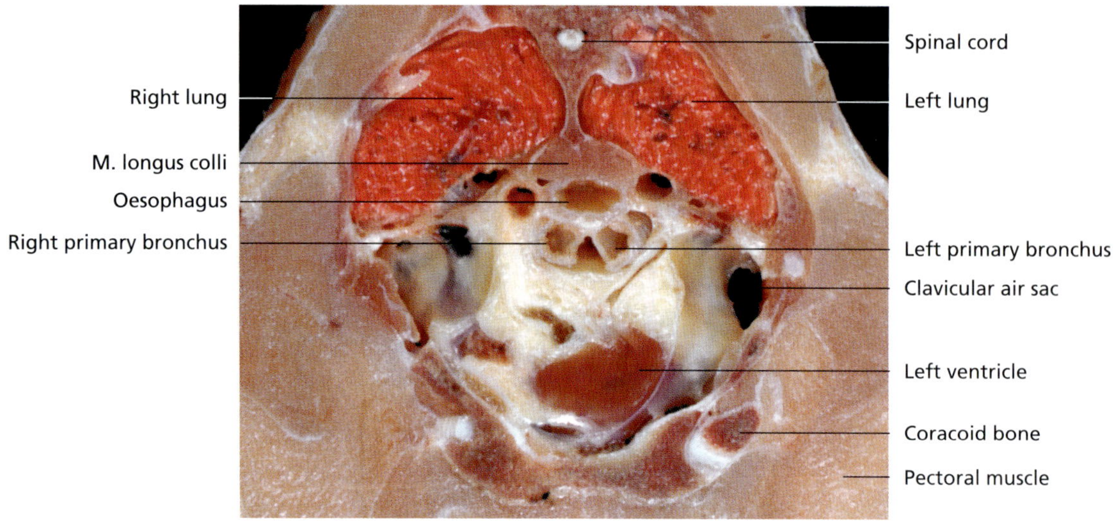

5.2 Transverse section at the level of the sternocoracoid joint (articulatio sternocoracoidea) of a chicken (cranial view). Courtesy of Professor Dr J. Ruberte, Barcelona.

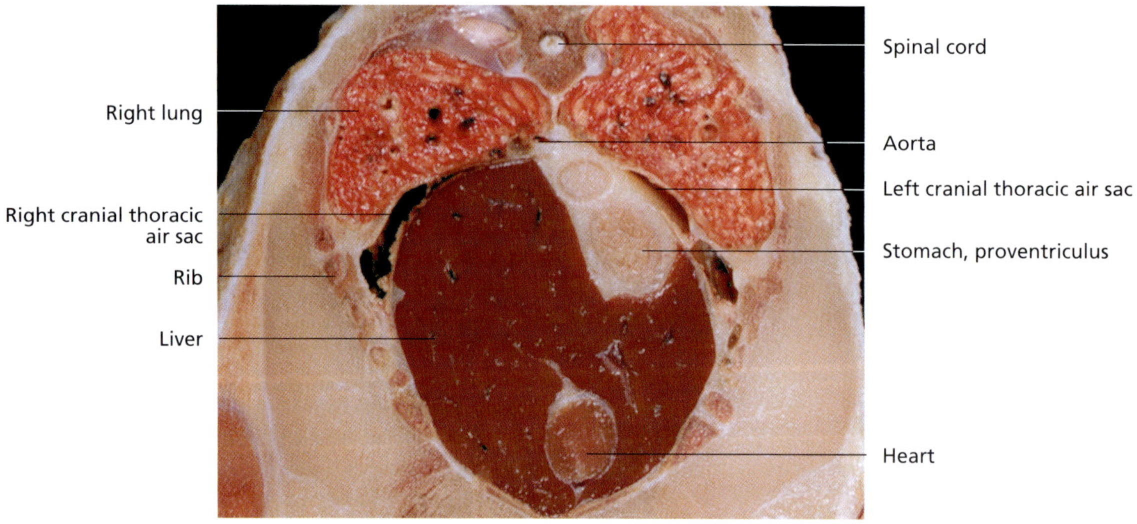

5.3 Transverse section at the level of the liver of a chicken (cranial view). Courtesy of Professor Dr J. Ruberte, Barcelona.

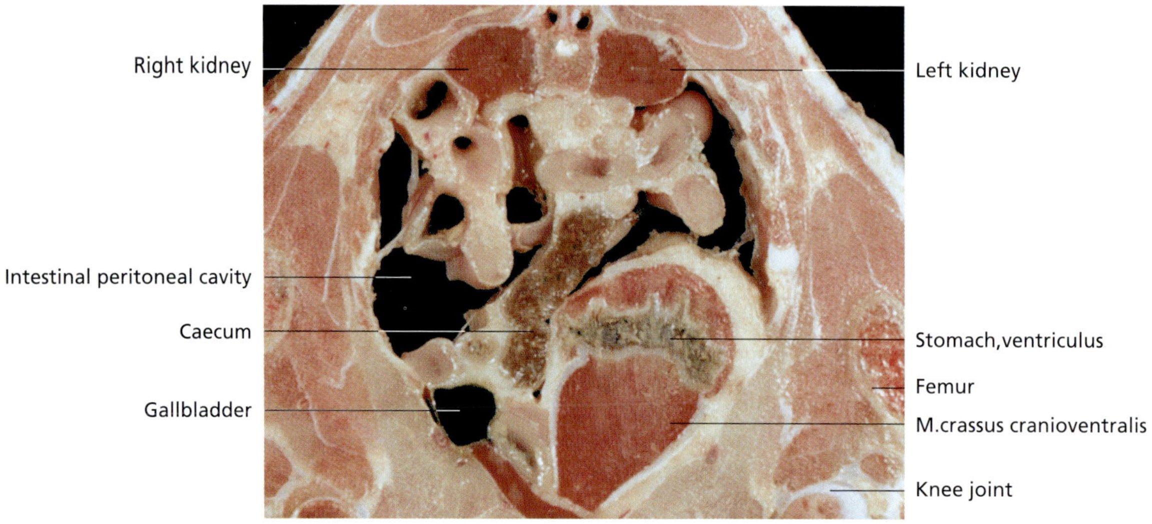

5.4 Transverse section at the level of the ventriculus of a chicken (cranial view). Courtesy of Professor Dr J. Ruberte, Barcelona.

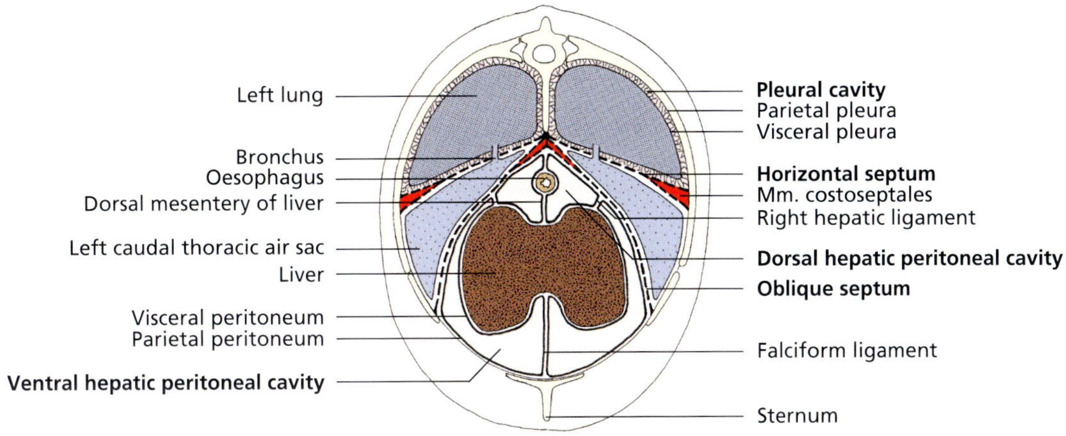

5.5 Section of the body of the chicken at the level of the liver (see Figure 5.1 section 1, caudal view), adapted from McLelland and King, 1970

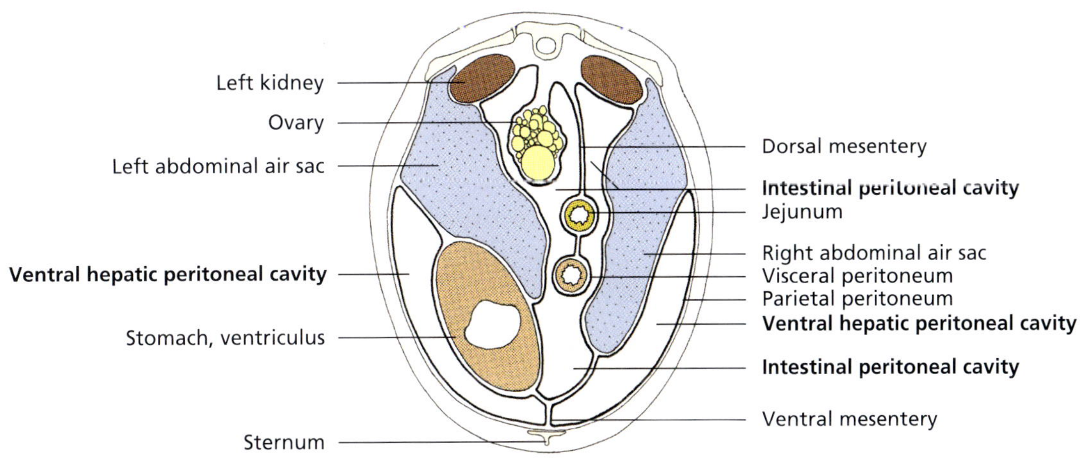

5.6 Section of the body of the chicken at the level of the ventriculus (see Figure 5.1 section 2, caudal view), adapted from McLelland and King, 1970.

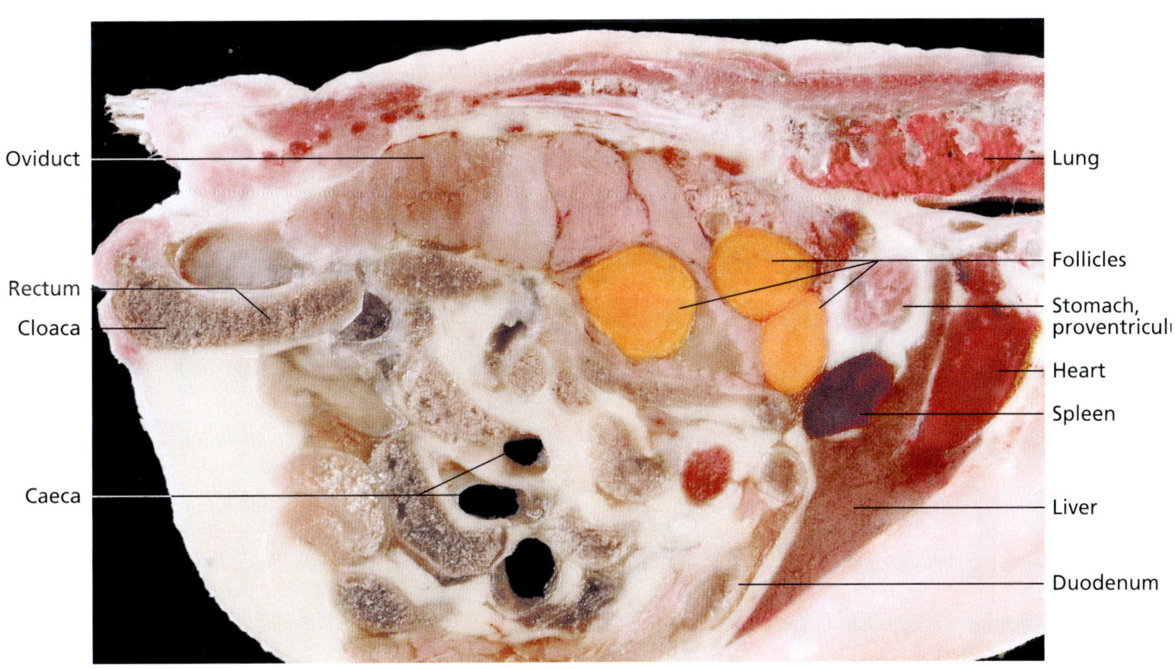

5.7 Median section of the body of a chicken (viewed from the right).

the thoracic wall and the horizontal septum. This septum is considered to be a derivative of the parietal pleura on the ventral surface of the lung. Dorsally, the ribs leave deep impressions in the lung parenchyma. Due to the attachment of the lung on all sides, the volume of the lung is remarkably constant between inspiration and expiration.

The hepatic peritoneal cavities (cava hepatica peritonei) are bounded caudally by the transverse posthepatic septum (septum posthepaticum), a double serosal layer enclosing the ventriculus on the left side of the body (Figure 5.6). The liver has a dorsal and ventral mesentery, the latter being considered the equivalent of the falciform ligament of mammals.

In addition, left and right hepatic ligaments, the **ligamentum hepaticum sinistrum** and **dextrum**, connect the liver with the oblique septum. Thus, four hepatic peritoneal cavities are formed (Figure 5.5). The two ventral components are considerably larger and can be visualised in dissected specimens by elevating the sternum. A connection sometimes exists between the left dorsal compartment and the intestinal peritoneal cavity.

The intestinal peritoneal cavity (cavum intestinale peritonei) (Figure 5.6), the largest of the peritoneal cavities, lies caudal to the posthepatic septum. In addition to the intestine, suspended by its dorsal mesentery (Figure

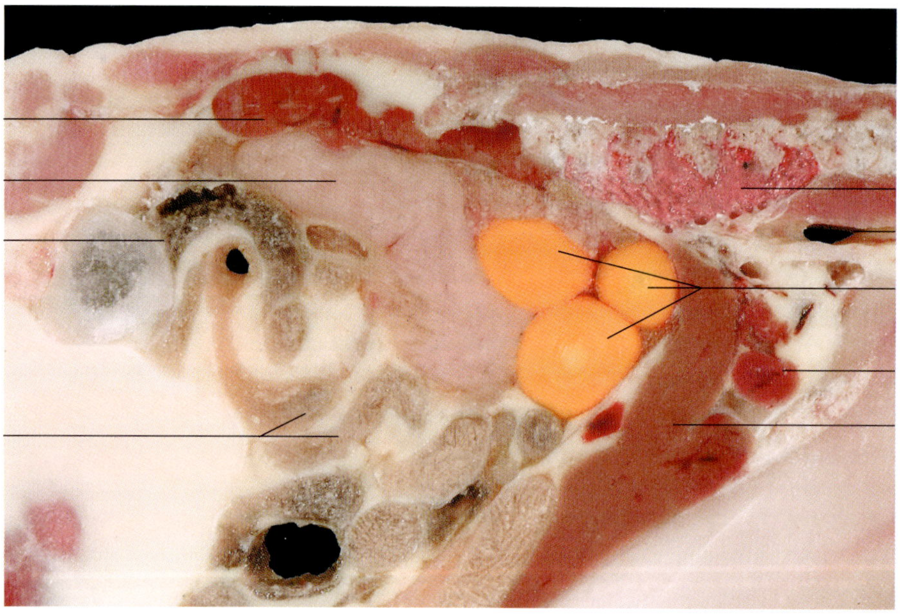

5.8 Paramedian section of a chicken (viewed from the right).

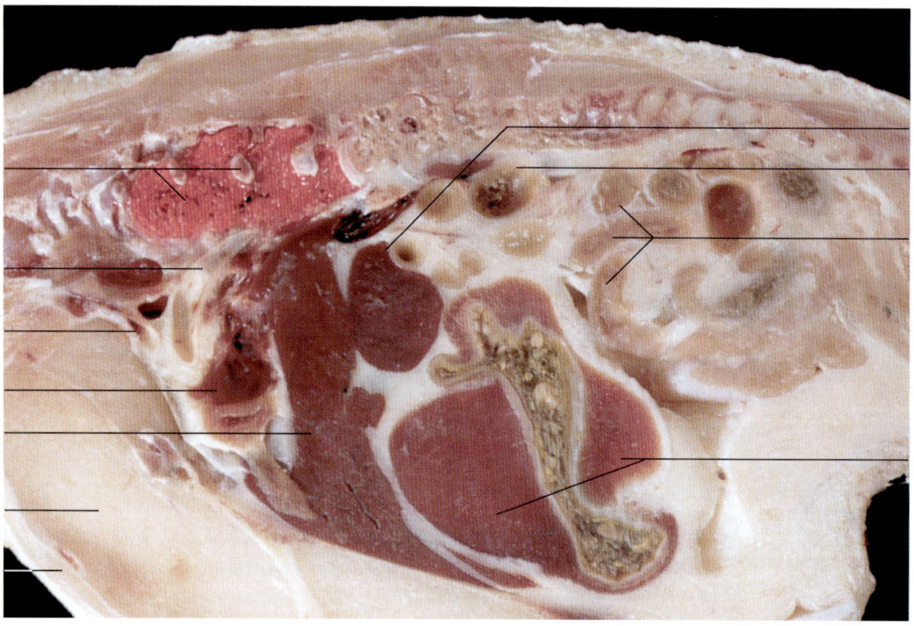

5.9 Right paramedian section of a chicken (viewed from the left). Courtesy of Professor Dr J. Ruberte, Barcelona.

5.6), the intestinal peritoneal cavity contains the ovary and oviduct, or the testes. The two abdominal air sacs project caudally into the intestinal peritoneal cavity. They extend between the intestinal loops and partially surround the testes and the kidneys.

The pericardial cavity (cavum pericardii) is similar in structure to that of mammals (see Chapter 11 'Cardiovascular system').

Clinical aspects

An understanding of the topographical anatomy of the body cavity, depicted in Figures 5.1 to 5.14 and 5.17, is indispensable in avian medicine and surgery for the following briefly outlined reasons.

Due to their common body cavity, with its network of air sacs, birds are excellent candidates for both endoscopic examination and endoscope-guided minimally invasive procedures (see Chapter 22 'Endoscopy'). Introduction of the endoscope into an air sac permits examination of the surrounding organs without the need for gaseous insufflation (as is required in humans and domestic mammals). This is described in more detail in Chapter 22 'Endoscopy'.

The **body cavity can be accessed surgically** through a region bounded by the caudal margin of the sternum (cranially) and the apex of the pubic bone (caudally). Laterally, and also dorsally, the accessible area is limited by the proximity of the kidneys. A midline approach is usually used. Common indications for surgical intervention in pet birds

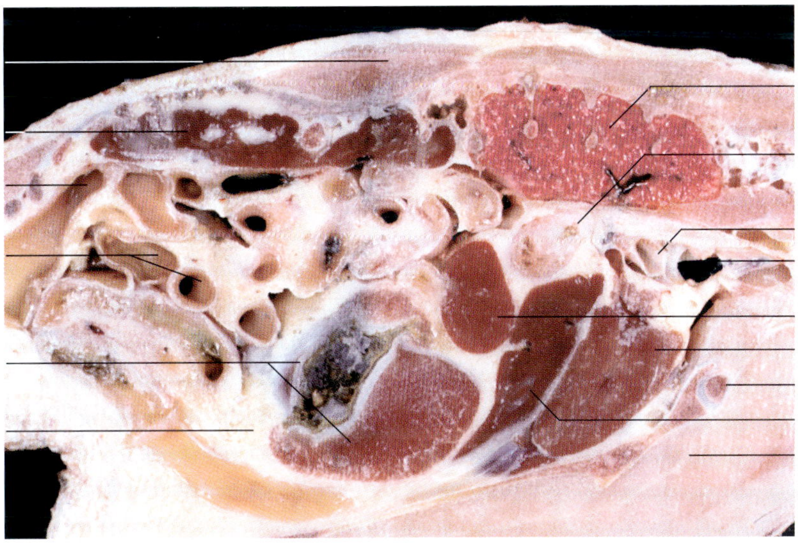

5.10 Left paramedian section of a chicken (viewed from the right). Courtesy of Professor Dr J. Ruberte, Barcelona.

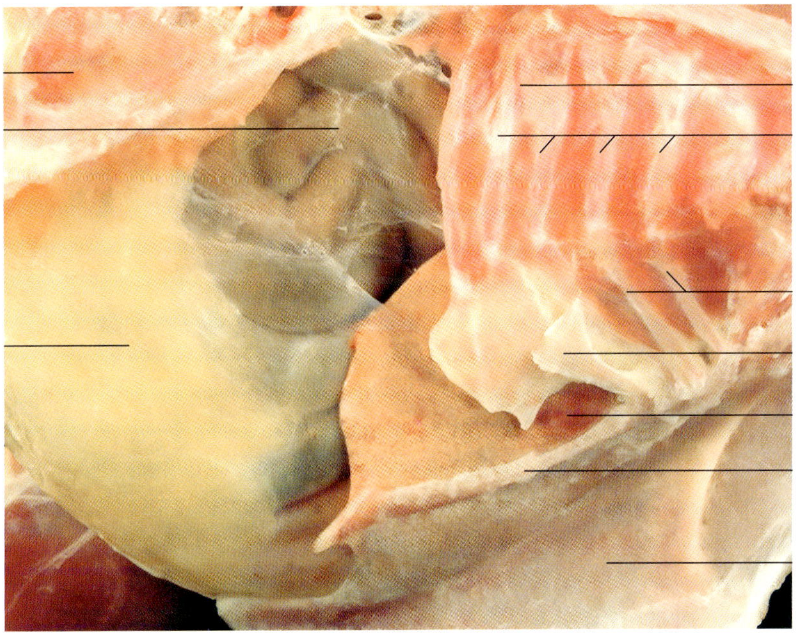

5.11 Coelomic cavity of a chicken, right lateral abdominal wall removed. Courtesy of Dr Annette Kaiser, Munich.

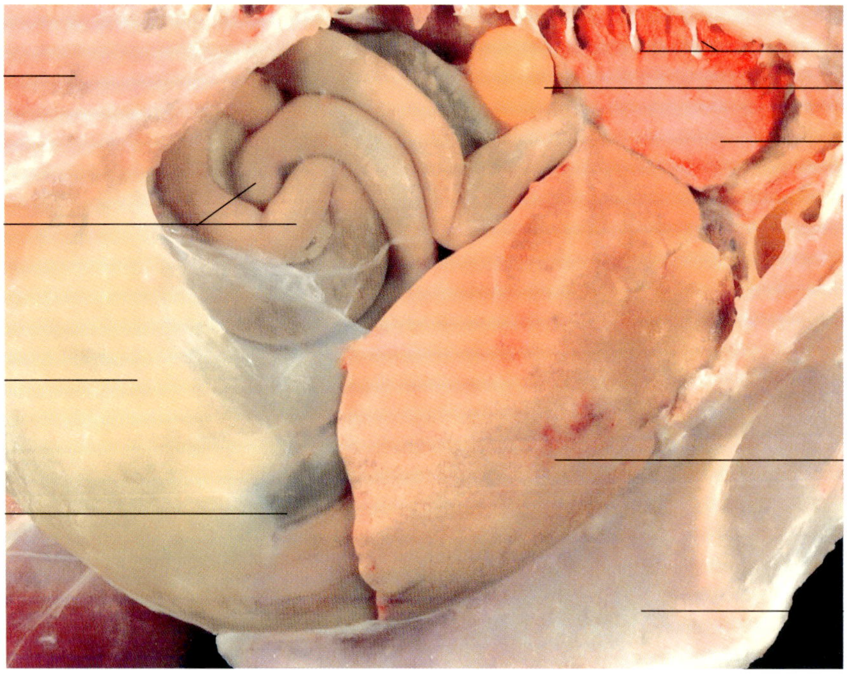

5.12 Coelomic cavity of a chicken, right lateral abdominal and thoracic wall removed. Courtesy of Dr Annette Kaiser, Munich.

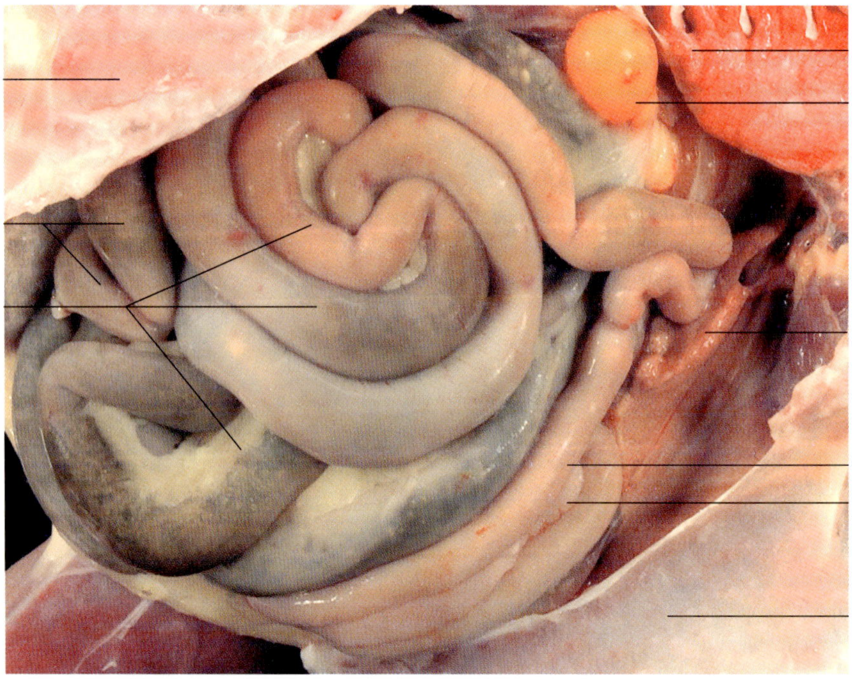

5.13 Coelomic cavity of a chicken, right lateral abdominal and thoracic wall and parts of the liver removed. Courtesy of Dr Annette Kaiser, Munich.

include the removal of eggs (combined, as appropriate, with ovariohysterectomy) in egg bound females or extraction of foreign bodies from the **stomach** or **intestine**.

When opening the body cavity of a bird in dorsal recumbency, the first typically encountered **clinically relevant anatomical reference point** is the superficially located, variably green-brown (according to degree of filling) cranial loop of the duodenum (see Figures 5.13 and 5.14). This encloses the usually pearlescent pancreas that, in relatively lean birds, may already be visible through the abdominal wall.

In laying female birds, the meandering coils of the porcelain-coloured **oviduct**, with its sub-serosal vascular plexus, is also usually seen lying superficially. Careful reflection of these structures is required to gain access to the deeper regions of the body cavity.

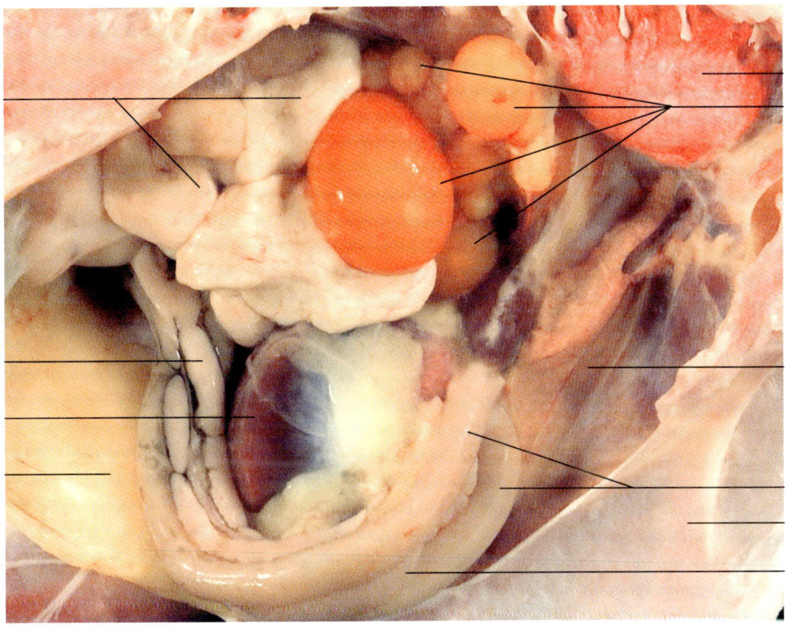

5.14 Coelomic cavity of a chicken, right lateral abdominal and thoracic wall, parts of the liver and intestine removed. Courtesy of Dr Annette Kaiser, Munich.

Lying deeper on the left side of the body cavity (right, from the surgeon's perspective), the ventriculus can serve as another **useful landmark**, particularly in granivorous species in which it is spherical and readily palpable.

For surgery in the vicinity of the kidneys or testes, the whole of the intestine must be reflected. Depending on species, and on the extent of the caudal margin of the sternum, the caudal part of the proventriculus usually represents the cranial limit of surgical access to the body cavity.

It is important to be aware that whenever the body cavity is opened (surgically or otherwise), the air sacs may be penetrated and inspiration becomes more difficult. Concomitant physiological disturbances of respiratory function associated with ventral recumbency also need to be considered, due to the pressure placed on the lungs by the viscera in this position.

Furthermore, during the inspiratory phase, in which the sternum and massive breast muscles are elevated, the entire weight of these structures rests on the relatively delicate respiratory musculature, particularly the intercostal muscles.

In general, therefore, it is imperative that the patient be transferred from ventral to sternal recumbency as soon as possible **after surgery is completed**.

While cranially positioned organs can usually be evaluated effectively using radiography (ventrodorsal and lateral views) (see Figures 5.15 and 5.16), distinguishing between the gastrointestinal tract and associated structures in the caudal body cavity is difficult without the use of contrast material. Contrast radiography is thus routinely employed for assessment of the stomachs and intestines (see Chapter 19 'Imaging techniques'). Orally administered barium sulphate suspension is typically used.

The ability to recognise normal anatomical relationships and to identify abnormal displacement of individual organs and organ systems within the body cavity is of considerable diagnostic importance in avian medicine. Altered abdominal topography can result from enlargement of individual organs with displacement of adjacent structures. This, in turn, can lead to other disease manifestations.

Imaging, particularly **contrast radiography**, is the method of choice for identifying abnormal anatomical

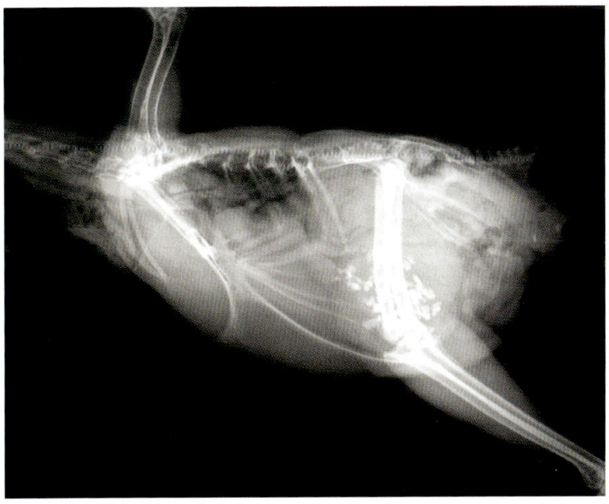

5.15 Radiograph of a chicken (Gallus gallus) (lateral view): discernible structures within the body cavity include the small lungs, the flattened heart resting on the sternum, the grit-filled ventriculus and the relatively radiolucent cloaca.

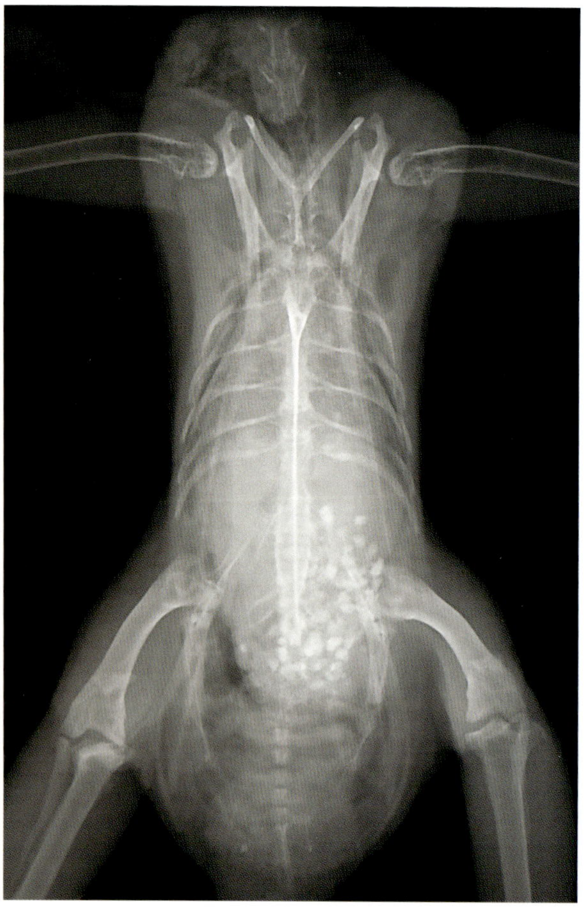

5.16 Radiograph of a chicken (Gallus gallus) (same patient as in Figure 5.15; ventrodorsal view): discernible structures within the body cavity include the small lungs, the liver lying caudal to the lungs and, at the level of the hip joints, the grit-filled ventriculus (the presence of grit is a normal finding in granivorous species). Individual components of the gastrointestinal tract and associated structures are difficult to distinguish without the use of contrast material.

relationships. The manner in which the contrast medium-filled gastrointestinal tract is displaced can be indicative of certain disease processes. **Ventral displacement**, for example, is suggestive of a space-occupying lesion associated with the kidney (kidney tumours, renal cysts), while **caudoventral dislocation** is consistent with a mass arising from the cranial region of the kidney, the gonads or the spleen.

Caudal or **caudodorsal** transposition of the intestine can result from a mass effect originating in the region of the stomach or the liver. In the latter case, the liver is pushed between the abdominal wall and the intestinal loops, displacing these dorsally and caudally.

Cranial displacement of the intestinal loops may be a consequence of cloacal obstruction. Malpositioning of viscera can also result from effusions into the body cavities and overinflation of the air sacs.

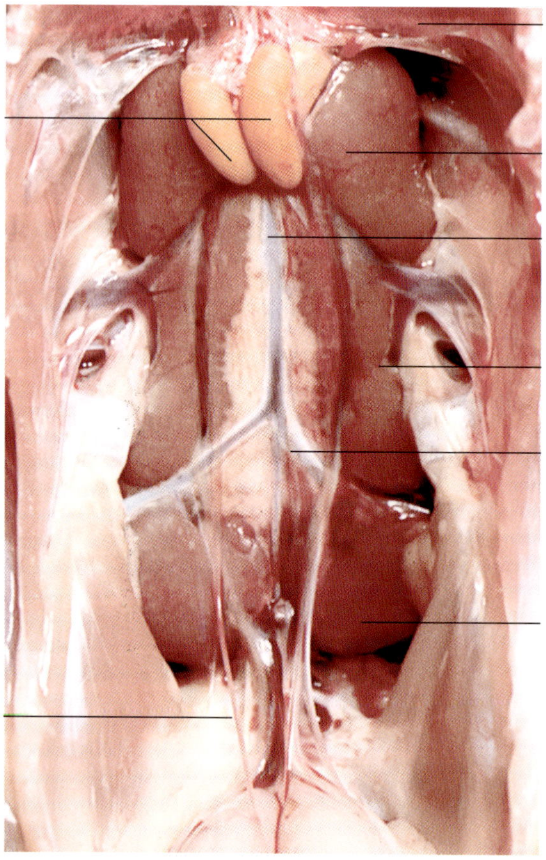

5.17 Organs associated with the dorsal wall of the coelomic cavity in a cockerel (Gallus gallus). Courtesy of Professor Dr J. Ruberte, Barcelona.

Tumours are a common cause of disease in birds. In the budgerigar (Melopsittacus undulatus), for example, they account for up to 12 per cent of patient presentations, with kidney tumours (and cysts) as well as tumours of the testes, spleen and liver being highly represented. A further species-specific example of a disease resulting in visceral displacement is **haemochromatosis** (iron storage disease) in mynas (Gracula religiosa), which causes enlargement of the liver and coelomic effusion.

In psittacines, dilatation of the proventriculus, ventriculus, and descending duodenum, together with excretion of undigested grain, are typical of virus-induced proventricular dilatation disease (PDD).

In conjunction with clinical pathology techniques, identification of abnormalities in the size and position of organs within the body cavity can also be useful in diagnosing other **infectious diseases**. Examples in chickens include **avian tuberculosis**, with granulomas occurring particularly in the liver and spleen, and lymphoid **tumours** associated with Marek's disease. Lymphoid leukosis may also result in tumours in various organs.

6

Digestive system (apparatus digestorius)

H. E. König, H.-G. Liebich, R. Korbel and C. Klupiec

The avian digestive system (Figure 6.1) is distinguished from that of mammals by the following features:

- the beak,
- lack of separation between the oral and pharyngeal cavities,
- the absence of teeth, lips and cheeks,
- the crop,
- division of the stomach into glandular and muscular components,
- the presence of two caeca,
- the cloaca.

Oral cavity (cavum oris) and pharynx

In contrast to mammals, in which the rostral oral cavity and caudally adjoining pharynx are clearly differentiated, the mouth and pharynx of birds constitute a combined cavity that is surrounded dorsally and ventrally by the beak.

This macroscopically and functionally common space is referred to as the **oropharynx**.

Beak, bill (rostrum)

The beak is a distinctive feature of the class Aves (see also Chapter 17 'Common integument'). Having evolved to suit the requirements of individual species, the beak varies considerably in shape (Figures 6.2ff.). While the beak is a particular avian adaptation for **feeding**, it also makes a significant contribution during flight as an **aerodynamic feature**.

The **maxilla** (premaxillary and maxillary bones) and **mandibles** form the bony foundation for the **rostrum maxillare** (upper beak) and **rostrum mandibulare** (lower beak). These bones are covered in a horny sheath, the

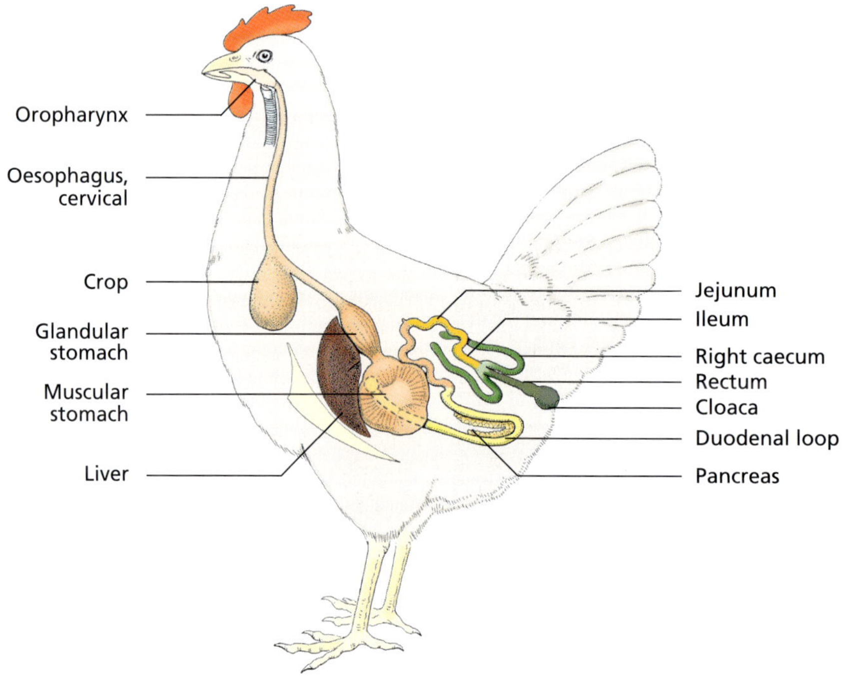

6.1 The digestive system of the chicken (schematic).

rhamphotheca (Figure 6.8). In many species this is continuously replaced, in a manner similar to a fingernail. Thus, if the beak of pet birds is not exposed to natural wear, appropriate trimming may be required.

6.2 Beak, oral cavity and tongue of a common buzzard (Buteo buteo). Courtesy of Dr S. Reese, Munich.

The dorsal median ridge of the upper beak is termed the **culmen**, while the equivalent ventral midline structure is referred to as the **gonys**. In many waterbird species, the tip of the upper and lower beak features a hard horny plate known as the **nail** (respectively, unguis maxillaris [see Figure 17.15] and unguis mandibularis). The **tomium** is the cutting edge of the upper and lower beak, where the inner and outer layers of horn come together. In several species, a sharp calcified projection is found on the culmen of full term chicks (Figure 6.8). Known as the 'egg tooth', this structure is used by young birds during hatching to break the egg membrane and sometimes also to penetrate the shell. The egg tooth is lost shortly after hatching (see Chapter 17 'Common integument').

The **external appearance of the beak** is species- and genus-specific (Figures 6.2 to 6.7). In Galliformes (e.g., chickens, turkeys, quail), the beak is pointed and hooked, whereas in Anseriformes (e.g., ducks, geese) it is flattened into a spoon-like shape. The rhamphotheca also exhibits variation. In granivores, for example, the outer covering of the beak is hard, while in ducks and geese it is soft

6.3 Muscovy duck (Cairina moschata) with leathery rhamphotheca and hard nail (containing the bill tip organ).

6.5 Female budgerigar (Melopsittacus undulatus) with characteristic brown cere.

6.4 Male budgerigar (Melopsittacus undulatus) with characteristic blue cere.

6.6 Male Indian peafowl (Pavo cristatus) with display feathers and slit-like nostrils.

6.7 Male elegant crested tinamou (Eudromia elegans) with bicoloured beak horn.

and leathery. This softer tissue is limited to the base of the upper beak in the chicken. In some species, the latter region is thickened forming a structure known as the **cere** (see Chapter 17 'Common integument'). Transversely oriented **lamellae** on the edges of the rhamphotheca of ducks and geese assist in the filtration of food.

In most avian species, the tip of the beak contains multiple aggregations of **sensory receptors** that form part of a complex sensory structure known as the **bill tip organ**. The sensory receptors are housed within so-called **'touch papillae'**. In the nail of the goose, these papillae have been observed at densities of up to 25 per square millimetre. Each papilla contains up to 40 receptors.

The cylindrical papillae are embedded within the keratinised tissue of the unguis maxillaris and the unguis mandibularis, with their tips extending to the free surface. They consist of a dermal core, surrounded by a soft horny (epidermal) coat. Each papilla is innervated predominantly by myelinated, but also some unmyelinated nerve fibres.

The sensory receptors within the papilla include **Herbst corpuscles** and **Grandry corpuscles**, which are thought to correspond functionally with the sensory Merkel's discs and Meissner's corpuscles of mammals.

The bill tip organ is used for selection and assessment of prehended foodstuffs, particularly in dabbling waterbirds, and plays an important role in plumage care. It is well developed in Anseriformes but is absent in pigeons and sparrows. In chickens, touch papillae are present only in the lower beak, although touch receptors are also found in the upper beak. With its highly developed bill tip organ, the beak of psittacines is both sensitive and powerful, making it useful for climbing.

Roof of the oral cavity and pharynx (oropharynx)

Birds lack a soft palate. Thus, there is no clear distinction between the roof of the mouth and the pharynx. Instead, the **palate** (palatum) forms the dorsal boundary of the combined cavities, or oropharynx. An elongated, sometimes oval **median cleft** (choana) in the palate connects the oropharynx with the left and right nasal cavities (Figure 6.9). The palate is covered by a non-glandular, often keratinised mucosa featuring **transverse ridges** (rugae palatinae) and **shallow grooves** (sulci palatini). Particularly in Galliformes (e.g., chickens) and Anatidae (e.g., ducks), the mucosa is studded with numerous caudally directed **papillae** (papillae palatinae). In Galliformes, a rostral, longitudinally oriented **median palatine ridge** (ruga palatina mediana) is also present (Figure 6.9). The papillae surround the choana and infundibular cleft (see below), some arranged in transverse rows, others scattered randomly. They have a **mechanical function** in transporting food and guarding against the ingestion of oversized foreign bodies.

The short infundibular cleft (rima infundibuli) lies caudal to the choana. Anatomically it is located in the pharynx,

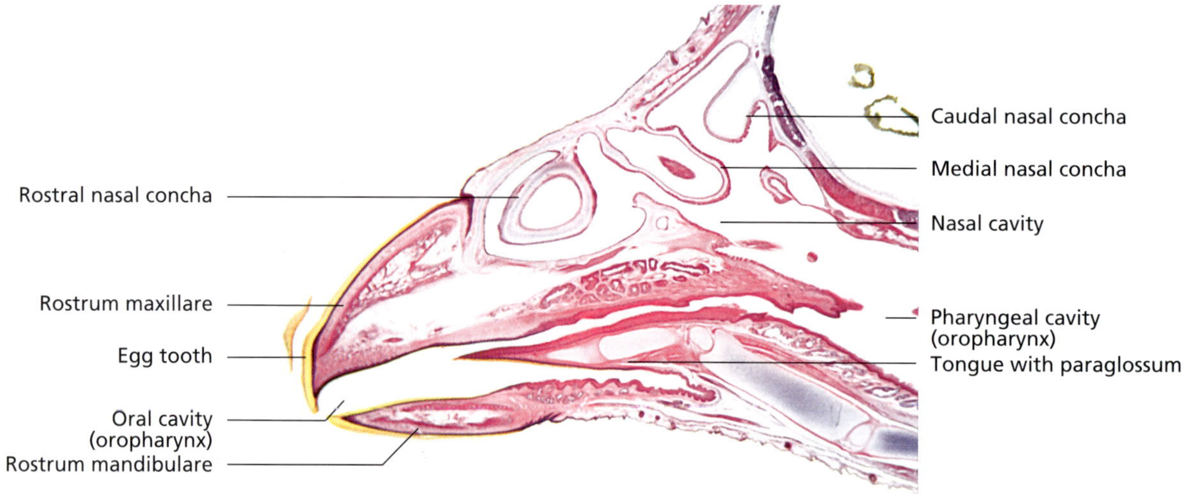

6.8 Paramedian section of the head of a chick with egg tooth.

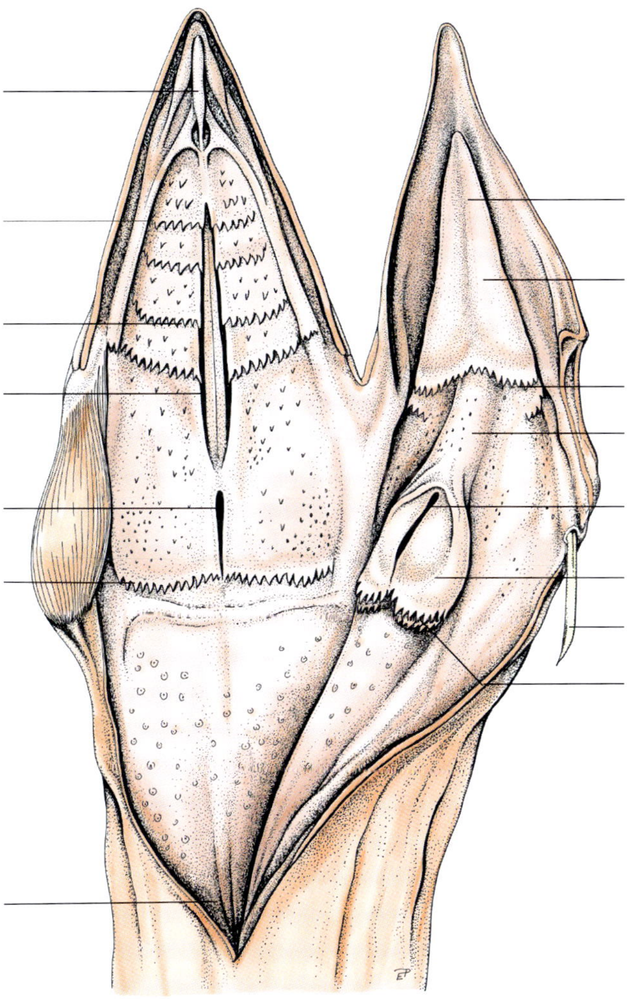

6.9 Oral cavity and pharynx of the chicken (opened and reflected; schematic), adapted from Dyce, Sack and Wensing, 2002.

which it connects with the middle ear (Figure 9.6). In birds, the left and right tuba auditiva join to form the unpaired **tuba auditiva communis**, which opens into a recess in the infundibular cleft. This pharyngotympanic connection is analogous to the tuba auditiva Eustachii in mammals. The non-glandular mucosa surrounding the infundibular cleft is richly endowed with subepithelial lymphatic tissue, referred to as the pharyngeal tonsil.

Floor of the oral cavity

The shape and development of the avian **tongue** (lingua) varies markedly, according to diet. Kolibris (hummingbirds) and insectivorous birds possess a very long and, where necessary, protrusible tongue. The tongue of psittacines is distinctively muscular, with that of lories exhibiting a brush-like tip for gathering nectar. In many avian species, particularly Galliformes, the tongue is pointed apically and broad at its base, with little if any muscle.

The tongue conforms to the shape of the lower beak. In chickens, a transverse row of caudally directed **lingual papillae** (papillae linguae) lies between the body (corpus linguae) and the **root** (radix linguae) of the tongue. In Anatidae (ducks), the caudal third of the tongue is thickened, forming a **torus linguae**.

The body of the tongue is supported by a bone, the **paraglossum** (= entoglossum), the intrinsic musculature being only rudimentary in most species, including chickens (Figure 6.8). Muscle features more prominently at the base of the tongue, into which extralingual muscle bundles radiate. Ventrally, the tongue is supported by a **keratinised plate** (cuticula cornificata lingualis).

In ducks and geese, the edges of the tongue are lined with spiny keratinised bristles that are directed towards the pharynx. These combine with the transverse ridges in the lower beak to assist in trapping small food particles during filter feeding of vegetable matter. Anatomically they are related to the **mechanical papillae** of the mammalian tongue.

Papillae containing tastebuds (papillae gustatoriae) are present on the dorsal surface of the tongue in most species, although these are usually sparse. Located at the centre of the papillae, the tastebuds collectively represent the gustatory organ. As in mammals, these papillae are innervated predominantly by branches of the glossopharyngeal nerve (IX). They also receive rami from the trigeminal nerve (V) and facial nerve (VII).

The avian **hyobranchial apparatus** (apparatus hyobranchialis) (Figures 2.13 and 2.14, Figure 7.5) differs markedly from the hyoid apparatus of mammals. It consists of the rostral, ventromedian paraglossum (which supports the body of the tongue), the basihyale (basibranchiale rostrale), the urohyale (basibranchiale caudale) and the paired, lateral horn-shaped cornu branchiale, formed by the ceratobranchiale and epibranchiale. The cornu branchiale is not attached to the base of the skull, extending instead to its caudolateral surface (Figure 2.17).

The muscles of the **hyobranchial apparatus** and the tongue are divided into two groups. Some originate from and also insert upon the hyobranchial bones, while others arise from the mandible and extend either to the hyobranchial apparatus or to a median strip of tendon in the intermandibular space.

Similarly to mammals, the tongue receives innervation from several **cranial nerves**. This includes sensory components of the trigeminal nerve (V), the chorda tympani (parasympathetic, motor) of the facial nerve (VII), the glossopharyngeal nerve (IX) and the exclusively motor hypoglossal nerve (XII). More detailed information pertaining to the hyobranchial apparatus and muscles of the tongue is reserved for the specialised literature.

Floor of the pharynx

A pronounced **laryngeal mound** (mons laryngealis) surrounds the slit-like glottis, the entrance to the laryngeal cavity. The surface of the laryngeal mound is covered with well-defined pharyngeal papillae (papillae pharyngeales), particularly at its caudal edge. The ducts of polystomatic salivary glands open at the base of the mound (Figures 6.8 to 6.10).

Salivary glands (glandulae salivariae)

An adequate supply of saliva is very important in birds, particularly in granivores. The **maxillary salivary gland** (glandula maxillaris) is located in the roof of the mouth (Figure 6.10). It empties via a duct, caudal to the bill tip organ of the upper beak. An abundance of tastebuds surrounds the duct orifice. At the angle of the mouth, the **glandula anguli oris** is also drained by a single duct.

The palate contains multiple openings of the **palatine glands** (glandulae palatinae). Numerous tastebuds surround these orifices. Openings of the manifold ducts of the **rostral**

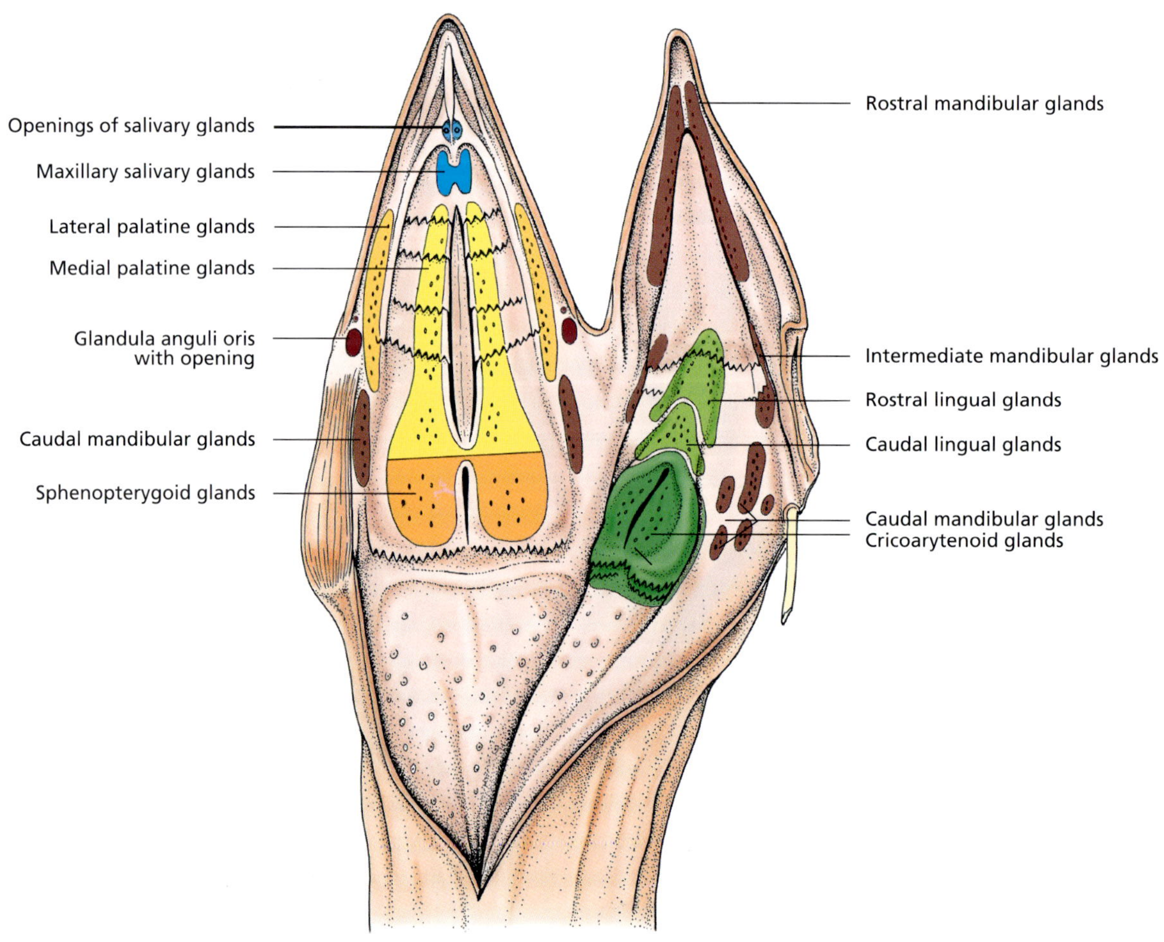

Fig 6.10 Salivary glands of the chicken (schematic), adapted from Dyce, Sack and Wensing, 2002.

and intermediate mandibular glands (glandulae mandibulares rostrales and intermediae) and **lingual glands** (glandulae linguales) are located in the floor of the mouth. Additional glands, located more caudally, empty into the oropharynx via several ducts. These include the sphenopterygoid glands (glandulae sphenopterygoideae), the cricoarytenoid glands (glandulae cricoarytenoideae) and the caudal mandibular glands (glandulae mandibulares caudales) (Figure 6.10).

Swallowing

After food is grasped by the beak, it is pressed up against the palate by the tongue. The particularly mucous saliva of chickens assists in holding the food in place. Following reflex closure of the choana, a rapid sequence of movements of the tongue propels the bolus towards the pharynx. This is accompanied by reflex closure of the infundibular cleft and glottis, aided by numerous lip-like mucosal projections lining the edges of these openings.

Abrupt peristaltic movements of the laryngeal mound facilitate transit of the bolus into the oesophagus. The swallowing process is further assisted by the many caudally directed papillae within the oropharynx, and by copious saliva.

Alimentary canal (canalis alimentarius)

The alimentary canal consists of the following components:

- **oesophagus**:
  - crop (ingluvies),
- **stomach (gaster)**:
  - proventriculus (pars glandularis),
  - ventriculus, gizzard (pars muscularis),
- **intestine (intestinum)**:
  - small intestine (intestinum tenue):
    - duodenum,
    - jejunum,
    - ileum,
  - large intestine (intestinum crassum):
    - caeca,
    - rectum,
- **cloaca**:
  - coprodeum,
  - urodeum and
  - proctodeum.

Oesophagus

The oesophagus is a flexible, thin-walled tube (Figure 6.15) that extends from the laryngeal mound to the proventriculus. Its **cervical component** (pars cervicalis) initially lies dorsal to the trachea. In the mid to lower cervical region, both the oesophagus and the trachea pass to the **right side of the neck**, contrasting markedly with the course of the trachea in mammals.

At the thoracic inlet, the oesophagus widens to form the **crop** that, in most species, lies ventrally. Thereafter the oesophagus is again positioned dorsal to the trachea and passes along the **ventral aspect of the lungs** and over the base of the heart. At the level of the third to fourth intercostal space, the oesophagus opens into the proventriculus. Within the thoracic region of the body cavity, the oesophagus is surrounded by components of the cervical, clavicular and cranial thoracic air sacs.

Crop (ingluvies)

The crop is formed by the dilation of the oesophagus immediately before its entry into the body cavity (Figures

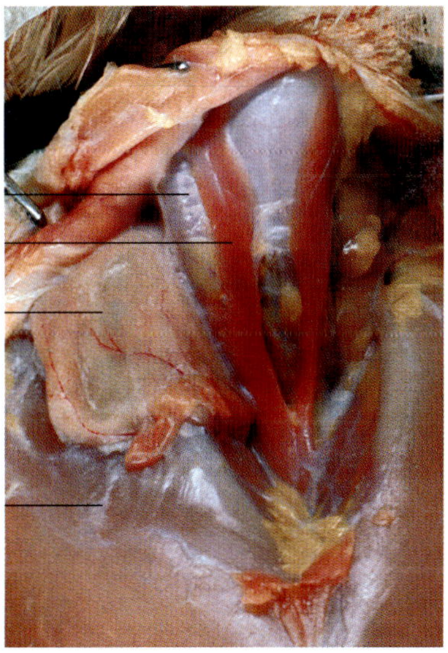

6.11 Crop of a chicken (ventral view).

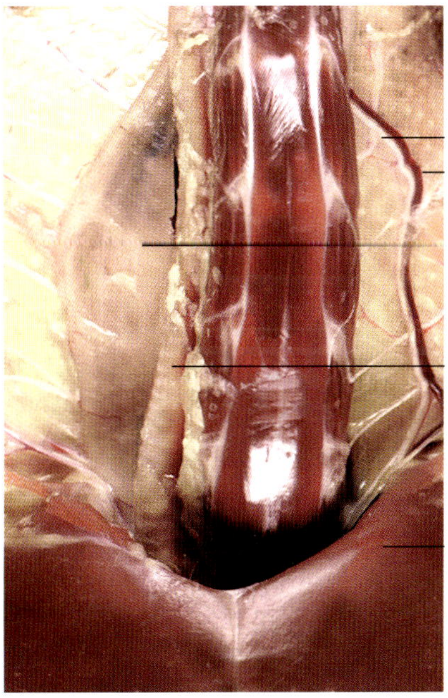

6.12 Crop (spindle shaped dilatation of the oesophagus) in a Mallard duck (ventral view).

6.11 and 6.12). A so-called 'crop channel' is located in its dorsal wall. The crop has several functions, including some species-specific specialisations. Generally, it permits temporary storage of ingesta as well as softening and pre-digestion of poorly digestible foodstuffs. Particularly in granivores, the crop is capable of considerable expansion. The highly developed crop of the hoatzin has a masticatory function.

The crop of ducks and geese is a simple spindle shaped dilatation of the oesophagus. In Columbiformes (pigeons and doves), the crop is divided into two large sacs. The columbiform crop also produces a substance known as 'crop milk', which is regurgitated and fed to nestlings. Crop milk consists of mucosal epithelial cells that have proliferated, become filled with lipid and subsequently desquamated.

The structure of the crop wall is similar to that of the oesophagus (Figure 6.15). It contains **mucous crop glands** (glandulae ingluviales) similar to those found in the oesophagus. In the chicken these are located in the vicinity of the crop channel, while in the pigeons they are limited to the fundus.

Forceful contractions of the muscles of the crop and crop channel propel the food into the stomach.

Stomach (gaster)

There are **three basic** types of avian stomach, reflecting the diet of different species:

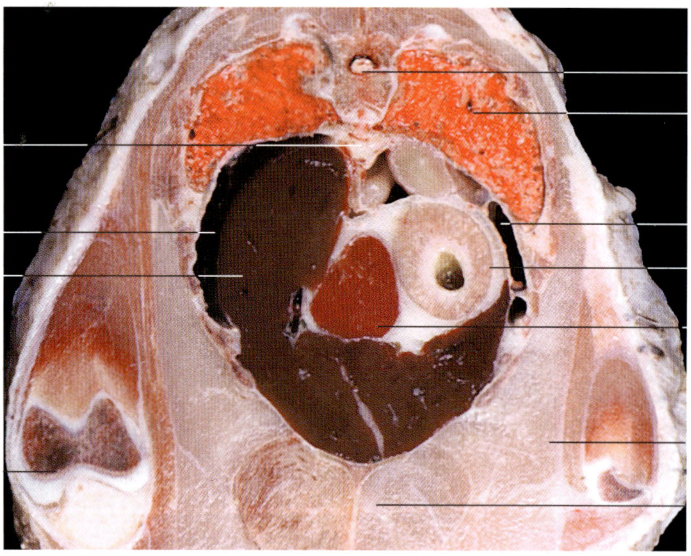

6.13 Transverse section of the body cavity of a chicken at the level of the proventriculus and adjacent organs. Courtesy of Professor Dr J. Ruberte, Barcelona.

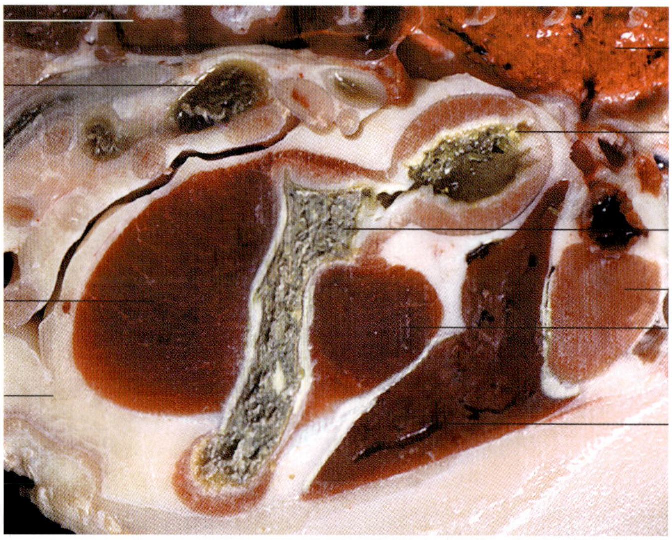

6.14 Paramedian section of the body cavity of a chicken at the level of the ventriculus and adjacent organs. Courtesy of Professor Dr J. Ruberte, Barcelona.

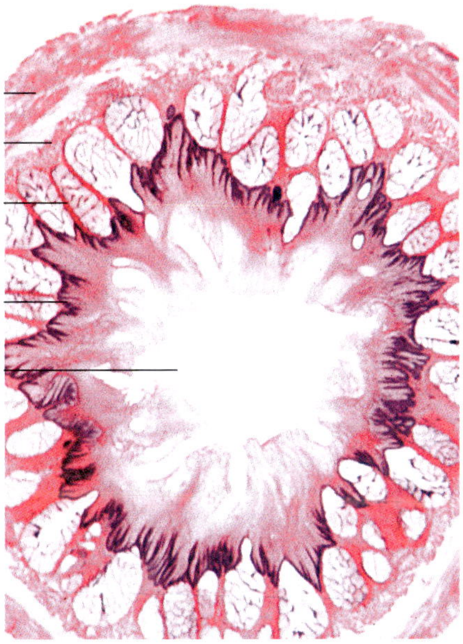

6.15 Histological section of the oesophagus of a chicken.

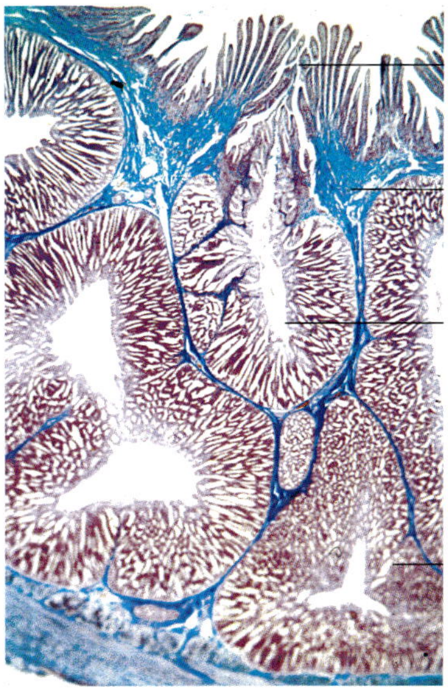

6.16 Histological section of the proventriculus of a chicken.

- carnivorous (fish- and meat-eating species),
- herbivorous, in which the stomach prepares the ingesta physically and chemically; in these species there is typically a clear distinction between the glandular and muscular stomachs, and
- intermediate (e.g., fruit- and nectar-eaters).

In seagulls and storks, which quickly ingest large quantities of animal protein, the stomach is a highly expandable, single sac-like structure with little muscle development. The stomach of grain- and plant-eating species, including the chicken, pigeon, goose and duck, is clearly divided into two compartments, the glandular proventriculus and the muscular ventriculus (Figures 6.17ff.). In fruit-eating species, such as tanagers (fruit-eating omnivores), the stomach is reduced to a rudimentary diverticulum.

Proventriculus (glandular stomach, pars glandularis)

The proventriculus continues from the oesophagus (Figures 6.17 and 6.18) without a clear anatomical boundary. It lies against the parietal surface of the liver. Together with the spleen, which lies to its left, the proventriculus is located in a pouch of the intestinal peritoneal sac. In most species, the mucosa is arranged in folds that, in the chicken, lie on prominent papillae (Figures 6.18ff.). The glands of the proventriculus are divided into:

- superficial proventicular glands (glandulae proventriculares superficiales) and
- deep proventricular glands (glandulae proventriculares profundae).

The walls of these tubular glands consist of secretory cells that, in contrast to the gastric glands of mammals, are morphologically indistinguishable from one another. These cells secrete pepsinogen, hydrochloric acid and a hydrogen carbonate-rich product, as well as intrinsic factor, required for absorption of Vitamin B_{12}. The loose connective tissue of the **tela submucosa** contains the submucosal nerve plexus (plexus nervorum submucosus; Meissner-Plexus) and numerous vessels. In the **tunica muscularis**, a well-developed inner circular layer (stratum circulare) is surrounded by a thinner outer longitudinal layer (stratum

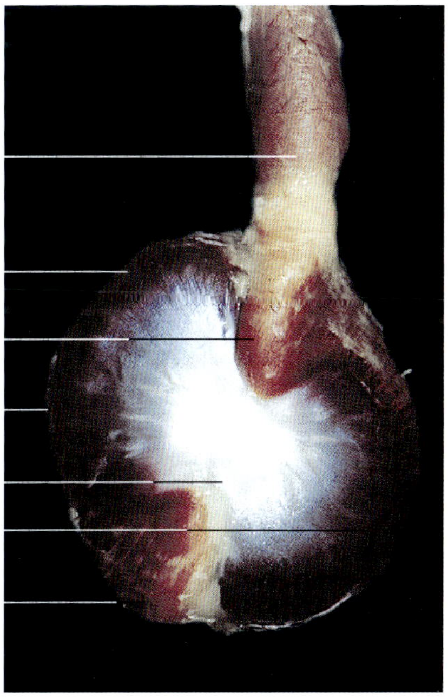

6.17 Proventriculus and ventriculus of a chicken.

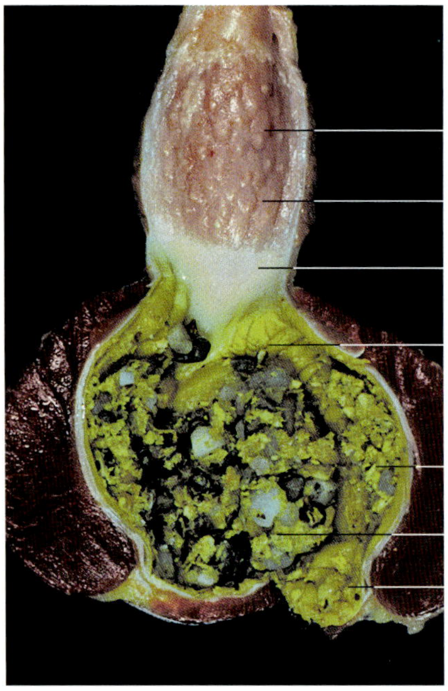

6.18 Proventriculus and ventriculus of a chicken (opened).

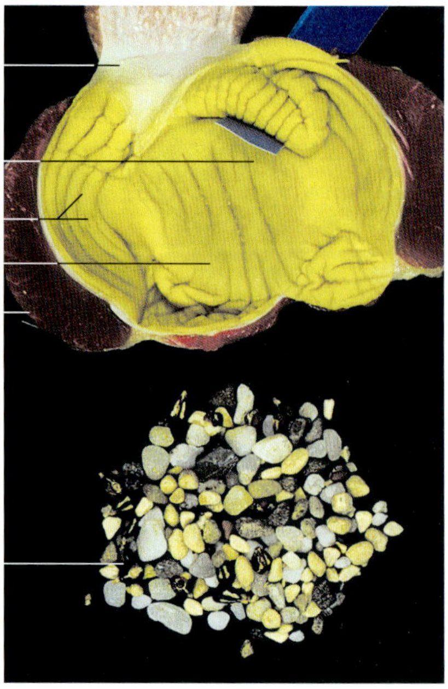

6.19 Ventriculus of a chicken (opened and grit removed).

longitudinale). The myenteric nerve plexus (plexus nervorum myentericus; Auerbach-plexus) lies between the two muscular layers. Externally, the proventriculus is lined by a single layered **tunica serosa**, which is continuous with the intestinal peritoneal sac.

Food only remains in the proventriculus for a short period. Rhythmic contractions force the ingesta into the ventriculus, where mixing and chemical breakdown of food commences.

The proventriculus is typically separated from the ventriculus by a narrow gastric isthmus (isthmus gastris). This transitional zone (zona intermedia) is usually free of mucosal elevations and folds, and is devoid of glands. In the chicken, the narrowing at the isthmus is attributable to a high concentration of elastic fibres and a relatively thin muscular layer.

Muscular stomach (ventriculus, pars muscularis)

The ventriculus, or 'gizzard', lies to the left of the intestinal peritoneal cavity, between the layers of the posthepatic septum. It is sometimes referred to as a **'masticatory organ'** as it replaces the function of the teeth. Features of the ventriculus include:

- the body (corpus ventriculi),
- two surfaces with a tendinous centre (facies tendineae),
- two blind sacs (saccus cranialis and saccus caudalis).

Shaped like a biconvex lens (Figures 6.17ff.), the ventriculus lies in the lower left quadrant of the body cavity. Much of its surface is covered by the left abdominal air sac, which adjoins the left ventral hepatic peritoneal cavity on the left side of the ventriculus. Part of the ventriculus is also in contact with the body wall and, as such, is retroperitoneal. The wall of the ventriculus has a typical layered structure (mucosa, submucosa, muscular tunic and serosa).

TUNICA MUCOSA GASTRIS

Most of the **gastric mucosa** is thrown into clearly visible **ventricular folds** (rugae ventriculi). These are absent near the tendinous centres (Figures 6.17ff.). The luminal surface of the ventriculus is covered by a greenish-yellow layer known as the **cuticle** (cuticula gastris). In some species, this acts as an abrasive surface for grinding food.

6.20 Stomach of an Indian runner duck (opened and emptied).

The cuticle consists of solidified secretions of **tubular glands in the lamina propria** (glandulae ventriculares). Secretory product is released onto the mucosal surface as compacted cylinders, or **columnae verticales**, that harden to form rods (Figures 6.21 and 6.22). Known as **koilin**, this solidified secretion comprises a keratin-like carbohydrate-protein complex. Koilin combines with the **matrix horizontalis** (softer secretions of the simple columnar gastric mucosal epithelium) to form a continuous layer of varying thickness. The compacted rods are seen as small processes that protrude beyond the surface of the cuticle.

The presence of a number of longitudinal ridges in the cuticle enhances its ability to break down food particles. Small stones and other foreign matter (grit) ingested with the food (Figures 6.18 and 6.19) further assist with the mechanical grinding of grain.

TUNICA MUSCULARIS GASTRIS

The wall of the muscular layer consists predominantly of smooth muscle tissue that can be divided macroscopically into four separate muscles (Figures 6.18 and 6.19). Their inner circular layer is usually more developed than the outer longitudinal layer. Based on the degree of muscle development in these layers, the muscles are characterised as thick (crassus) or thin (tenuis), and are further identified by their anatomical position as:

- m. crassus caudodorsalis,
- m. crassus cranioventralis,
- m. tenuis craniodorsalis and
- m. tenuis caudoventralis.

The strong **m. crassus caudodorsalis** and **m. crassus cranioventralis** extend from one **tendinous centre** (centrum tendineum) to the other, forming the dorsal and ventral borders of the ventriculus (Figure 6.17). Lying between the thick muscles are the weaker **m. tenius craniodorsalis** and **m. tenius caudoventralis**. These pass over the cranial and caudal blind sacs, ending likewise at the tendinous centres.

In the chicken, in which the ventriculus is highly differentiated, complete contraction of the muscular wall can generate pressures of up to 100–200mm Hg. Particularly in grain-feeders, the ventriculus is the site of mechanical breakdown of food. Digestion of protein also increases in this portion of the stomach. The ground gastric contents pass from the ventriculus into the pylorus through the **ventriculopyloric ostium** (ostium ventriculopyloricum), situated adjacent to the opening between the proventriculus and ventriculus. The **cuticle** terminates at this point (Figures 6.19 and 6.20).

Intestine (intestinum)

In most birds, the intestine is shorter in relative terms than in domestic mammals (Figures 6.25ff.). Among avian species, the intestine is longer in grain- and grass-feeders than in carnivores. **Villi are present in all segments of the intestine**. Chemical digestion and absorption of nutrients takes place in the small intestine. The caecum is responsible for breakdown of cellulose, and reabsorption of water occurs in the rectum and cloaca.

Gut-associated lymphatic tissue (GALT)

As in mammals, the immune system of the digestive tract of birds comprises diffuse lymphatic tissue that extends

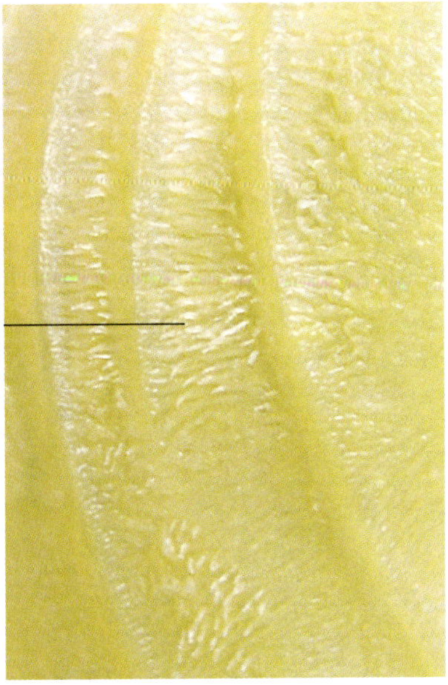

6.21 Cuticle on the luminal surface of the ventriculus of a chicken. Courtesy of Dr Sergio Donoso E. Chillan, Chile.

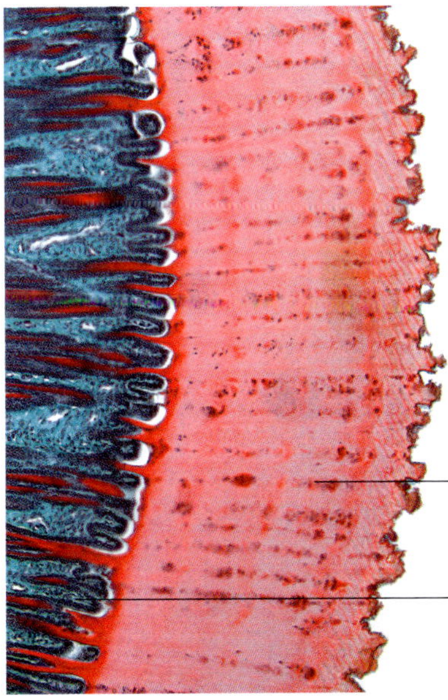

6.22 Histological section of the cuticle of the ventriculus of a chicken, with regions of varying hardness.

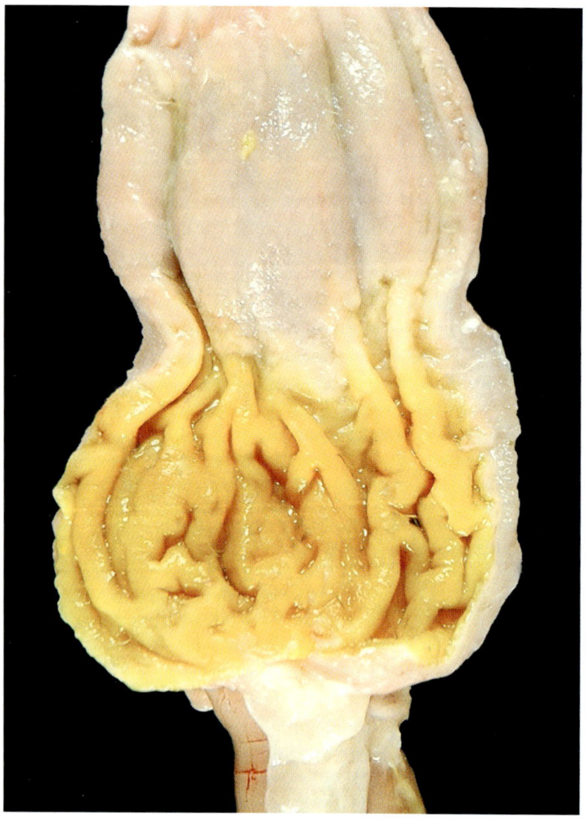

6.23 View of the opened proventriculus and ventriculus of a common buzzard (Buteo buteo). Compared with granivorous bird species, the wall of the ventriculus is thinner and the transition between the two stomachs is less clearly demarcated.

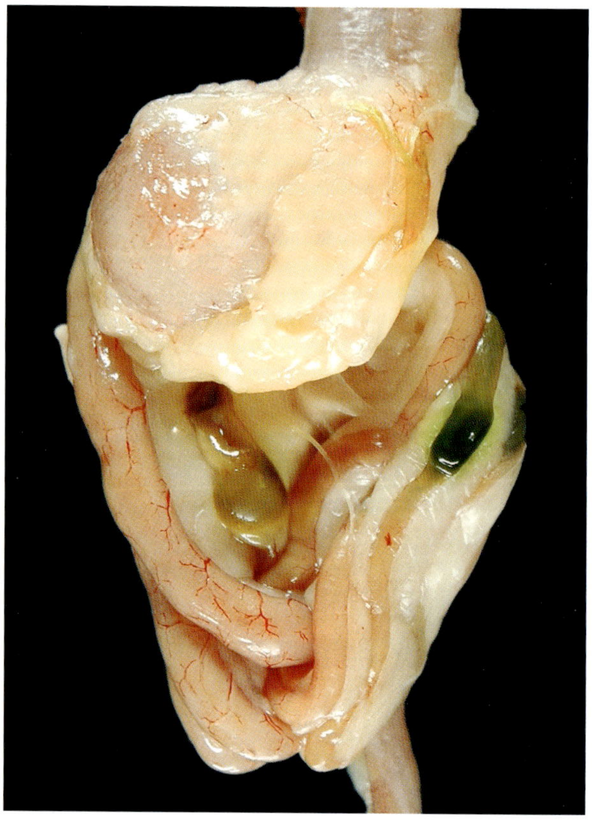

6.24 External view of the ventriculus, descending duodenum, spleen (brown) and additional duodenal and jejunal loops in the common buzzard (Buteo buteo). Note the abundant intra-abdominal fat deposition, seen in late summer and autumn.

throughout the entire mucosa of the stomach and intestine. This so-called **mucosa-associated lymphatic tissue** (**MALT**) includes scattered lymphatic tissue as well as clusters or aggregations of lymph nodules referred to as tonsils.

In both birds and mammals, the MALT of the gut is referred to as **gut-associated lymphatic tissue** (**GALT**). Particularly in chickens, aggregated lymph nodules known as Peyer's patches are most conspicuous near the Meckel's diverticulum, in the distal jejunoileum and at the entrance to the caecum (Figure 6.32).

These tonsil-like structures are involved in presenting ingested antigens to the immune system. **Discrete lymph nodes are absent in the gastrointestinal tract of birds**.

Small intestine (intestinum tenue)

As in mammals, the small intestine (Figures 6.25ff.) is comprised of the:

- duodenum,
- jejunum,
- ileum.

The boundaries between these intestinal segments are poorly differentiated. In view of the lack of a morphological distinction between the jejunum and the ileum, these segments are sometimes referred to collectively as the '**jejunoileum**'.

DUODENUM

The duodenum (Figures 6.25ff.) begins at the **ostium ventriculo-pyloricum** of the ventriculus. Topographically, therefore, the **stomach exit** is closely associated with its entrance (Figure 6.20). The duodenum forms a U-shaped **ansa duodeni** consisting of a descending portion, the **pars descendens**, and an ascending component, the **pars ascendens**. Lying between these segments is the pancreas. In contrast to mammals, the ascending duodenum typically accommodates the openings of three pancreatic ducts and two bile ducts, the **ductus hepatoentericus** and the **ductus cysticoentericus** (Figure 6.26). These ducts open at or near the **papilla duodeni**.

JEJUNUM AND ILEUM

Located just cranial to the cranial mesenteric artery, immediately ventral to the vertebral column, the flexura duodenojejunalis forms the junction between the duodenum and the jejunum. The jejunum and ileum are arranged in loops occupying the right caudal quadrant of the body cavity.

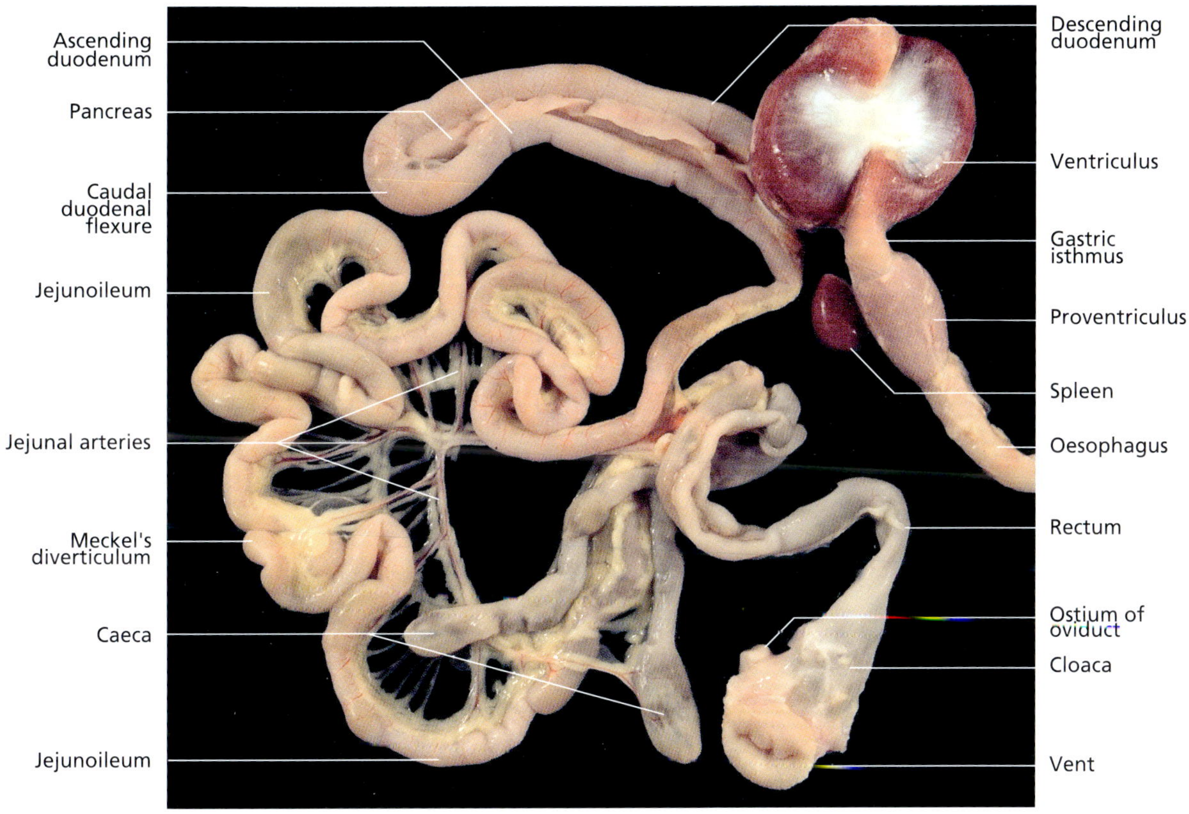

6.25 Gastrointestinal tract of a chicken with proventriculus, ventriculus and intestinal loops (separated). Courtesy of Dr Annette Kaiser, Munich.

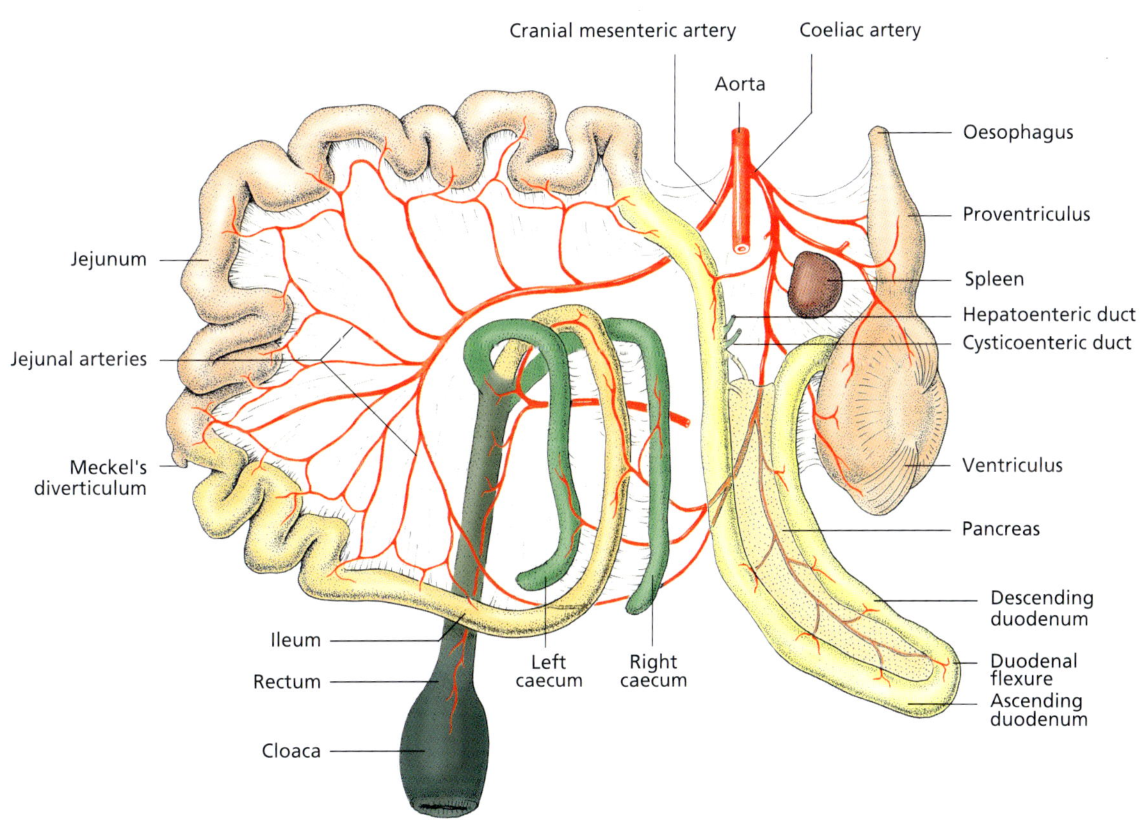

6.26 Gastrointestinal tract and associated arteries of the chicken (schematic), adapted from McLelland, 1975.

The **Meckel's diverticulum** (diverticulum vitellinum), an embryonic remnant of the yolk stalk, is located on the ansa axialis (an intestinal loop in the middle of the jejunoileum), opposite the longest middle branch of the cranial mesenteric artery (Figures 6.25, 6.26 and 6.31). This rudimentary structure is often considered to represent the boundary between the jejunum and ileum, although there are no discernible morphological differences upon which to base such a distinction. The presence of the Meckel's diverticulum is variable. It is absent in 40 per cent of chickens and pigeons, 20 per cent of ducks and 10 per cent of geese.

In ducks, geese and pigeons, the final loop of the **ileum** is termed the **supraduodenal loop** (ansa supraduodenalis) on account of its location, dorsal to the duodenum. In the chicken, the jejunum and ileum are arranged in garland-like coils, in which an ansa supraduodenalis is not discernible.

Large intestine (intestinum crassum)

The large, or terminal, intestine includes the **caeca** and the **rectum** (Figures 6.25ff.).

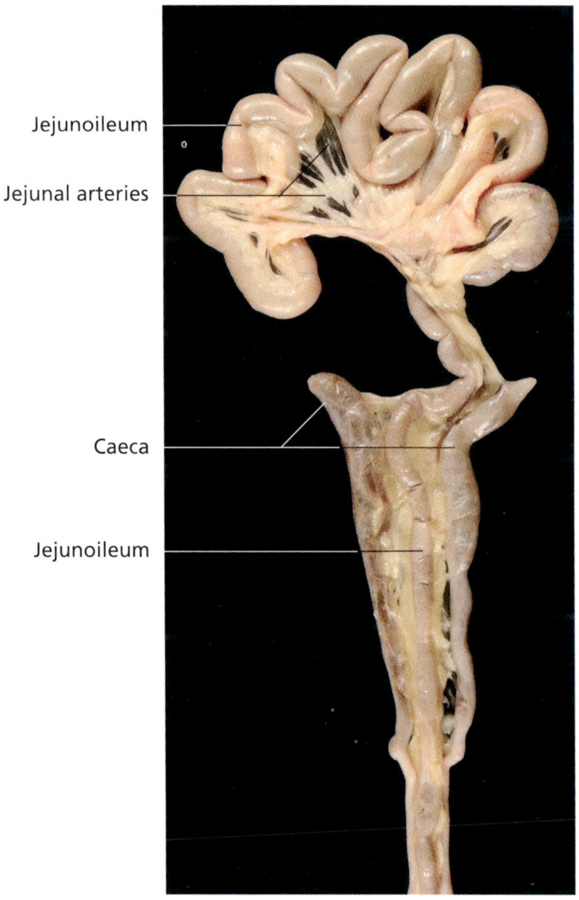

6.27 Segment of intestine of a chicken. Courtesy of Dr Annette Kaiser, Munich.

CAECA

In contrast to mammals, domestic poultry have two large caeca. They begin at the transition between the ileum and the rectum and are connected to the terminal ileum by a well-defined **ileocaecal ligament** (lig. ileocaecale). Each caecum communicates with the rectum via an **ostium caeci.** A muscular sphincter is present at the **base** (basis caeci) of each caecum.

The caeca are particularly well developed in the chicken. At its base, each caecum is richly endowed with lymphatic tissue, sometimes referred to as the **caecal tonsils** (Figure 6.32). The **caecal body** (corpus caeci) is thin-walled and often ampulliform, while the **apex** (apex caeci) may be pointed or vesicular. In herbivorous and frugivorous species, the caeca are the site of digestion of plant polysaccharides such as cellulose.

In pigeons, the caeca are short and rudimentary. Caeca are absent in parrots and in several carnivorous species. Loons and herons have an unpaired caecum, while the caeca of the ostrich communicate with the rectum via a single opening.

RECTUM

The rectum is the final, straight segment of the intestine that passes to the cloaca (Figures 6.25ff.). Until recently, it was referred to as the colon. In the ostrich, duck and goose, the transition to the cloaca is marked by an annular mucosal fold, the **plica rectocoprodealis**.

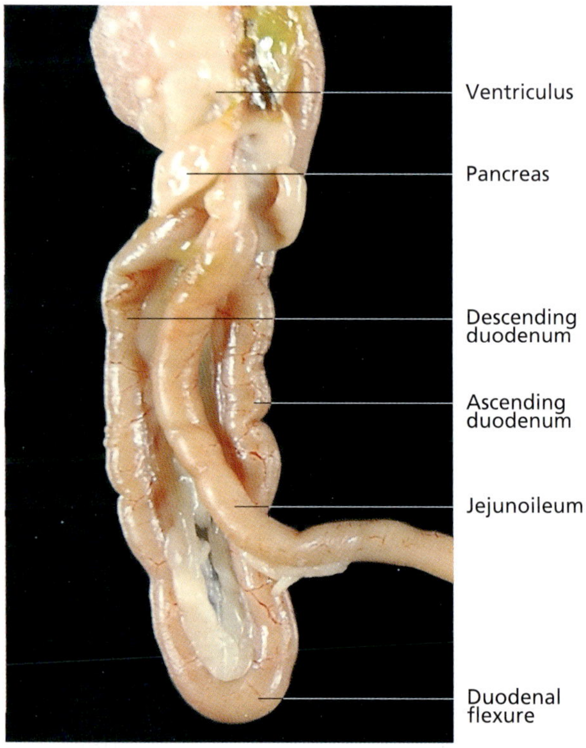

6.28 Stomach and intestine of a common buzzard.

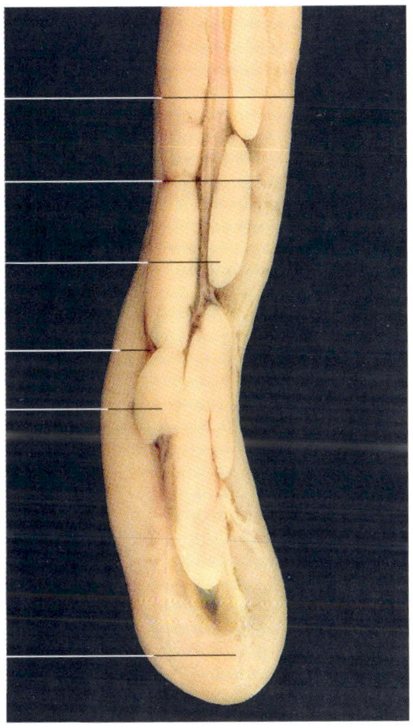

6.29 Segment of duodenum and pancreas of a chicken, Courtesy of Dr Annette Kaiser, Munich.

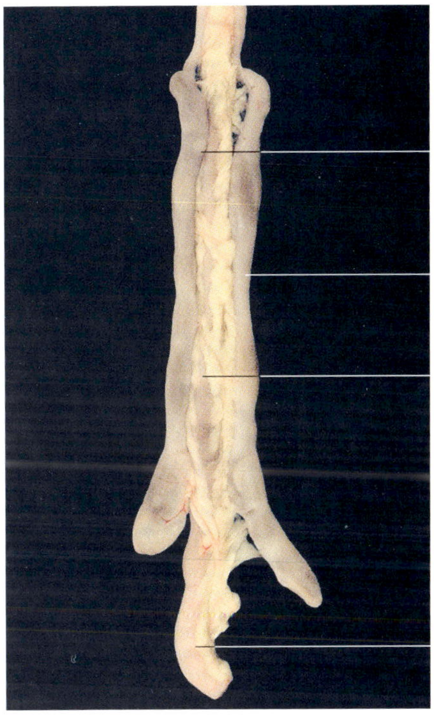

6.30 Caeca of a chicken with jejunoileum, Courtesy of Dr Annette Kaiser, Munich.

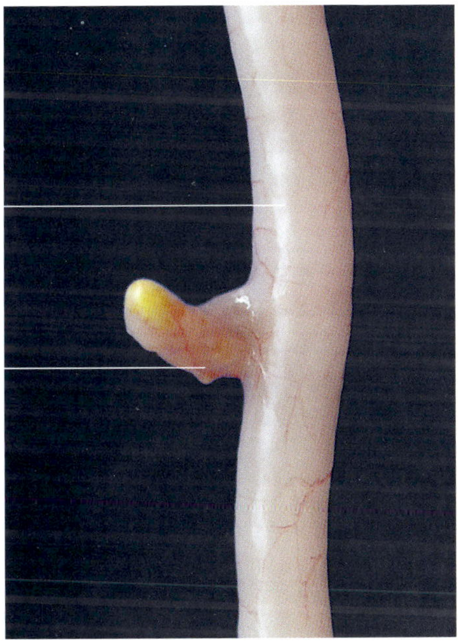

6.31 Diverticulum vitellinum (Meckel's diverticulum) of a 14 day old chicken, Courtesy of Dr Sergio Donoso E., Chillan, Chile.

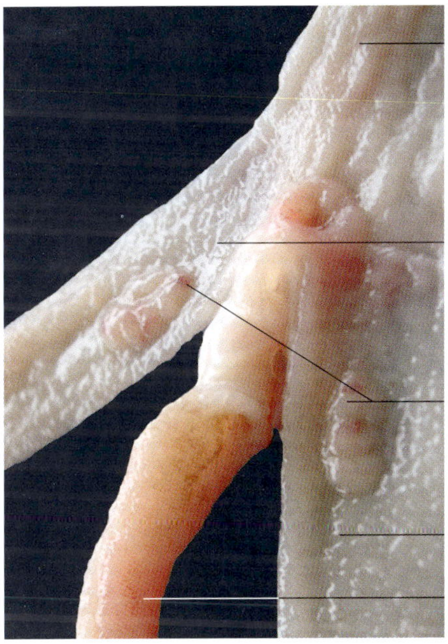

6.32 Base of the caeca of a chicken (opened), Courtesy of Dr Sergio Donoso E., Chillan, Chile.

Cloaca

The cloaca is the common excretory passage for the digestive and urogenital systems (Figures 6.25, 6.26 and 6.38). In the chicken, it is 2.5cm long and 2–2.5cm wide. Two mucosal folds divide the cloaca into three sections:

- coprodeum,
- urodeum and
- proctodeum.

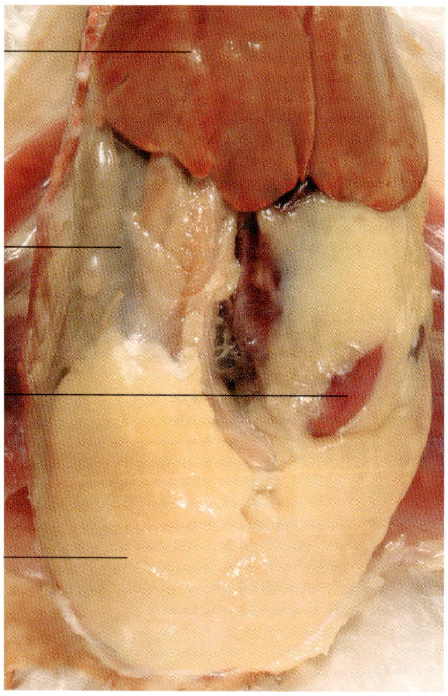

6.33 Organs of the peritoneal cavity of a chicken, with fat accumulation in the intestinal peritoneal cavity (ventral view). Courtesy of Dr Annette Kaiser, Munich.

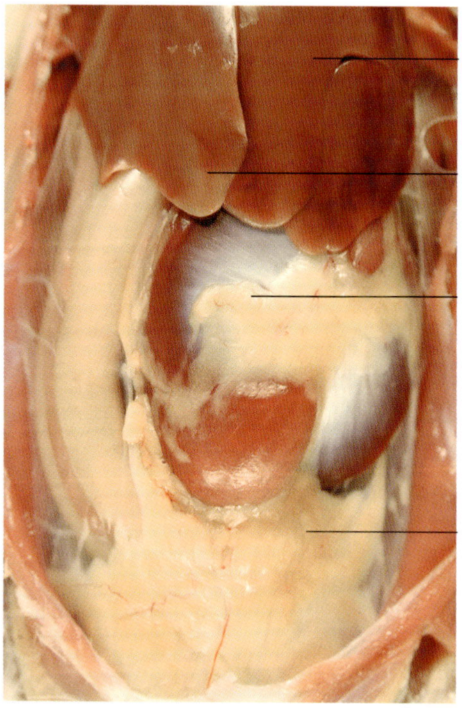

6.34 Organs of the peritoneal cavity of a chicken (ventral view after partial exposure the intestinal peritoneal cavity) . Courtesy of Dr Annette Kaiser, Munich.

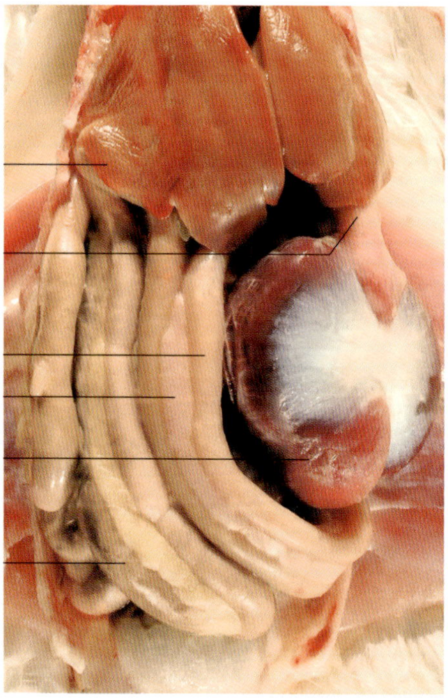

6.35 Organs of the peritoneal cavity of a chicken (ventral view, intestinal peritoneal cavity exposed). Courtesy of Dr Annette Kaiser, Munich.

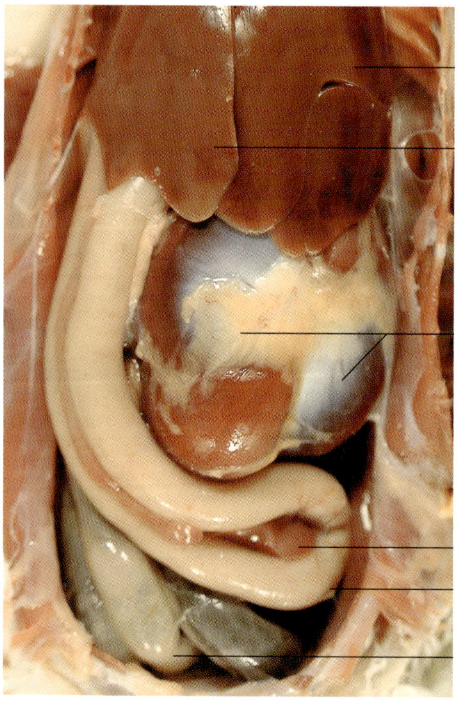

6.36 Organs of the peritoneal cavity of a chicken (ventral view, intestinal peritoneal cavity exposed; detailed view). Courtesy of Dr Annette Kaiser, Munich.

The rectum empties into the coprodeum. This joins the urodeum, into which the urinary and reproductive tracts open. The proctodeum is the terminal section of the cloaca. Schematic illustrations of the individual cloacal segments are shown in Chapter 9, 'Male genital organs' (Figures 9.1, 9.7 and 9.8).

The chicken lacks a plica rectocoprodealis, the rectum merging with the coprodeum without a distinct boundary.

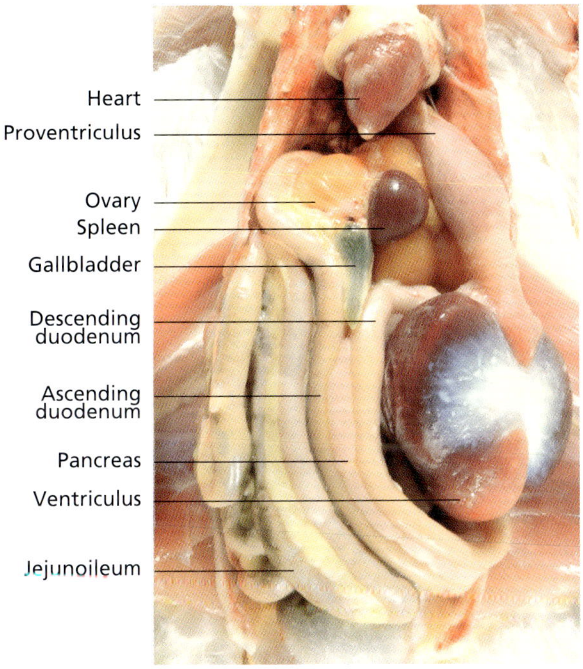

6.37 Organs of the peritoneal cavity of a chicken (ventral view, intestinal peritoneal cavity exposed and liver removed). Courtesy of Dr Annette Kaiser, Munich.

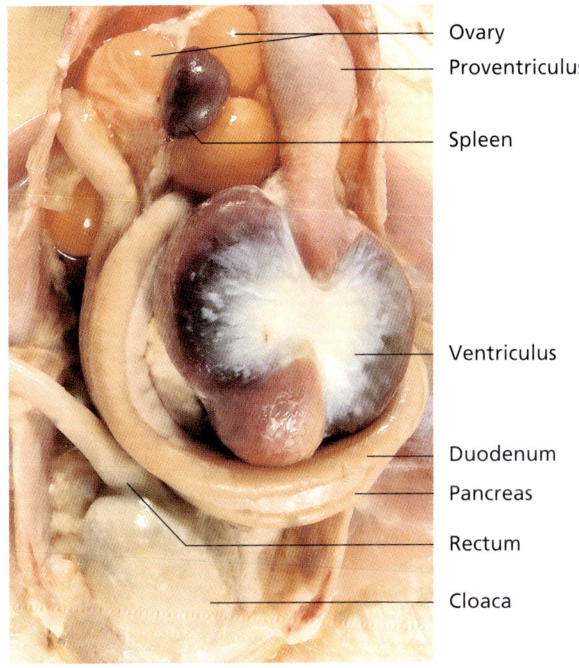

6.38 Organs of the peritoneal cavity of a chicken (ventral view, liver removed and rectum and cloaca reflected). Courtesy of Dr Annette Kaiser, Munich.

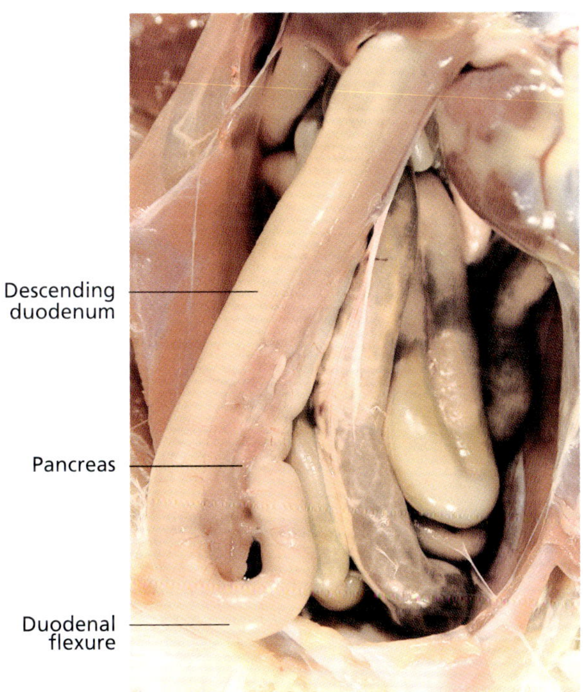

6.39 Intestinal loops of a chicken (ventral view, organs reflected). Courtesy of Dr Annette Kaiser, Munich.

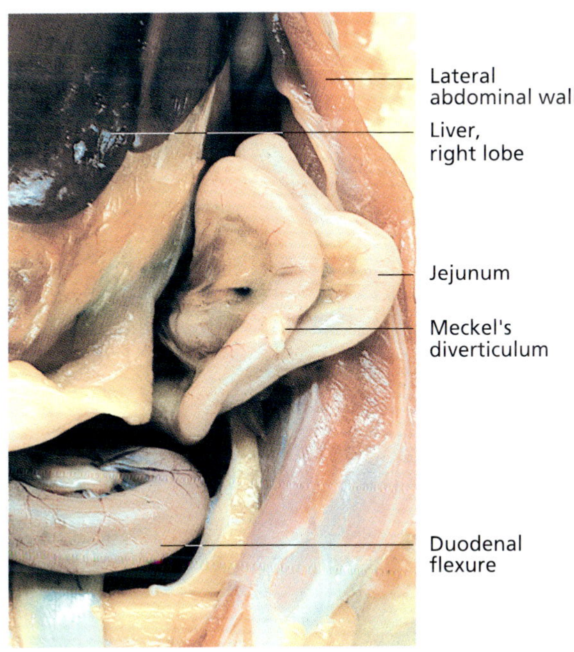

6.40 Intestinal loops of a chicken (detailed ventral view, organs reflected).

In the coprodeum, the villi are particularly broad, becoming shorter caudally.

The dorsal wall of the **urodeum** contains the openings of the ureters. In the male, the deferent ducts open on conical papillae adjacent to each ureteral orifice. In females, the left side of the urodeum receives the left oviduct. The right, vestigial oviduct, which sometimes appears as a vesicular fluid filled structure next to the rectum, terminates in a nondescript recess in the urodeum.

In the final section, the **proctodeum**, the rectal lining transitions to a non-glandular mucosa that is continuous with the external skin. The opening of the **cloacal bursa**

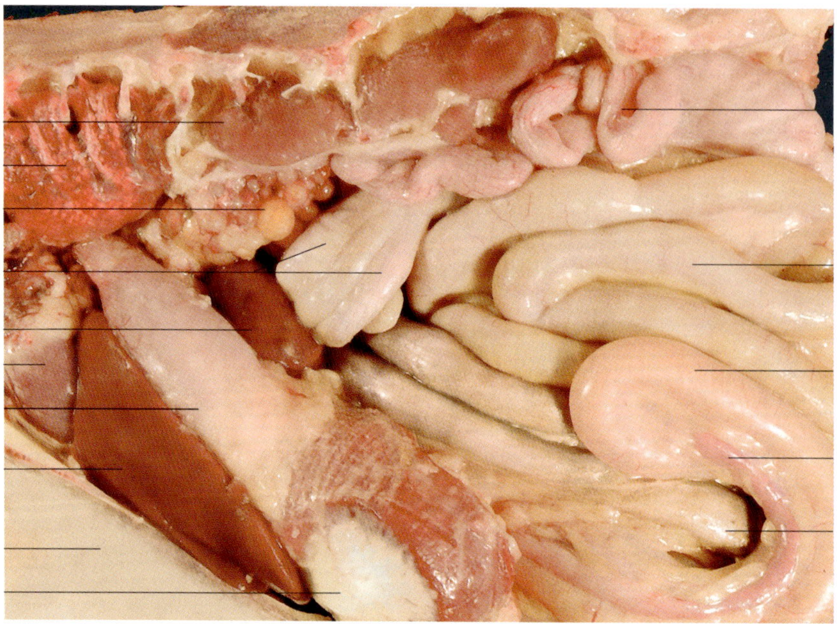

6.41 Organs of the body cavity of a chicken, left body wall removed (lateral view). Courtesy of Dr Annette Kaiser, Munich.

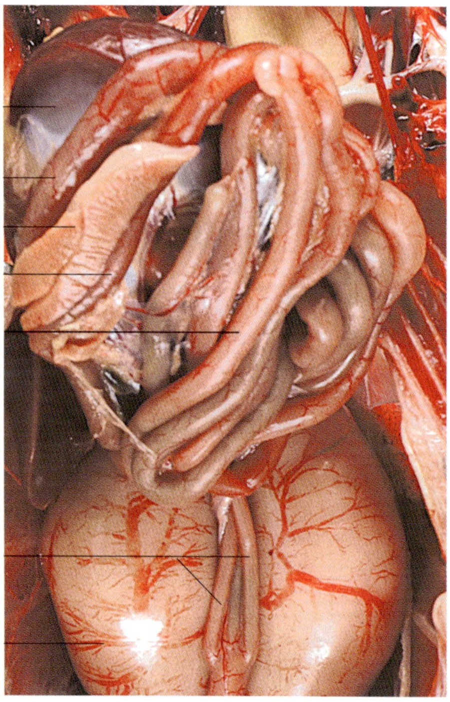

6.42 Intestine of a male Indian runner duck (ventral view).

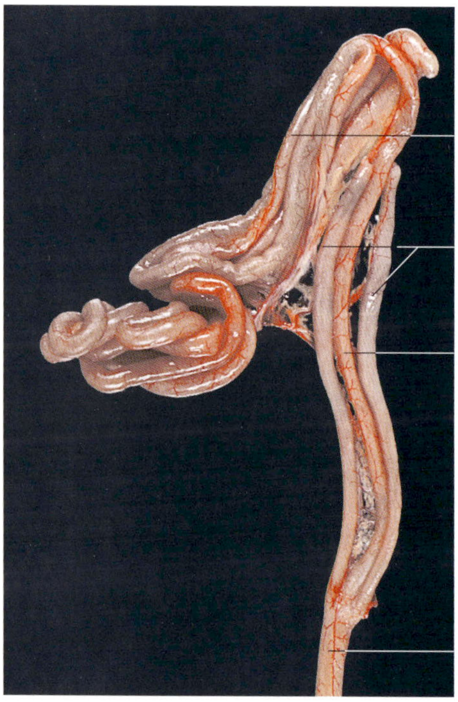

6.43 Intestine of a male Indian runner duck (ex situ).

(bursa cloacalis; bursa of Fabricius) (see Chapter 12 'Immune system and lymphatic organs') is located in the dorsal wall of the proctodeum. **Glands** (glandulae proctodeales) are present in the dorsolateral wall. In males, the floor of the proctodeum houses the **copulatory organ** (phallus).

At the external opening of the cloaca, referred to as the **vent** (ventus), there is a dorsal and ventral **lip** (labium venti dorsale and ventrale), each of which contains **glands** (glandulae venti). In the chicken, the lips contain numerous sensory Herbst corpuscles. **Cloacal muscles** (mm. cloacales) allow the cloaca to expand for copulation, egg laying and defaecation. A **muscular sphincter** (m. sphincter cloacae) surrounds the vent (Figure 6.44).

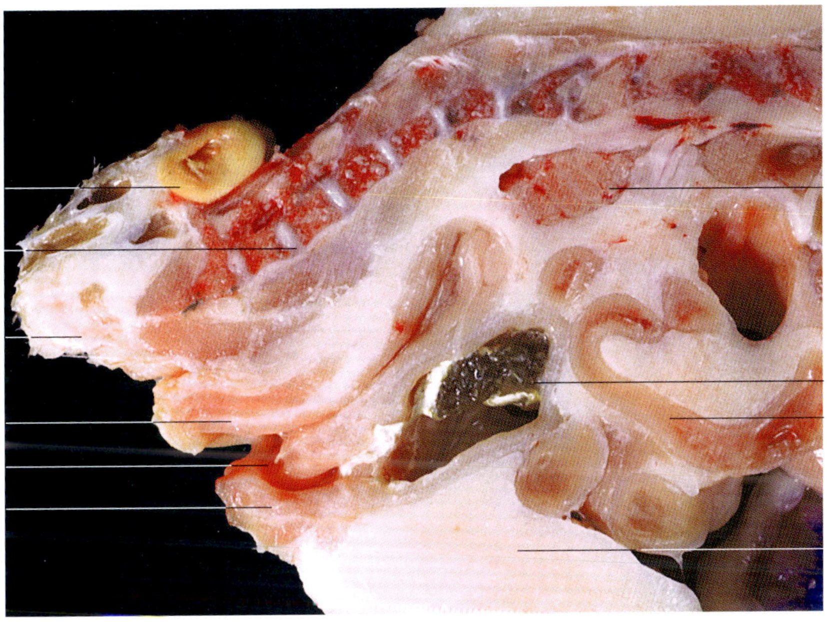

6.44 Longitudinal section of the cloaca with organs of the caudal body cavity of a chicken. Courtesy of Professor Dr J. Ruberte, Barcelona.

Glands associated with the alimentary canal

As in mammals, the intestine is connected to the liver (the largest gland in the body) and the pancreas (Figures 6.45ff.), both of which are derived from a common embryonic region, the hepatopancreatic ring.

Liver (hepar)

The liver of chickens is conspicuously large. Surrounded by the hepatic peritoneal sac, it covers a large portion of the median trabecula of the sternum and its sides are in contact with the sternal ribs (Figures 6.45, 6.48 and 6.49). The cranioventral segments of both hepatic lobes surround the

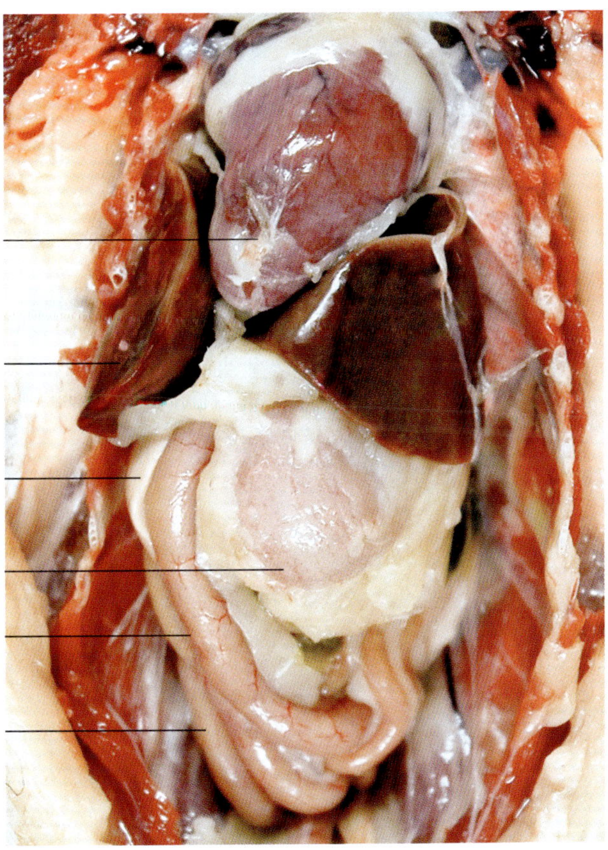

6.45 Topography of the body cavity of a common buzzard (Buteo buteo) including the heart, both lobes of the liver, the ventriculus and the duodenal loops. As shown here, pericardial fat and intra-abdominal fat depots are particularly well developed in wild predatory birds in late summer and autumn. Both serve as reserves for the winter months when prey is scarce. Endoscopy is contra-indicated in the presence of extensive fat deposition.

pericardium, forming a deep **cardiac impression** (impressio cardiaca) in the liver. The visceral surface of the liver is in contact dorsally with the lungs and, on the left, with the proventriculus, ventriculus and spleen. On the right, the visceral surface is associated with the duodenum. In males, the right caudal margin lies adjacent to the right testicle. The caudal vena cava passes through the right lobe of the liver.

The colour of the liver of adult animals is red-brown to light brown. Its consistency varies from soft (chicken and pigeon) to firm (duck and goose). At hatching, the liver has a yellow hue, resulting from carotenoid pigments in egg yolk lipids that enter the liver in the final days of incubation.

Cranial and caudal incisures (incisura interlobaris cranialis, incisura interlobaris caudalis) divide the liver into a **left lobe** (lobus sinister hepatis) and a **right lobe** (lobus dexter hepatis) (Figures 6.46 and 6.47). The parenchymal bridge connecting the two lobes is the pars interlobaris. Caudal to the hepatic porta, small intermediate (also referred to as dorsal) processes project from the left and right lobe. These vary according to species (right intermediate process in ducks and geese; left intermediate process in chickens, ducks and geese). A papillary process may also be present (absent in pigeons).

The left lobe is further divided into caudodorsal and caudoventral (also referred to as medial and lateral) parts.

Porta hepatis

In contrast to mammals, the hepatic porta receives its **nutritional blood supply** from two vessels, the left and right hepatic arteries (a. hepatica sinistra, a. hepatica dextra), which arise from corresponding branches of the **coeliac artery** (a. coeliaca) (Figures 6.46 and 6.47).

The **functional supply** to the liver also differs from that of mammals, consisting of **two (right and left) hepatic portal veins** (v. portalis hepatica dextra and v. portalis hepatica sinistra). Typically constituting the larger vessel, the right hepatic portal vein carries blood from the small intestine, caeca, rectum, cranial portion of the cloaca, pancreas and the spleen. The left, usually smaller vein receives blood from the vv. proventriculares and the vv. gastricae of the proventriculus and ventriculus (see Chapter 11 'Cardiovascular system').

Blood is drained from the left lobe, pars interlobaris and right lobe of the liver by, respectively, the **v. hepatica sinistra**, **v. hepatica media** and the **v. hepatica dextra**. Blood enters these vessels from the **central veins** (vv. centrales) of the **hepatic lobules** (lobuli hepatici). The efferent vv. hepaticae drain into the **caudal vena cava** (v. cava caudalis) shortly after it emerges from the right lobe on the parietal aspect of the liver.

Attachments of the liver

The surface of the liver is covered with a tunica serosa, underlain by a thin stratum fibrosum. Double-layered serous lamellae anchor the liver within the intestinal hepatic cavity. Additional fixation is provided by the **hepatic ligaments** (ligamenta hepatica; extensions of the oblique septum), the **duodenohepatic ligament** (lig. duodenohepaticum) and the **falciform ligament** (lig. falciforme hepatis) (Figure 5.5).

Gallbladder (vesica fellea)

The gallbladder lies on the visceral surface of the right lobe of the liver. It is absent in most species of pigeon and parrot.

Bile from each liver lobe is drained by a **hepatic duct** (ductus hepaticus dexter, ductus hepaticus sinister). The two ducts pass towards the hepatic porta and unite to form the **common hepatoenteric duct** (ductus hepaticoentericus communis). This continues to the duodenum (Figures

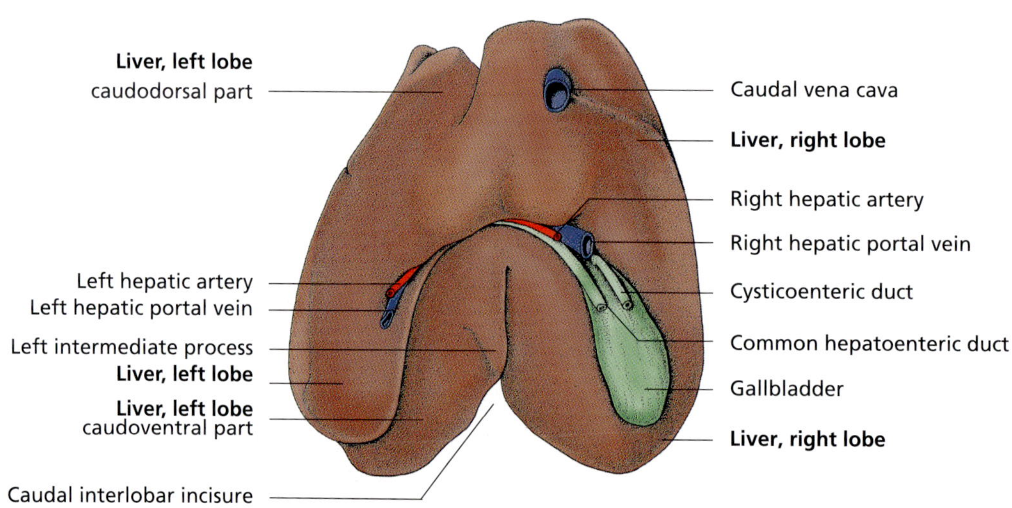

6.46 Visceral surface of the liver with hepatic porta in the chicken (schematic), adapted from Vollmerhaus and Sinowatz, 2004.

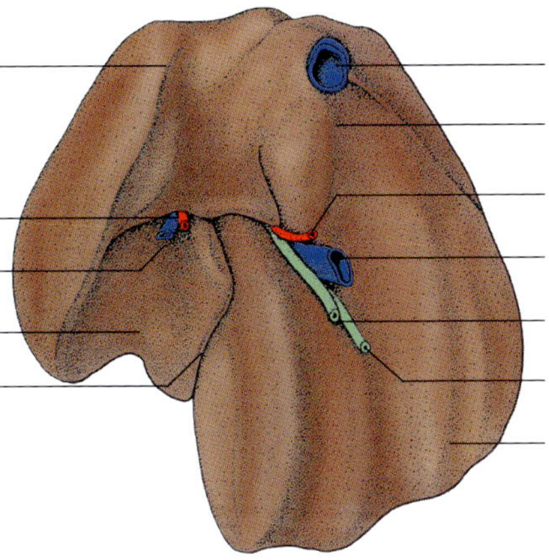

6.47 Visceral surface of the liver with hepatic porta in the pigeon (schematic), adapted from Vollmerhaus and Sinowatz, 2004.

6.46 and 6.47) and is the functional equivalent of the common bile duct of mammals. In Galliformes (e.g., chickens) and Anseriformes (e.g., ducks), in which the gall bladder is present, the right hepatic duct sends a branch, the **hepatocystic duct** (ductus hepatocysticus), to the gallbladder. Bile is thence carried by the **cysticoenteric duct** (ductus cysticoentericus) to the duodenum (Figure 6.26).

In birds lacking a gallbladder, the branch of the right hepatic duct opens directly into the duodenum as the right hepatoenteric duct.

The structure of the wall of the gallbladder is generally similar to that of mammals.

Pancreas

The pancreas lies within the mesoduodenum between the two limbs of the duodenum (Figures 6.25, 6.29 and 6.35). It is usually pale yellow to pink in colour, reaching lengths of up to 140mm in chickens, ducks and geese, and up to 80mm in pigeons. The pancreas consists of three lobes:

- dorsal lobe (lobus pancreatis dorsalis),
- ventral lobe (lobus pancreatis ventralis),
- splenic lobe (lobus pancreatis lienalis).

The secretory product of the **exocrine component** of the pancreas (Figure 6.51) empties into the ascending duodenum by up to three ducts, the **ductus pancreaticus dorsalis, ventralis** and **tertius**. As is the case in mammals, the exocrine component of the pancreas constitutes one of only two serous glands in birds, the other being the parotid salivary gland.

The endocrine tissue of the pancreas consists of pancreatic islets, or islets of Langerhans (Figure 6.51). As in mammals, the islets produces glucagon (A cells), insulin

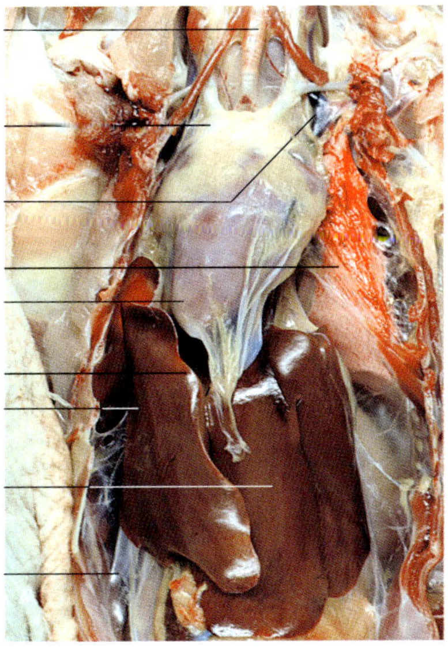

6.48 Liver with heart and pericardium in a chicken (ventral view).

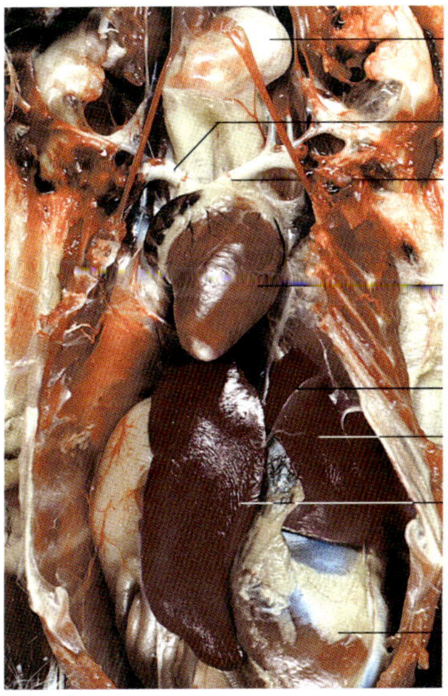

6.49 Liver, stomach and heart of a male Indian runner duck (ventral view).

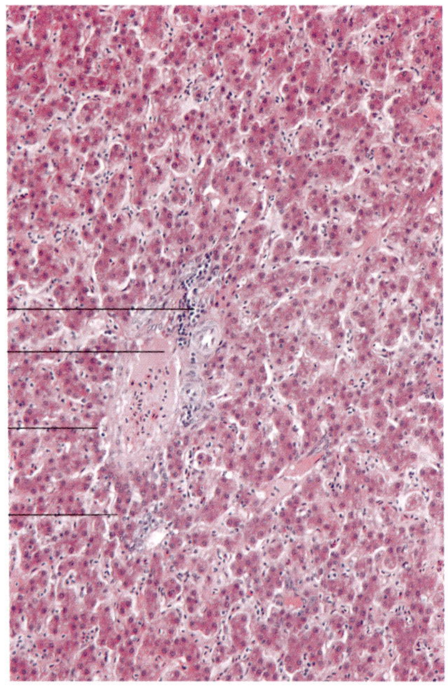

6.50 Histological section of the liver of a chicken with sparse interlobular fibrous tissue.

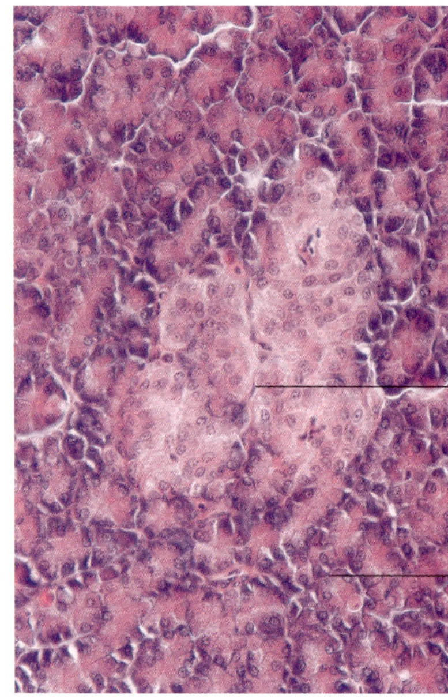

6.51 Histological section of the pancreas of a chicken.

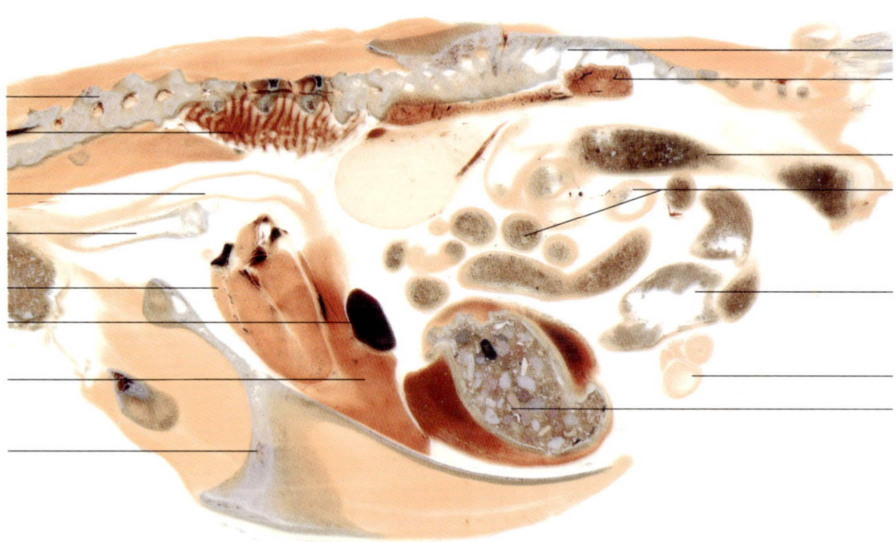

6.52 Sheet plastinate of the body cavity of a chicken (right paramedian section, viewed from the left). Courtesy of Dipl.-Biol. Martin Kobienia, Munich.

(B cells) and the inhibitory hormone somatostatin (D cells). A further cell type (PP- or F-cell) produces pancreatic polypetide. The islets of Langerhans are largest and most numerous in the splenic lobe.

Clinical aspects

The preferred method for clinical examination of the digestive system of birds is **contrast radiography** (Figures 6.54ff.; see also below). Other imaging modalities such as computed tomography (see Chapter 19 'Imaging techniques') are also used. Contrast radiography is particularly useful for detecting abnormalities in the **time taken for contrast material to pass into the distal gastrointestinal segments** (e.g., increased transit time associated with gastroparesis in parrots, Figures 6.68 and 6.69; reduced transit time due to foreign body obstruction, Figures 6.62 to 6.65).

In budgerigars (Melopsittacus undulatus), the colour of the cere at the base of the beak can be used in determining the sex of this otherwise monomorphic (lacking visible external sexual dimorphism) species. Predominantly blue in males, the cere of females is brown due to superficial keratisation. Although widely used among bird-owners and veterinarians, this method is not infallible. Its accuracy is approximately 80 per cent in blue-toned birds (so-called

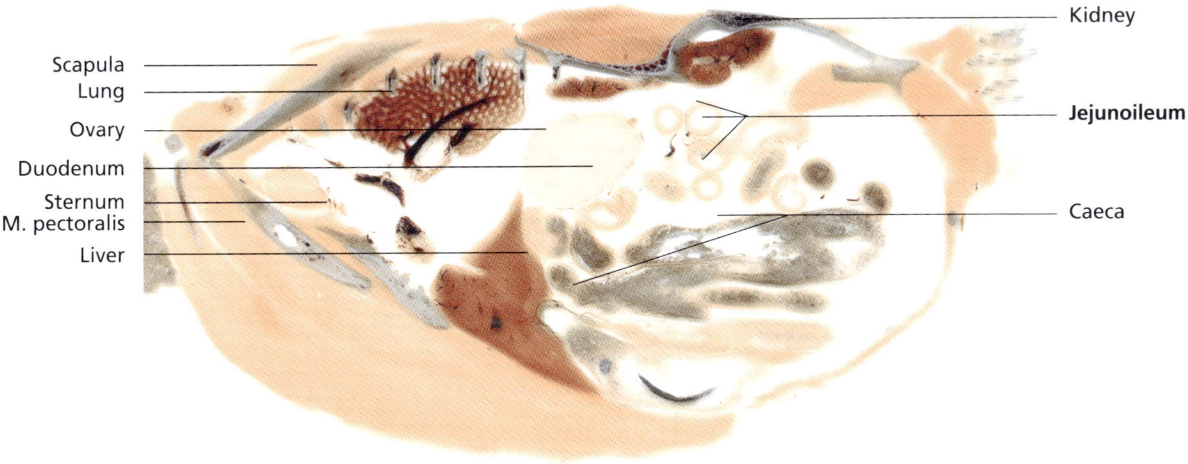

6.53 Sheet plastinate of the body cavity of a chicken (right median section, viewed from the left). Courtesy of Dipl.-Biol. Martin Kobienia, Munich.

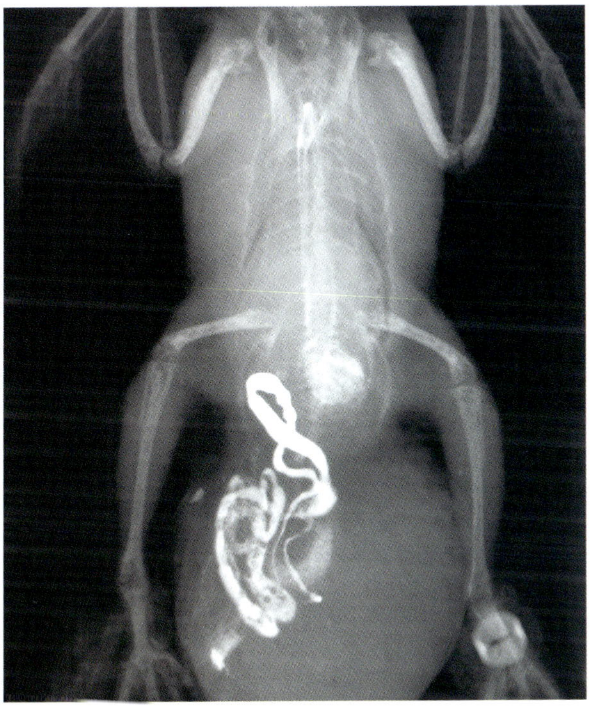

6.54 Contrast radiograph (ventrodorsal view) of a high grade abdominal hernia with prolapse of the intestinal loops resulting from tumour-induced hyperoestrogenism in a black-cheeked lovebird (Agapornis nigrigenis).

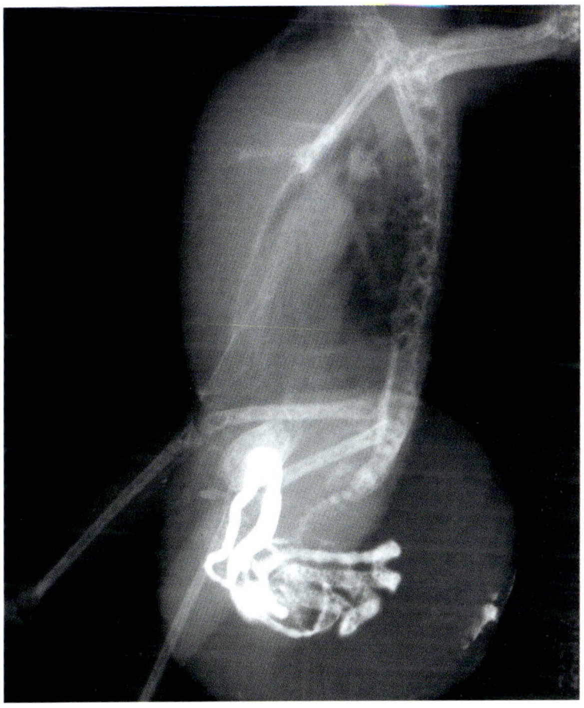

6.55 Contrast radiograph (lateral view) of an abdominal hernia in a black-cheeked lovebird (Agapornis nigrigenis; see Figure 6.54) with complete prolapse of the intestinal loops.

'opaline' budgerigars) and only 60 per cent in yellow birds ('lutinos').

In male budgerigars, a **change in the colour of the cere** from blue to brown is suggestive of hyperoestrogenism induced by a **Sertoli cell tumour** (testicular tumour).

In older females, hormonally induced hyperkeratosis of the cere, with possible horn formation, is not uncommon. A **pumice-like hyperkeratotic cere** with multiple macroscopically visible bore-holes is a characteristic sign of infestation with **scaly face mite** (Cnemidocoptes pilae).

Beak trimming (trimming of the tip of the upper beak) is still sometimes used in poultry production to reduce feather pecking and cannibalism. However, this practice is highly controversial and is limited to trimming of the horny coat (rhamphotheca) of the tip of the beak of chickens and the soft surface of the bill of ducks. In some jurisdictions the procedure is prohibited.

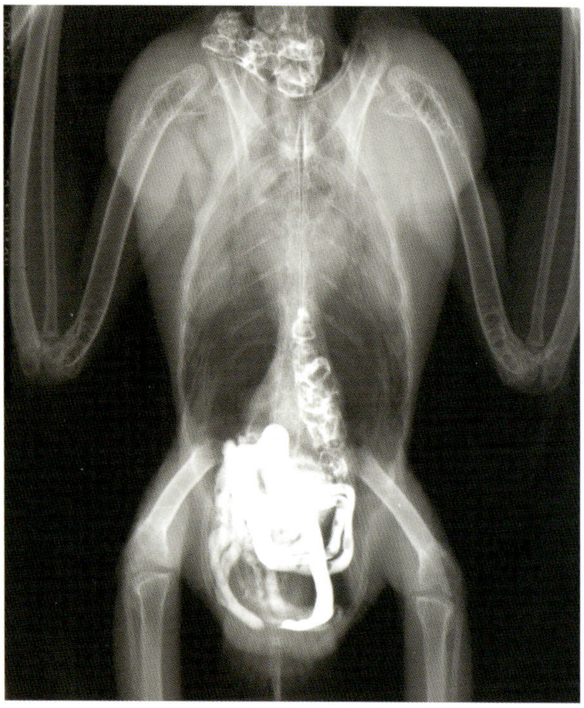

6.56 Contrast radiograph (ventrodorsal view) showing the anatomical relationships of the digestive tract in a white cockatoo (Cacatua alba) two hours after administration of contrast material. Remnants of contrast material are visible in the crop and proventriculus after filling of the ventriculus and intestinal loops.

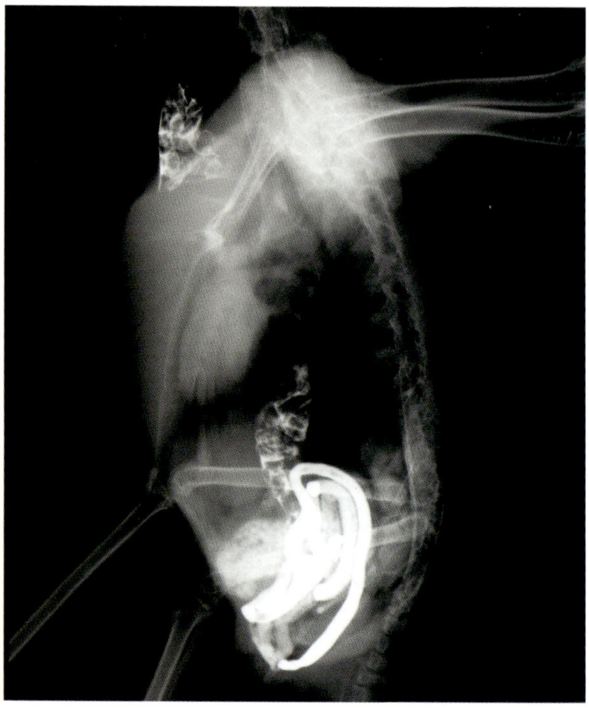

6.57 Contrast radiograph (lateral view, see Figure 6.56) showing the anatomical relationships of the digestive tract in a white cockatoo (Cacatua alba). Contrast material has entered the distal intestinal segments, with remnants still visible in the ventriculus.

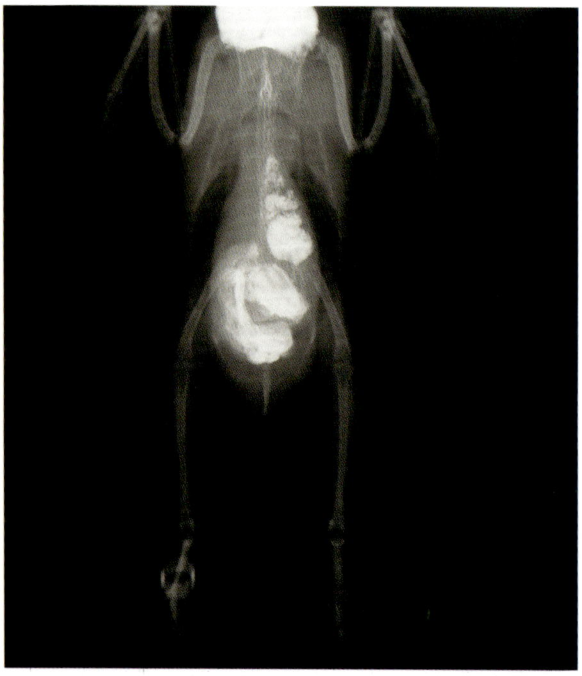

6.58 Contrast radiograph (ventrodorsal view) of a budgerigar (Melopsittacus undulatus) with 'megabacteriosis' (mycotic infection caused by Macrorhabdus ornithogaster). Typical interruption of the stream of contrast material is evident in the isthmus between the proventriculus and ventriculus.

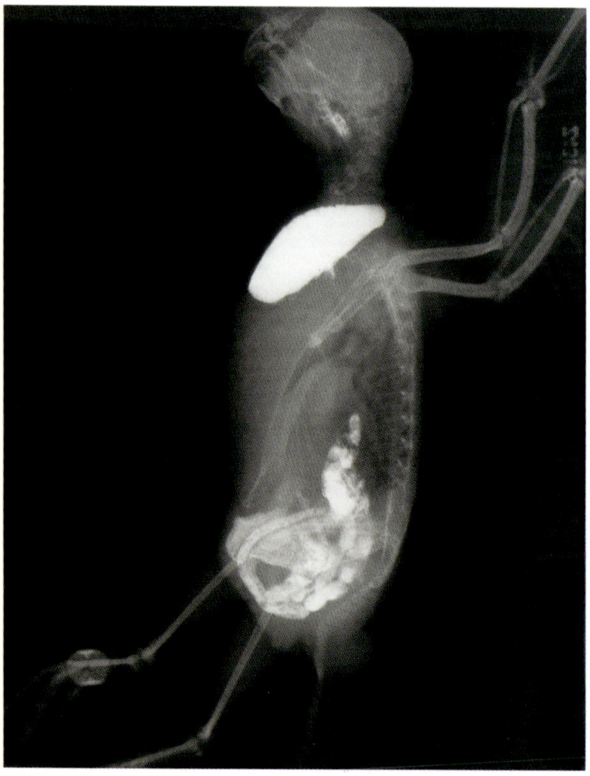

6.59 Contrast radiograph (lateral view) of a budgerigar (Melopsittacus undulatus) with 'megabacteriosis' (see Figure 6.58). The crop is massively distended with millet (surrounded by contrast material) as a result of polyphagia.

Given that the rhamphotheca of the chicken measures only a few millimetres in thickness (Figure 6.8), one can easily appreciate the difficulty associated with conducting beak trimming to acceptable animal welfare standards. In waterfowl, the leathery coating of the bill is even thinner and humane trimming is virtually impossible. The risk of damage to the highly sensitive bill tip organ, and the associated pain, render this procedure very difficult to justify.

Traumatic injuries involving **fractures of the upper beak** and underlying bone carry a poor prognosis, particularly with more proximally located lesions. In view of the complex and protracted healing process, and the usually adverse outcome in cases where appropriate fixation is not possible, such injuries are often an indication for euthanasia. The inability of the patient to feed by themselves during the healing period and the impracticality of long-term supported feeding (e.g., due to capture stress) represent additional complicating factors.

The connection between the nasal cavity and the oropharynx (choana) can be utilised in small birds for administering medication by dropper via the nostrils, reducing the likelihood of stress-induced adverse reactions.

In large parrots, the presence of visible masses between the mandibles, with associated malposition of the tongue, is a characteristic sign of **metaplastic enlargement of the sublingual glands**, typical of **vitamin A deficiency**.

The normal anatomical relationships of the components of the intestinal tract are shown in Figures 6.56 and 6.57. Two imaging planes are required for representation of spatial relationships. Displacement or **prolapse of the intestinal loops** (abdominal hernia; see Figures 6.54 and 6.55) can result from space occupying lesions in the body cavity or from hormonal imbalance; for example, hyperoestrogenism. In ornamental cage birds, raptors and pigeons, common infectious diseases of the intestinal tract include **bacterial and mycotic infections (candidiasis)** of the crop as well as parasitism (e.g., trichomoniasis; see Figures 6.66 and 6.67). **Dilatation of the stomachs** and the proximal duodenum is observed in association with proventricular dilatation disease (neuropathic gastric dilatation) in parrots (Figures 6.68 and 6.69). This is one of the most important diseases of this group of birds.

Hourglass-shaped constrictions or interruptions of the stream of contrast material in the **region of the isthmus** between the proventriculus and ventriculus are characteristic of a mycotic infection (Macrorhabdus ornithogaster). Particularly common in small parrots, this disease was previously erroneously referred to as 'megabacteriosis' (Figures 6.58 and 6.59).

Disease associated with **gastrointestinal foreign bodies** occurs in both captive (Figures 6.62 and 6.63) and wild birds (Figures 6.64 and 6.65). As well as causing mechani-

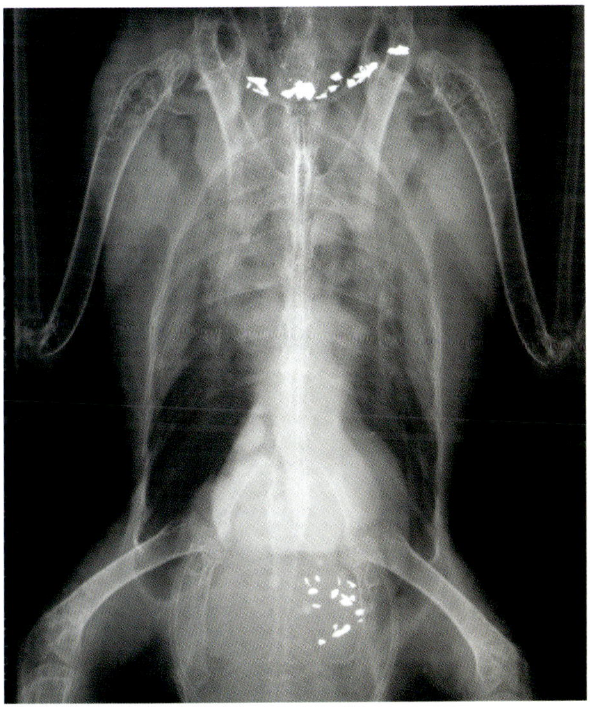

6.60 Radiograph (ventrodorsal view) of an African grey parrot (Psittacus erithacus) with lead intoxication. Lead particles (curtain weight cord, components of a Tiffany lamp and a stained glass frame) are present in the crop and ventriculus.

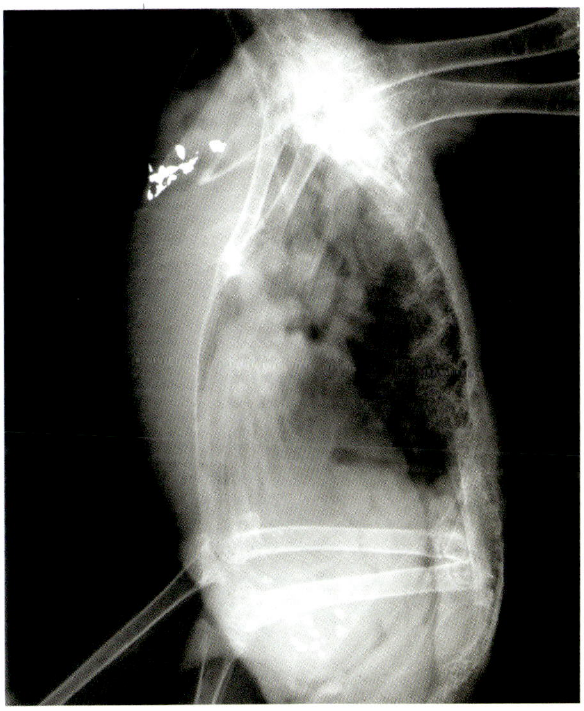

6.61 Radiograph (lateral view) of an African grey parrot (Psittacus erithacus) with lead intoxication (see Figure 6.60). Signs indicative of intoxication include atony and dilatation of the intestinal loops in the caudal body cavity as well as marked enlargement of the kidneys.

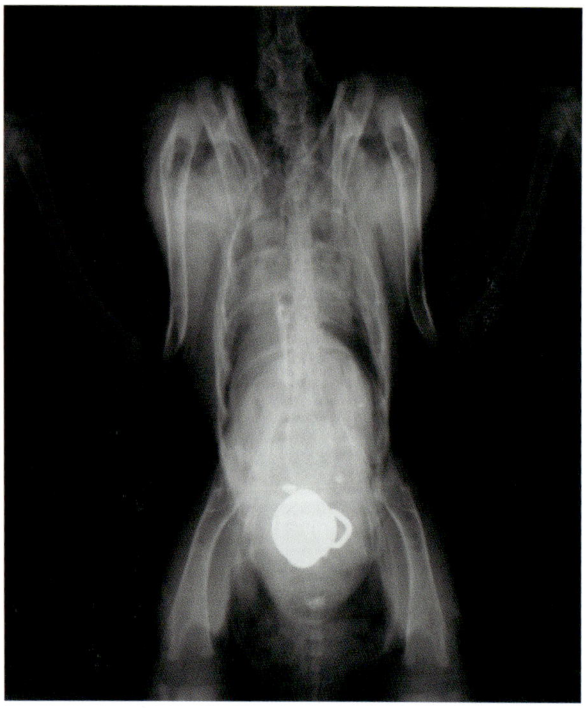

6.62 Radiograph (ventrodorsal view) of a tame Indian peafowl chick (Pavo cristatus). A foreign body is present in the gizzard. The chick had swallowed a metal bell attached to the wrapping of a chocolate Easter bunny.

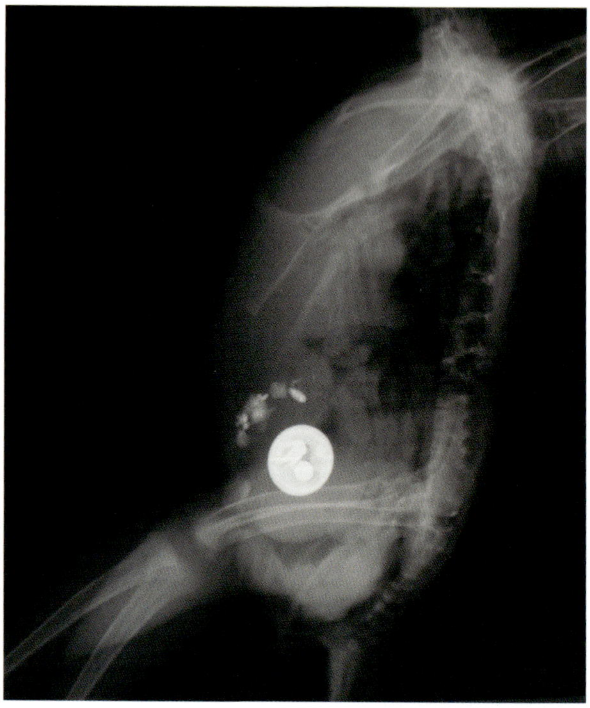

6.63 Radiograph (lateral view) of an Indian peafowl chick (Pavo cristatus; see Figure 6.62). The size of the foreign body in the ventriculus necessitates surgical removal.

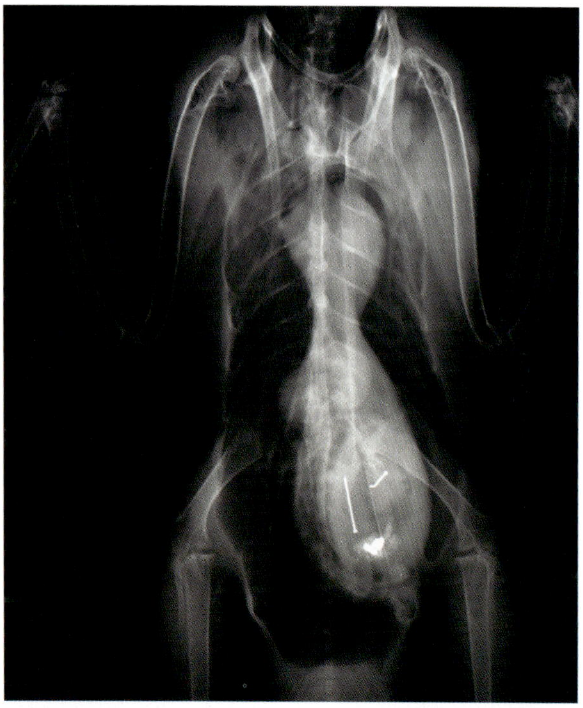

6.64 Radiograph (ventrodorsal view) of a runner duck (Anas platyrhynchos domesticus). Needles are visible in the proventriculus and descending duodenum.

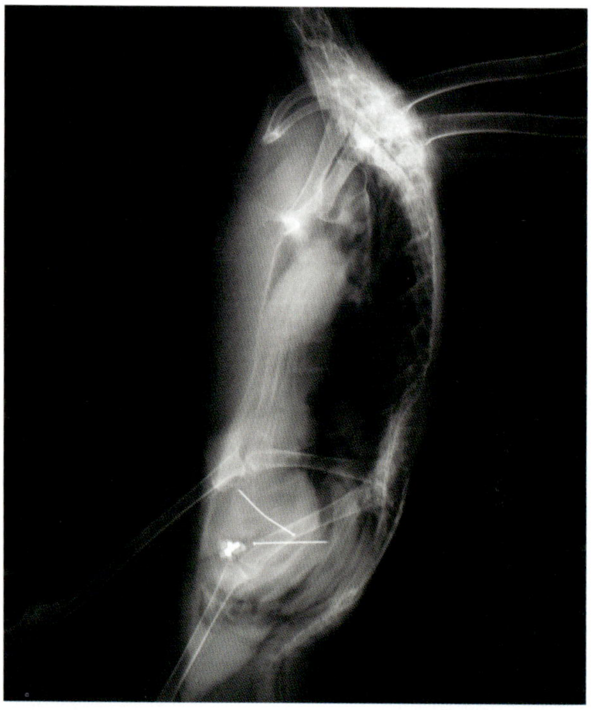

6.65 Radiograph (lateral view) of a runner duck (Anas platyrhynchos domesticus; see Fig 6.64). Lead pieces originating from fishing equipment are present in the ventriculus.

cal disturbances, foreign bodies can lead to rapidly fatal intoxications, particularly when they contain lead or zinc (Figures 6.60 and 6.61).

A pathognomonic finding in **pancreatic disease** of budgerigars is the excretion of light ochre coloured faeces that dry rapidly to resemble a solid foam. This results from inadequate digestion of protein.

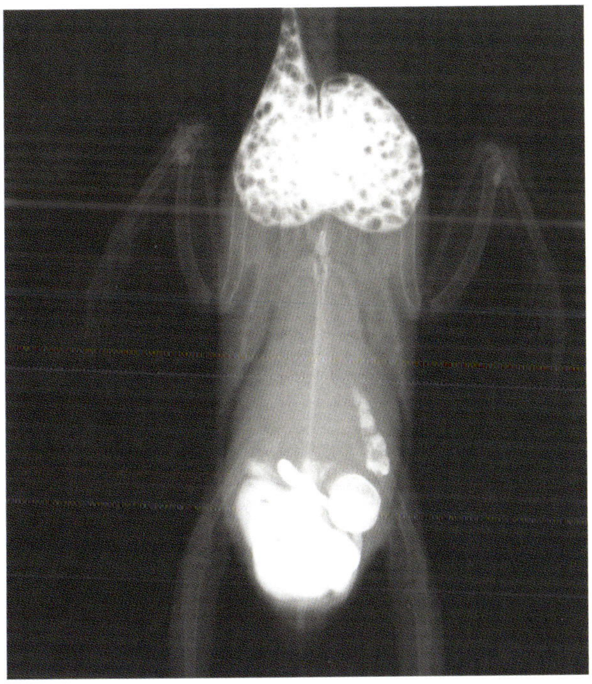

6.66 Contrast radiograph (ventrodorsal view) of a budgerigar (Melopsittacus undulatus) with trichomoniasis. The crop is dilated and contains gas accumulations.

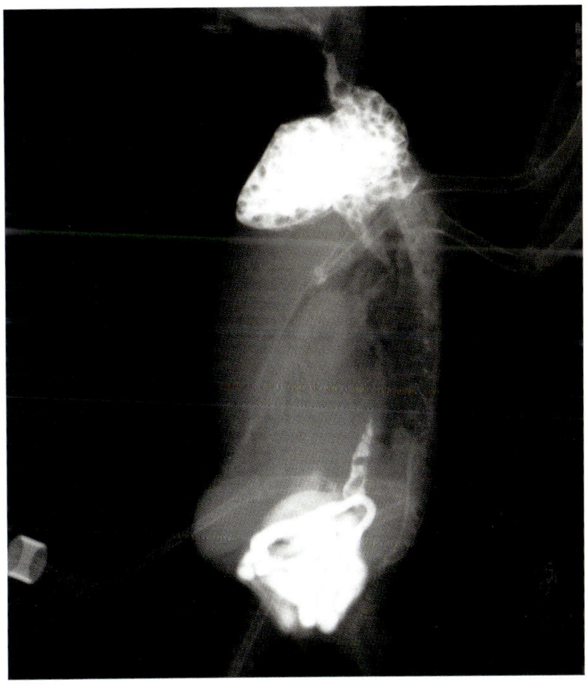

6.67 Contrast radiograph (lateral view) of a budgerigar (Melopsittacus undulatus) with trichomoniasis (see Figure 6.66). The crop is dilated and contains gas accumulations.

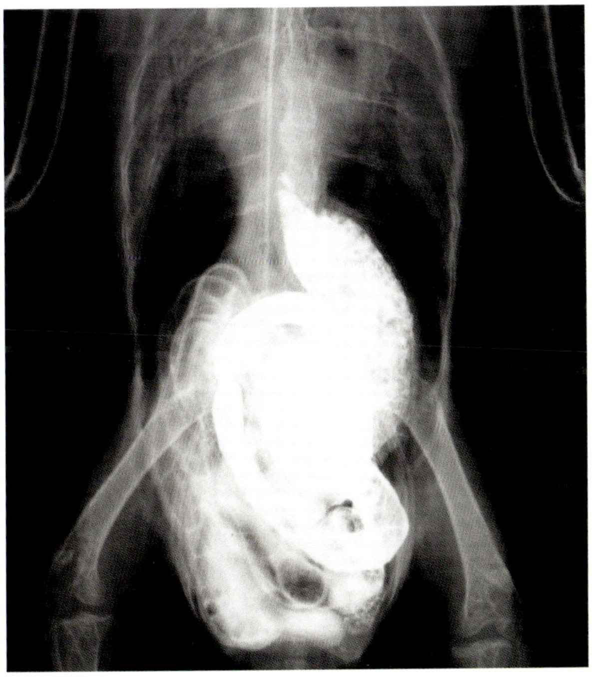

6.68 Contrast radiograph (ventrodorsal view) of an African grey parrot (Psittacus erithacus) with proventricular dilatation disease. The proventriculus and ventriculus are markedly distended as a result of damage to intramural nerve ganglia.

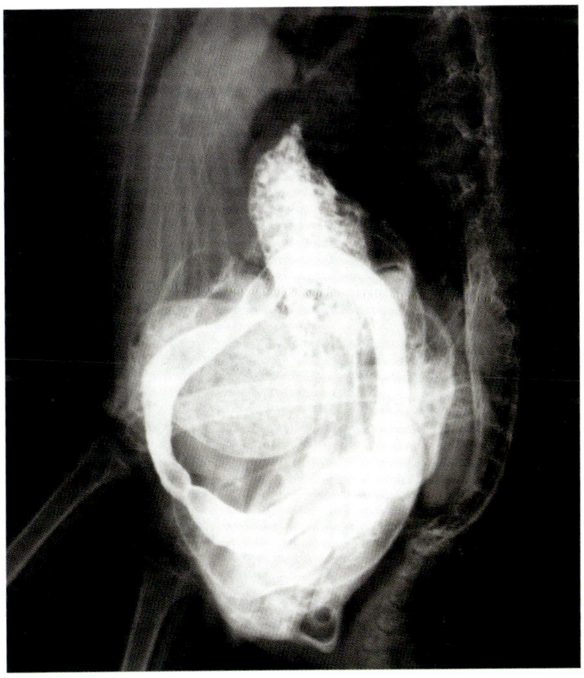

6.69 Contrast radiograph (lateral view) of an African grey parrot (Psittacus erithacus) with proventricular dilatation disease (see Figure 6.68), five hours after administration of contrast material. Dilatation of the stomachs has resulted in increased transit time.

7

Respiratory system (apparatus respiratorius)

H. E. König, M. Navarro, G. Zengerling and R. Korbel

The following features distinguish the respiratory system of birds from that of mammals:

- the presence of both a larynx and a syrinx,
- ossification of the tracheal rings,
- a relatively constant lung volume,
- the absence of pleura post-hatching and
- the presence of air sacs (Figure 7.1).

Nasal cavity (cavum nasi)

The nasal cavity is situated to the left and right of the median **nasal septum** (septum nasale).

The position of the **nostrils** (nares) varies considerably between species. In the chicken they are located at the base of the beak. The nares can also be surrounded by feathers and may be tubular in structure.

In some species, including the chicken and turkey, a cornified plate known as the **operculum** projects from the dorsal border of the nares. In pigeons, the operculum is covered by the fleshy **cere** (see Chapter 17 'Common integument'). The left and right ceres may coalesce dorsally and, in many breeds of pigeon, are quite pronounced. In the duck and goose, the nasal septum is perforated by a small opening at the level of the nares.

Most birds have three **nasal conchae** (conchae nasales) (Figure 7.2). In contrast to mammals, they are arranged in a rostrocaudal, rather than dorsoventral, sequence.

They are composed of the:

- rostral nasal concha (concha nasalis rostralis),
- middle nasal concha (concha nasalis media),
- caudal nasal concha (concha nasalis caudalis).

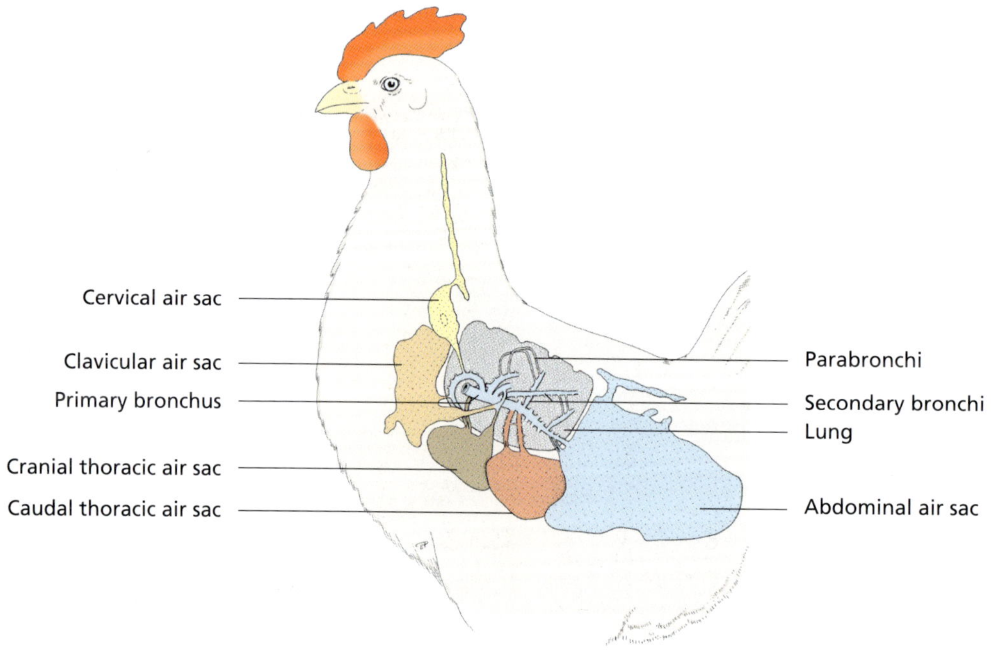

7.1 Relationship between the air sacs and the bronchial system in the chicken (schematic).

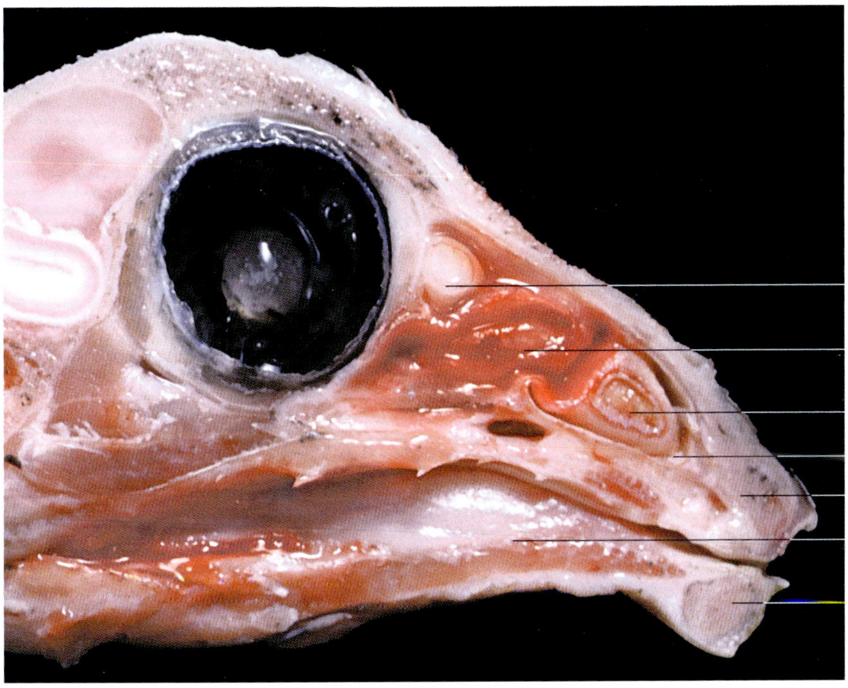

7.2 Nasal conchae of a chicken (paramedian section).

In the chicken, a **cartilaginous lamella** arises from the ventral border of the nostril, in front of the rostral nasal concha. The **nasolacrimal duct** (ductus nasolacrimalis) opens into the nasal cavity between the rostral and middle nasal conchae. Its course is described in Chapter 15 'The eye' (Figure 15.43).

The cranial portion of the **nasal cavity** (nasal vestibule; regio vestibularis) is lined with non-glandular mucosa. This transitions caudally, in the **respiratory region** (regio respiratoria), into a pseudostratified ciliated epithelium containing goblet cells (respiratory epithelium). In the **olfactory region** (regio olfactoria) the epithelium contains neurosensory cells (olfactory epithelium) (Figures 7.2 to 7.4). Its histological structure is similar to that of mammals.

In chickens and waterbirds, the typically yellowish **olfactory region** consists of a small circumscribed area on the caudal nasal concha and the caudal nasal septum. The avian olfactory mucosa, like the olfactory bulb, is usually limited in extent and function. Development of the olfactory apparatus is generally greater in fish- and meat-eating birds than in grain-eating species.

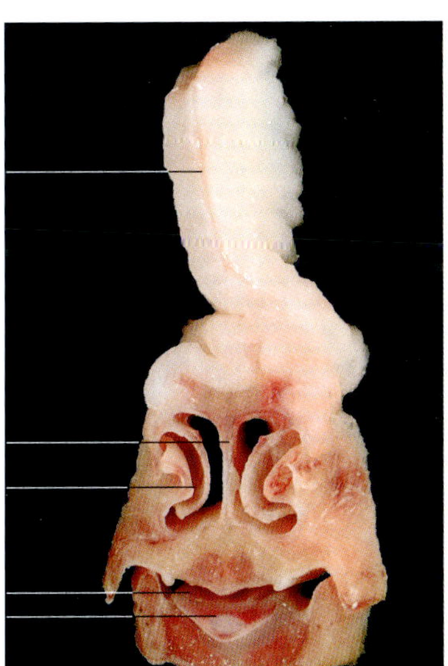

7.3 Transverse section of the nasal cavity of a chicken at the level of the rostral nasal concha.

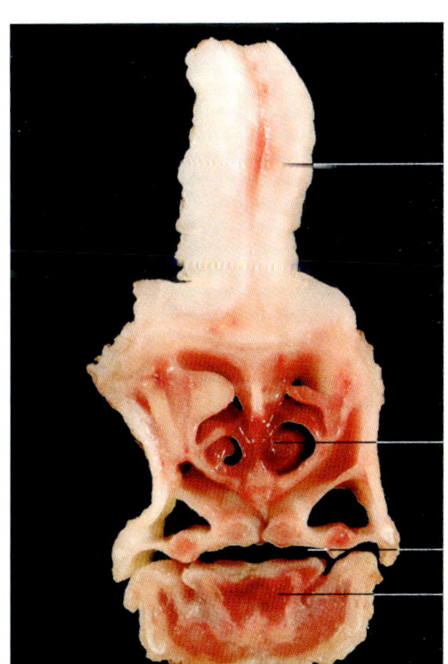

7.4 Transverse section of the nasal cavity of a chicken at the level of the middle nasal concha.

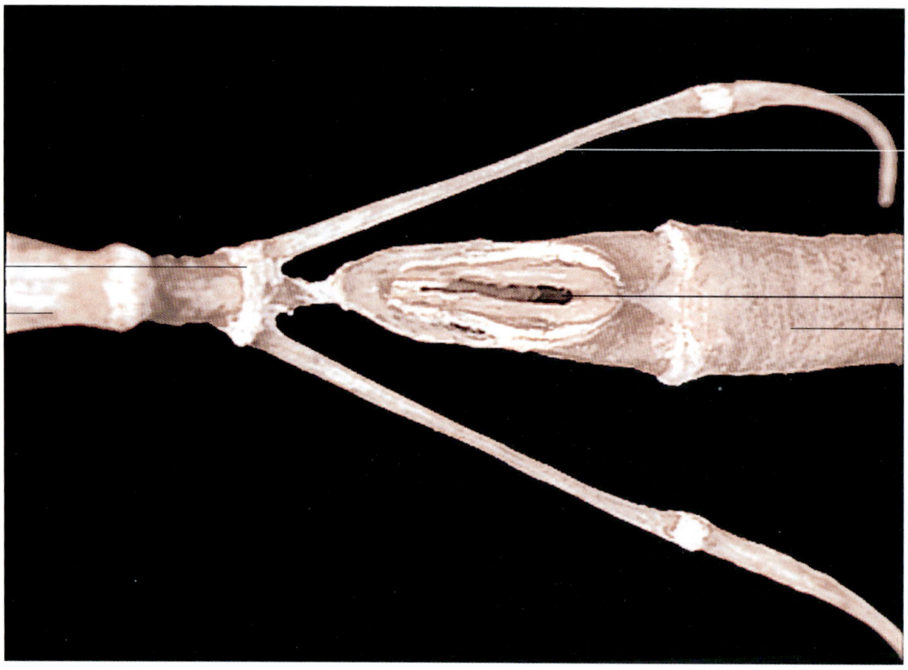

7.5 Anatomical relationships of the larynx, hyobranchial apparatus and tongue of a chicken.

The nasal cavity communicates with a single paranasal sinus. Termed the **infraorbital sinus** (sinus infraorbitalis), this occupies a relatively large, triangular space situated immediately under the skin, rostroventral to the eye. It is surrounded almost entirely by soft tissue. Near the nasal angle of the eye, the infraorbital sinus communicates with the cavity of the caudal nasal concha. The infraorbital sinus is clinically significant and can be accessed for paracentesis (see Chapter 18 'Clinical examination').

In the genus Ara, the sinus is extensively subdivided. Several pouches have been described, extending in front of, ventral to and behind the eye (pars pre-, infra- and postorbitalis), medial to the mandible and deep into the cervical region.

The nasal vestibule receives the secretions of the **nasal gland** (glandula nasalis; absent in pigeons), which serve to humidify the nasal opening. In most birds, the nasal gland consists of a **lateral** and **medial** lobe, each with its own duct (see Chapter 15 'The eye'). Only the medial lobe is present in the chicken, its caudal portion lying over the dorsal surface of the eyeball. The **duct** empties by a slit-like opening on the nasal septum near the rostral nasal concha. In sea birds, as in some coastal reptiles, the nasal gland secretes a concentrated salt solution.

Larynx

The larynx (Figures 7.5 to 7.7) presents as a conspicuous mound in the ventral oropharynx, caudal to the tongue. Two rows of caudally directed conical **papillae pharyngeales** line its caudal margin. A longitudinally oriented slit-like **laryngeal opening**, or **glottis**, is located in the midline of the laryngeal mound (Figure 7.6). The larynx is supported by the **laryngeal cartilages** (cartilagines laryngis) (Figure 7.7). These are the:

- cricoid cartilage (cartilago cricoidea),
- procricoid cartilage (cartilago procricoidea),
- arytenoid cartilage (cartilago arytaenoidea).

The **cricoid cartilage** is shaped like a 'sugar scoop' (Figure 7.7). Caudally, the two wing-like ends of the 'scoop' curve dorsally and articulate in the dorsal midline with the small, median comma-shaped **procricoid cartilage**. The **arytenoid cartilage** is paired. Its shape resembles a tuning fork with its tines directed caudally. The ventral tine, the body of the arytenoid, articulates with the procricoid.

7.6 Glottis of a common buzzard (Buteo buteo), Courtesy of Professor Dr Daniel Gonzalez-Acuna, Chillan, Chile.

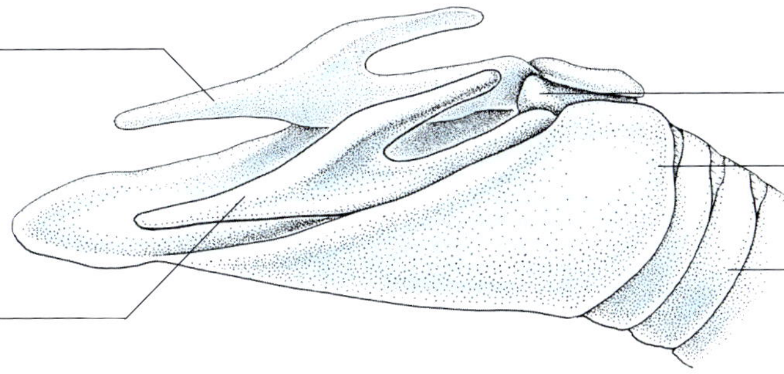

7.7 Laryngeal cartilages of the chicken (schematic), adapted from Ghetie, 1976.

Two laryngeal muscles, the **m. dilatator** and the **m. constrictor glottidis**, are responsible for opening and closing the glottis. Elevation of the larynx is achieved by the action of the m. cricohyoideus, while contraction of the mm. tracheales draws the larynx ventrally. The primary function of the larynx is to prevent access of foreign matter to the deeper airways through reflex closure of the glottis. It does not contribute to phonation.

Trachea

The trachea begins at the caudal end of the cricoid cartilage. In the chicken, the upper portion is located in the midline of the cervical region. Its course then continues, together with the oesophagus, on the right side of the neck. The trachea regains its median position upon entering the thoracic inlet (Figure 7.9). In some species, such as swans, cranes, spoonbills and birds of paradise, the trachea is particularly long and wound into coils that lie between the skin and the breast muscles, or within the sternum itself.

The trachea is supported by a series of cartilaginous rings. Except in pigeons, these tend to become ossified (Figures 7.7ff.). The **tracheal rings** resemble a signet ring, with the expanded portion alternately forming the left and right half of each subsequent ring (Figure 7.8). Considerable species variation exists in the number of tracheal rings (120 in the chicken). The rings gradually decrease in diameter towards the body cavity.

The **tracheal mucosa** is lined by pseudostratified ciliated epithelium including tall columnar cells, narrow basal cells and goblet cells. Functioning as small intra-epithelial glands, the mucin secreting goblet cells form small crypts in the mucosa. The epithelium is underlain by a lamina propria containing seromucous glands, lymphoid follicles and diffuse lymphoreticular tissue.

The band-like **tracheal muscles** (mm. tracheales) extend along the length of the trachea. These are the:

- m. tracheolateralis,
- m. cleidohyoideus,
- m. sternotrachealis and
- m. cleidotrachealis.

The **m. tracheolateralis** arises from the syrinx and passes along the lateral trachea to the cricoid cartilage.

Beginning at the clavicle, the **m. cleidohyoideus** courses cranially and attaches to the cricoid cartilage and the hyobranchial apparatus. The **m. sternotrachealis** has its origin at the craniolateral process of the sternum and inserts laterally at the caudal end of the trachea. Its functional continuation is the **m. tracheolateralis** (Figure 7.9).

The **m. cleidotrachealis** (formerly the m. ypsilotrachealis) arises from the clavicle and inserts on the trachea, cranial to the site of attachment of the m. sternotrachealis.

Due to its length, the internal air resistance of the trachea is relatively high. This is counteracted by its comparatively large diameter. Therefore, relative to body weight, the air resistance of the trachea of birds and mammals is similar. However, the dead space within the avian trachea is approximately four times that of a mammal of comparable size. Birds thus have a much lower respiratory rate than mammals.

Syrinx

The syrinx is located at the level of the bifurcation of the trachea into the primary bronchi (see below) (Figures 7.8

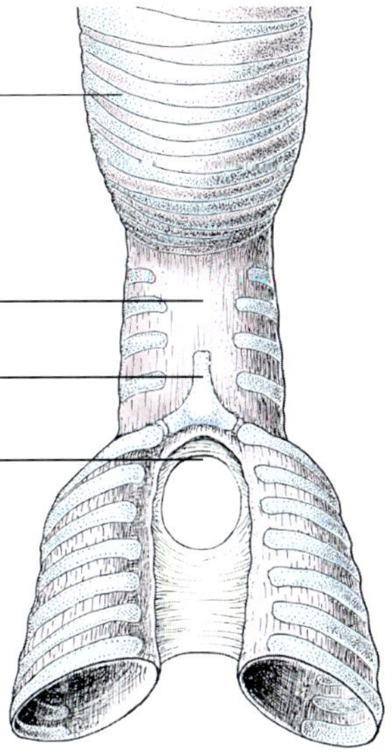

7.8 Syrinx of the chicken (schematic), adapted from Ghetie, 1976.

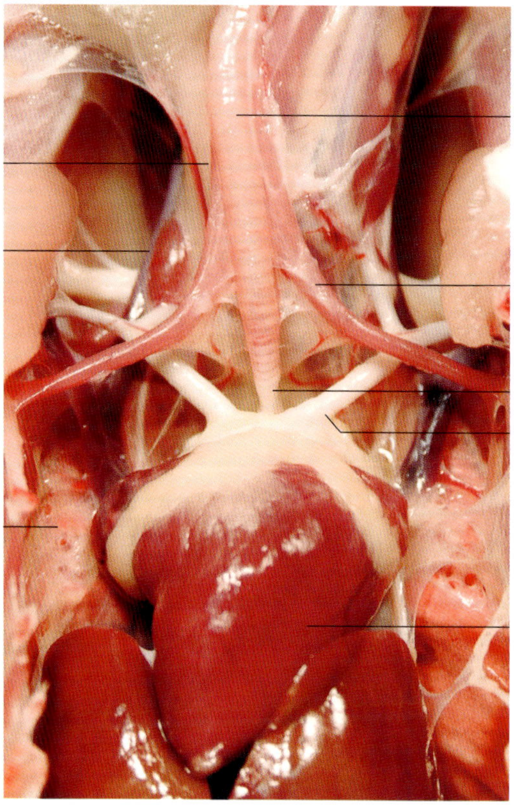

7.9 Anatomical relationships of the trachea and syrinx in a chicken.

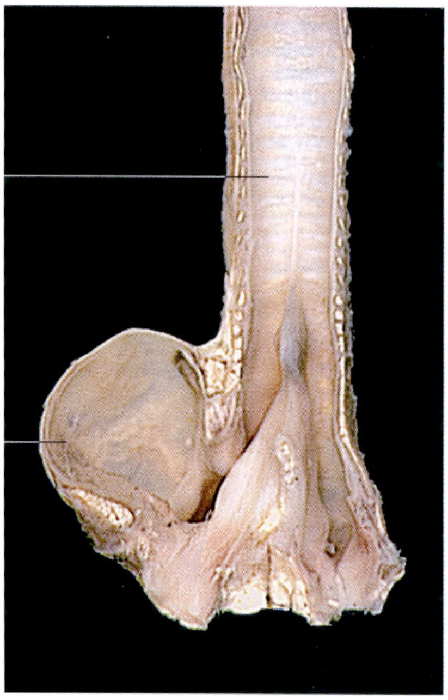

7.10 Trachea and syrinx of a male Indian runner duck (opened).

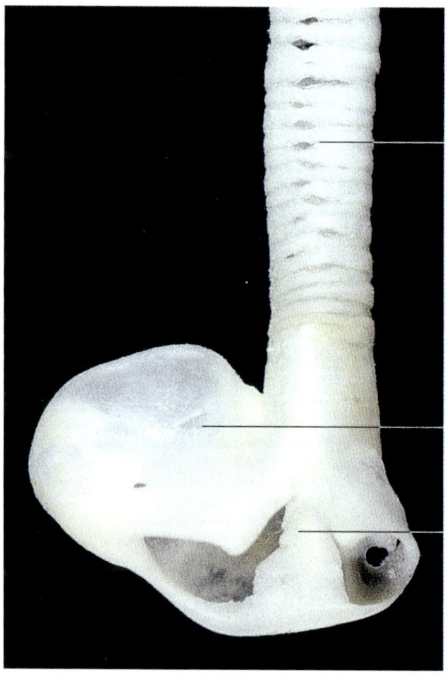

7.11 Trachea and syrinx of a male Indian runner duck (dorsal view).

to 7.11). In the chicken, the last four tracheal rings are considered to be part of the syrinx. The subsequent rings are no longer complete. Instead, they are joined at one or both ends to a median bridge known as the **pessulus**. Extending cranially from the pessulus is a mucosal fold, the **membrana semilunaris**.

Together, the cartilaginous components of the syrinx form the **tympanum**. Left and right **lateral tympaniform membranes** (membrana tympaniformis lateralis) extend from the tympanum to the lateral side of the bronchial cartilages. Paired **medial tympaniform membranes** (membrana tympaniformis medialis) pass from the pessulus to the (incomplete) medial aspect of the bronchial cartilages.

Elastic connective tissue pads known as **labia** project from the membranes into the lumen of the syrinx. During phonation, the membranes and labia function in a similar manner to the **vocal folds** of the mammalian larynx. **Syringeal muscles** are present in song birds and **absent** in domestic poultry.

In males of various breeds of duck, the syrinx is profoundly modified by the presence of a dilated compartment, the bony **syringeal bulla** (bulla syringis), that extends from its lateral side (Figures 7.10 and 7.11). The bulla is divided into a large and a small cavity and is believed to act as a **resonance chamber**.

Lung (pulmo)

The **left and right avian lungs** (pulmo sinister and pulmo dexter) occupy a dorsal position, either side of the vertebral column. They are **not lobed** (Figures 7.12ff.). The ribs are deeply embedded in the dorsomedial portion of the lungs, forming distinctive **impressions** (sulci costales) that separate the lung tissue into segments known as **tori intercostales** (Figures 7.13 and 7.14).

The cranial margin of the lungs is located at the level of the first rib. Caudally, the lungs extend to or beyond the last rib. On its ventral surface, or **facies septalis**, each lung

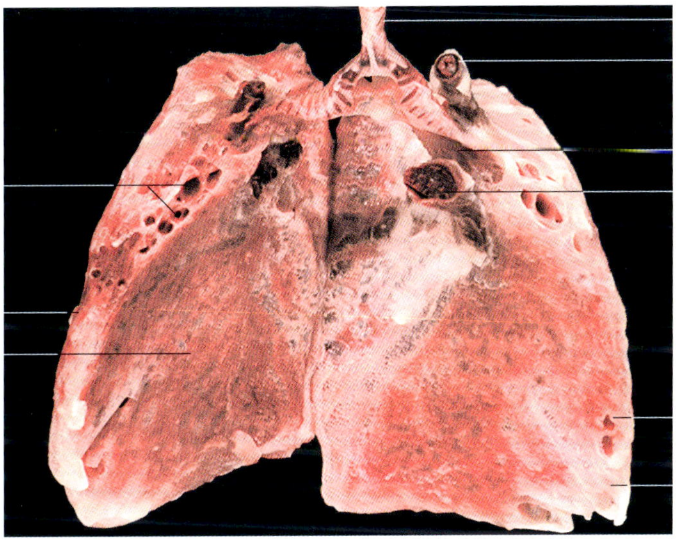

7.12 Right and left lung of a chicken with pulmonary hilus (ex situ, ventral view). Courtesy of PD Dr J. Maierl, Munich.

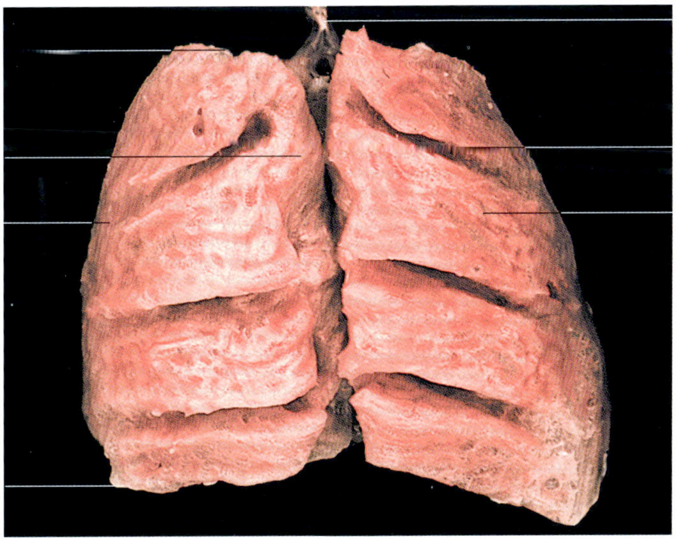

7.13 Right and left lung of a chicken (ex situ, dorsal view). Courtesy of PD Dr J. Maierl, Munich.

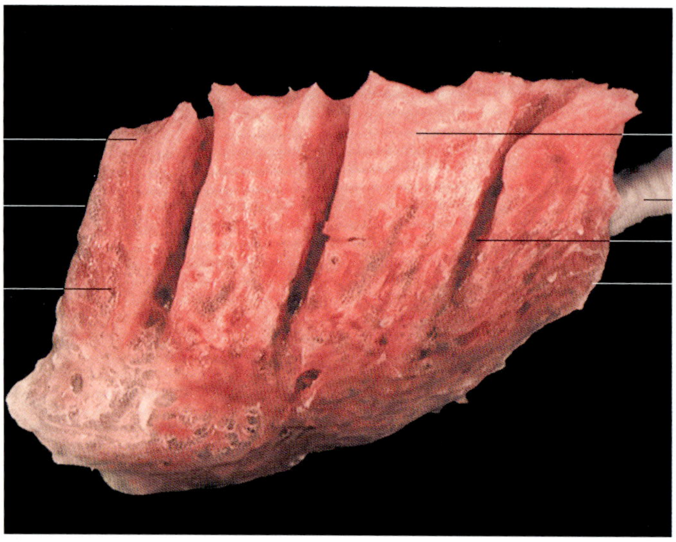

7.14 Right lung of a chicken (ex situ, lateral view). Courtesy of PD Dr J. Maierl, Munich.

has a **hilus** (hilus pulmonalis) through which blood vessels and the primary bronchi enter. The ventral surface is fused with the **horizontal septum** (see Chapter 5 'Body cavities') and contains openings that communicate with the air sacs (Figure 7.12).

The lung of the chicken is approximately rectangular, while that of the goose and duck is more triangular in shape. Apart from small embryonic remnants, the lung has **no associated pleura**.

Connective tissue attachments to the **ribs**, **vertebrae** and the **horizontal septum** prevent the lung from collapsing (Figures 7.15 and 7.16). The thin walled bronchi and air capillaries remain permanently open to the passing air. Relative to body weight, the lungs of birds are similar in weight to those of mammals. However, the **volume** of the avian lung is just one-tenth that of a comparably sized mammal.

Bronchial system and gas exchange

The divisions of the **bronchi** are as follows:

- two primary bronchi (bronchi primarii),
- secondary bronchi (bronchi secundarii),
- parabronchi and
- air capillaries (pneumocapillares).

The **primary bronchi** are also referred to as **first-order bronchi**. They penetrate the **horizontal septum** and pass through the lung to its caudal margin, where they open into the abdominal air sacs. The walls of the primary

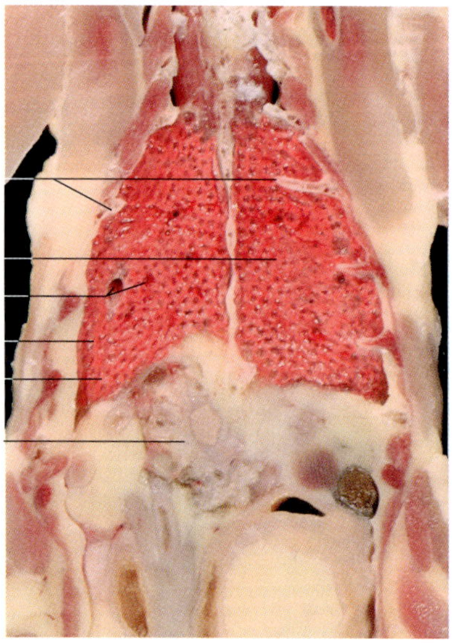

7.15 Horizontal section of a chicken, ventral to the vertebral column (dorsal view).

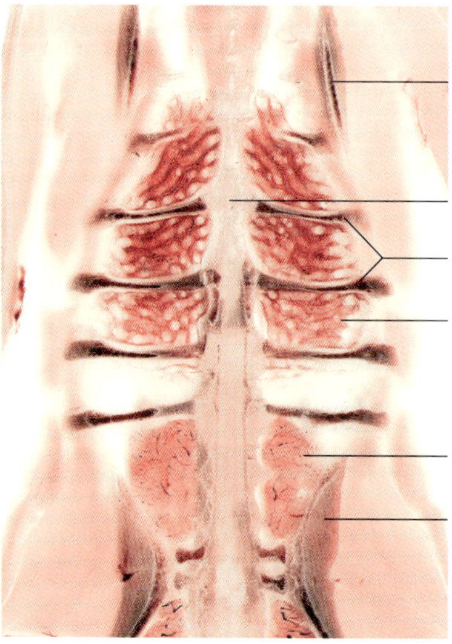

7.16 Horizontal section of a chicken at the level of the vertebral column (sheet plastinate, dorsal view). Courtesy of Dipl.-Biol. M. Kobienia, Munich.

bronchi contain **incomplete C-shaped rings of cartilage** that are absent from all subsequent bronchial divisions. In the primary bronchi, the lumen is surrounded by **respiratory epithelium**, underlain by elastic and collagen fibres, seromucous glands and lymphoreticular tissue. The **smooth muscle** of the primary bronchi is mostly circular. An adventitia of loose connective tissue is present.

The **secondary bronchi**, or **second-order bronchi**, are given off by the primary bronchi. According to the direction in which they pass, they are grouped into:

- laterodorsal secondary bronchi,
- 7–10 mediodorsal secondary bronchi,
- 4–7 lateroventral secondary bronchi and
- 4 medioventral secondary bronchi.

The four regions of the lung ventilated by their respective secondary bronchi are of different phylogenetic 'ages'. The laterodorsal sector is present in more highly developed birds and is thus termed the 'new lung' or **neopulmo.** All birds have the remaining sectors, referred to as the 'old lung' or **paleopulmo**.

The secondary bronchi are interconnected by **parabronchi**, or **third-order bronchi**. These are the **functional units of the avian lung**. Parabronchi arising from secondary bronchi within the dorsal paleopulmo meet and anastomose with their ventral counterparts in the interior of the lung.

The parabronchi are arranged in a parallel array of elongated tubules, from which they derive their alternative name of 'air pipes'. In most species their diameter is around 0.5mm (1–1.5mm in the chicken). Individual parabronchi are separated by **interparabronchial septa** (septa interparabronchialia) composed of connective tissue. Interparabronchial blood vessels pass through the septa (Figure 7.17). Due to the arrangement of the septa, the parabronchi appear hexagonal in transverse section.

Parabronchi have several distinctive features:

- they anastomose with one another,
- their walls contain chambers called atria,
- they contain gas exchange units and
- their diameter is uniform within species.

Internally, the parabronchi are lined with **simple squamous epithelium**. From the lumen, numerous small air chambers known as **atria** bulge outwardly into the so-called mantle of the parabronchus (Figure 7.17). The atria are lined with squamous to cuboidal epithelium containing **lamellated osmiophilic bodies** (surfactant) for reduction of surface tension. The epithelium is surrounded by muscle cells and elastic fibres.

Several funnel-shaped **infundibulae** open from the atria and radiate into the mantle. These give rise to an anastomosing three-dimensional network of tubular **air capillaries** (pneumocapillares).

The diameter of the air capillaries varies with species from about 3μm to 10μm. Due to the high surface tension within these small calibre tubes, their diameter remains relatively constant. The air capillaries are intimately intermeshed with a dense network of blood capillaries, permitting **gas exchange** to take place across the **blood–gas barrier**.

The avian blood–gas barrier is considerably thinner than that of mammals. It consists of three elements:

- the endothelial cells of the blood capillaries,
- the fused basal membranes of the blood and air capillaries and
- the epithelium of the air capillaries.

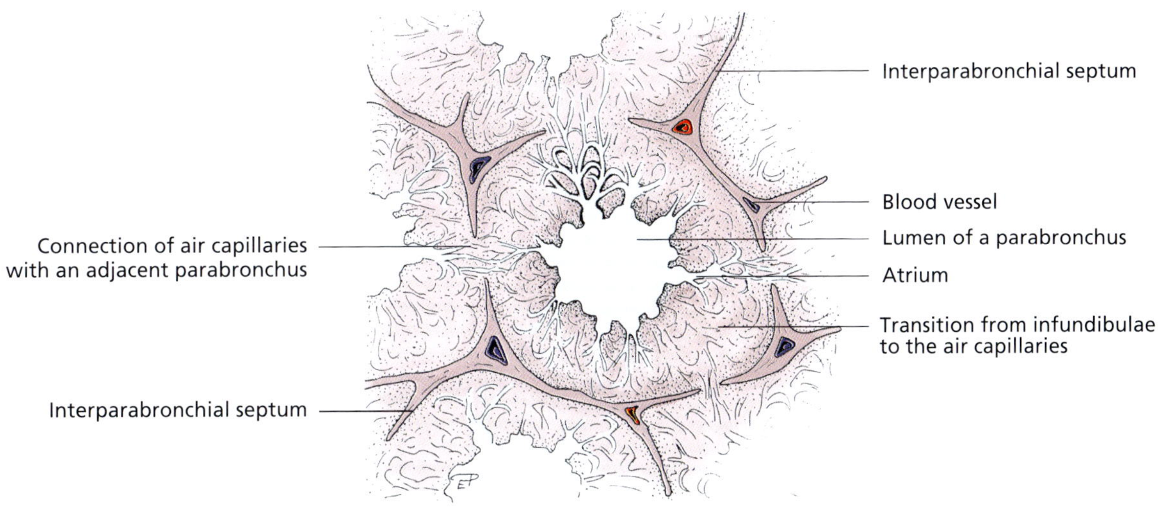

7.17 Parabronchi of the chicken (schematic).

Relative to body weight, the surface area for gas exchange in birds is around ten times greater than in mammals. In the chicken, it constitutes $18 cm^2/g$ of body weight.

Air sacs (sacci pneumatici, sacci aerophori)

The air sacs (also referred to in Chapter 5 'Body cavities') are **thin-walled deformable** cavities attached to the lungs. They provide mechanical ventilation of the lungs by acting as a bellows. The air sacs are joined by connective tissue with adjacent organs or muscles, but can also be partially covered with a tunica serosa. Their walls contain collagen and elastic fibres as well as smooth muscle cells. By penetrating the bones, the air sacs also serve to pneumatise the skeleton.

In the chicken there are **eight air sacs**, of which two are unpaired and three are paired (Figures 7.18ff.). The unpaired air sacs develop embryonically as paired structures, fusing at hatching to give rise to:

- a cervical air sac (saccus cervicalis),
- a clavicular air sac (saccus clavicularis).

Located more caudally are the paired:

- cranial thoracic air sacs (saccus thoracicus cranialis),
- caudal thoracic air sacs (saccus thoracicus caudalis),
- abdominal air sacs (saccus abdominalis).

With the exception of the abdominal air sacs, which are connected directly to the **primary bronchi**, the air sacs communicate with the **secondary bronchi**. This occurs mainly at the ventral margin of the lung and around the hilus. All of the connections between the secondary bronchi and the air sacs involve penetration of the **horizontal septum** (Figure 7.12).

The air sacs are divided into two groups, based on the movement of air during inspiration and expiration:

- **cranial air sacs**:
  - cervical air sac,
  - clavicular air sac and
  - cranial thoracic air sac,
- **caudal air sacs**:
  - caudal thoracic air sac and
  - abdominal air sac.

The **cervical air sac** consists of a median chamber lying over the oesophagus, and two elongated diverticulae that extend cranially into the vertebral canal and the transverse canal of the cervical vertebrae (Figures 7.18 and 7.20).

The **clavicular air sac** is capacious and complex. As well as enveloping the heart, the great vessels at the base of the heart and the syrinx, it penetrates the humerus and extends between the muscles of the pectoral girdle (Figures 7.18 and 7.20).

The **cranial and caudal thoracic air sacs** are located between the horizontal and oblique septa. They have no diverticulae (Figures 7.18 and 7.19). The caudal thoracic air sacs are small in the chicken and absent in the turkey. Air sac perfusion anaesthesia (APA; see Chapter 20 'Handling, restraint and anaesthesia') is usually performed via the left caudal thoracic air sac.

The **abdominal air sacs** project around the abdominal viscera (Figures 7.20 and 7.21). Their volume far exceeds that of the other air sacs, with the right being larger than the left. Dorsally they lie against the kidneys, as well as the testes in the male. Diverticulae extend as far as the hip joint and penetrate the synsacrum and the ilium. The abdominal air sacs play a key role in the mechanical ventilation of the lungs.

Movement of air through the lung–air sac system is brought about by raising and lowering of the caudal margin of the sternum, and by movement of the ribs (see Chapter 20 'Handling, restraint and anaesthesia'). The muscles contributing to **inspiration** are the:

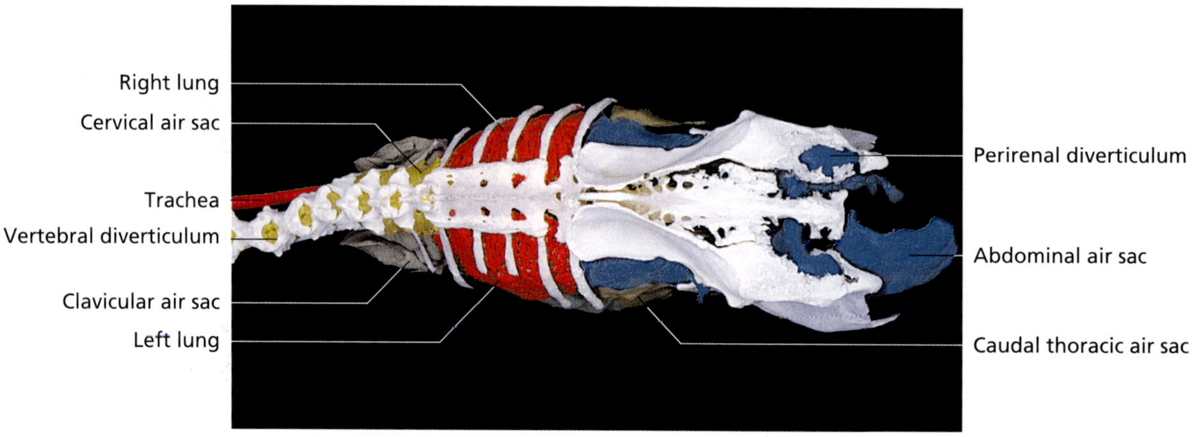

7.18 Air sacs of a chicken, corrosion cast (dorsal view). Courtesy of Professor Dr J. Ruberte, Barcelona.

- mm. levatores costarum,
- m. scalenus,
- mm. intercostales externi.

The muscles of **expiration** are the:

- mm. intercostales interni,
- muscles of the abdominal wall.

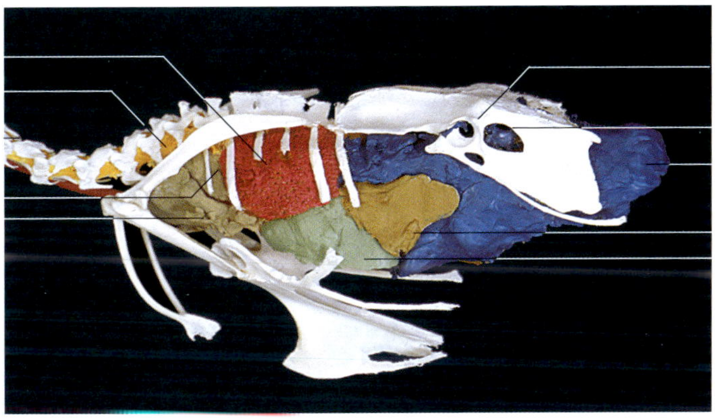

7.19 Air sacs of a chicken, corrosion cast (lateral view). Courtesy of Professor Dr J. Ruberte, Barcelona.

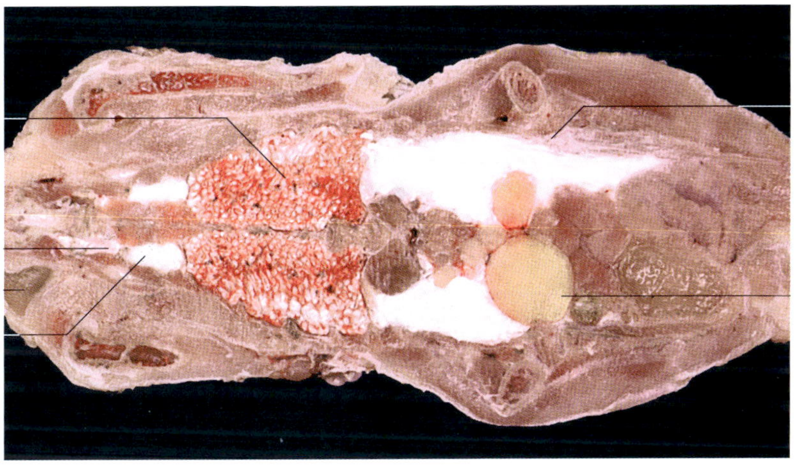

7.20 Horizontal section of the body of a chicken at the level of the lungs (air sacs injected). Courtesy of Professor Dr J. Ruberte, Barcelona.

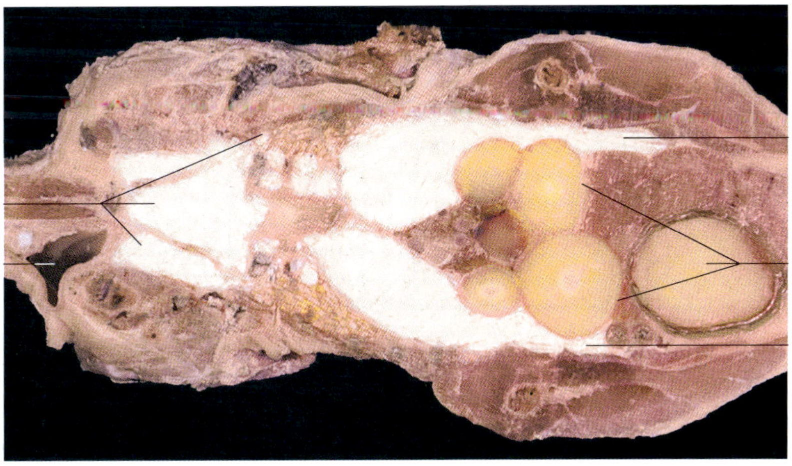

7.21 Horizontal section of the body of a chicken, ventral to the lungs (air sacs injected). Courtesy of Professor Dr J. Ruberte, Barcelona.

During inspiration, air passes into the caudal and cranial air sacs. The air reaching the caudal air sacs is mainly fresh, having travelled directly through the primary bronchi. Air entering the cranial air sacs, on the other hand, has passed through the parabronchi and is therefore partly used (see Figure 20.3).

During expiration, air is expelled from the cranial air sacs through the primary bronchi and the trachea. The relatively fresh air in the caudal air sacs passes through the parabronchi, where it participates in gas exchange, before ultimately being expelled through the trachea.

Clinical aspects

Due to its specific anatomical characteristics, particularly the air sacs, the **avian respiratory system** constitutes a substantial target for **infection** or compromise by **husbandry-related diseases**. At the same time, these anatomical features provide clinicians with a range of highly efficient options for diagnostic and therapeutic intervention. Indeed, some of these procedures can only be performed in avian patients (see below) and are well-suited to the need for rapid diagnosis and prompt instigation of treatment.

One of the features that distinguish the **trachea** of birds from that of other species is the presence of complete tracheal rings. When anaesthetising birds, **non-cuffed 'Cole'-style tubes** should therefore be used for **intubation**, as the inflation of a cuff may result in pressure necrosis of the delicate tracheal mucosa or even rupture of the tracheal rings. When intubating some species of zoo birds (e.g., flamingos), it is also important to be aware that the bifurcation of the trachea into the primary bronchi occurs relatively close to the larynx (in the upper or mid-cervical region). Intubation of one bronchus results in **ventilation of only one lung**, which may lead to inadequate anaesthesia.

Surgical transection of the lateral tympaniform membrane of the syrinx was once used for 'devoicing' birds (e.g., to prevent crowing in roosters). In some cases the membrane was replaced with stainless steel mesh. This procedure is now considered unethical on animal welfare grounds and is widely prohibited.

Due to the dynamics of air flow around the **pessulus**, this is a common site of **fungal granuloma** formation. Granulomas can occur in isolation, without associated pathology elsewhere in the respiratory tract. The resulting disease presentation, referred to as 'isolated' **syringeal** or **tracheal mycosis,** is characterised by sudden onset of pronounced inspiratory noise and signs of asphyxiation (dyspnoea with open-mouthed inspiration). In these cases,

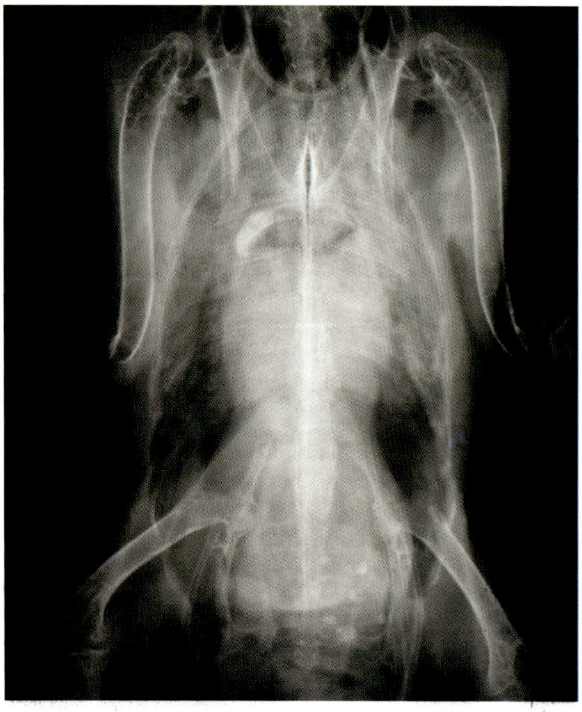

7.22 Radiograph (ventrodorsal view) of an African grey parrot (Psittacus erithacus) with aspergillosis. Abnormal findings include nodular opacities (fungal granulomas) in the lungs and typical 'bridge-like' opacities in the caudal thoracic air sacs. 'Air trapping' is seen in the right caudal thoracic air sac (on left of picture, refer to text).

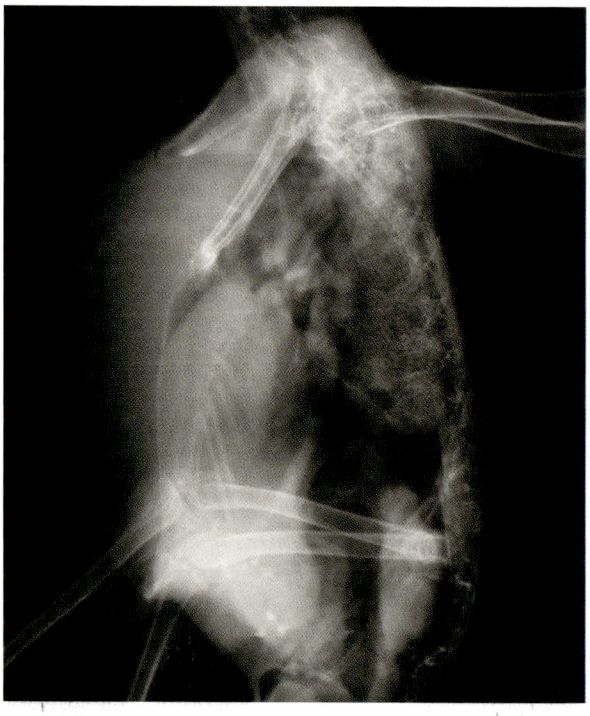

7.23 Radiograph (lateral view) of an African grey parrot (Psittacus erithacus, same patient as in Figure 7.22) with aspergillosis, showing nodular opacities in the lung and typical 'bridge-like' opacities in the region of the cranial and caudal thoracic air sacs. Pronounced, mycotoxin-induced renomegaly is a supportive radiographic finding in cases of aspergillosis.

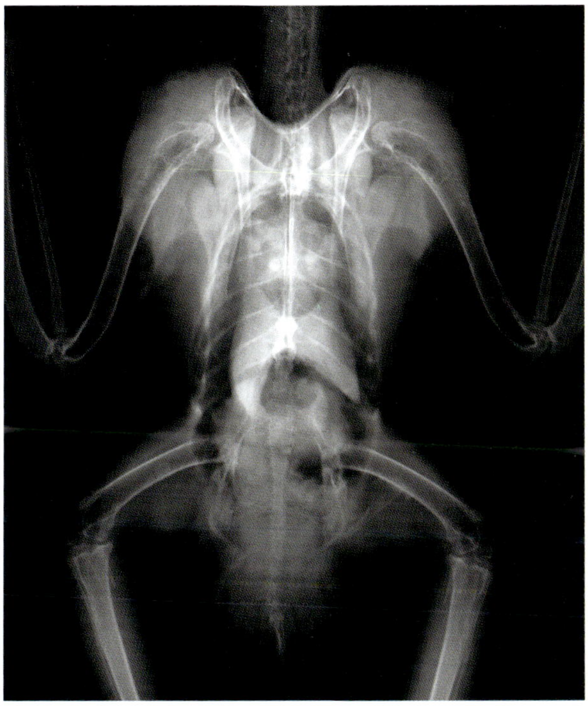

7.24 Radiograph (ventrodorsal view) of a peregrine falcon (Falco peregrinus) with severe aspergillosis. Abnormal findings include asymmetry of the thoracic air sacs, bilateral air trapping in the region of the caudal thoracic air sacs and hepatomegaly. The relatively large heart is typical for a trained raptor.

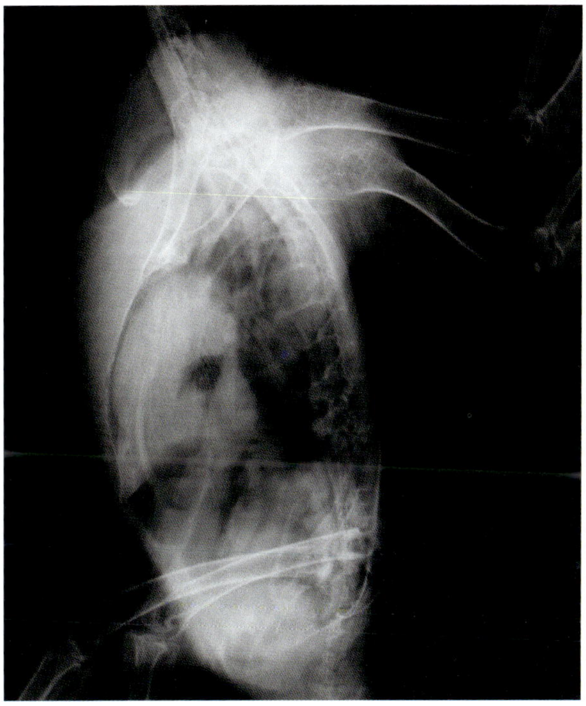

7.25 Radiograph (lateral view) of a peregrine falcon (Falco peregrinus) with aspergillosis (see Figure 7.24) showing nodular lesions (aspergillomas) in the lungs, opacities in the region of the thoracic air sacs and marked renomegaly.

the anatomy of the air sacs facilitates life-saving intervention. Prompt re-establishment of pulmonary ventilation can be achieved by opening an air sac, ideally the left **caudal thoracic air sac,** permitting normal, unrestricted breathing.

By fixing a flexible catheter in the opening, this process of 'air sac ventilation' or 'air sac perfusion' can be maintained for up to 72 hours, with regular monitoring of catheter patency. While the **trachea is temporarily bypassed**, granulomas or other obstructions can be removed with the aid of an endoscope.

The principle of air sac perfusion is also used in anaesthesia, whereby the mixture of carrier and anaesthetic gas is delivered retrograde through the caudal thoracic air sac and expelled through the trachea. This permits unimpeded surgical access to the head as well as the upper respiratory and digestive tracts. In addition, continuous unidirectional perfusion of the respiratory apparatus creates a CO_2 washout-effect, with suppression of the breathing stimulus. The resulting (reversible) apnoea, and thus complete immobilisation of the patient, is typically observed 8–10 seconds after commencing perfusion. These factors (surgical access, patient immobilisation) make this an ideal anaesthetic technique for ophthalmic microsurgery (e.g., cataract operations).

Some birds (e.g., pelicans) have a system of **subcutaneous air pockets** that can extend over the entire body and may feel like 'bubble wrap' when the bird is handled. These air pockets communicate with the lung-air sac system and play an important part in thermal insulation and achieving lift. They should not be confused with **subcutaneous emphysema**.

Emphysema is a common consequence of **air sac perforation** resulting from (cat) bite injuries (Figure 7.29) with subsequent leakage of air into the subcutaneous tissue. This is treated using paracentesis or by placing an intracutaneous drain.

In central Europe, **respiratory aspergillosis** is the most common disease of the lung–air sac system, accounting in some areas for more than 50 per cent of all clinical presentations (including large psittacines, penguins and various raptors such as gyrfalcons). The anatomy of the respiratory tract, including the presence of blind air sacs, favours the deposition of fungal spores. Other factors that promote infection in this geographic region include:

- pre-existing damage to the respiratory epithelium caused by inappropriate husbandry conditions (insufficient humidity, e.g., when heating is used in winter),
- inadequate nutrition (hypovitaminosis A) and
- prolonged antibiotic therapy.

In addition to the aforementioned tracheal or syringeal mycoses, aspergillus infection may result in **nodular lesions** of the lung (Figures 7.26 and 7.27) and **plaque-like growths** on the walls of the air sacs (Figure 7.28). The associated clinical presentation is typically chronic, including recurrent dry respiratory noise. Orthogonal radiographic projections of the air sacs reveal typical 'bridge-like' opacities in the cranial and caudal thoracic air sacs (Figures 7.22 and 7.23). The prognosis is usually poor.

Mycotoxin-induced **renomegaly** is a **supportive radiographic finding** in diagnosing aspergillosis. Hyperinflation of the air sacs, known as 'air trapping', may also occur due to interference with the normal outflow of air. In severe cases, this can be fatal. The disturbance of air flow by mycotic lesions frequently involves disruption of the aerodynamic valving system in the lung-air sac apparatus. Consisting of a series of 'physiological' valves (air vortices) – **as opposed to anatomical structures** – the aerodynamic valving system ensures the even distribution of air throughout the lung and air sacs in the normal bird.

The importance of promptly returning surgical patients to sternal recumbency in the post-operative period, to avoid air sac compression and respiratory depression, is addressed in Chapter 5.

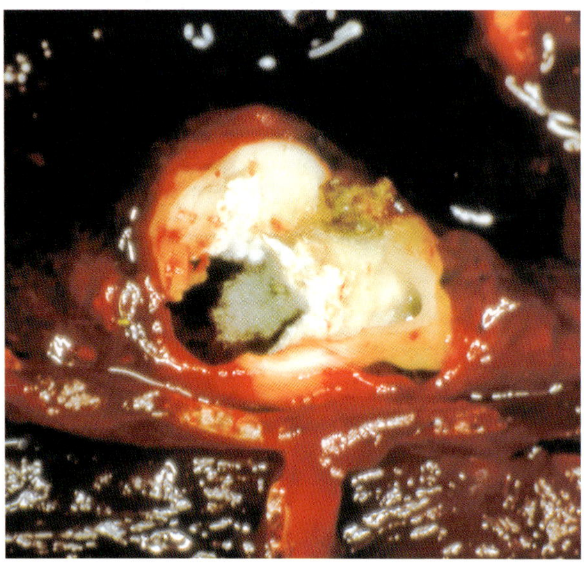

7.26 Opened pulmonary fungal granuloma with fungal elements (centrally) surrounded by a fibrin layer in a peregrine falcon (Falco peregrinus).

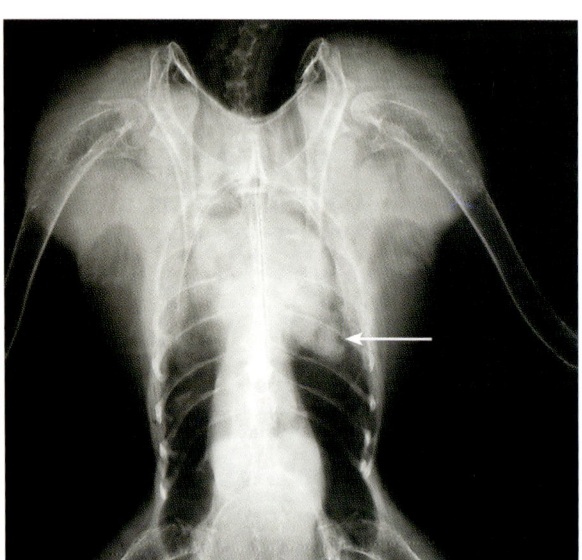

7.27 Radiograph (ventrodorsal view) of a peregrine falcon (Falco peregrinus) with aspergillosis. Arrow indicates a pulmonary fungal granuloma (see Figure 7.26).

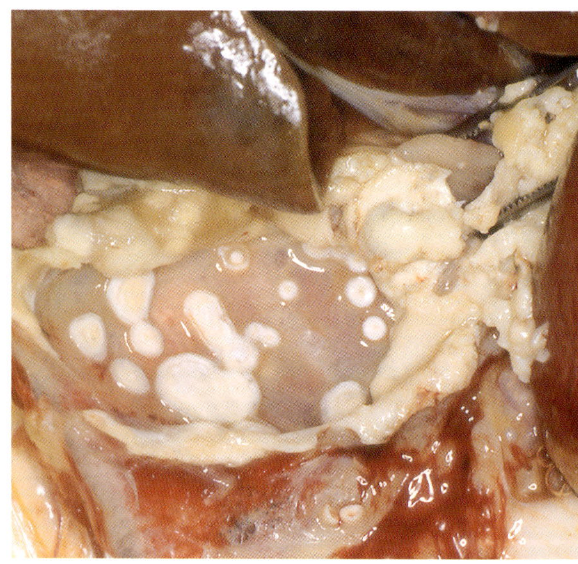

7.28 Extensive nodular and plaque-like fungal lesions on the wall between the cranial and caudal thoracic air sacs in a blue-fronted Amazon (Amazona aestiva).

7.29 Common blackbird (Turdus merula) with extensive dorsal subcutaneous emphysema following air sac perforation due to a cat bite.

8

Urinary system (organa urinaria)

A. Carretero, H. E. König, H.-G. Liebich and R. Korbel

Kidney (nephros, ren)

Structure of the kidney

Considerable differences exist between birds and mammals with respect to the excretion of urine, and the anatomy of the organs in which urine is produced and conveyed.

Birds are **uricotelic**, excreting toxic nitrogenous compounds (particularly **ammonia**) in the form of a considerably less harmful metabolic product (uric acid) in the urine. Birds lack the enzyme urate oxidase (uricase) and are therefore unable to convert urate into allantoin. They share this metabolic characteristic with humans, reptiles, many amphibians and with the Dalmatian dog, in which uricase appears to be biologically unavailable for urate conversion.

As a result, birds too can develop articular and visceral forms of gout.

A urinary bladder (vesica urinaria) is absent in birds. This reduces bodyweight, and thus facilitates flight. Urinary components are excreted together with the faeces.

Certain features also distinguish the renal vasculature of birds from that of mammals, the main difference being the presence of an avian **renal portal system**. This additional, 'downstream' capillary bed receives venous blood

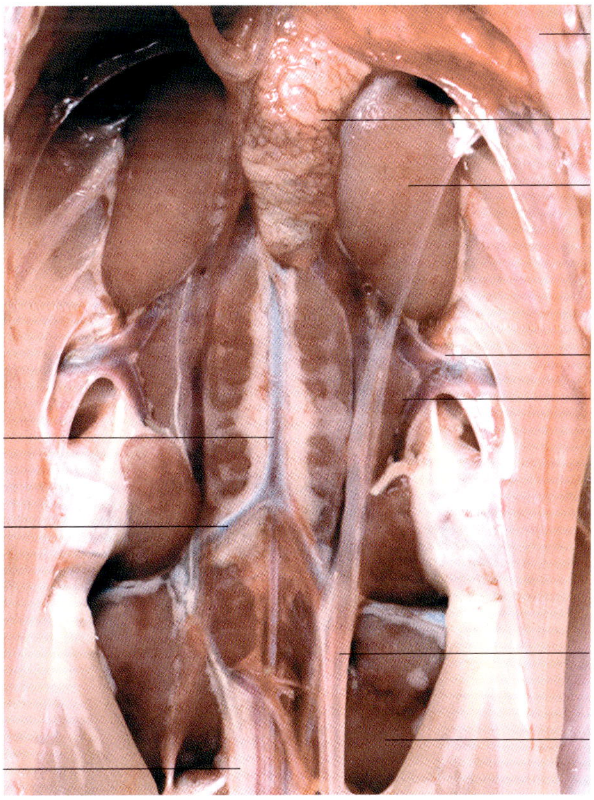

8.1 Kidneys and ureters in a chicken (ventral view).

from other parts of the body, permitting augmented filtration of blood (see Chapter 11 'Cardiovascular system'). Mammals are the only vertebrates in which this vascular arrangement is lacking.

The avian kidneys consist of three **renal divisions** (divisiones renales) (Figure 8.1). In most bird species, these are connected with one another by parenchymal bridges.

The divisions are distinguishable macroscopically as the:

- cranial renal division (divisio renalis cranialis),
- middle renal division (divisio renalis media) and
- caudal renal division (divisio renalis caudalis).

The kidneys are embedded dorsally in excavations of the **synsacrum**.

As in mammals, the **adrenal glands** lie medial to the cranial pole of the kidneys. They are relatively large in birds. In both genders, the gonads are situated adjacent to the kidneys and adrenal glands (unlike most of their mammalian counterparts, male birds have internal testes). The ventral surface (facies ventralis) of both kidneys is in contact with the paired **abdominal air sac** (saccus pneumaticus abdominalis) and with the wall of the **intestinal peritoneal cavity**. By definition, therefore, the kidneys are **retroperitoneal** (Figure 8.7).

In adult Galliformes, such as the chicken, the combined renal divisions are approximately 70mm long, 20mm wide and 15mm thick.

Particularly in Galliformes, **renal lobules** (renculi) may be visible macroscopically as small dome-shaped bulges (diameter 1–2mm in chickens) on the surface of the kidneys (Figure 8.2). Based on this feature, the avian kidney has been compared by some authors with the multi-lobar kidney of the ox; in birds, however, the lobules are not as clearly demarcated as the renal lobes in the ox, and not all lobules reach the surface of the kidney.

The **external iliac artery** and **vein** pass between the cranial and middle renal divisions, while the middle division is separated from the caudal division by the **ischiadic artery** and **vein** (Figure 8.1). Further detail regarding the vascular supply of the kidney is provided in Chapter 11 'Cardiovascular system'.

Branches of the **lumbar** and **sacral plexus** also lie between the renal divisions.

Structure of renal lobules

The composition of the avian kidney may be considered from both a functional and a structural perspective.

In functional terms, the kidney can be divided into **renal lobes** (lobi renales) and **renal lobules** (lobuli renales) based on the branching pattern of the ureter.

Renal lobes are thus defined as a portion of medulla that drains into secondary branches of the ureter, plus the region of the cortex that is drained by that medullary tissue. Each lobe is composed of several **renal lobules**. Individual lobules (Figure 8.2) drain into tertiary branches

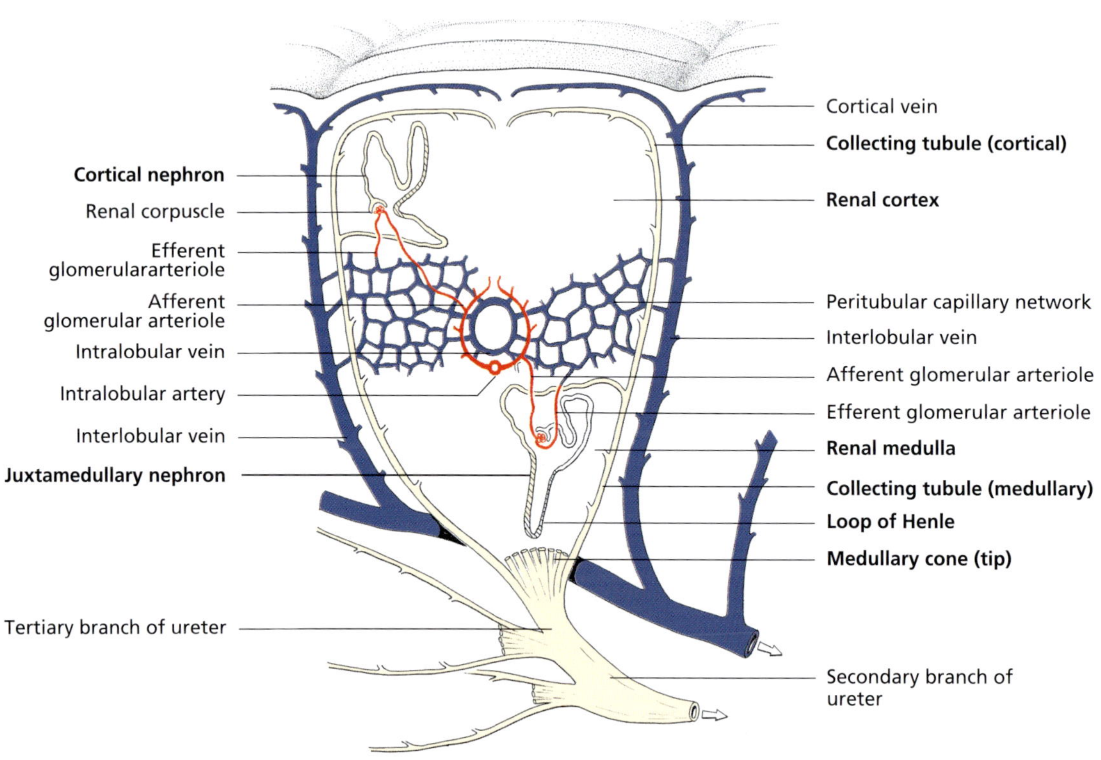

8.2 Renal lobule of the chicken (schematic), adapted from King and McLelland, 1978.

of the ureter, which then combine to form the aforementioned secondary ureteral branches.

Renal lobules are encircled by **interlobular veins** (venae interlobulares) and **collecting tubules** (tubuli colligentes) (Figure 8.2). Each lobule includes both medullary tissue (analogous to the medullary pyramid of mammals) and the cortical tissue that it drains (Figure 8.2). The medullary component of each renal lobule consists of cone-shaped bundles of **medullary collecting tubules** (tubuli colligentes medullares) enclosing blood vessels and loops of Henle of juxtamedullary nephrons.

Passing through the centre of each lobule are an efferent vein of the renal portal system (intralobular vein) and an afferent artery that supplies the lobule (intralobular artery).

In most bird species, the branching of the **intralobular vasculature** is comparable with that of mammals. The **intralobular artery** gives off **afferent arterioles** (arteriolae glomerulares afferentes) that each ramify to form a rete mirabile known as the **glomerulus**. Post-glomerular **efferent arterioles** (arteriolae glomerulares efferentes) give rise to capillaries that anastomose with the capillaries of the **renal portal system**, together forming the **peritubular capillary network** that surrounds the renal tubules (Figure 8.2).

In **histological section** (Figures 8.3ff.), renal lobes and lobules are seen at various levels, such that the medullary and cortical zones appear intermingled. In birds that have a high capacity for water conservation, the medullary regions (and thus the number of loops of Henle) are particularly well-developed, with each medullary region draining only a relatively small area of cortex. This arrangement presumably allows for production of more concentrated urine.

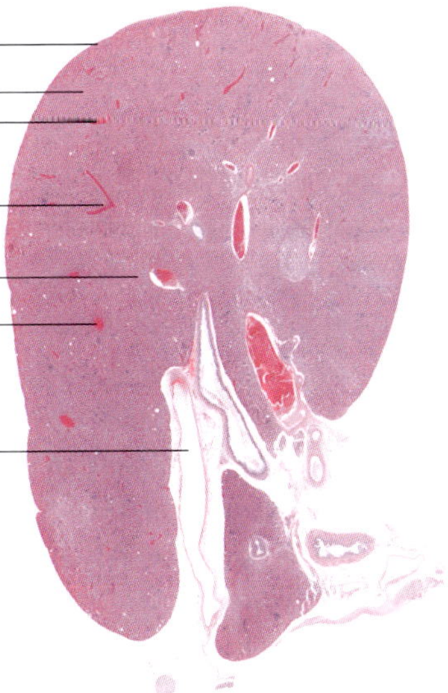

8.3 Histological section of the kidney of a chicken.

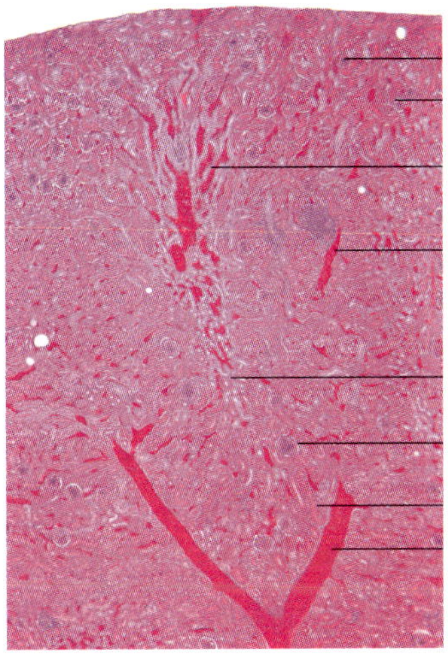

8.4 Histological section of the kidney of a chicken (detail).

RENAL CORPUSCLE (CORPUSCULUM RENIS, MALPIGHIAN BODY) AND NEPHRON

As in mammals, the renal corpuscle consists of a **glomerulus** (capillary tuft arising from the afferent glomerular arteriole) and a double-walled **capsule** (capsula glomeruli, Bowman's capsule) (Figures 8.5 and 8.6).

The **visceral layer of Bowman's capsule** (paries internus) lies against the capillary loops. At the **vascular pole** (polus vascularis) the visceral layer transitions into the outer **parietal layer** (paries externus), consisting of a single layer of flattened epithelial cells. The cells of the visceral layer, known as **podocytes**, possess branching processes (primary, secondary and tertiary) that extend to the basal membrane and form part of the barrier responsible for **glomerular ultrafiltration**.

The **renal corpuscles** of birds are smaller, but more numerous than those of mammals. Differences are also observed in the tubules. In particular, the length and diameter of the loop of Henle is reduced in birds.

Morphologically, there are two types of nephron: **cortical nephrons** and **juxtamedullary (or medullary) nephrons.** Cortical nephrons are greater in number (up to 90 per cent of nephrons) and have smaller renal corpuscles. The loop of Henle is short or absent. Juxtamedullary nephrons are characterised by more substantial renal corpuscles and well-defined loops of Henle that penetrate deep into the medulla (Figure 8.2).

TUBULES AND COLLECTING DUCTS

The morphology and nomenclature of the tubules of the avian nephron is the subject of ongoing clarification. Terms and descriptions presented here are based on the *Nomina Anatomica Avium* (2nd edition).

The convoluted tubules of the cortical nephrons consist of a **proximal tubule** (tubulus proximalis; site of glucose,

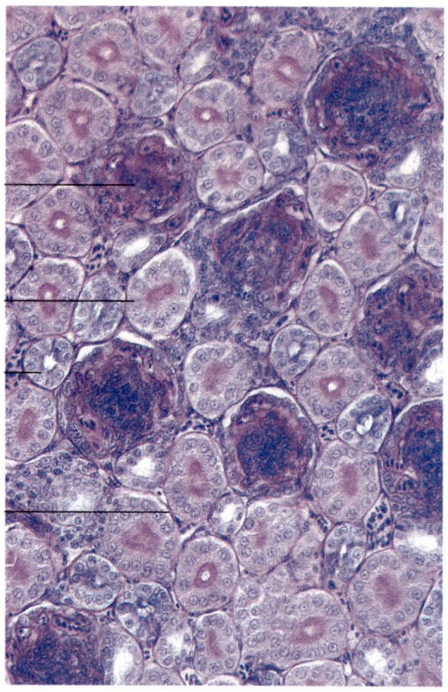

8.5 Histological section of the kidney of a chicken.

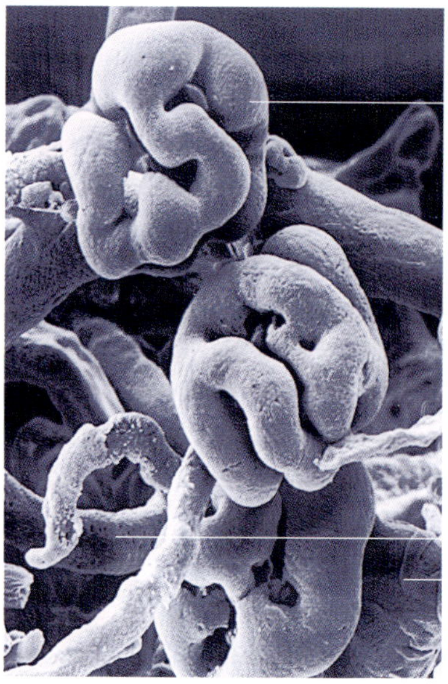

8.6 Capillary loops in the kidney of a chicken.

amino acid and electrolyte reabsorption), a short and variable **intermediate section** (tubulis intermedius) and a **distal tubule** (tubulus distalis). A clearly defined loop of Henle is lacking. Folding of the basal membrane of the cuboidal epithelial cells in the proximal tubule (as seen in mammals) is also absent in birds.

In contrast, the juxtamedullary nephron features a distinct loop of Henle, also referred to as a **medullary loop** (ansa nephrica). The tubules of these nephrons are comprised of the proximal convoluted tubule (tubulus contortus proximalis), proximal straight tubule (tubulus rectus proximalis), thin tubule (tubulus attenuatus), straight distal tubule (tubulus rectus distalis) and convoluted distal tubule (tubulus contortus distalis). The descending limb of the loop of Henle is composed of the straight proximal tubule, thin tubule and the initial segment of the straight distal tubule. The ascending limb consists of the remainder of the straight distal tubule, which then continues as the distal convoluted tubule. Structurally, these juxtamedullary nephrons resemble the nephrons of mammals.

The distal convoluted tubules open into the peripherally located **collecting tubules**. At the tip of the medullary cone, the **medullary collecting tubules** (tubuli colligentes medullares) from several renal lobules unite to form a **secondary branch** of the ureter. As described above, the various lobules that contribute to each secondary branch

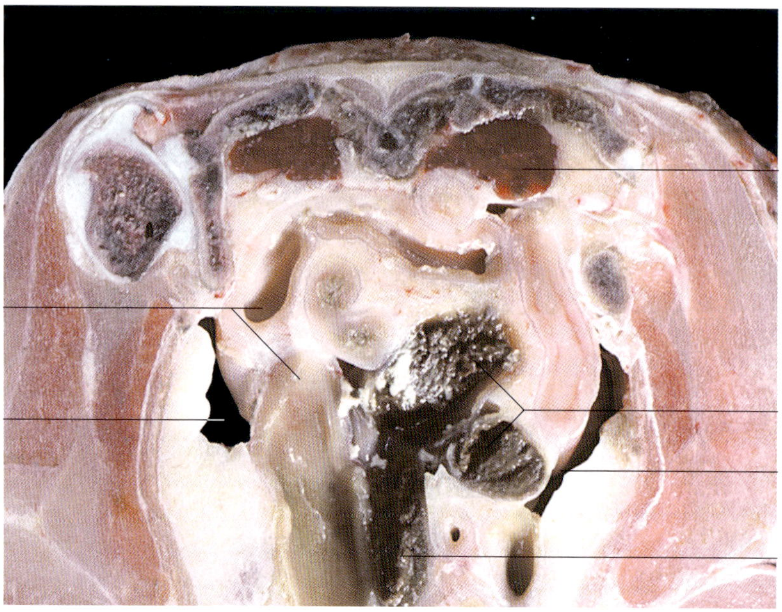

8.7 Transverse section of the body of a chicken at the level of the caudal renal division.

constitute a **renal lobe**. In this sense, despite the absence of renal calyces, there is some organisational similarity with the multi-lobar bovine kidney. The secondary ureteral branches drain into **primary branches** that ultimately open into the **ureter**.

JUXTAGLOMERULAR APPARATUS (APPARATUS JUXTAGLOMERULARIS)

The components of the avian juxtaglomerular apparatus are analogous to those of mammals, comprising a **macula densa**, **epithelioid (juxtaglomerular) cells** and **extraglomerular mesangial cells (Goormaghtigh cells)**.

As in mammals, the **juxtaglomerular apparatus** is located at the vascular pole of the renal corpuscle. The cells of the **macula densa** are **chemosensitive**, monitoring the **sodium concentration** in the blood in the afferent arteriole, and in the primary urine in the distal tubule. Falling sodium concentrations in the distal tubule (and thus the primary urine) stimulate the release of the proteolytic enzyme **renin**, which converts the circulating hepatocyte-derived plasma protein **angiotensinogen** into the **decapeptide angiotensin I**.

Angiotensin I is converted into the active octapeptide **angiotensin II** by pulmonary angiotensin converting enzyme (ACE). Angiotensin II triggers the release of the steroid hormone **aldosterone** from the **zona arcuata** of the **adrenal cortex**. Aldosterone drives the **reabsorption of sodium**, followed by **water**, in the distal tubule of the nephron. Through this system, the kidney of birds (like that of mammals) influences **blood volume** and thereby **blood pressure**, as well as **glomerular filtration rate** and **tubular reabsorption**.

Urine formation

Due to the generally shorter and thinner structure of the loops of Henle, birds have a relatively limited capacity for concentrating urine.

Uric acid formed as the end product of **purine metabolism** is excreted as a 2–3.5 per cent **colloidal solution**. Precipitation of uric acid, as may be seen for example in the Dalmatian dog, is thus prevented. **Mucin- and mucopolysaccharide-containing material** secreted by the epithelium of the ureter blocks the aggregation of small precipitates into larger urate crystals.

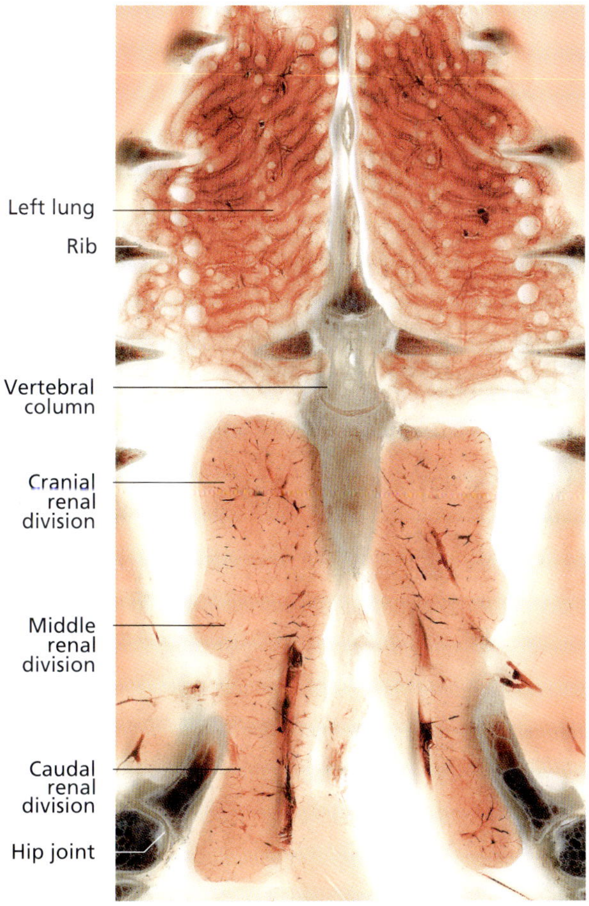

8.8 Horizontal section of the kidney of a chicken at the level of the hip joint (sheet plastinate). Courtesy of Dipl.-Biol. M. Kobienia, Munich.

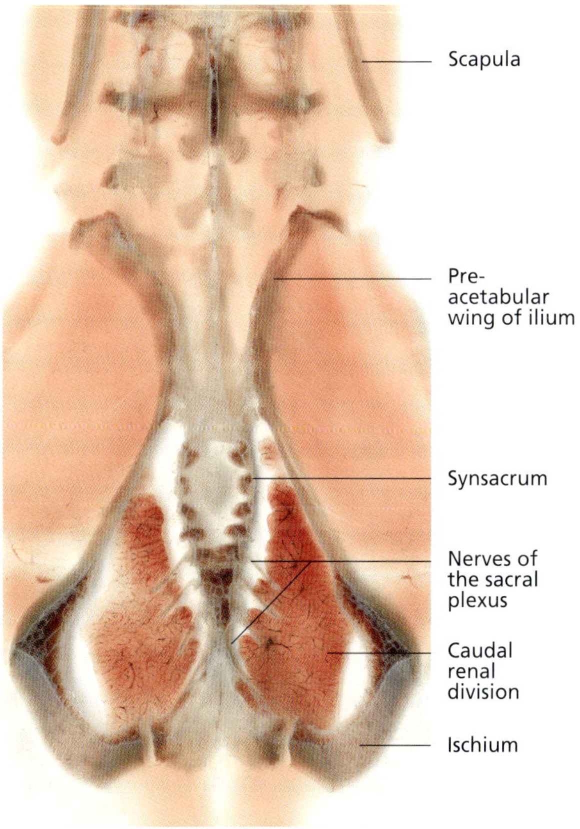

8.9 Horizontal section of the kidney of a chicken at the level of the dorsal vertebral column (sheet plastinate). Courtesy of Dipl.-Biol. M. Kobienia, Munich.

A viscous, mucous stringy urine is ultimately deposited by the ureter into the urodeum of the cloaca. There is no facility for storing urine as birds do not have a urinary bladder. Urine can be moved by retroperistalsis into the rectum where water is reabsorbed. In most species, urine and faeces are expelled together. The urine component consists of a liquid portion and a white paste consisting of concentrated uric acid (Figure 18.6).

Ureter

In most species, the ureter emerges from the ventral surface of the kidney in the region of the **middle renal division** (Figures 8.1 and 8.3). The ureter of the chicken is formed from the union of approximately 17 primary branches, each of which drains 5–6 secondary branches.

The avian ureter, like that of mammals, is lined by a specialised form of pseudostratified (polygonal) epithelium (epithelium transitionale). In the extra-renal ureter, the lamina propria underlying the epithelium increases in thickness. This, together with a well-developed network of filaments in the cytoplasm of the apical epithelial cells, serves to protect the deeper structures from the effects of the hypertonic urine.

The loose connective tissue of the **lamina propria** is supported by bundles of elastic fibres and may contain infiltrates of lymphoreticular cells (B and T cells). The **tunica muscularis** is thickened along the extra-renal portion by circular and longitudinal smooth muscle bundles. Reinforcement is provided near the cloaca by an additional longitudinal muscle layer. The ureter enters the dorsolateral wall of the cloaca at an acute angle, opening into the **urodeum** at the **ureteric ostium** (ostium cloacale ureteris) (see Chapter 9 'Male genital organs', Figure 9.7).

In most bird species, as in mammals, the relatively long intramural course of the ureter (within the wall of the urodeum) serves to prevent the retrograde flow of urine: as the urodeum fills, the lumen of the ureter is automatically closed off. This mechanism aids in guarding against ascending infections of the urinary tract.

Clinical aspects

Due to the anatomical relationships of the avian urinary and digestive tracts, excreta originating from the intestine and the kidney exit the body together through the cloaca. Many bird-owners are not aware of this and often mistakenly describe **polyuria** as diarrhoea. Polyuria is characterised by a pool of urine surrounding a faecal component that is, depending on species, relatively firm (in granivores, e.g., budgerigars) or semi-liquid (in raptors and soft feeders such as mynas). In the case of diarrhoea, liquid faeces are surrounded by a normal amount of urine. An awareness of this distinction is important when inspecting cage faeces as part of a physical examination.

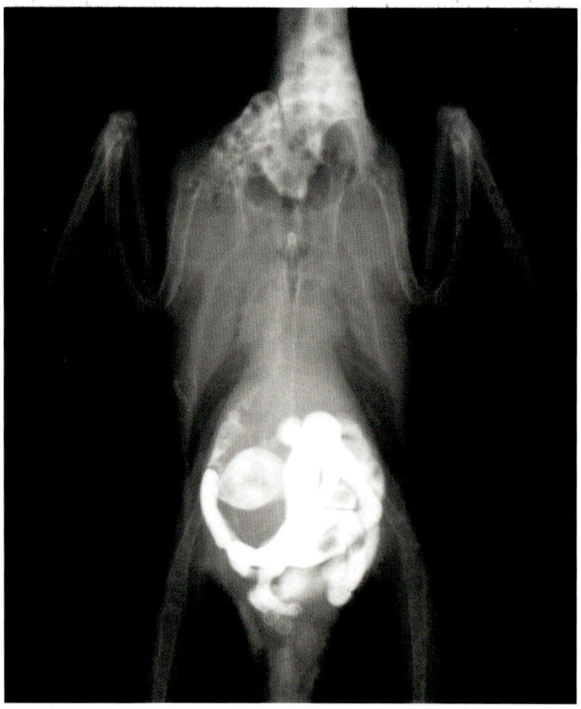

8.10 Contrast radiograph (ventrodorsal view) of a budgerigar (Melopsittacus undulatus; same patient as Figure 8.11) with a kidney tumour causing congestion of the liver and dislocation of the intestine and ventriculus.

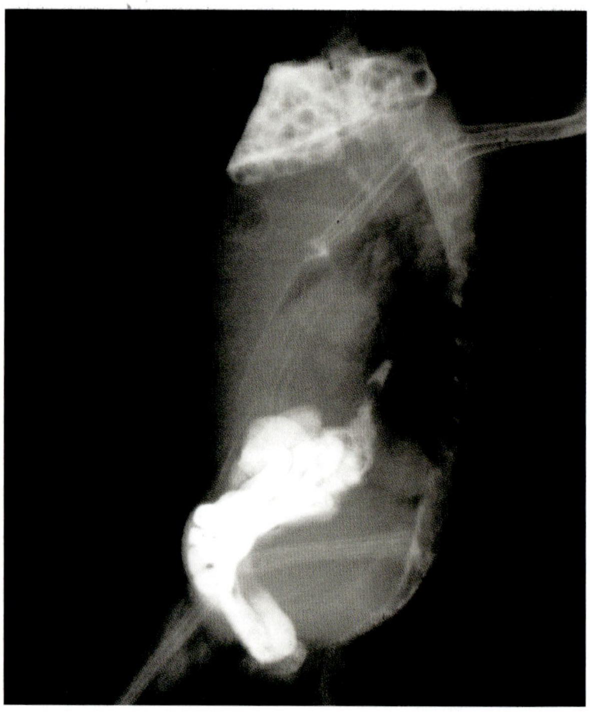

8.11 Contrast radiograph of a budgerigar (Melopsittacus undulatus) with a kidney tumour showing typical ventral displacement of intestinal loops filled with contrast material.

Stress associated with examination may induce a **transient polyuria**, while a **physiological polyuria** occurs shortly before egg deposition. **Discolouration of urine**, for example by blood components (**haematuria**), can be a typical sign of certain diseases (e.g., lead poisoning, Figure 6.61). It is important to note, however, that the colour of excreted urine may also be influenced by diet (e.g., strongly coloured fruits), potentially mimicking a pathological finding.

The kidney has very efficient mechanisms for reabsorbing water. Some birds, particularly species originating from dry climates – including the budgerigar (Melopsittacus undulatus) – may therefore have low water intake requirements. This is sometimes misinterpreted by owners as a sign of illness. For the same reason, it is difficult to achieve accurate and effective therapeutic doses when delivering medications via the drinking water. Thus, treatment of individual birds should generally be carried out using injectable products (see Chapter 21 'Medication and blood collection techniques'). When treating flocks (e.g., commercial poultry), on the other hand, dosing by means of the drinking water is indicated for practical reasons.

The highest incidence of kidney disease is seen in the budgerigar (Melopsittacus undulatus). In this species, **kidney tumours** or **fluid filled renal cysts** account for up to 10–15 per cent of all patient presentations. These are usually diagnosed using radiography (lateral view), whereby ventral displacement of contrast medium-filled intestinal loops is observed (see Ch. 19 'Imaging techniques') (Figure 8.11). Tumours with a diameter of 1cm or more can be palpated by experienced examiners (see Ch. 18 'Clinical examination'). However, these should not be confused with the muscular stomach (gizzard, ventriculus), which is palpable as a freely movable spherical structure in the left caudal quadrant of the body cavity.

In the absence of evidence of a fracture or luxation, **uni- or bilateral lameness in the hindlimbs** of budgerigars can be indicative of a kidney tumour. Enlargement of the kidney places pressure on the lumbosacral plexus between the kidney and the dorsally located synsacrum, giving rise to neural deficits. Kidney tumours may become quite substantial, potentially filling the body cavity and displacing all of the internal organs. Associated pressure on the lungs often results in dry respiratory sounds and dyspnoea.

Renal tumours cannot be surgically removed, as the intricate topographic association of the kidney with blood vessels and nerves makes surgical access difficult. The external iliac artery and vein, as well as branches of the lumbar plexus, pass between the cranial and middle renal divisions, and the ischiadic artery and vein and branches of the sacral plexus traverse the region between the middle and caudal divisions.

Pathological precipitation of uric acid crystals in the renal parenchyma (**renal gout**) can be caused by various factors, including genetic predisposition. Depending on their consistency, urate precipitates can only be identified radiographically in around 10–15 per cent of cases (Figures 8.12 and 8.13). Endoscopy is usually more useful for establishing a diagnosis (see Chapter 22 'Endoscopy').

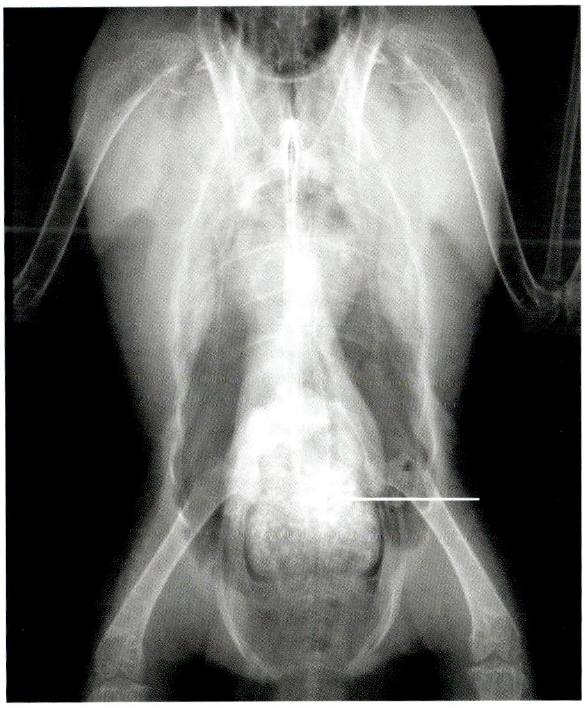

8.12 Radiograph (ventrodorsal view) of an African grey parrot (Psittacus erithacus) with renal gout (arrow).

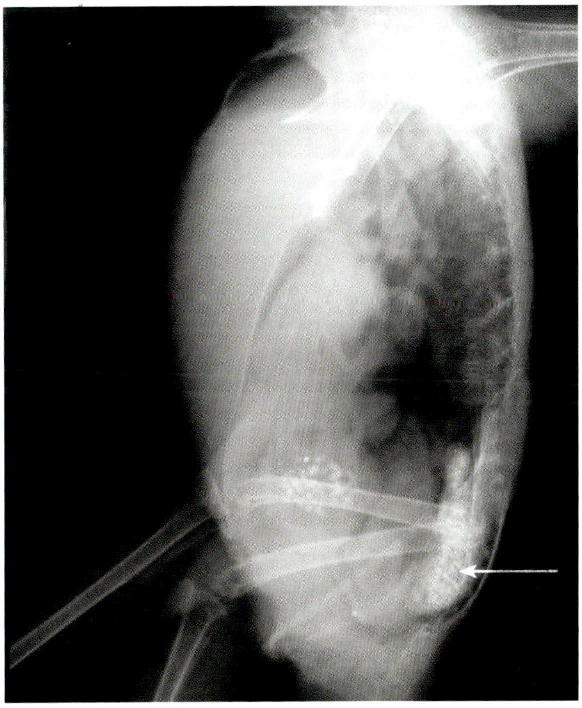

8.13 Radiograph (lateral view) of an African grey parrot (Psittacus erithacus) with renal gout (arrow) showing typical miliary opacities.

Laboratory findings indicative of kidney disease include elevated blood urate (**hyperuricaemia**). Severe kidney disease occasionally leads to excretion of uric acid by the crop mucosa resulting in decreased pH and inflammation (ingluvitis). Clinical signs include vomiting, and a crop swab may reveal the presence of urate crystals.

9

Male genital organs (organa genitalia masculina)

H. E. König, H. Bragulla, H.-G. Liebich and R. Korbel

The sexes are separate in birds, the reproductive apparatus consisting of either male or female genital organs. However, many avian species, particularly Psittaciformes, do not exhibit phenotypic sexual dimorphism. Definitive gender determination (e.g. pre-purchase examination of parrots) therefore often requires endoscopy (see Chapter 22 'Endoscopy') or DNA testing.

The male genital organs are comprised of the (Figure 9.1):

- testes,
- epididymides,
- deferent ducts (ductus deferentes),
- copulatory organ (phallus) and
- accessory organs of the phallus.

Testis (orchis)

The testes of birds, like those of mammals, are paired. However, as in elephants and cetaceans, the testes **do not undergo descent** (descensus testis). Both testes remain in the common body cavity, located – with some species variation – in the vicinity of the kidney. Accordingly, birds also **lack a scrotum**.

The potentially deleterious impact of the high avian body temperature (up to 41.5 °C) on the developing spermatozoa within the internally located gonads is prevented by the elaboration of a richly branching **venous plexus** around the testes. This vascular network, similar to that seen in elephants, serves as a heat exchange system for cooling of the testes, thus circumventing the need for

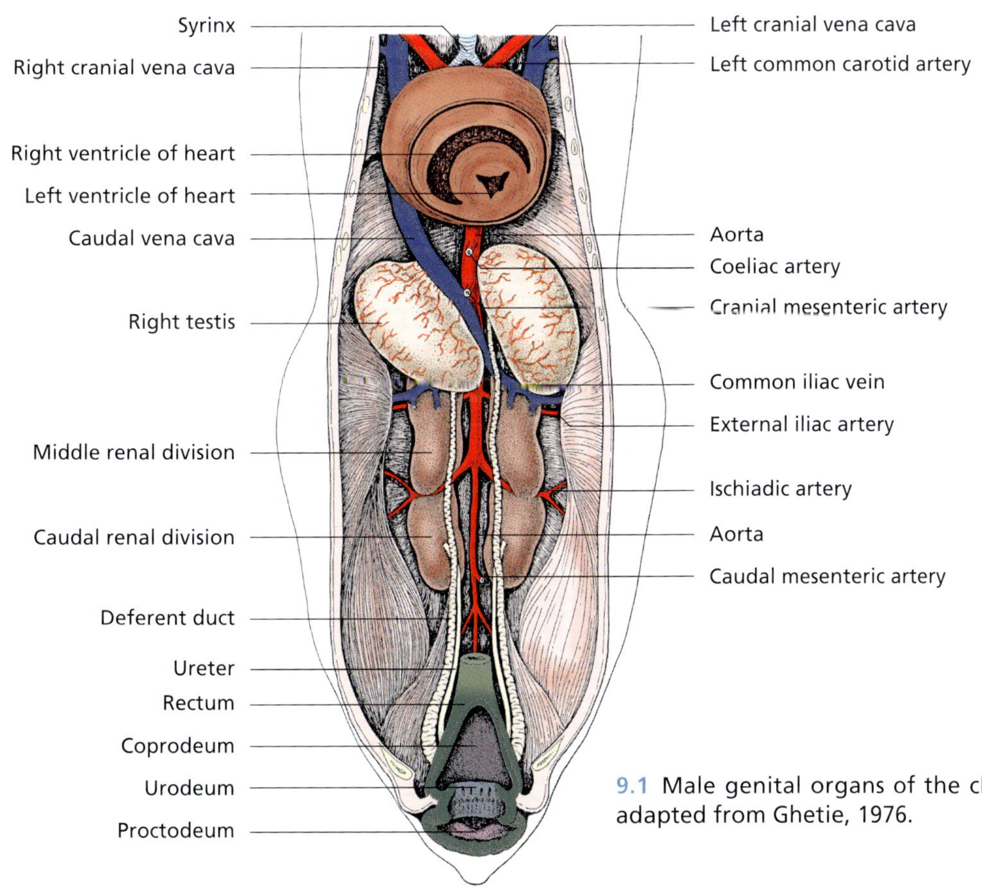

9.1 Male genital organs of the chicken (schematic), adapted from Ghetie, 1976.

exteriorisation of the testes for the purposes of thermoregulation. It also assists in distinguishing between the kidneys and the testes during endoscopic examination of the coelomic cavity.

The **size** and **development** of the testes exhibits marked variation associated with season, climate, age and breed.

Outside the breeding season, the testes of male Galliformes are approximately the size of a cherry pit. During the mating period, they grow to many times this size, reaching 60mm in length and 30mm in thickness in male Galliformes, and up to 80mm by 45mm in drakes (Figures 9.2 and 9.3). In some wild bird species the **weight** of the testes can increase by up to 1,000-fold.

A double layer of serosa, the **mesorchium**, attaches the epididymal border of each of the paired testes to the dorsal wall of the body cavity, near the kidneys.

In most bird species, the testes are located high in the body cavity, between the lungs and the cranial renal division (divisio renalis cranialis), adjacent to the adrenal glands, aorta and the caudal vena cava (Figure 9.2).

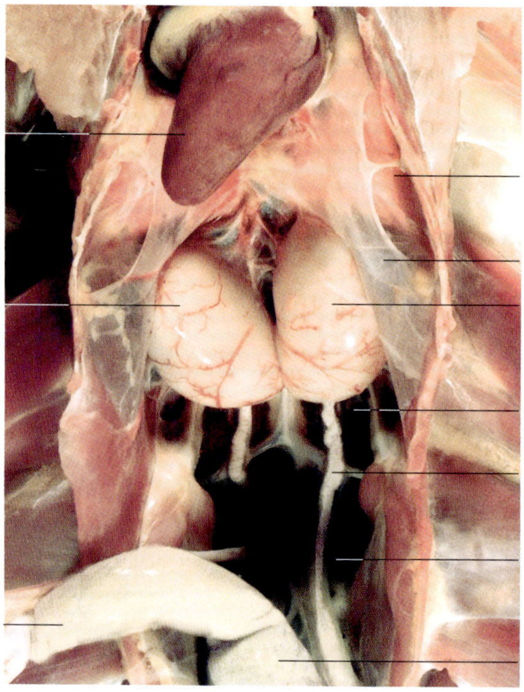

9.2 Testes of a sexually active male chicken (ventral view, intestine reflected).

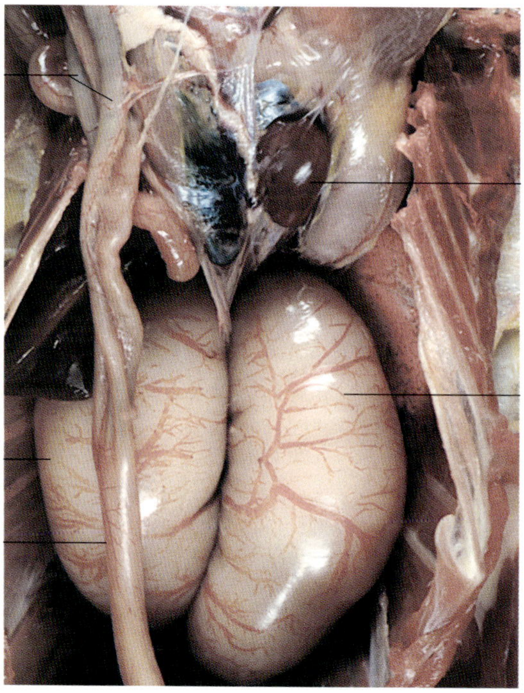

9.3 Testes of a sexually active male Indian runner duck (ventral view, intestine reflected).

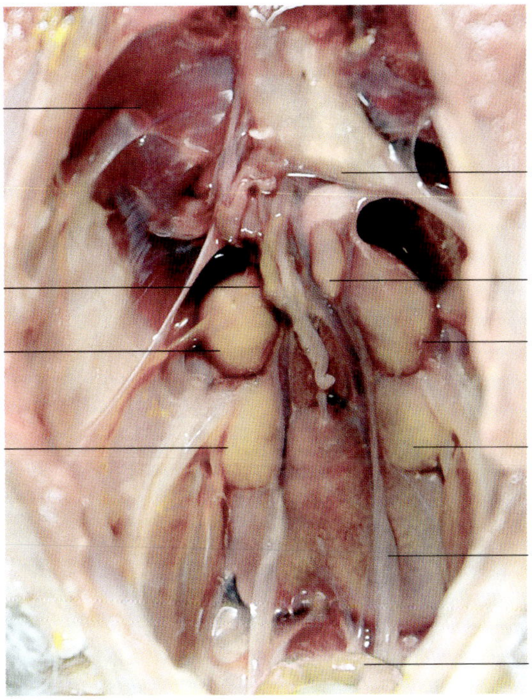

9.4 Testes of a male African grey parrot (intestine reflected).

The increase in testicular size during the breeding season brings the gonads into contact with the thoracic and abdominal air sacs. This has an additional cooling effect. **Viewed endoscopically**, the testes typically appear yellowish-white. Some species, including cockatoos, have strongly pigmented testes that appear *in situ* as black, elongated ovoid structures lying alongside the divisions of the kidney (see Chapter 22 'Endoscopy', Figure 22.12). As such, they serve as highly specific anatomical landmarks when examining the body cavity in these species.

Subcapsular testicular veins (vv. testiculares externae), typically three in number, pass over the **surface of the testes**. They exhibit considerable species-dependent branching. Numerous connecting branches (rami communicantes) extend between these veins forming the **rete venosum** that participates in **thermoregulation of the testicular parenchyma**.

Structure of the testis

Unlike the male gonads of mammals, the internally positioned avian testes are covered laterally, medially and ventrally by the **peritoneum** (a single layer of serosa representing an invagination the wall of the **intestinal peritoneal cavity**) (see Figure 5.6).

Also contrasting with mammals, the underlying **tunica albuginea** is relatively thin in most bird species. In Galliformes, it contains muscle cells that confer a limited degree of contractility. At the beginning of the breeding season, the tunica albuginea undergoes marked fibrogenesis. Under the influence of somatotropin and 17-beta-oestradiol, a proliferation of fibroblasts results in the formation of a new layer of fibrous tissue. Consequently, the tunica appears temporarily to consist of two layers.

The tunica albuginea gives off delicate bundles of connective tissue that, unlike the well-developed septa of mammals, do not divide the testes into lobules. Accordingly, there is no testicular mediastinum.

In the parenchyma of the avian testes, the highly coiled **convoluted seminiferous tubules** (tubuli seminiferi convoluti) follow a tortuous course within the loose interstitial connective tissue. They anastomose extensively, forming a network that is more delicate and complex than in mammals. During the breeding season, the tubules increase in length in proportion with the increase in size of the testes, up to a total of 250m in chickens.

The intertubular connective tissue of the testis contains androstenone-producing **Leydig cells** and, particularly in cockatoo species, pigment-forming **melanocytes**.

The wall of the seminiferous tubules is composed of a membrana propria (or lamina limitans) and a germinal (spermatogenic) epithelium (epithelium spermatogonicum) comprising spermatogenic cells and sustentacular (supporting, Sertoli) cells.

The **membrana propria** consists of a basal membrane and loose connective tissue incorporating delicate bundles of elastic fibres and numerous contractile smooth muscle cells. This tissue layer forms a structural component of the diffusion barrier referred to as the **blood-testis-barrier**.

In juvenile and sexually quiescent adult males, the spermatogenic epithelium consists of a single layer of cells, in

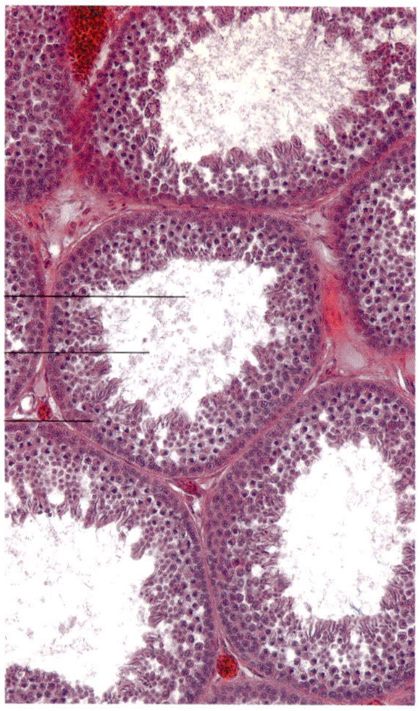

9.5 Histological section of the testis of a male chicken showing seminiferous tubules in transverse section.

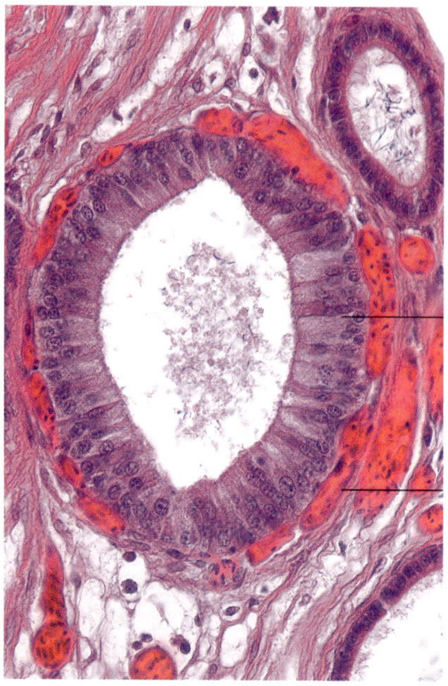

9.6 Histological section of the epididymal duct of a male Indian runner duck.

which the spermatogenic cells and sustentacular cells are in contact with the basal membrane.

In male Galliformes, the spermatogenic and sustentacular cells become more numerous from the fifth week post-hatching and the epithelial layer increases in height. With the onset of **sexual maturity**, at about 16 to 24/26 weeks in the cock (depending on breed), the epithelium becomes stratified. This is associated with the mitotic division of spermatogonia and formation of spermatozoa. The **duration of spermatogenesis**, up to delivery of spermatozoa into the lumen of the convoluted seminiferous tubule, is usually no more than 12 days in the chicken.

From the convoluted seminiferous tubules, spermatozoa pass into the efferent components of the testicular tubular system, the short **straight seminiferous tubules** (tubuli seminiferi recti). In most species, these are lined by cuboidal epithelium. The straight tubules open into a series of anastomosing channels of varying diameter, the **rete testis**.

In birds, the rete testis is located on the medial aspect of the testis and is divided into intratesticular, intracapsular and extracapsular components. The last of these opens into the **proximal efferent ductules** (ductuli efferentes proximales) from which sperm are conducted into the epididymis.

Epididymis

The epididymis lies against the dorsomedial surface of the testis, extending along approximately two-thirds of the testicular border. This anatomical relationship is consistent across species of the class Aves, in contrast to the interspecies variation seen in mammals.

Consisting largely of the convoluted epididymal duct, the epididymis reaches a length of only 3–4mm in sexually active chickens. Unlike the mammalian epididymis, which consists of a head, body and tail, the epididymis of birds is not divided into segments.

The previously described proximal efferent ductules narrow to form **distal efferent ductules** (ductuli efferentes distales) (approximately 70 in the cock). These empty via **connecting ductules** (ductuli conjugentes) into the **epididymal duct** (ductus epididymis).

Efferent ducts enter the epididymis along its entire length. This differs from the arrangement in mammals, in which efferent ducts of the testes typically open only into the head of the epididymis. The epithelium of the efferent ductules is simple columnar (similar to mammals).

The **epididymal duct** is lined by stereociliated pseudostratified columnar epithelium surrounded by loose connective tissue that, in some bird species, is heavily invested with smooth muscle cells.

Along its length, the epididymal duct gradually increases in thickness. At the caudal pole of the epididymis it opens into the deferent duct.

Deferent duct (ductus deferens)

The deferent duct (Figures 9.1 and 9.2) follows a strongly meandering course, occupying a retroperitoneal position ventromedial to the kidney. At the level of the middle renal division, it crosses to the lateral aspect of the ureter, which it then accompanies on its passage to the cloaca. Caudally, its lumen may be considerably expanded.

The ductus deferens is lined by pseudostratified epithelium with sparse stereocilia and few secretory cells.

The terminal **straight segment** of the ductus (pars recta ductus deferentis) opens into a dilated section, the ampulliform **receptaculum ductus deferentis**. While the term 'ampulla' is appropriate in a descriptive sense, this structure is not homologous with the ampulla ductus deferentis of mammals, having no secretory cells in its walls. The ductus deferens opens at the **ostium ductus deferentis** located on the conical **papilla ductus deferentis** (particularly prominent in chickens) in the urodeum (Figures 9.7 and 9.8).

Within the wall of the urodeum, near the receptaculum, there is an **arterial network** (rete mirabile arteriosum) arising from the pudendal artery. Referred to as the **vascular body of the phallus** (corpus vasculare phalli), this structure contributes to intumescence of the phallus. It is particularly well-developed in species with a protrusible copulatory organ (phallus protrudens).

The function of the ductus deferens, as in mammals, is to convey mature sperm. By virtue of its highly convoluted arrangement, the ductus deferens traverses a distance of only around 10cm in Galliformes, despite being at least 60cm long in these species.

In male chickens, the white milky semen contains approximately 3.5 million sperm per microlitre. The volume of ejaculate is 0.5–1ml in chickens and 2–5ml in the ostrich. These values are comparable with those seen in the bull.

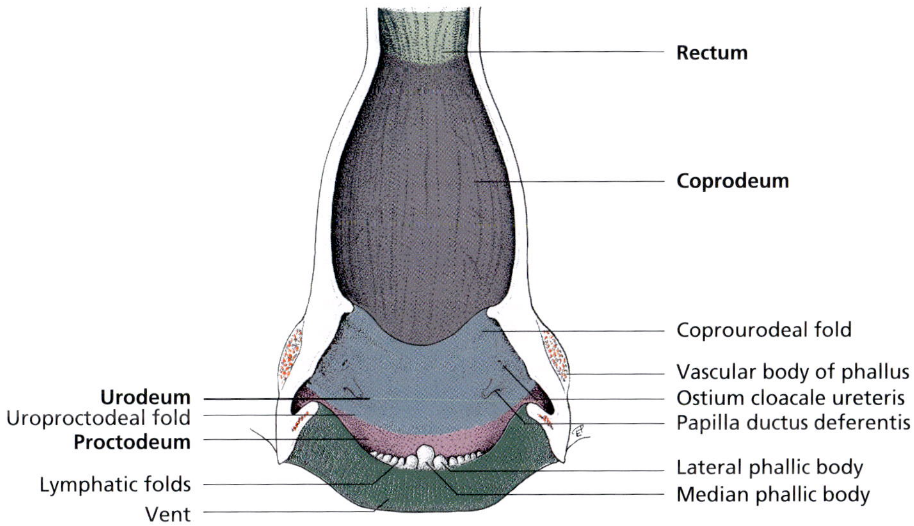

9.7 Cloaca of the male chicken (schematic), adapted from Waibl and Sinowatz, 2004.

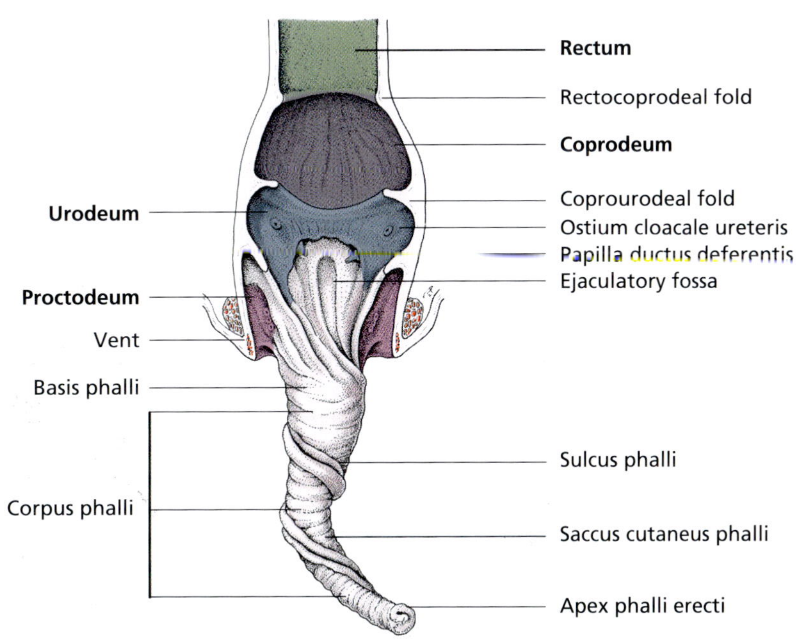

9.8 Cloaca of the male Indian runner duck with phallus protruded (schematic), adapted from Komarek, 1969.

Phallus (penis, phallus masculinus)

The phallus of the male bird is a **component of the cloaca**. Among modern bird species there are two types of phallus:

- non-protrusible (phallus nonprotrudens) and
- protrusible, intromittent (phallus protrudens).

The **non-protrusible phallus**, seen for example in chickens, is composed of:

- an unpaired median phallic body (corpus phallicum medianum) flanked by
- paired lateral phallic bodies (corpora phallica lateralia)

In chickens the median phallic body is visible in day-old chicks. It is rounded in males and conical in females. This subtle difference allows experienced operators to sex chicks at a very young age (see Chapter 18 'Clinical examination').

Ducks, geese and ratites have a **protrusible phallus**. In the gander, the tumescent phallus can reach 60–80mm in length. Terminology relating to the complex structure of the protrusible phallus has undergone extensive revision. Synonyms for the terms used in the following highly simplified description can be found in the *Nomina Anatomica Avium* (2nd edition).

The components of the protrusible phallus include the:

- base (basis phalli),
- body (corpus phalli),
- phallic sacs (saccus cutaneus phalli and saccus glandularis phalli),
- flexura phalli (non-erecti)/apex phalli (erecti).

The phallus arises as the **basis phalli** in the ventral wall of the cloaca, where it rests in a trough-like plate of fibrocartilage, the corpus fibrocartilagineum. It incorporates a **lymphatic cistern** (cisterna lymphatica basis phalli) that is partly divided into left and right components. This cistern continues as a narrow chamber (cisterna lymphatica corporis phalli) that extends around the cutaneous and glandular phallic sacs (saccus cutaneous phalli and saccus glandularis phalli). These sacs form the hollow interior of the body (corpus) of the phallus. The sacs are arranged 'in series', with the more proximal cutaneous sac leading into the more distal glandular sac. In the non-tumescent phallus, the body is completely invaginated and the junction between the cutaneous and glandular phallic sacs, the **flexura phalli**, is curved.

During intumescence, **lymph** derived from the vascular body of the phallus (described above) fills the lymphatic cisterns to bring about erection. This results in eversion of the cutaneous phallic sac through the **ostium sacci cutanei phalli** on the floor of the proctodeum. The glandular sac is not everted (it comes to lie within the cutaneous sac). As a result, the exteriorised flexura phalli becomes the tip (apex) of the erect phallus. A **phallic sulcus** (sulcus phalli) spirals around the free part of the erect phallus**.**

Erection of the non-protrusible phallus also occurs due to engorgement with lymph, although in these species the erect phallus protrudes only slightly, if at all, from the cloaca.

Accessory structures of the phallus

The accessory structures of the phallus include the:

- vascular body of the phallus (corpus vasculare phalli),
- elastic ligament of the phallus (lig. elasticum phalli),
- m. retractor phalli.

The **vascular body of the phallus** consists of a capillary tuft (originating from the **pudendal artery**) that is intricately intermingled with lymphatic vessels. Fluid passes from blood capillaries into the interstitium, from which it enters the lymph vessels. The lymph passes through two ducts into the lymphatic cisterns of the phallus. As described above, filling of these cisterns is responsible for erection, aided by contraction of the m. sphincter cloacae.

During detumescence, the phallus is returned to its invaginated state by the action of the usually striated **m. retractor phalli** in the ventral wall of the cloaca, with assistance from the **elastic ligament of the phallus** (highly developed in ducks), which runs through the centre of the erect phallus. In waterfowl, lymph is pumped from

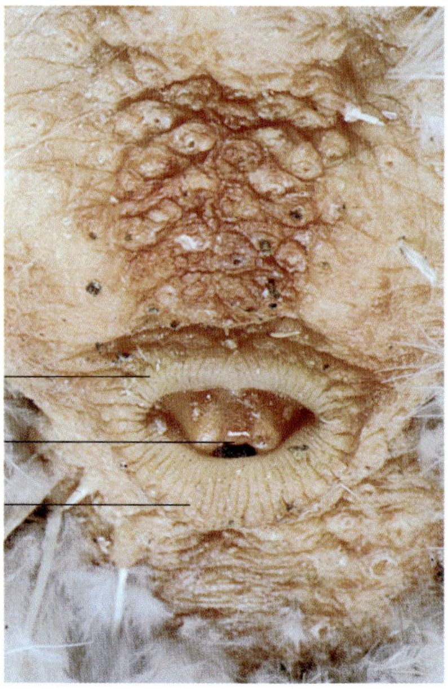

9.9 Vent of a male chicken.

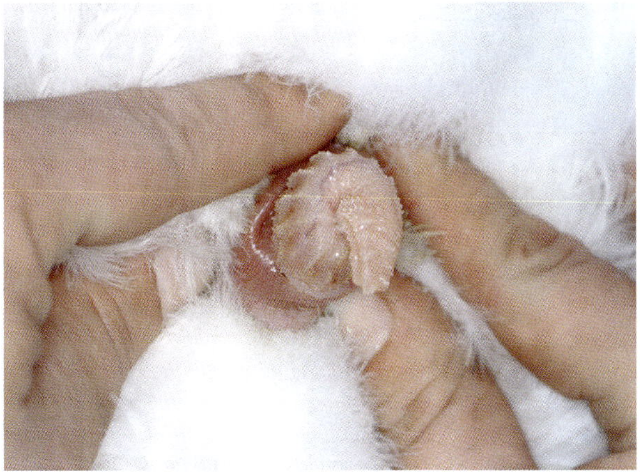

9.10 Sexing of a male goose (gander) by protruding the spiral phallus through the vent. The index and middle fingers are used to apply pressure between the cloaca and the pubic bone. Routine use of this technique for sexing monomorphic species requires considerable skill.

the phallus by two **lymph hearts** (cor lymphatica), each located above the transverse process of the first free caudal vertebra.

During ejaculation, semen flows over the surface of the phallus. In the drake and gander, the edges of the phallic sulcus close during ejaculation to form a tube.

Clinical aspects

Sexing of birds plays an important role in commercial poultry production, captive bird-breeding and conservation programs (e.g., for raptors, zoo and wild birds). Financial considerations are the most common reason for gender determination (e.g., selection of breeding pairs, separation of males and females in broiler production). Targeted selection of breeding pairs is also an important aspect of breeding programmes used in the management of hunting birds, zoo birds and certain wild bird species. While pet bird-owners may wish to know the sex of a bird for breeding purposes, they are often motivated simply by personal interest.

As the testes are internal, little **phenotypic sexual dimorphism** is evident in young birds. Most psittacines never exhibit phenotypic differences, and those observed in birds of prey may be very subtle.

Sexing of **day-old chicks** is usually performed by specialist operators and is based on inspection of the median phallic body, or median tubercle, in the cloaca. Feather morphology is also used in commercial operations for sexing of chickens ('feather sexing'). **Anseriformes** can be **sexed** by manually protruding the spiral shaped phallus through the vent (Figure 9.10). This technique involves the application of an appropriate amount of pressure on either side of the cloaca, with the bird restrained in dorsal recumbency. It requires considerable experience and skill.

In the late 1970s, endoscopy was adopted as a technique for sexing **parrots** (psittacines) (see Chapter 22 'Endoscopy'). Males are identified endoscopically by visualisation of the generally porcelain coloured, spherical or bean-shaped appearance of the testes. An important exception is the cockatoo, which has pigmented testes that appear black on endoscopic examination.

To ensure that the testes are differentiated from other structures, such as the inflection point in a loop of intestine, the branching testicular veins (usually three) should also be visualised. In addition, care must be taken to distinguish the testes from secondary or tertiary ovarian follicles, which are similarly rounded and may appear porcelain-coloured under endoscopic illumination. However, there are usually multiple follicles and they lack the superficial vascular network of the testes.

The **ductus deferens** is another useful anatomical feature for endoscopic identification of male birds. Its convoluted appearance allows it to be distinguished from the straight and potentially urine-filled (thus white) ureter, and from the oviduct of juvenile females (see Chapter 10 'Female genital organs').

In breeding birds, errors in gender determination may have significant financial consequences and can result in litigation. Thus, additional methods of sexing should also be considered. The current method of choice is DNA testing (see Chapter 17 'Common integument'), which can be carried out using commercially available kits.

Several other 'traditional' methods are also used in sexing. Due to the substantial increase in size of the testes during the breeding season, the male gonads can sometimes be **visualised** on **radiographs**, situated cranioventral to the cranial renal division. However, small inactive testes are usually not visible and can be confused with other structures located in the same area (e.g., the spleen). This method is therefore unreliable.

Examination of the **external cloacal structures** can be useful in **sexing** canaries (Serinus canariae). Appearing as cone-shaped projections from the vent in the male, these are undetectable in the female. The distance between the pubic bones, usually greater in laying females, can also be assessed.

Further **species-specific phenotypic differences** include the **colour of the cere** (see Chapter 6 'Digestive system' and Chapter 17 'Common integument'), **iris colour** (see Chapter 15 'The eye'), **plumage colour** and **feather morphology** (see Chapter 17 'Common integument') and **body size** (see Chapter 24 'Falconry and raptor medicine').

Collection of semen for artificial insemination is practised to a relatively limited extent in captive breeding programmes. Semen is obtained by massaging the cloacal region. Due to a lack of effective preservation techniques, artificial insemination is almost exclusively performed using fresh semen.

Among birds of prey, semen is generally collected from 'manned' birds (see Chapter 24 'Falconry and raptor medicine') that are strongly imprinted on humans, or caged birds that exhibit aggression towards their mate and express sexual behaviour towards humans. These birds are no longer accepted by their peers. Semen is collected using latex copulation hats worn on the head of the collector.

Diseases of the male genital organs include testicular tumours (high incidence in budgerigars, Melopsittacus undulatus), which are diagnosed using contrast radiography (Figure 9.11). These are frequently Sertoli cell tumours. Depending on their size they can result in dyspnoea and hormonally induced changes in the cere (see Chapter 6 'Digestive system' and Chapter 17 'Common integument').

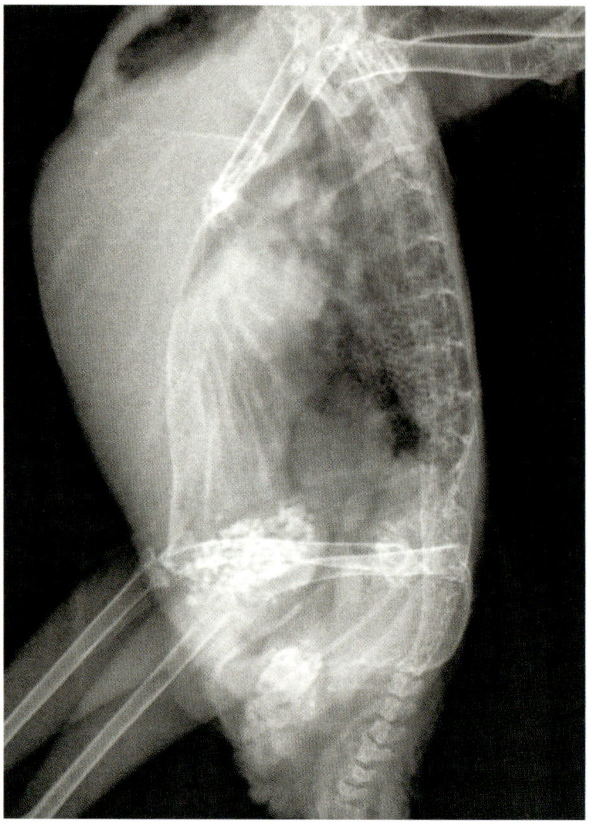

9.11 Contrast radiograph of a budgerigar (Melopsittacus undulatus) with a testicular tumour located cranioventral to the kidneys. There is typical ventral displacement of contrast medium-filled intestinal loops.

10

Female genital organs (organa genitalia feminina)

H. E. König, I. Walter, H. Bragulla and R. Korbel

The composition and development of the female avian genital apparatus differs substantially from that of mammals. Although paired symmetrically positioned ovaries and oviducts are present during embryonic development, only the **left ovary and oviduct** develop to **functional maturity** in most avian species. Those on the right side of the body rapidly **regress** after hatching, remaining throughout life as rudimentary structures that may be filled with fluid. This one-sided development presumably represents an adaptation that serves to facilitate flight by reducing bodyweight.

In several birds of prey and in the kiwi, only the right oviduct undergoes involution, while both ovaries develop fully. Complete development of both ovaries and oviducts has also been reported in some species. To date, however, definitive evidence for ovulation of oocytes from both the left and right ovaries has only been obtained in the female goshawk.

Avian embryogenesis begins in the maternal oviduct. At the time of oviposition, the embryo within the egg is still at a relatively early stage of differentiation (**ovipary**). Development is subsequently completed during the

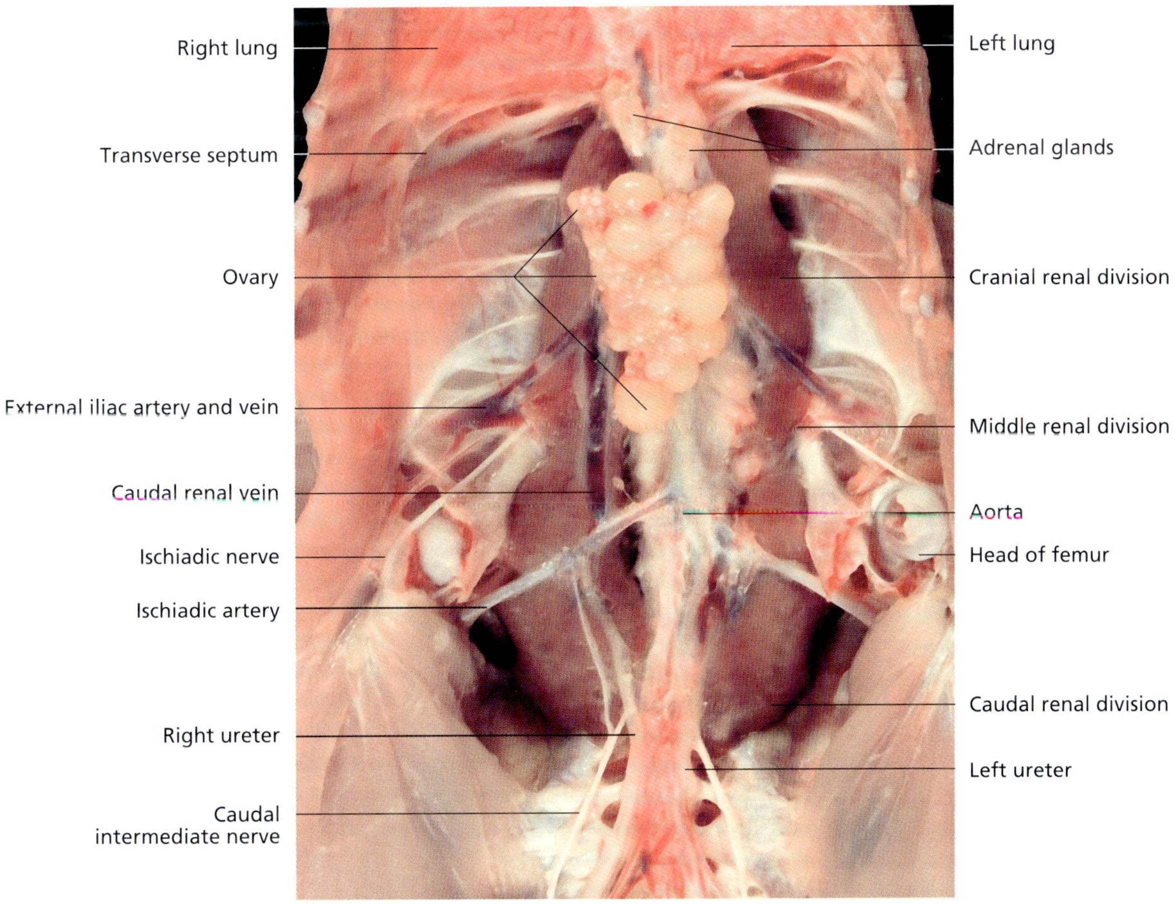

10.1 Ovary of a female chicken, intestines removed (ventral view). Courtesy of Dr Annette Kaiser, Munich.

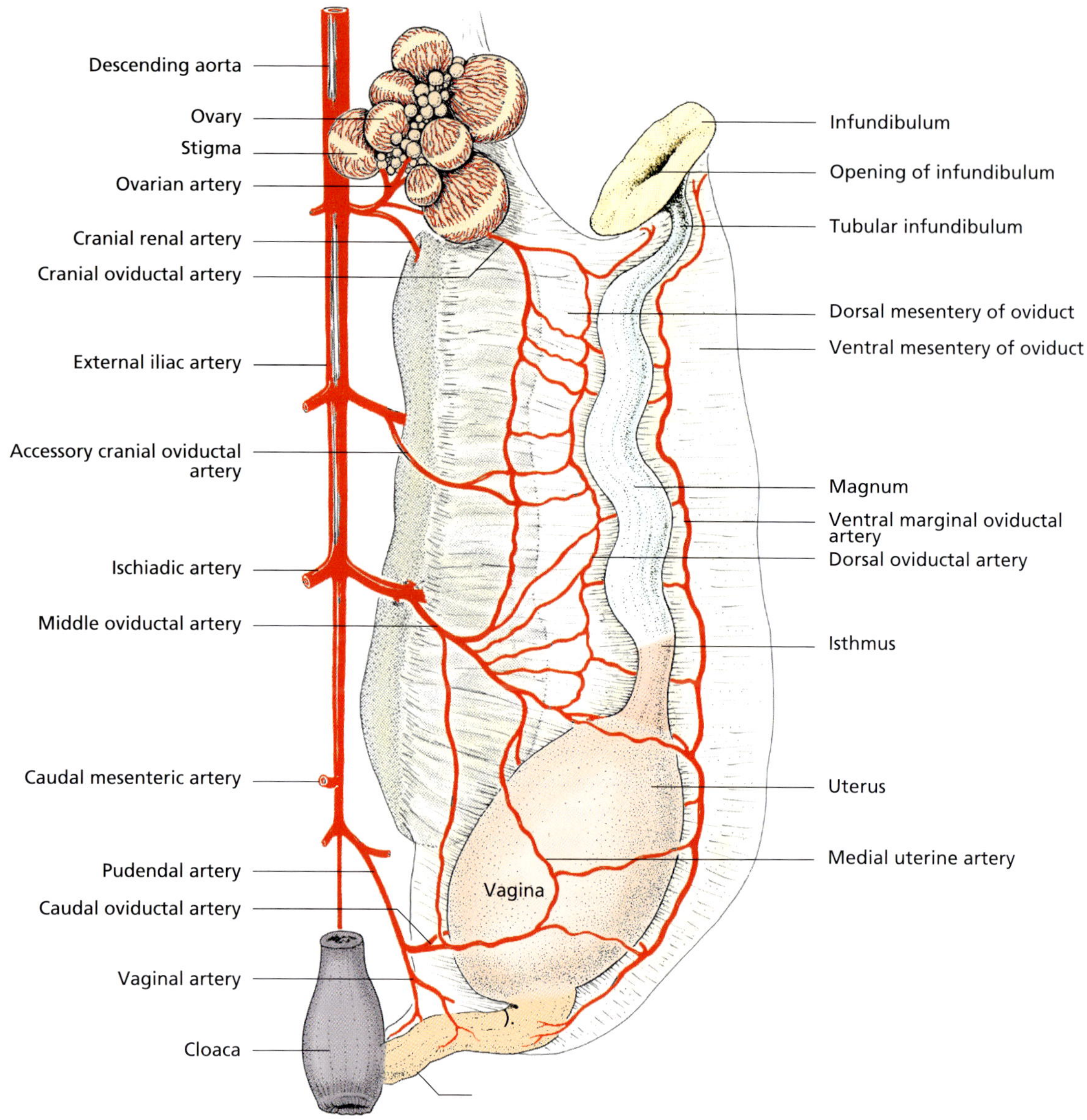

10.2 Vascular supply of the female genital organs of the chicken (schematic), adapted from Ghetie, 1976.

incubation period, which varies in length according to species.

The female genital organs are richly supplied with blood, particularly during the laying period. An overview of the vascular supply is provided in Figure 10.2.

Ovary (ovarium)

In genetically predetermined female embryos (note that in birds the female is heterogametic), large numbers of gametes migrate from the right to the left ovary in the first days of embryonic development.

From **day seven**, the **left ovary** begins to take on its definitive form, while the medulla of the right ovary retains only a few undifferentiated germ cells and oocytes.

The ovary is located craniodorsally in the **intestinal peritoneal cavity**. It is attached to the body wall by a short **mesovarium** (a few millimetres long in chickens) and lies against the caudal margin of the left lung, the left adrenal gland, the cranial pole of the left kidney, the aorta and the caudal vena cava. The infundibulum of the oviduct extends to the caudal end of the ovary (Figures 10.2 and 10.6).

In the juvenile and non-laying mature female chicken, the ovary is a compact, roughly triangular structure, measuring approximately 15–20mm by 10mm and weighing around 0.5g. Its surface has a finely granular appearance. As the ovarian follicles mature prior to laying, the ovary increases in just a few days to a size of 110mm by 70mm, reaching a weight of more than 60g (Figure 10.6).

At the **time of hatching**, the ovary consists of a **cortex** (cortex ovarii) and a **medulla** (medulla ovarii). The cortex contains the ovarian follicles, comprising oocytes surrounded by follicular epithelium. At **sexual maturity**, the macroscopic separation between cortex and medulla becomes less distinct and is eventually completely obliterated (Figure 10.9).

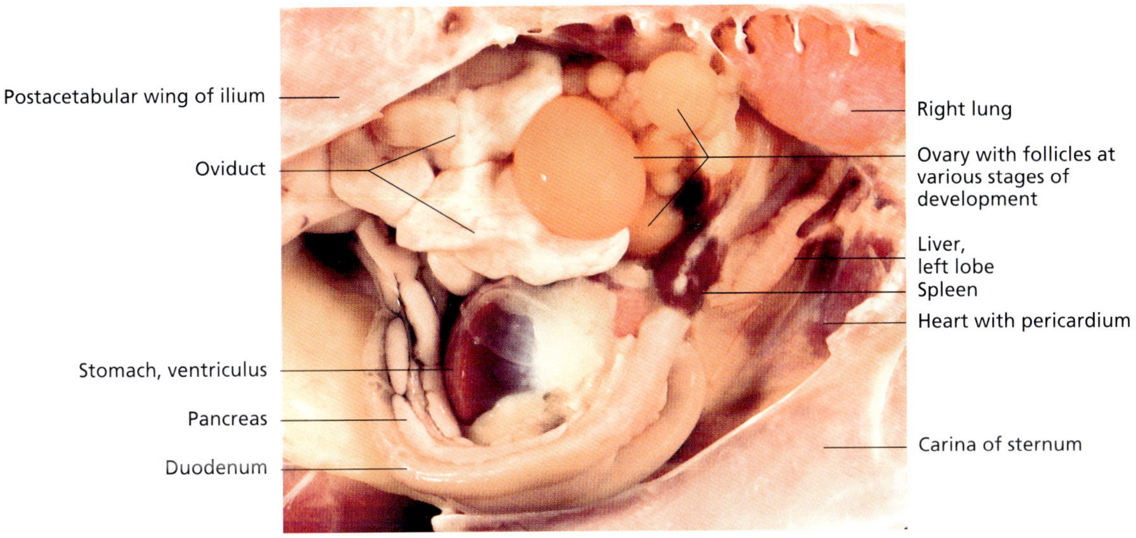

10.3 Ovary, oviduct and parts of the gastrointestinal tract of a female chicken (right aspect, superficial view). Courtesy of Dr Annette Kaiser, Munich.

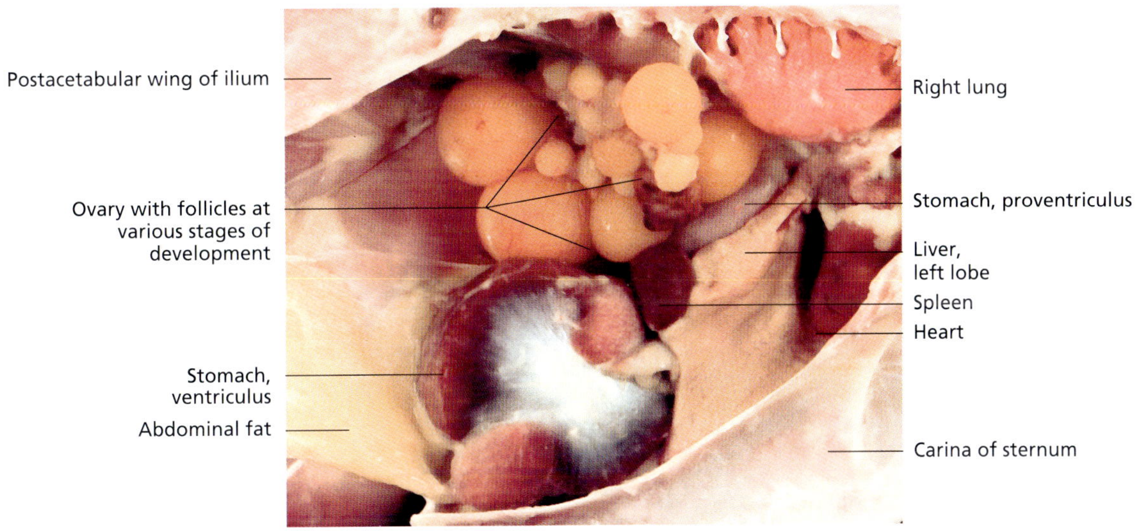

10.4 Ovary and stomach of a female chicken (right aspect, deep view). Courtesy of Dr Annette Kaiser, Munich.

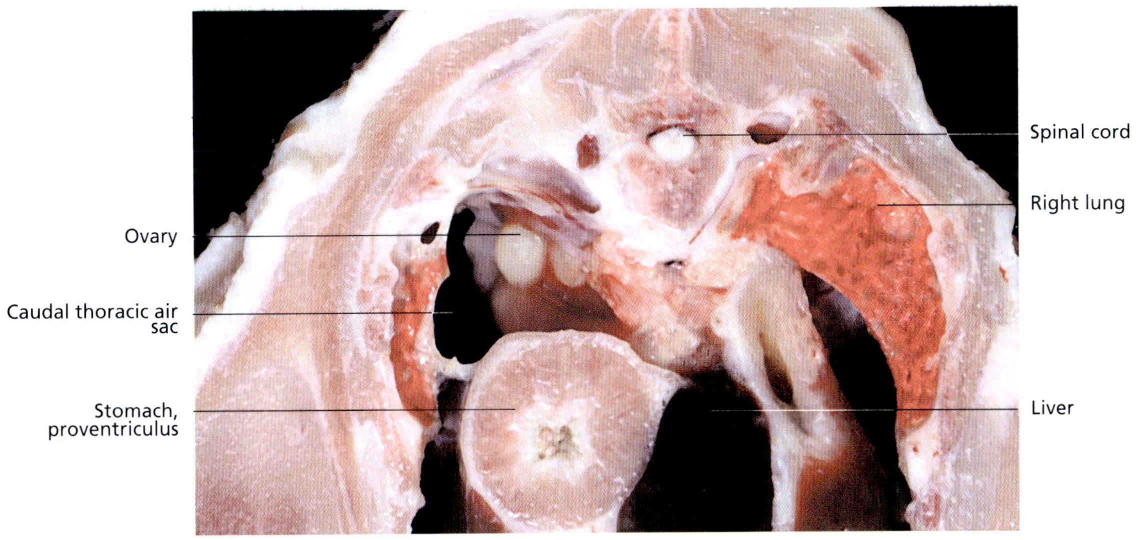

10.5 Transverse section of a female chicken at the level of the ovary. Courtesy of Professor Dr J. Ruberte, Barcelona.

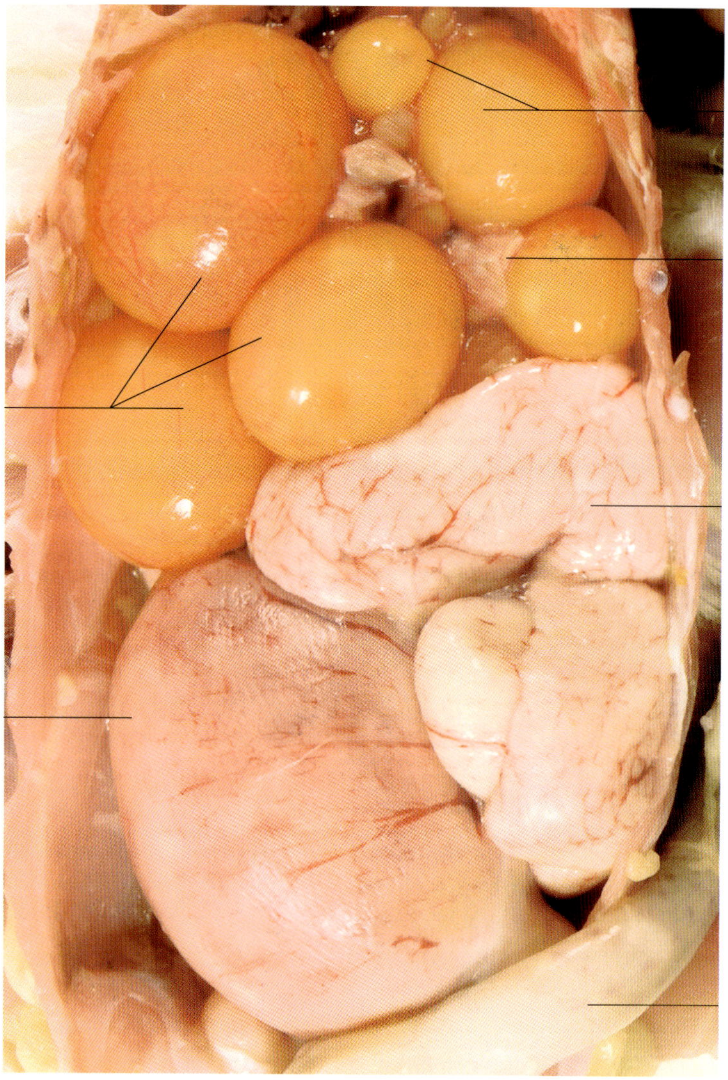

10.6 Genital apparatus of a laying female chicken (ventral view, gastrointestinal tract removed). Courtesy of Dr Annette Kaiser, Munich.

During the laying period, follicles in the **ovarian parenchyma** grow to varying sizes. At a given point in time, most follicles have a diameter of approximately 5mm. Others mature fully, reaching up to 40mm in diameter.

The **mature oocyte of birds** is the **largest female gamete in the animal kingdom** (Figures 10.3, 10.4, 10.6 and 10.8). At this stage of development, the surrounding follicular **wall consists of several layers**. This pre-ovulatory follicle is equivalent to the tertiary, or Graafian, follicle in mammals. It is connected to the ovary by a peduncle into which blood vessels, nerves and smooth muscle cells are drawn. The follicular wall, which has undergone increasing vascularisation and innervation during the development of the oocyte, also contains smooth muscle (Figure 10.7). Positioned meridionally in the follicle wall is a pale and relatively avascular region known as the **stigma** (Figure 10.8).

Endoscopic visualisation of ovarian follicles can be used for gender determination in monomorphic avian species (see Chapter 22 'Endoscopy').

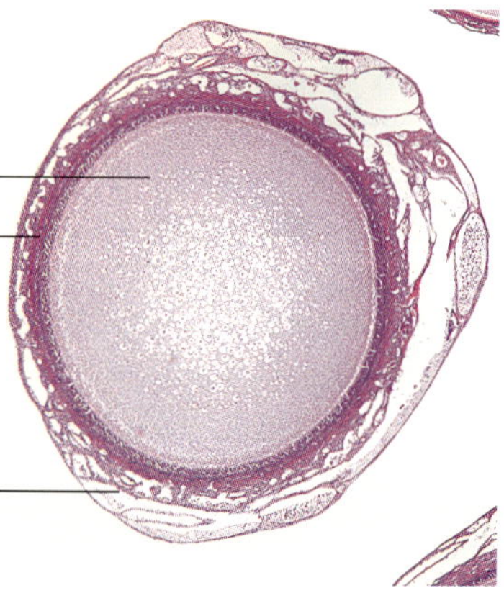

10.7 Histological section of a follicle of a female chicken.

Oogenesis

The development and maturation of the polylecithal (yolk-rich) avian oocyte begins, as in mammals, early

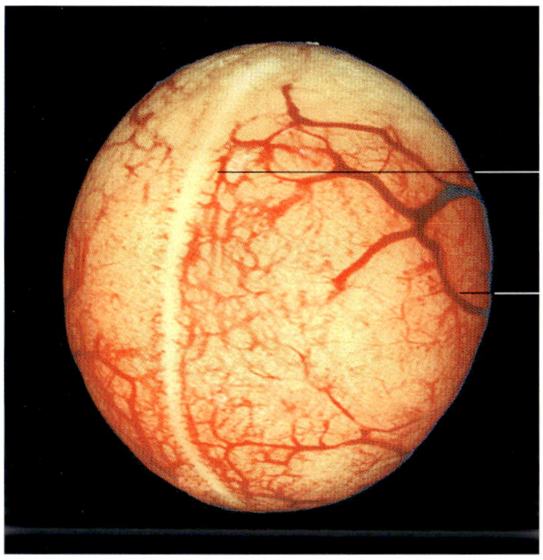

10.8 Mature follicle of a female chicken. Courtesy of PD Dr S. Reese.

in embryogenesis. Primordial germ cells migrate from the embryonic yolk sac to the gonadal area where they differentiate into oogonia. The oogonia undergo repeated mitotic division. By the time they reach the prophase of their first meiotic division they are considered primary oocytes. They remain in the diplotene stage of the prophase of meiosis until shortly before ovulation.

During this phase, the oocyte increases in size through the uptake of large amounts of yolk (**vitellus**) into its cytoplasm (**vitellogenesis**). The gel-like yolk consists of species-specific lipids and soluble proteins.

Yolk formation takes place in three phases of different length. During the first phase, which may last several years, the oocytes undergo a modest increase in size. In the subsequent phase, the oocyte grows markedly in volume over a period of 8–10 months, reaching 4mm in diameter in chickens. The third phase is associated with a substantial accumulation of yolk, with the oocyte reaching its typical final size of around 40mm in the chicken and 20mm in the pigeon. Taking approximately 14 days, this stage is the distinguishing developmental feature of the polylecithal avian oocyte.

The completion of the **first meiotic division** occurs just a few hours prior to ovulation, resulting in a **secondary oocyte** and production of the first polar body. In contrast to mammals, a second division of this polar body has not been observed in birds.

Ovulation occurs under the influence of the peptide hormone luteinising hormone (LH) produced by the adenohypophysis. The secondary oocyte is released by rupture of the follicular wall along the stigma. Spermatozoa, if present, penetrate the oocyte around 15 minutes after

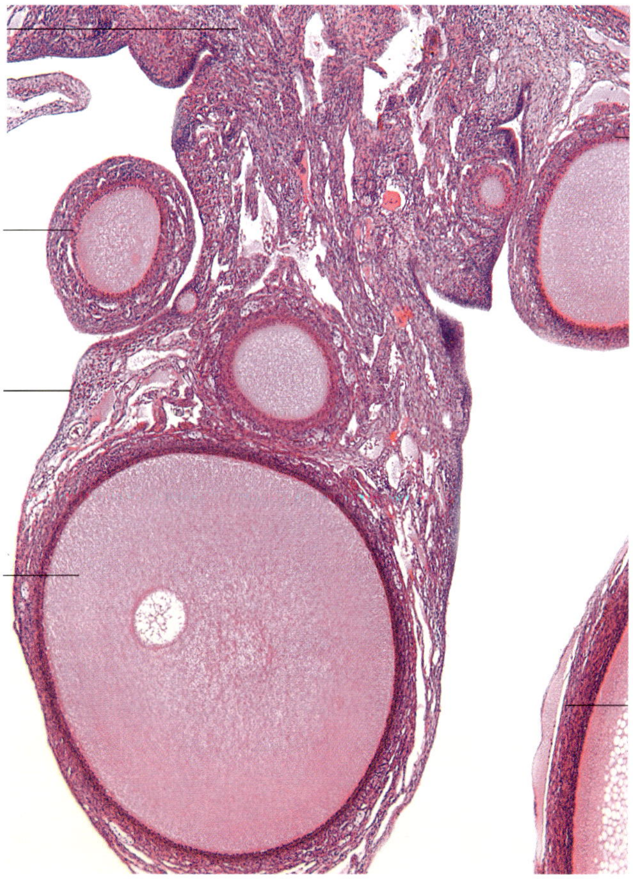

10.9 Histological section of the ovary of a female chicken.

ovulation. This is followed, within the infundibulum of the oviduct, by the **second phase of meiosis** (resulting in a mature female gamete, or ovum, and a second polar body) and fertilisation. In contrast to mammals, **polyspermy** (penetration of the oocyte by more than one spermatozoon) may occur but only one spermatozoon fuses with the nucleus of the oocyte.

As indicated above, the female is **heterogametic** in birds. The oocyte contains either a Z or W chromosome, whereas all avian spermatozoa contain the Z chromosome. Gender is therefore determined **prior to fertilisation**.

Each subsequent ovulation takes place about half an hour after an egg is laid. Not all oocytes enter the infundibulum. At the beginning and the end of the laying period, some ovulated oocytes pass into the coelomic cavity where they are quickly resorbed. Occasionally these may undergo concretion and persist for some time. This phenomenon is not clinically significant.

Immediately after ovulation, the wall of the follicle regresses. Within six days, only an insignificant vestige remains and eventually this disappears altogether. A corpus luteum similar to that of mammals is **not formed**. As in mammals, **oestrogens** are produced by endocrine cells of the thecal cells of the follicle and by the zonae parenchymatosae of the ovary. **Androgens** are produced by interstitial ovarian cells. The post-ovulatory follicle is capable of producing **progesterone**, thus functionally resembling the mammalian corpus luteum. Oestrogens are responsible for stimulating the production of yolk by the liver. The yolk is transferred via the bloodstream to the ovarian follicles.

Oviduct (oviductus)

The oviduct (Figures 10.2, 10.3, 10.6 and 10.10) of birds consists of the:

- infundibulum,
- magnum,
- isthmus,
- uterus and
- vagina.

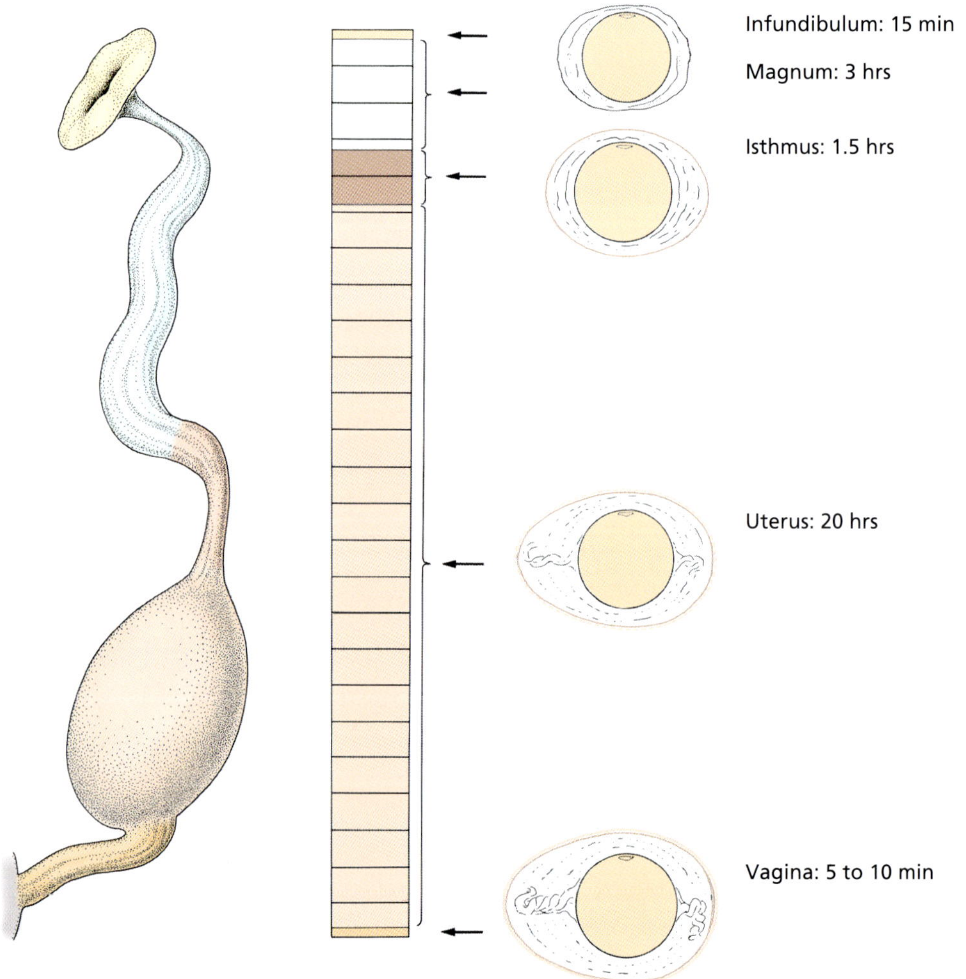

10.10 Development of the egg of the chicken in the segments of the oviduct (schematic).

As with the ovary, only the left oviduct develops fully in most birds. In the non-laying female chicken, the oviduct is an inconspicuous tube within the common body cavity. During the laying period, the oviduct increases considerably in size, forming loops that fill the caudodorsal portion of the coelomic cavity, within the intestinal peritoneal cavity. Towards the end of the laying period, it reaches 65mm in length and 75g in weight. During brooding and moulting, the oviduct is once again considerably foreshortened.

The oviduct adds successive layers to the developing egg. In the chicken, the passage of the egg through all of the segments of the oviduct takes approximately 25 hours (Figure 10.10).

The mesentery of the oviduct consists of dorsal and ventral components (Figure 10.2). In the chicken, the **dorsal mesentery** (lig. dorsale oviductus) is approximately 3cm long. It arises on the dorsolateral wall of the body cavity at the level of the last rib and passes caudal to the left kidney, descending gradually towards the cloaca.

The ventral mesentery (lig. ventrale oviductus) extends from the ventral infundibulum to the ventral surface of the vagina (Figure 10.2).

In common with other hollow viscera, the wall of the oviduct consists of a:

- **tunica muscularis**:
  - stratum circulare and
  - stratum longitudinale;
- **tunica serosa**:
  - serosal epithelium (epithelium serosae) and
  - lamina propria (lamina propria serosae);
- **tunica mucosa**:
  - mucosal epithelium (epithelium mucosae) and
  - lamina propria (lamina propria mucosae);
- **tela submucosa**.

Initially the epithelium is simple and flat, transitioning through cuboidal to columnar and pseudostratified columnar. The columnar cells comprise endoepithelial glandular cells, superficial ciliated cells and basal cells. The distal segments contain regions of pseudostratified columnar epithelium that subsequently become reduced in thickness.

The lamina propria contains **glands** along most of its length. These vary considerably in structure, number and density in the different segments of the oviduct.

Mucosal folds are developed to a greater or lesser degree throughout the oviduct. In chickens, the height and thickness of the folds varies distinctly from one segment of the oviduct to another. The folds are arranged in a gentle spiral such that the egg is turned around its longitudinal axis as it passes through the oviduct.

The circular and longitudinal muscle layers produce **peristaltic contractions** that assist in **transporting the egg**, as well as antiperistaltic contractions that **convey sperm** in the opposite direction.

Infundibulum

The infundibulum consists of a funnel-shaped proximal section and a tubular distal portion. Its **opening** (ostium infundibulare) (Figure 10.2) is approximately 80mm wide in the chicken. In contrast to mammals, the infundibular opening is surrounded by relatively few fimbriae

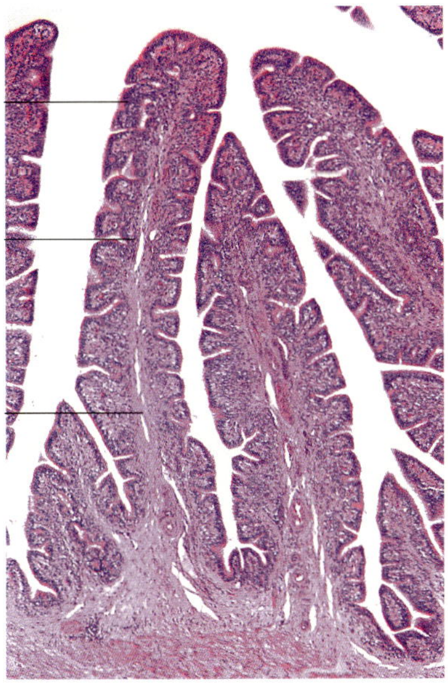

10.11 Histological section of the magnum of the oviduct of a female chicken.

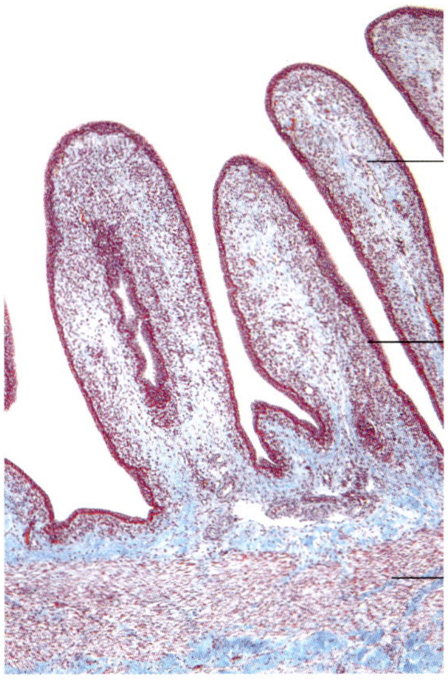

10.12 Histological section of the uterus of a female chicken.

(fimbriae infundibulares). The thin wall of the funnel is thrown into shallow primary and secondary folds, the latter being particularly pronounced in chickens. Initially the infundibulum is non-glandular. Toward the caudal portion of the funnel, alveolar invaginations known as **fossae glandulares infundibuli** appear in the lamina propria. In the subsequent, tubular section of the infundibulum, the glands increase in size and complexity forming tubular **glandulae tubi infundibulares**.

The wall of the funnel contains smooth muscle, giving it contractile properties that aid in the uptake of oocytes after ovulation. In the **tubular section**, the wall of the infundibulum is thicker and features more prominent primary and secondary mucosal folds. **Fertilisation** of the oocyte by the spermatozoa occurs in this segment.

Transit of the oocyte through the infundibulum takes around 15–20 minutes in the chicken (Figure 10.10), but can be considerably faster or slower in other species. During this time, glycoproteins and phospholipids secreted by the glands are laid down around the oocyte to form the **chalaziferous layer**. This inner dense layer of albumen later forms the twisted **chalazae** that suspend the yolk as it rotates about its longitudinal axis (Figure 10.13).

Magnum

In all bird species, the magnum is the longest and broadest segment of the female genital tract. In the chicken it reaches a length of 34cm. Similar to the uterus of the pig, it follows a looping course (resembling that of the small intestine). In this segment, the epithelium transitions from pseudostratified columnar to a single layer of mostly columnar cells. The mucosa is arranged in substantial folds (up to 22mm deep, without secondary folds) that are richly endowed with coiled (and, in the chicken, extensively branched) tubular glands (glandulae magni), forming a substantial secretory apparatus.

The glands produce ovalbumin, ovotransferrin and ovomucoid. These hygroscopic proteins form the main component of the albumen, to which water is added in the uterus. The time spent by the oocyte (or zygote) in the magnum is approximately **three hours** (Figure 10.10).

Isthmus

The isthmus is clearly distinguishable macroscopically from the surrounding segments of the oviduct. It is approximately 10cm long in the chicken. The commencement of the isthmus is marked by the translucent, non-glandular **pars translucens isthmi**, in which the mucosa is completely devoid of folds.

Further distally, the mucosa becomes thicker, is thrown into longitudinal folds and contains numerous **tubular glands** (glandulae isthmi). The folds are shallower than in the magnum and are associated with secondary folds of varying depth. The oocyte (or zygote) passes through the isthmus in around **1.5 hours** (Figure 10.10).

The glands of the isthmus are similar to those of the magnum. Their secretory product, comprising particularly stable sulphur-containing keratin-type proteins, is unique to this segment of the oviduct and forms the **inner and outer shell membranes**. The **air cell** later forms in the space between these membranes, at the blunt end of the egg. More albumen is also added in the isthmus.

Uterus (metra)

The uterus is sometimes also referred to as the 'shell gland'. It continues from the isthmus without any obvious macroscopic demarcation and is about 8cm long in the chicken. Initially tubular, the uterus expands into a pouch-like segment. The muscular tunic is well-developed. Longitudinal folds in the mucosa are intersected by circular folds, giving rise to leaf-like lamellae.

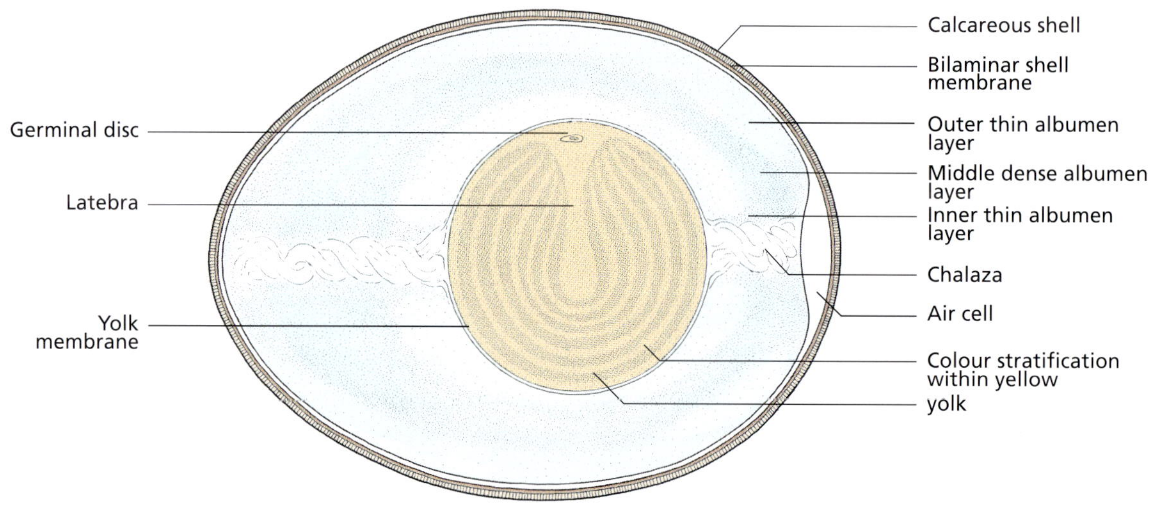

10.13 Longitudinal section of the egg of a chicken (schematic), adapted from Waibl and Sinowatz, 2004.

Branching tubular **uterine glands** (glandulae uterinae) are similar morphologically and functionally to those of the isthmus, although they are even more tightly packed in the lamina propria. The final component of the albumen is laid down in the uterus, and the addition of a large amount of water substantially '**plumps up**' the hygroscopic mix of proteins that make up the completed 'egg white'.

The egg spends **around 20 hours** in the uterus (Figure 10.10), considerably longer than in any other segment of the oviduct. Most of this time is occupied by the formation of the calcareous shell from calcium carbonate and other calcium salts. The organic matrix of the shell is produced from secretions of the columnar epithelial cells.

The thin, organic outermost layer of the egg, known as the **cuticle** (cuticula), is also derived from the uterus.

Vagina

At the junction between the uterus and the vagina, the already strong circular muscle thickens to form the **m. sphincter vaginae**. The vagina is approximately 8cm long and folded upon itself into a sigmoid shape. Its muscular wall is well-developed throughout its length.

The vaginal mucosa has a simple ciliated columnar epithelium and is arranged in narrow primary and secondary folds. Near the m. sphincter vaginae, the lamina propria contains the branching **tubular utero-vaginal sperm host glands** (tubuli spermatici or fossulae spermatici) that serve as storage sites for sperm.

These **reservoirs** are remarkable in that they can house viable sperm for some weeks, allowing a female chicken to lay fertilised eggs for up to two weeks after mating (the closest equivalent is the bitch, in which sperm can be stored for a week).

The time taken for the egg to pass through the vagina is highly variable with an average range of **5–10 minutes**.

Structure of the avian egg

From an evolutionary perspective, the egg represents an ingenious strategy for the development and nourishment of young in flying animals. The delivery of offspring as eggs, and the development of the embryo outside the body of the female, permits birds to reproduce with a minimum increase in body weight, leaving their capacity for flight relatively unencumbered.

The avian oocyte is the **largest female gamete in the animal world**. Prior to ovulation, the yolk-rich oocyte (like the ovum of mammals) is surrounded by the **theca folliculi**, consisting of an outer theca folliculi externa and an inner theca folliculi interna. The oocyte lies within the theca folliculi interna and is surrounded by a thin **inner yolk membrane** (lamina perivitellina), which rests on the **cytolemma of the oocyte** (cytolemma ovocyti).

The **germinal disc** (discus germinalis; blastodisc) is a round, pale 1–2mm area on the **surface of the oocyte**. Based on this arrangement, in which (as in amphibians) the embryo-forming cytoplasm is concentrated at one pole (animal pole) and the yolk accumulates at the other (vegetal pole), the avian oocyte is classified as **telolecithal**. After fertilisation, cleavage begins at the animal pole with the formation of blastomeres. Cleavage is incomplete as it is impeded by the large mass of yolk. This partial type of division occurring at the animal pole is referred to as **meroblastic discoidal cleavage**.

During the initial divisions in the fertilised ovum, the blastodisc differentiates into the **area pellucida** and **area opaca**. A pendulous strand of 'white yolk' extends from the blastodisc into the 'yellow yolk' to form the **latebra** (Figure 10.13).

Upon ovulation, the oocyte, surrounded by the inner yolk membrane, enters the infundibulum of the oviduct. Within the oviduct, an outer yolk membrane is laid down and a series of layers is added (Figure 10.13; see also above).

The **albumen** component consists of the:

- stratum chalaziferum (chalazae),
- inner thin albumen,
- middle dense albumen,
- outer thin albumen.

The **egg membranes** (membranae testae) comprise the:

- inner shell membrane and
- outer shell membrane.

In all species, the avian egg has one relatively blunt and one more pointed pole. At the blunt end of the egg, the two egg membranes separate to form the air cell. The calcareous shell (testa) consists of:

- an inner organic stratum mammillarium,
- an organic stratum spongiosum incorporating a palisade layer (stratum vallatum) of calcium carbonate crystals (more than 40µm wide),
- an organic outer layer (cuticle; cuticula).

The **cuticle** is a semi-permeable barrier that has an important role in preventing bacterial penetration of the egg. It should not be removed by washing. Pores (c. 10µm wide) on the surface of the shell, covered only by the cuticle, lead into air canaliculi that extend to the outer shell membrane. These tiny air passages allow the diffusion of gases and water vapour.

Clinical aspects

For a description of the sexing of monomorphic birds (i.e., those lacking phenotypic sexual dimorphism), refer also to Chapter 9 'Male genital organs'.

When using **endoscopy for sex determination**, the **ovary** is the most important anatomical landmark for identifying female animals. It is associated with the cranial pole of the kidney and is developed to a varying degree, according to the age and reproductive status of the animal (see Chapter 22 'Endoscopy'). Depending on species, the ovary can be identified from the age of eight to ten weeks.

In the vast majority of birds, only the **left ovary** and **oviduct** are fully developed. Exceptions include various birds of prey, including the common kestrel (Falco tinnunculus), and some species commonly kept in zoos (e.g., the scarlet ibis [Eudocimus ruber]), in which both ovaries are fully formed. As a result of pigmentation, the ovary of some species (e.g., cockatoos [Cacatua spp.]) may appear black.

In juvenile animals that have not begun to lay, the **oviduct** is an inconspicuous, straight, semi-transparent tubular structure. When **laying has commenced**, the oviduct manifests as pale, porcelain-coloured meandering loops. The increase in the size of the oviduct in **laying birds** is substantial, with a concurrent increase in blood flow through the larger longitudinally oriented ventral marginal oviductal artery and the smaller circularly coursing middle oviductal arteries. These vessels have a considerable influence on the superficial appearance of the oviduct and can be an important means of distinguishing between the oviduct and loops of intestine during endoscopy and laparotomy. Intestinal loops are typically darker in colour (depending on the colour of the contents) and their associated arterial supply assumes a predominantly circular pattern.

When enlarged by the presence of mature follicles, the ovary may be visible **radiographically**, cranioventral to the cranial renal division. However, this is not a reliable means of determining gender due to the presence of other structures in this region that may have a similar radiographic appearance (e.g., the spleen).

Additional phenotypic features used in sexing are described in Chapter 9 'Male genital organs'.

During oviposition, the **pointed pole of the egg** usually emerges first. This process is associated with a more or less complete **physiological prolapse** of the cloaca. In hens kept for egg production, this can be a stimulus for excessive pecking (cannibalism) by other animals.

A **prolapse** that persists after the egg is deposited is considered to be pathological. It should be noted, however, that cloacal prolapse can also have other causes, including hyperoestrogenism, excessively intense lighting and cloacal papillomas. Surgical reduction may be required in some cases (e.g., via purse string suture).

Chicks penetrate the egg shell using the so-called 'egg tooth' (see Chapter 6 'Digestive system') located on the upper beak. The chick's head and neck move repeatedly in an anti-clockwise direction until the shell is opened (Figure 10.15).

The **air cell** between the inner and outer shell membranes at the **blunt end of the egg** is of practical significance, in terms of both the **production of eggs for human consumption** and the management of **breeding programs** (e.g., commercial flocks, pet birds, zoo species and birds of prey). After oviposition, the egg cools (from the body temperature of the hen to the ambient environmental temperature) and the volume of its contents decreases (evaporation of water in the presence of low ambient humidity). This causes air to be drawn into the egg through the pores in the shell, resulting in an **increase in the size of the air cell**. Micro-organisms present on the surface of the shell can enter the egg by the same mechanism, particularly if the cuticle is damaged (e.g., by washing). This type of contamination can reduce the shelf-life of the egg or result in horizontal (egg-borne) infection of the chick prior to hatching. Refrigerated storage of eggs can, contrary to the intended outcome, lead to a reduction in shelf-life as the low ambient temperature and humidity can actually promote penetration of the egg by micro-organisms present on the surface of the shell. Furthermore, it is inappropriate to store the eggs with their blunt end down, as the air cell migrates to the pointed pole, encouraging further intake of air and other materials.

Egg-binding is one of the most important **disorders of the female reproductive tract** (Figure 10.14). In addition to excessive egg size, either in absolute or relative terms, egg-binding has multiple causes including calcium deficiency and infectious disease. Initial diagnosis is based on palpation, although radiography should also be performed for confirmation and more detailed assessment. In simple cases, treatment can be undertaken by removing the egg per viam naturale (via the 'natural passage', i.e., the cloaca). It is important, however, to avoid breaking the egg as this can result in injury and incomplete removal of the egg.

If the bird has been egg-bound for a prolonged period, radiographic findings may include roughening of the surface of the shell, resulting from inflammation.

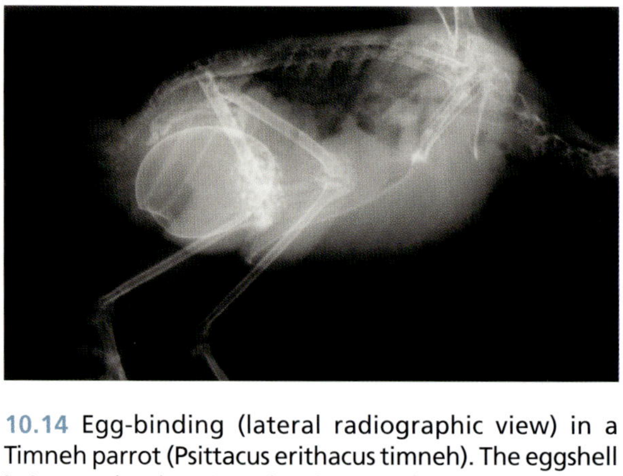

10.14 Egg-binding (lateral radiographic view) in a Timneh parrot (Psittacus erithacus timneh). The eggshell is incompletely mineralised, its surface is roughened and it is partly collapsed. This indicates that the bird has been egg-bound for a prolonged period.

10.15 Freshly hatched chick (Araucana) with eggshell. The green colouring of the shell is normal and results from accumulation of products of metabolism of blood components. Green-shelled eggs are marketed as a specialty product, although their contents are essentially the same as those of conventional eggs.

In these cases, surgical removal of the egg by **laparotomy** is required. If extensive inflammation and adhesions are present, which can predispose to recurrence, hysterectomy should also be performed. This involves removal of the oviduct – with transection of the ventral and dorsal ligaments of the oviduct – from as far cranially as possible (ideally at the junction between the tubular infundibulum and the magnum) to a point just before the oviduct opens into the cloaca. Double ligation is required to reduce bleeding and to seal off the cloacal lumen.

It is particularly important to remove all of the hormonally active uterus to break the feedback loop acting on the ovary, which would otherwise continue to produce oocytes that may enter the coelomic cavity. The ovary itself is difficult and dangerous to remove because of its very short ligament and the close association of its blood supply with the major vessels.

A multitude of infectious diseases can also affect the female reproductive tract, resulting in depigmentation of eggs, weak or incomplete eggshells, calcification defects or very small, yolkless eggs. These include infectious bursitis, avian encephalomyelitis and egg drop syndrome.

11

Cardiovascular system (systema cardiovasculare)

J. Ruberte, H. E. König, R. Korbel and C. Klupiec

The avian circulatory system is distinguished from that of mammals by the following features:

- the aortic arch lies to the right of the median plane,
- there are three renal arteries on each side of the body,
- the presence of a renal portal system,
- the presence of two cranial vena cavae,
- the presence of two hepatic portal veins.

Avian blood also has characteristic features including:

- nucleated red blood cells,
- heterophils (functional equivalent of neutrophils),
- thrombocytes (functional equivalent of platelets).

Heart (cor)

The heart occupies a **midline position** in the cranial portion of the body cavity, partially surrounded on both sides by the **liver** (Figures 11.2 and 11.3). Relative to body size, birds have a larger heart than mammals. The **septum** separating the right and left sides of the heart is fully formed at hatching. Small perforations present in the interatrial septum during embryogenesis (Figure 11.1) close after the bird has hatched.

The heart is surrounded by the **pericardium**, which is similar in structure to that of mammals (Figure 11.2).

The pericardium is composed, from exterior to interior, of the following layers:

- peritoneum,
- fibrous pericardium (pericardium fibrosum) and
- parietal serous pericardium (pericardium serosum parietale).

At the great vessels at the base of the heart (basis cordis), the parietal serous pericardium reflects onto the myocardium to become the epicardium (or visceral serous pericardium). This gives rise to the **pericardial cavity** (cavum pericardii). The pericardium is joined to the dorsal surface of the sternum by the **sternopericardial ligament** and to the ventral mesentery of the liver by the **hepatopericardial ligament**.

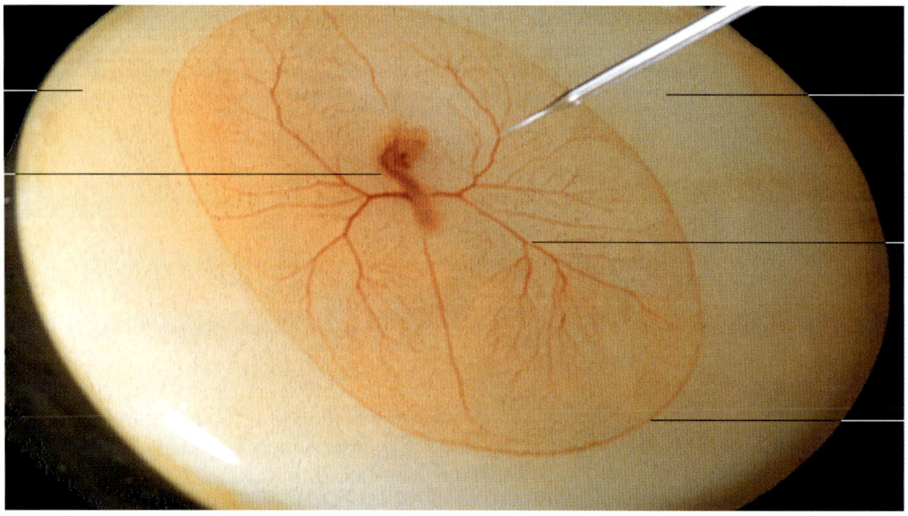

11.1 Blastoderm of a chicken with vitelline vessels. Courtesy of Professor A. Carretero, Barcelona.

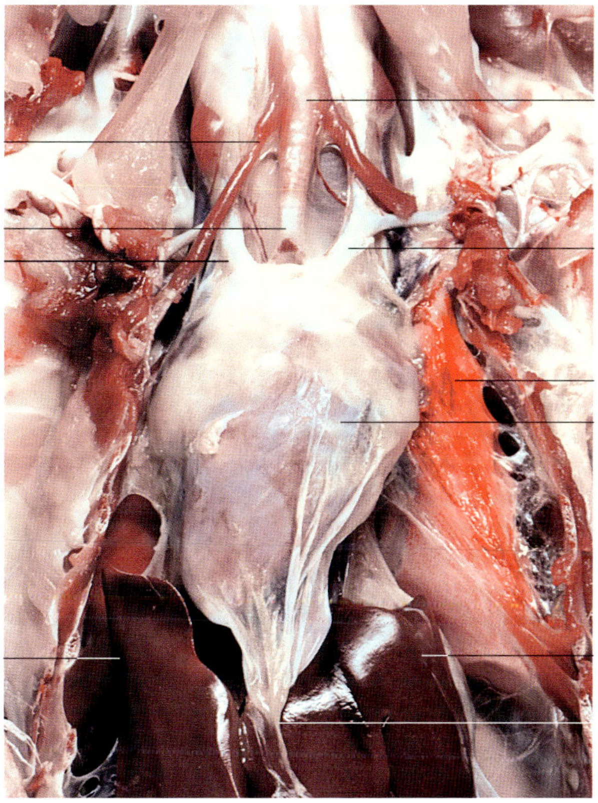

11.2 Heart and pericardium of a chicken (ventral view).

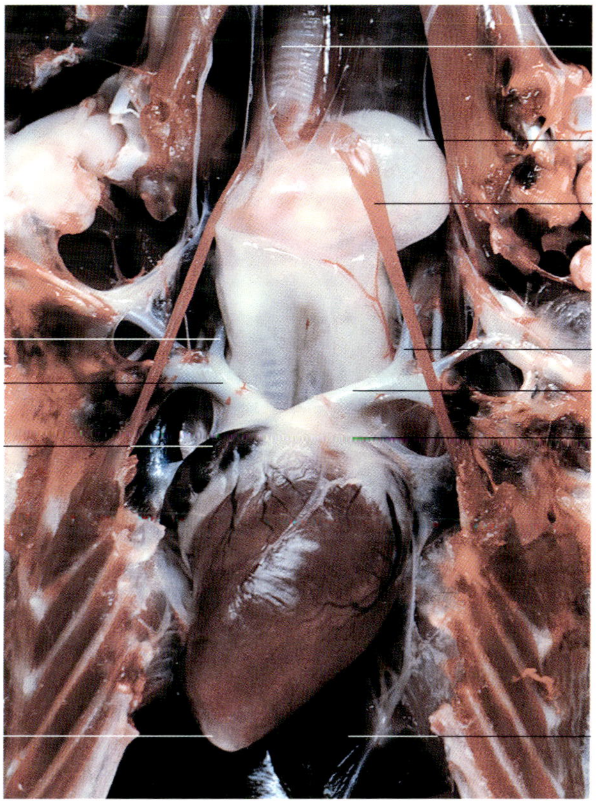

11.3 Heart (pericardium removed) in a male Indian runner duck (ventral view).

Dorsally, the heart is related to the **lungs** and the **horizontal** and **oblique septa**. Cranioventrally, the base of the heart and the great vessels are embedded in the clavicular air sac. Opening of the pericardial sac reveals the fat-filled **coronary groove** (sulcus coronarius) (Figures 11.3ff.)

The **longitudinal paraconal** and **subsinuosal ventricular grooves** (sulcus interventricularis paraconalis and subsinuosus) are indistinct and do not reach the apex of the heart. In chickens and ducks, a **small fat depot** is typically present at the **apex**.

The **chambers of the heart** are similar to those of mammals. Manifesting as a triangular shelf of muscle, the **atrioventricular valve** (valva atrioventricularis dextra) is situated between the **right atrium** (atrium cordis dextrum) and the **right ventricle** (ventriculus cordis dexter) (Figure 11.5). It has no **chordae tendinae**.

The **left atrioventricular valve**, composed of three indistinctly defined cusps, is located at the **left atrioventricular opening** (ostium atrioventriculare sinistra). Its associated **chordae tendinae** arise from three flat **papillary muscles** (mm. papillares) (Figure 11.6). Both the aortic valve and the valve of the pulmonary trunk consist, as in mammals, of three **semi-lunar cusps**.

Clinical procedures relating to the heart (e.g., cardiac puncture) are described in Chapter 21 'Medication and blood collection techniques'.

Blood vessels of the heart

In common with their mammalian counterparts, domestic birds have **right** and **left coronary arteries** (a. coronaria dextra and a. coronaria sinistra). In contrast to mammals, however, the right coronary artery is considerably larger

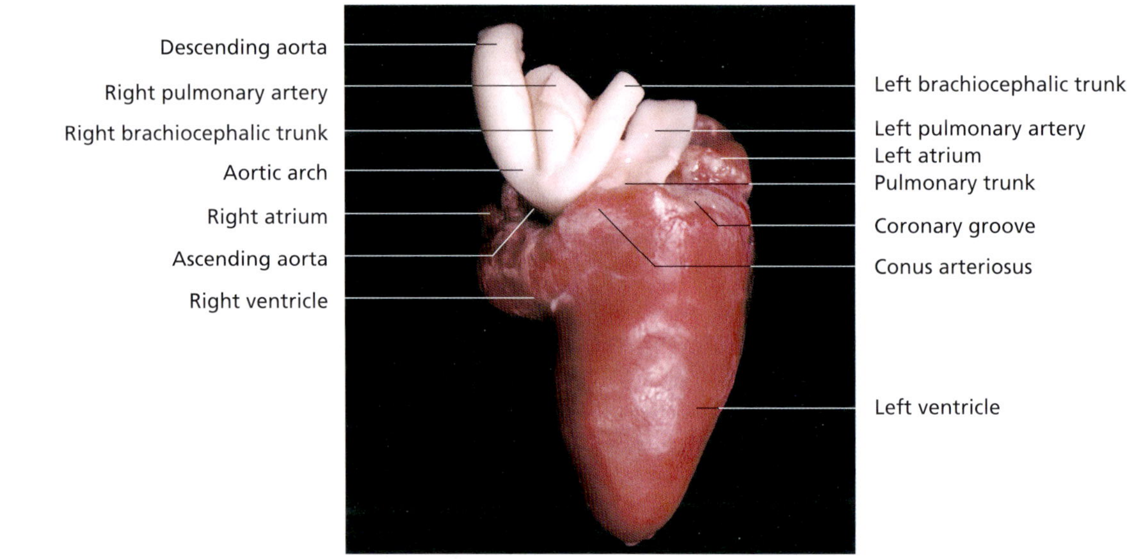

11.4 Heart of a chicken (right aspect).

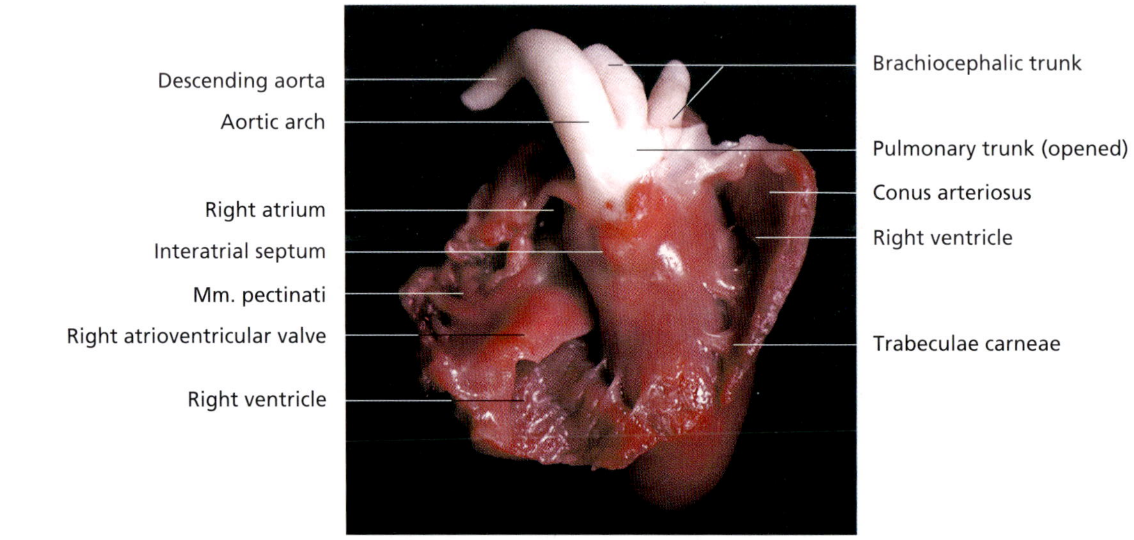

11.5 Heart of a chicken, right atrium and ventricle opened.

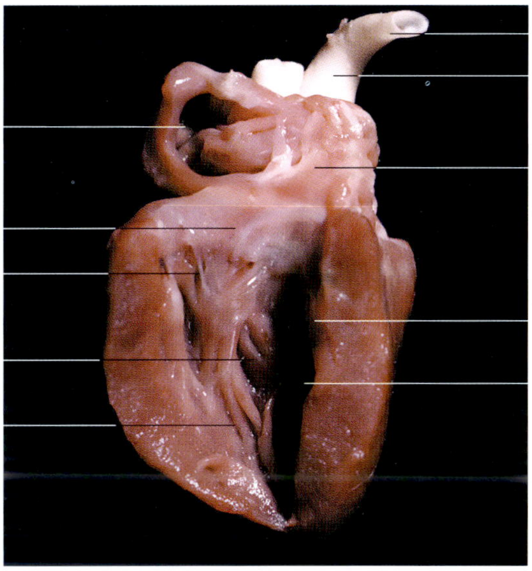

11.6 Heart of a chicken, left atrium and ventricle opened.

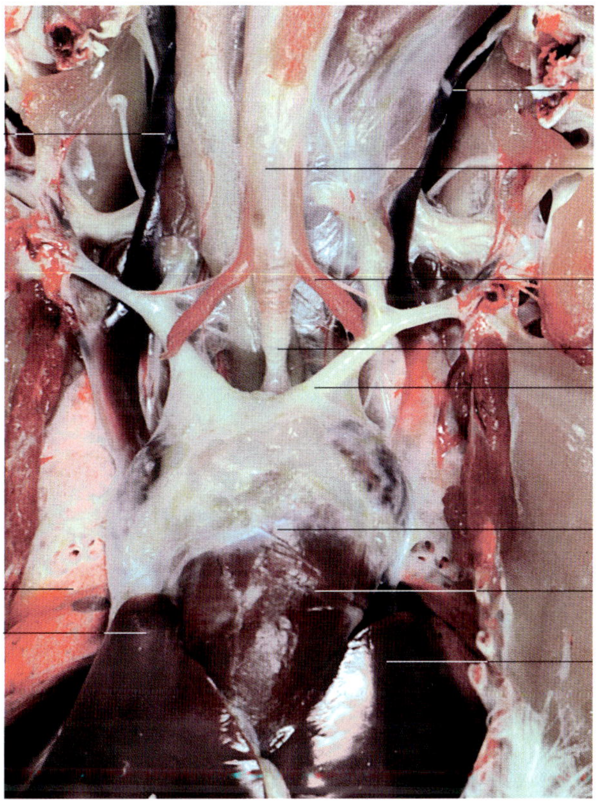

11.7 Heart of a chicken, clavicular air sac opened.

than the left in the chicken. It supplies the bulk of the septum between the right and left heart and the walls of the cardiac chambers. The less substantial left coronary artery principally supplies the basal segments of the wall of the left ventricle and the septum.

Generally the cardiac veins do not accompany the arteries. Apart from the smallest vv. cardiacae minimae, the veins of the heart empty into the right atrium.

Conduction system of the heart

In birds, the conducting system of the heart consists of the following components:

- sinoatrial node (nodus sinuatrialis),
- atrioventricular node (nodus atrioventricularis) and the atrioventricular bundle (fasciculus atrioventricularis),
- truncobulbar node (nodus truncobulbaris) and the truncobulbar bundle (fasciculus truncobulbaris).

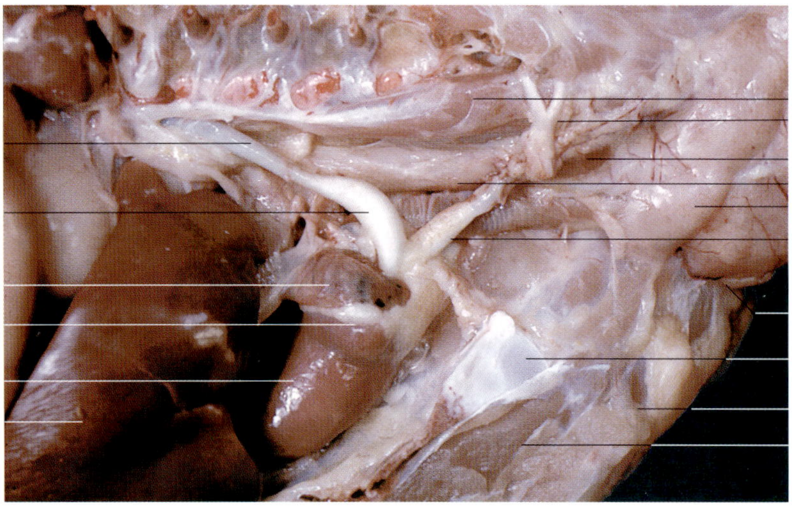

11.8 Heart of a chicken (right aspect).

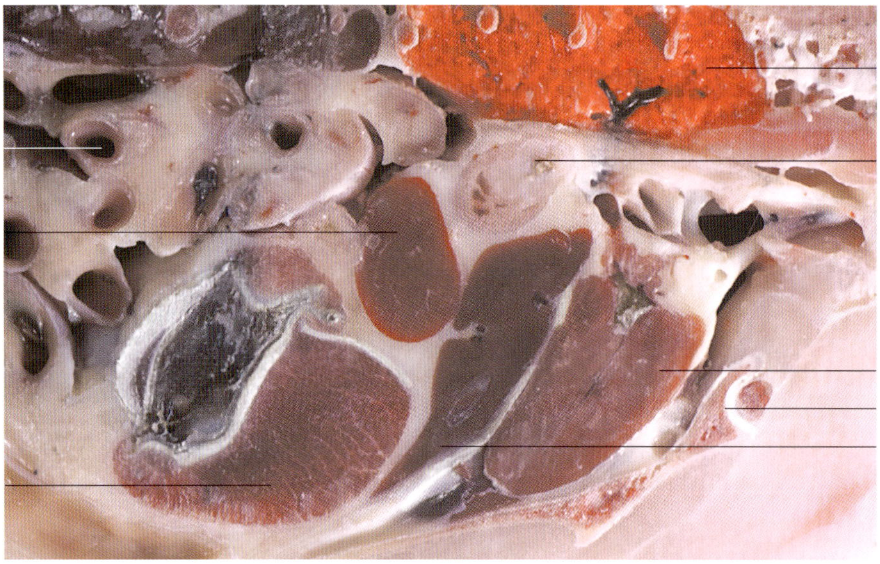

11.9 Right paramedian section at the level of the heart in a chicken (viewed from the right).

The relatively inconspicuous **sinoatrial node** (nodus sinuatrialis) lies adjacent to the opening of the right cranial and caudal vena cava. Somewhat larger, the **atrioventricular node** (nodus atrioventricularis) is located near the ostium of the left cranial vena cava. The **atrioventricular bundle** (fasciculus atrioventricularis) extends from the atrioventricular node and, as in mammals, divides into a **crus dextrum** and **crus sinistrum**. Covered in endocardium, the crura course within the **ventricular septum** toward the apex of the heart and ramify as **Purkinje fibres.** In birds, an **additional limb** extends from the atrioventricular node, passing around the right atrioventricular ostium to the base of the heart where it meets the **truncobulbar node** (nodus truncobulbaris).

As a third discrete node, the **truncobulbar node** is peculiar to the avian heart. It is situated under the origin of the **left coronary artery** and is probably involved in regulating the function of the muscular right atrioventricular valve. Ventrally, the truncobulbar node continues as the **truncobulbar bundle** (fasciculus truncobulbaris), which unites with the atrioventricular bundle just before its bifurcation. The **function of the avian heart** is also regulated by sympathetic fibres of the autonomic nervous system and by fibres of the vagus nerve.

Pulmonary vessels

The **pulmonary trunk** (truncus pulmonalis) arises from the right ventricle. Shortly thereafter it divides into the **left** and **right pulmonary arteries** (a. pulmonalis sinistra and dextra) (Figure 11.14).

After penetrating the horizontal septum, the pulmonary arteries divide within the lung, coursing independently of the bronchial tree. The blood capillaries form a network that is closely associated with the air capillaries (see

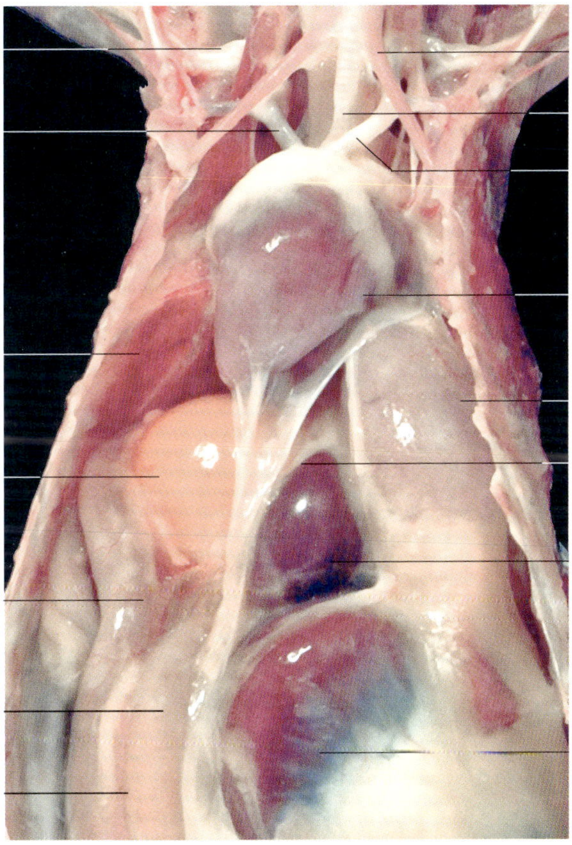

11.10 Heart of a chicken (sternum and liver removed, ventral view). Courtesy of Dr Annette Kaiser, Munich.

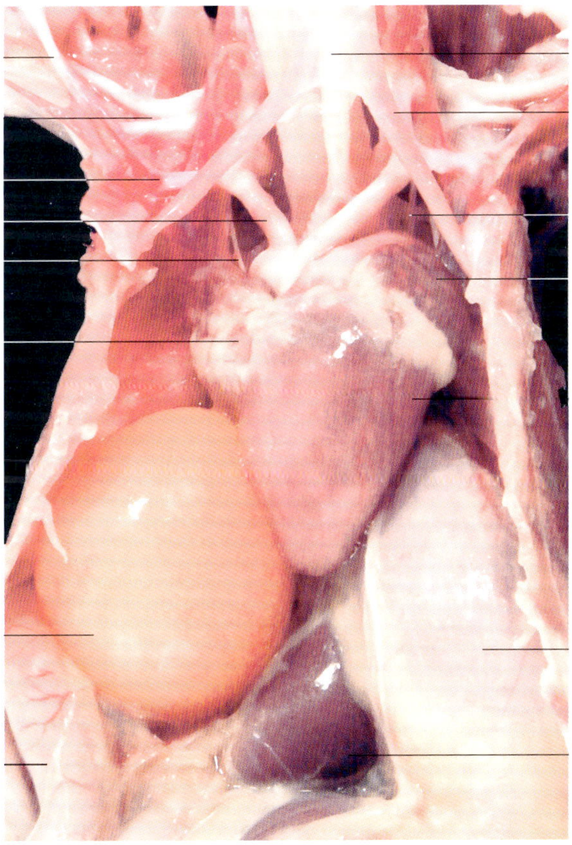

11.11 Heart of a chicken (sternum, liver and pericardium removed, ventral view). Courtesy of Dr Annette Kaiser, Munich.

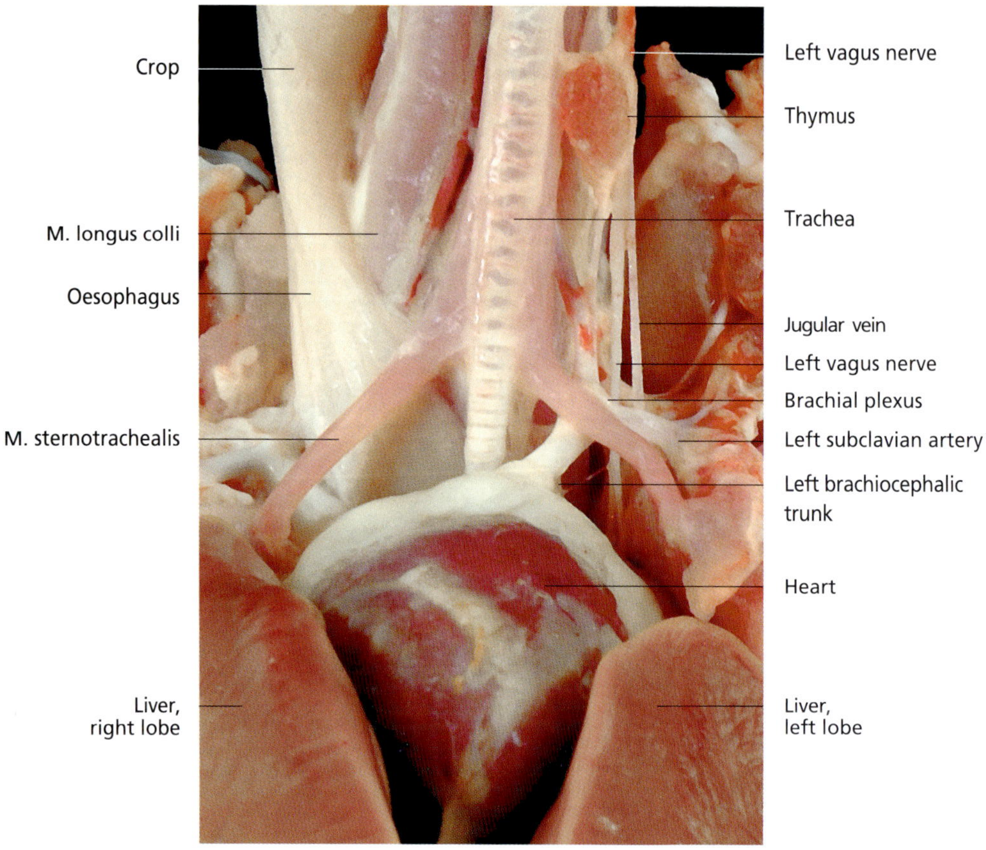

11.12 Anatomical relationships of organs and vessels lying cranial to the heart in a chicken (ventral view). Courtesy of Dr Annette Kaiser, Munich.

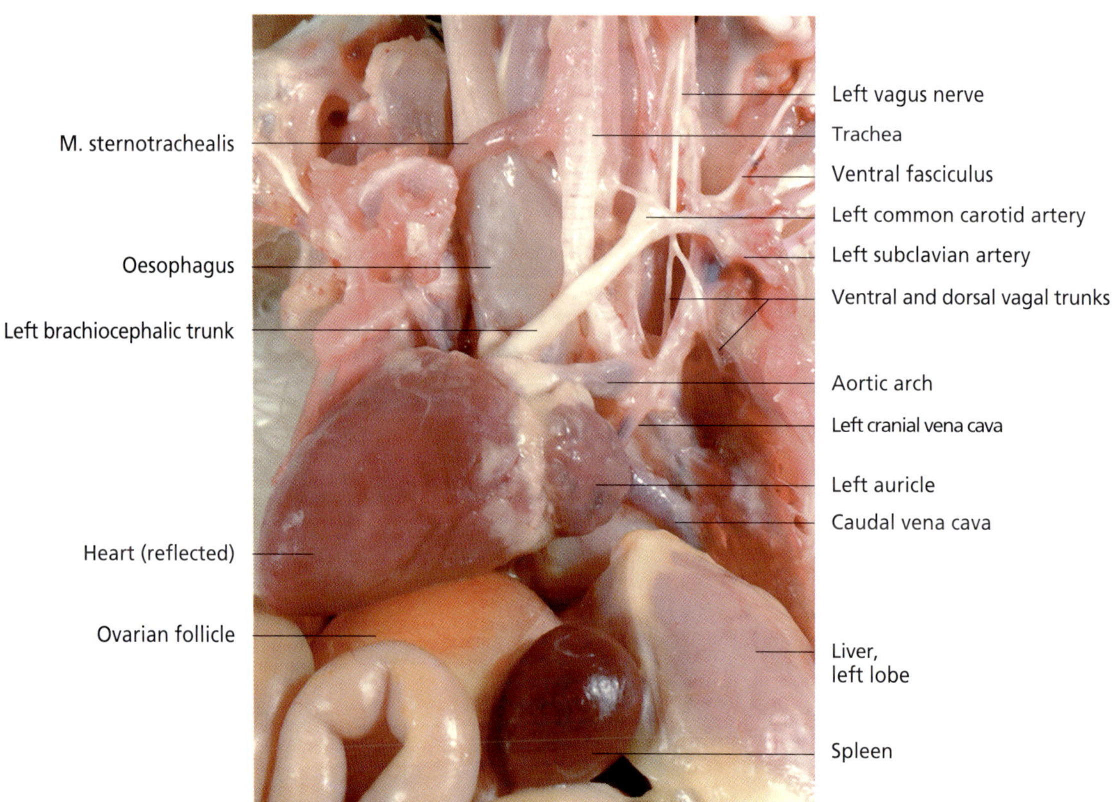

11.13 Anatomical relationships of organs and vessels at the base of the heart in a chicken (ventral view, heart reflected to the right side of the body). Courtesy of Dr Annette Kaiser, Munich.

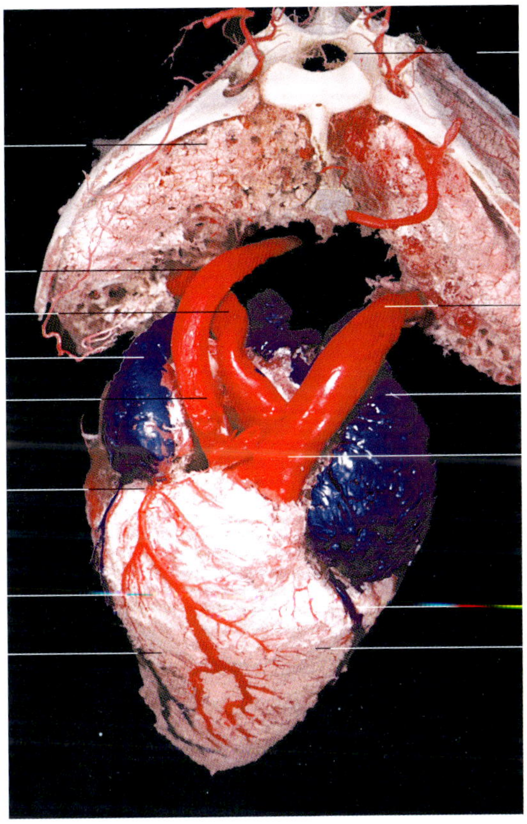

11.14 Heart and lung of a chicken (corrosion cast of blood vessels, cranial view).

Chapter 7 'Respiratory system'). Oxygenated blood enters the pulmonary venous system and drains into **right and left pulmonary veins** (v. pulmonalis dextra and sinistra). The pulmonary veins empty via separate ostia into the **left atrium**. Delivery of pharmacological agents via intrapulmonary injection is described in Chapter 21 'Medication and blood collection techniques'.

Systemic arteries

The aorta originates from the left ventricle. Located at the aortic ostium, the three-cusped **aortic valve** (valva aortae) prevents retrograde flow of blood during diastole. Aortic sinuses are located above each cusp. Two of these contain the openings of the coronary arteries. In contrast to mammals, the avian aorta is formed **embryologically** from the **right fourth aortic arch** and the **right dorsal aorta**. The **ascending aorta** (aorta ascendens) and **aortic arch** (arcus aortae) thus lie on the right side of the body.

In avian anatomical nomenclature, the **descending aorta** (aorta descendens) is *not* divided into thoracic and abdominal components because of the **absence of the diaphragm** in birds. The **left** and **right brachiocephalic trunks** (truncus brachiocephalicus sinister and dexter) arise together from the **ascending aorta** (Figures 11.3 to 11.5, 11.7 and 11.8). Due to the considerable size of the **brachiocephalic trunks**, the ascending aorta appears to undergo a three-way division (two brachiocephalic trunks and the aortic arch) (Figure 11.15).

In addition to **segmental arteries** (aa. intersegmentales) that supply the body wall and tail region, the descending aorta (Figure 11.8) gives rise, from cranial to caudal, to the following vessels (Figure 11.15):

- **coeliac artery (a. coeliaca)** (unpaired),
- **cranial mesenteric artery (a. mesenterica cranialis)** (unpaired),
- **cranial renal artery (a. renalis cranialis)**:
  - testicular artery (a. testicularis) (paired) or
  - ovarian artery (a. ovarica) (left only);
- **external iliac artery (a. iliaca externa)**,
- **ischiadic artery (a. ischiadica)**:
  - middle renal artery (a. renalis media),
  - caudal renal artery (a. renalis caudalis);
- **caudal mesenteric (a. mesenterica caudalis)** (unpaired),
- **internal iliac artery (a. iliaca interna)**:
  - caudal lateral artery (a. lateralis caudae),
  - pudendal artery (a. pudenda) and
- **caudal median artery (a. mediana caudae)** (continuation of the aorta).

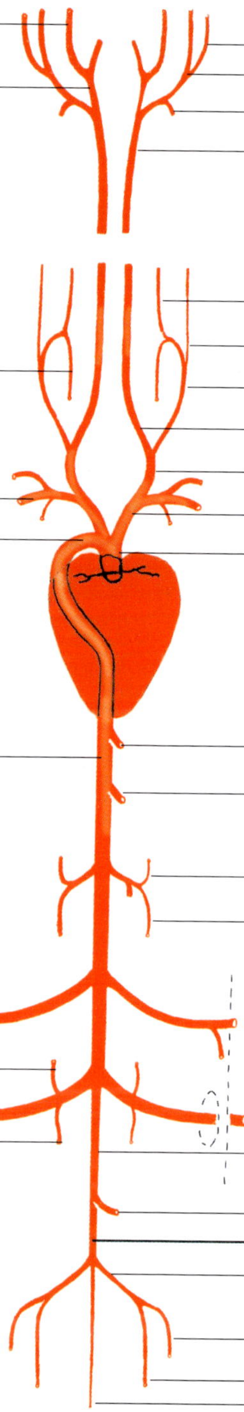

11.15 Principal systemic arteries in the male chicken (schematic; ventral view), adapted from King and McLelland, 1978.

Brachiocephalic trunk (truncus brachiocephalicus)

The brachiocephalic trunks supply the wings, particularly the flight muscles, the neck and the head. Both trunks (Figure 11.3 to 5, 7 & 15) give off the **common carotid artery** (a. carotis communis) and continue as the **subclavian artery** (a. subclavia). The short **common carotid artery** extends to the origin of the vertebral trunk (truncus vertebralis), then continues as the **internal carotid artery** (a. carotis interna). The common carotid artery gives rise to a vessel that supplies the oesophagus and trachea (a. oesophagotracheobronchialis) as well as the caudal thyroid artery (a. thyroidea caudalis).

The left **vertebral trunk** detaches a limb to the crop, as well as a delicate vessel that accompanies the vagus nerve (a. comes [nervi] vagi). On both sides, the vertebral trunk eventually divides into the **ascending vertebral artery** (a. vertebralis ascendens) (Figure 11.17) and **descending**

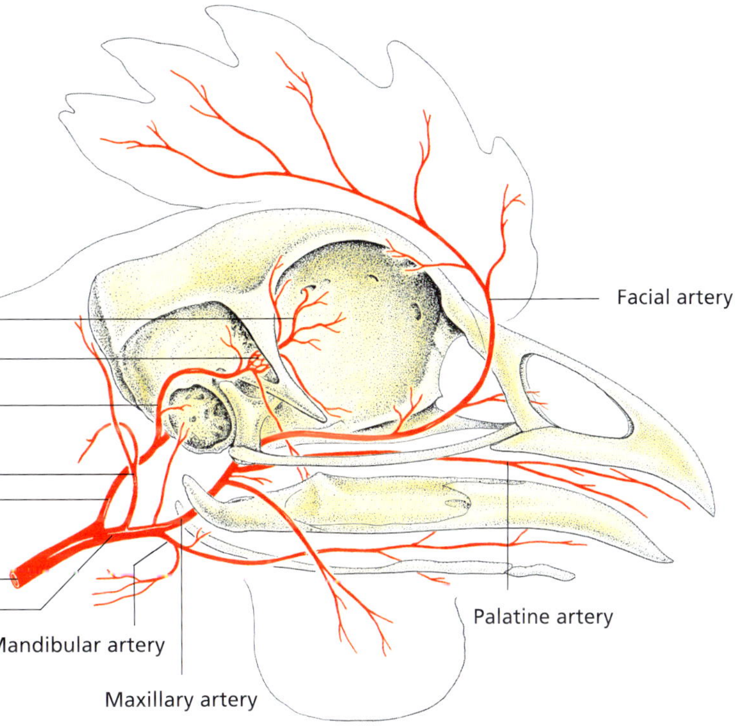

11.16 Principal arteries of the head in the chicken (schematic), adapted from Ghetie, 1976.

vertebral artery (a. vertebralis descendens) (Figure 11.15). These supply the cervical vertebrae, the cranial thoracic vertebrae and the spinal cord in these regions.

The following is an overview of the branches of, and regions supplied by, the brachiocephalic trunk:

- **common carotid artery (a. carotis communis)**:
  - **internal carotid artery (a. carotis interna)** (head),
  - **vertebral trunk (truncus vertebralis)** (vertebral column, spinal cord):
    - a. comes nervi vagi (thyroid gland, oesophagus),
    - ascending vertebral artery (a. vertebralis ascendens),
    - descending vertebral artery (a. vertebralis descendens);
- **subclavian artery (a. subclavia)**:
  - **axillary artery (a. axillaris)** (wing) and
- **pectoral trunk (truncus pectoralis)** (breast muscles).

The **internal carotid artery** provides the entire blood supply of the head (Figures 11.15ff.). It divides into the following branches:

- **A. carotis interna**:
  - **external carotid artery (a. carotis externa)**:
    - occipital artery (a. occipitalis),
    - mandibular artery (a. mandibularis),
    - maxillary artery (a. maxillaris);
  - **cerebral carotid artery (a. carotis cerebralis)**:
    - external ophthalmic artery (a. ophthalmica externa).

The left and right **internal carotid arteries** run alongside each other on the ventral surface of the cervical vertebrae. Ventrally they are covered by the m. longus colli. In some species, the two vessels become fused. After detaching the **external carotid artery** (a. carotis externa) near the head (Figure 11.16), the internal carotid artery goes on to divide into the **cerebral carotid artery** (a. carotis cerebralis) (Figure 11.18) and the **external ophthalmic artery** (a. ophthalmica externa).

The **external carotid artery** supplies the external regions of the head. At the base of the skull, it gives off the occipital artery (a. occipitalis) before dividing into the larger **maxillary artery** (a. maxillaris) and the smaller **mandibular artery** (a. mandibularis) (Figures 11.15ff.).

The **maxillary artery** (Figures 11.16ff.) courses rostrally, ventral to the joints of the mandible. A branch of the maxillary artery, the **facial artery** (a. facialis), passes around the ventral and rostral margins of the eye before ascending to subdivide within the comb. The continuation of the maxillary artery extends to the palate, where it converges with its opposite number to form the unpaired **median palatine artery** (a. palatina mediana) that supplies the tip of the upper beak. Blood to the lower beak

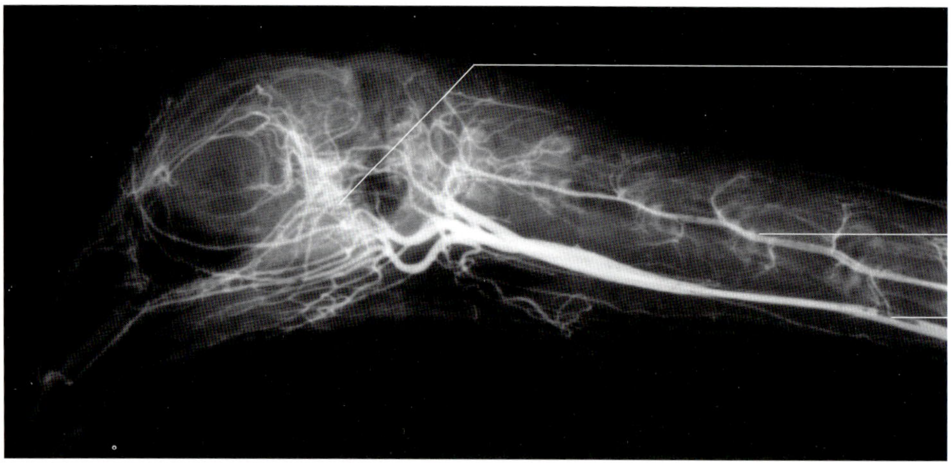

11.17 Angiogram of the head and neck of a chicken.

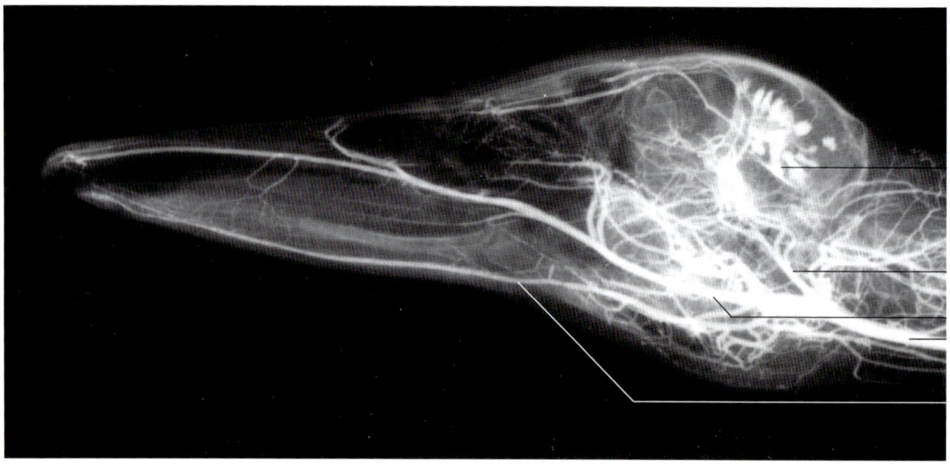

11.18 Angiogram of the head of a Muscovy duck (Cairina moschata).

and tongue is supplied by the **mandibular artery** (Figures 11.16 and 11.18).

After passing caudal to the external acoustic meatus, the **external ophthalmic artery** (Figure 11.16) approaches the orbit from a caudolateral direction and ramifies into the **rete mirabile ophthalmicum**. Branches of the rete supply the eye, its adnexa and the third eyelid (see Chapter 15 'The eye').

The **cerebral carotid artery** (Figures 11.16 and 11.18) traverses the carotid canal of the sphenoid bone to enter the cranial cavity. Caudal to the hypophysis, it joins its counterpart on the opposite side to form the **intercarotid anastomosis** (anastomosis intercarotida) and gives off arteries to the brain. Its branching pattern resembles that of the arterial circle of mammals. A detailed description of these vessels is beyond the scope of this text.

The **subclavian artery** (Figure 11.19) supplies the flight muscles and the wings. It forms the lateral continuation of the **brachiocephalic trunk,** sending branches to the clavicle, coracoid and sternum. Distal to the origin of the axillary artery, the subclavian artery continues as the **pectoral trunk** (truncus pectoralis). This large artery eventually divides into the cranial, middle and caudal **pectoral artery** (a. pectoralis). The middle branch is absent in the chicken.

- **subclavian artery (a. subclavia)**:
  - sternoclavicular artery (a. sternoclavicularis),
  - internal thoracic artery (a. thoracica interna);
- **axillary artery (a. axillaris)**:
  - subscapular and supracoracoid arteries (a. subscapularis and a. supracoracoidea),
  - brachial artery (a. brachialis):
    - deep brachial artery (a. profunda brachii),
    - dorsal circumflex humeral artery (a. circumflexa dorsalis humeri),
    - bicipital artery (a. bicipitalis),
    - radial artery (a. radialis),
    - ulnar artery (a. ulnaris);
- **pectoral trunk (truncus pectoralis)**:
  - cutaneous thoracoabdominal artery (a. cutanea thoracoabdominalis),
  - cranial pectoral artery (a. pectoralis cranialis) and caudal pectoral artery (a. pectoralis caudalis).

A branch of the pectoral trunk, the **external thoracic artery** (a. thoracica externa), extends to the breast muscles (Figure 11.19). The superficial **cutaneous thoracoabdominal artery** (a. cutanea thoracoabdominalis) passes to the skin of the incubation (brood) patch and the abdomen. It is important to note that the m. pectoralis is particularly well vascularised. Intramuscular injection should therefore be given into the underlying m. supracoracoideus (formerly m. pectoralis profundus).

The **axillary artery** (a. axillaris), the principal artery of the wing, is smaller than the pectoral trunk. It gives off two branches that divide within the m. subscapularis and m. supracoracoideus.

The **brachial artery** (a. brachialis) is the prolongation of the axillary artery. It gives off the **deep brachial artery** (a. profunda brachii) that supplies the caudal muscles of the proximal wing. After the origin of the dorsal circumflex humeral artery (a. circumflexa humeri dorsalis) and

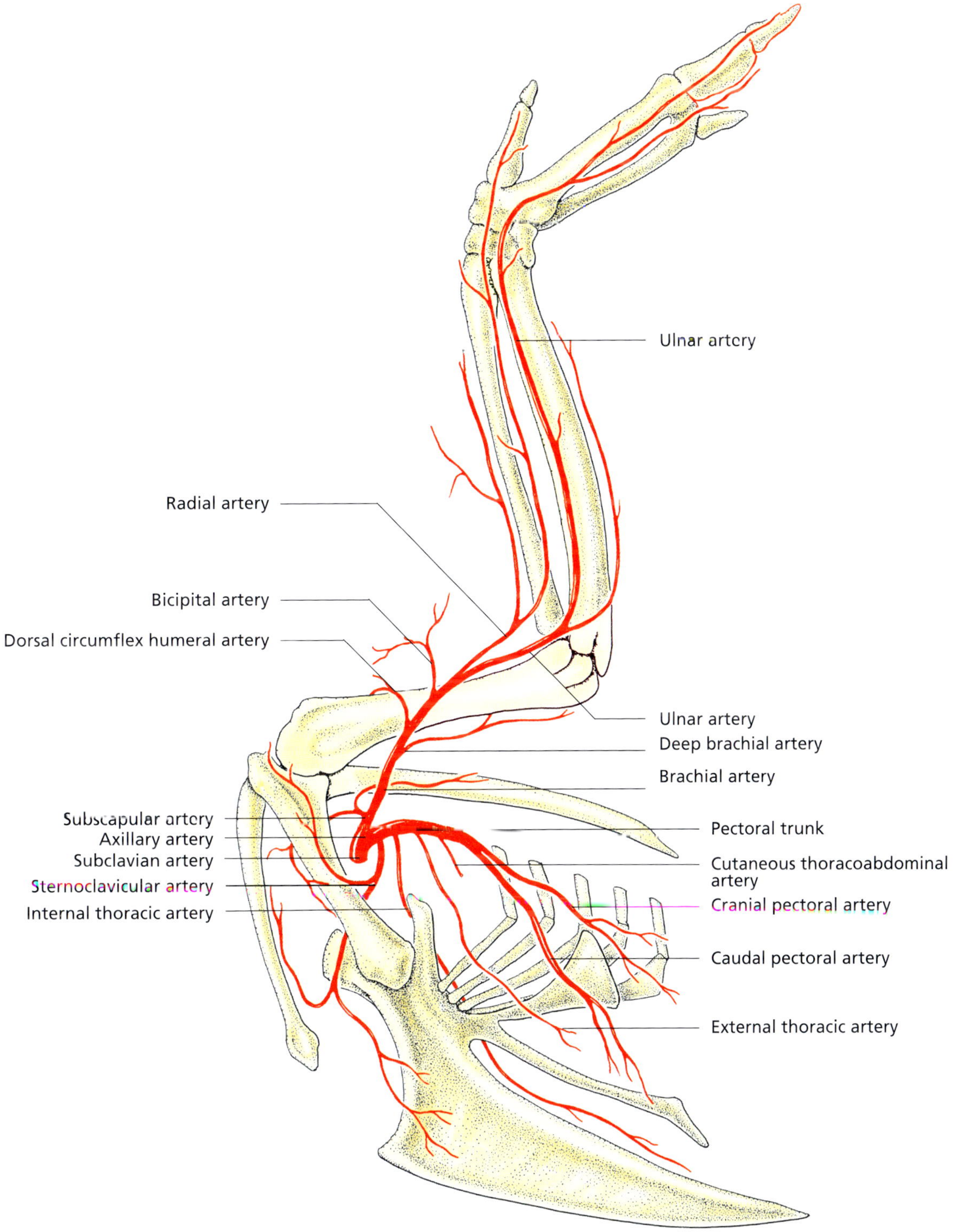

11.19 Principal branches of the subclavian artery in the chicken (schematic), adapted from Ghetie, 1976.

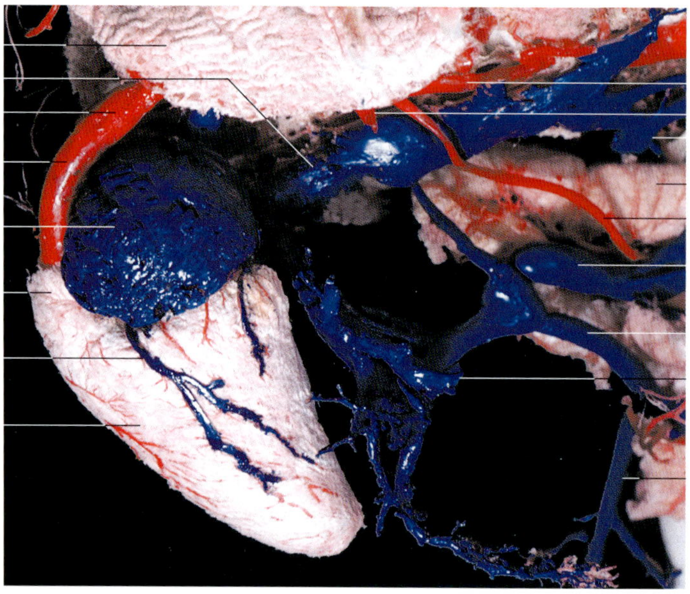

11.20 Blood vessels of the cranial body cavity of a chicken (corrosion cast).

the **bicipital artery** (a. bicipitalis), the brachial artery divides into the smaller **radial artery** (a. radialis) and larger **ulnar artery** (a. ulnaris). In some species (e.g., common buzzard), the ulnar artery may be surrounded by a network of veins. The radial and ulnar arteries supply the distal segments of the wing.

Visceral branches of the descending aorta

The **coeliac artery** (a. coeliaca) (Figures 11.15 and 11.20) provides blood to the glandular and muscular stomachs, the duodenum, ileum, caeca, pancreas and spleen.

It gives off the following branches:

- **oesophageal artery (a. oesophagealis),**
- **dorsal proventricular artery (a. proventricularis dorsalis),**
- **left branch (ramus sinister):**
  - ventral proventricular artery (a. proventricularis ventralis),
  - left gastric artery (a. gastrica sinistra),
  - ventral gastric artery (a. gastrica ventralis);
- **right branch (ramus dexter):**
  - right hepatic artery (a. hepatica dextra),
  - duodenojejunal artery (a. duodenojejunalis),
  - right gastric artery (a. gastrica dextra),
  - pancreaticoduodenal artery (a. pancreaticoduodenalis),
  - ileocaecal artery (a. ileocaecalis).

The region supplied by the coeliac artery is thus considerably more extensive in birds than in mammals.

The **cranial mesenteric artery** (a. mesenterica cranialis) (Figure 11.15) arises from the descending aorta immediately caudal to the coeliac artery at the level of the sixth rib. It supplies the jejunum, terminal duodenum, ileum and the apices of the caeca (see Chapter 6 'Digestive system', Figure 6.26). Branches of the cranial mesenteric artery include the:

- duodenojejunal artery (a. duodenojejunalis),
- jejunal arteries (aa. jejunales, c. 20) and
- ileal arteries (aa. ileae).

Of the approximately 20 jejunal arteries, one represents the continuation of the course of the cranial mesenteric artery. Where present, the Meckel's diverticulum is located near the distal end of this artery (see Chapter 6, Figure 6.26).

The **caudal mesenteric artery** (a. mesenterica caudalis) arises from the descending aorta after the latter has given off the **ischiadic artery** (a. ischiadica) (Figure 11.23). It divides into a cranial branch, supplying the rectum, and a caudal branch that passes to the cloaca and bursa of Fabricius.

Renal arteries

The only paired visceral branches of the descending aorta are the **cranial renal arteries**. In the male, the cranial renal artery originates together with the testicular and adrenal arteries. In the female, ovarian and cranial oviductal arteries are only present on the left side, originating in common with the left cranial renal artery. The **middle** and **caudal** renal arteries arise from the **ischiadic artery** (Figure 11.23).

Arteries of the pelvic limb

The pelvic limb is supplied by two large vessels arising from the **descending aorta:** the **external iliac artery** (a. iliaca externa) and **ischiadic artery** (a. ischiadica) (Figures 11.22 and 11.23)**.** These vessels leave the aorta at the level of the middle and caudal renal division respectively.

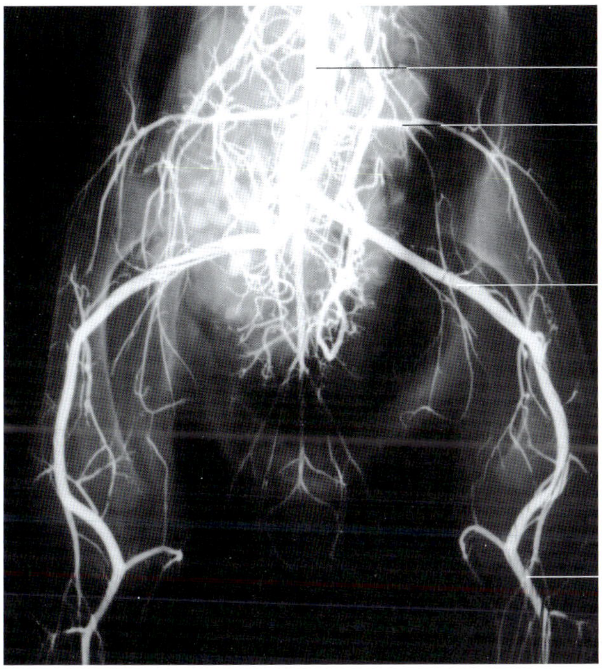

11.21 Angiogram of the caudal body cavity and pelvic limbs of a chicken.

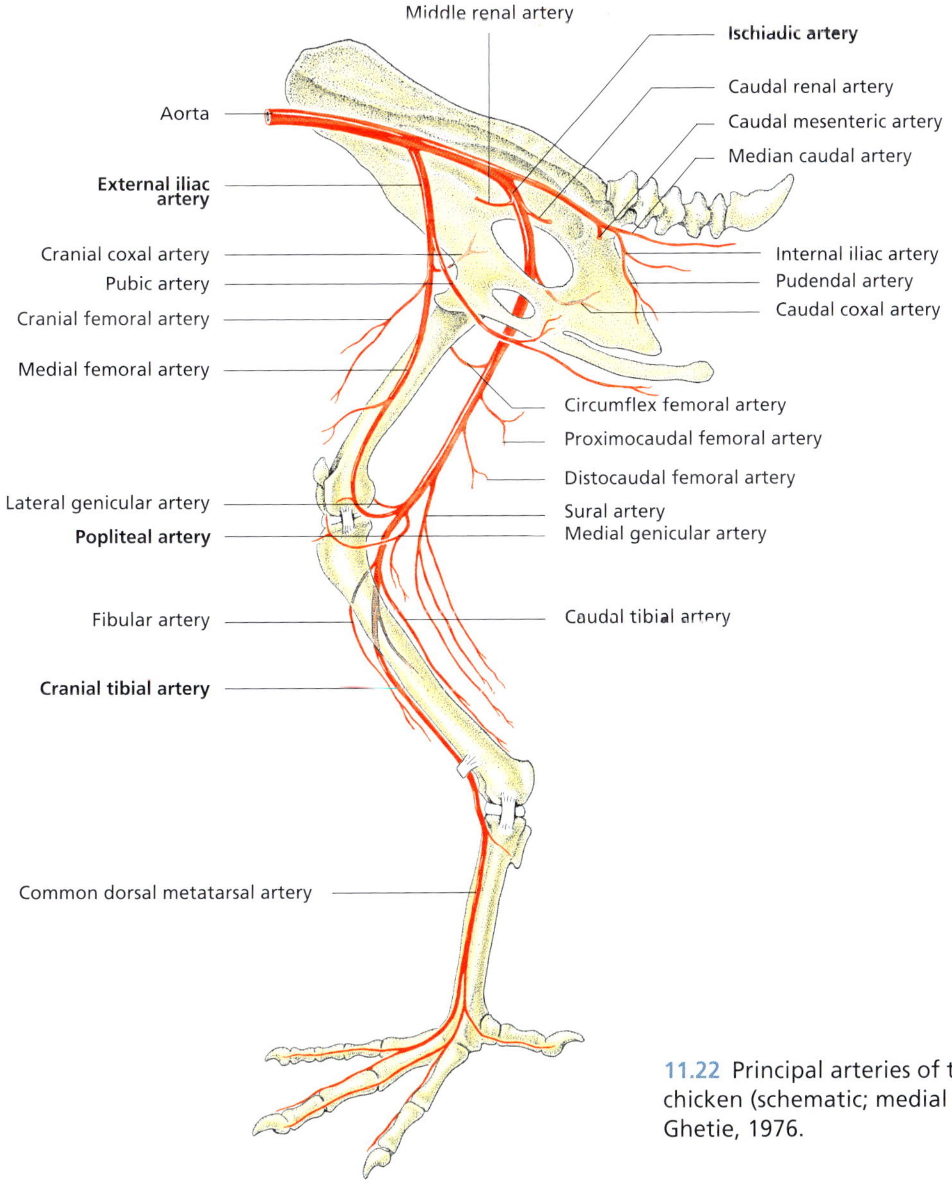

11.22 Principal arteries of the pelvic limb of the chicken (schematic; medial view), adapted from Ghetie, 1976.

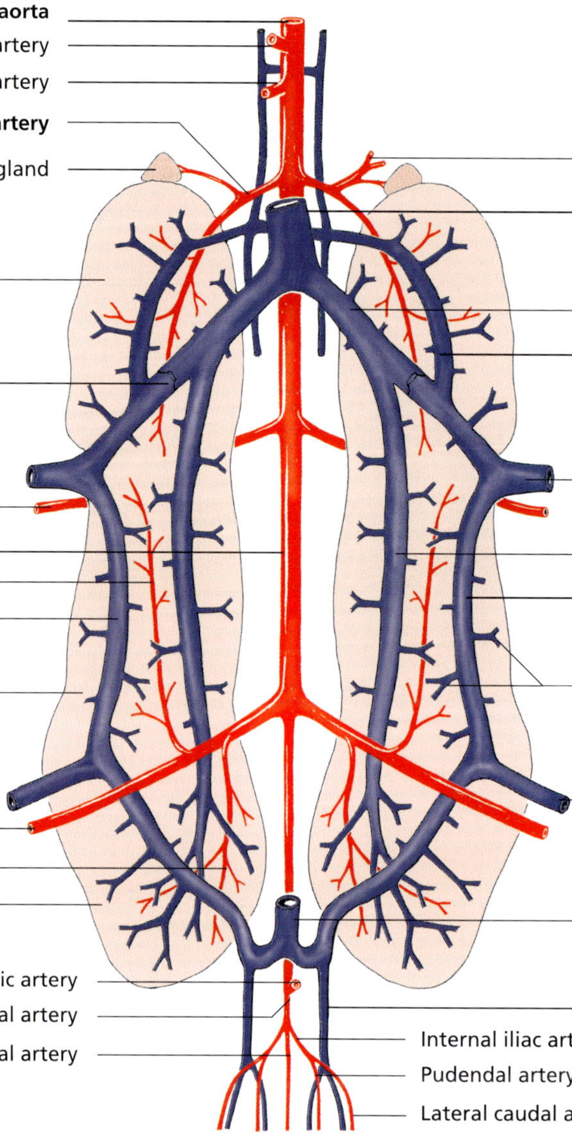

11.23 Vascular supply of the kidneys and pelvis in the chicken (schematic), adapted from Rickart-Müller, 1968.

The external iliac is the smaller of the two arteries of the pelvic limb. It passes cranial to the acetabulum to reach the thigh (Figure 11.22).

The branches of the external iliac artery are the:

- **pubic artery (a. pubica)**,
- **medial femoral artery (a. femoralis)**:
  - cranial femoral artery (a. femoralis cranialis),
  - cranial coxal artery (a. coxae cranialis).

The external iliac artery and its continuation, the **medial femoral artery** (a. femoralis medialis) supply the caudal abdominal musculature as well as the cranial and medial muscles of the thigh.

Ischiadic artery

The ischiadic artery, the larger of the vessels supplying the pelvic limb, passes caudally, dorsal to the hip joint (Figures 11.22 and 11.23). It traverses the ilioischiadic foramen to reach the thigh. In the caudomedial thigh, the vessel runs parallel to the ischiadic nerve.

The branches of the ischiadic artery are the:

- middle renal artery (a. renalis media),
- middle oviductal artery (a. oviductalis media),
- caudal renal artery (a. renalis caudalis),
- caudal coxal artery (a. coxae caudalis),
- obturator artery (a. obturatoria),
- trochanteric artery (a. trochanterica),
- circumflex femoral artery (a. circumflexa femoris),
- proximocaudal femoral artery (a. femoralis proximocaudalis),
- distocaudal femoral artery (a. femoralis distocaudalis),
- sural artery (a. suralis).

- **popliteal artery (a. poplitea)**:
  - lateral genicular artery (a. genicularis lateralis),
  - medial genicular artery (a. genicularis medialis),
  - medial tibial artery (a. tibialis medialis),
  - caudal tibial artery (a. tibialis caudalis),
  - fibular artery (a. fibularis);
- **cranial tibial artery (a. tibialis cranialis)** and
- **common dorsal metatarsal artery (a. metatarsalis dorsalis communis)**.

Arteries of the pelvic region

After detaching the **caudal mesenteric artery**, the descending aorta divides into the paired **internal iliac artery** (a. iliaca interna) and the **median sacral artery** (a. sacralis mediana). The latter continues as the **median caudal artery** (a. caudae mediana) (Figure 11.23). Shortly after its origin, the internal iliac artery gives off the lateral caudal artery (a. caudae lateralis) and continues as the **pudendal artery** (a. pudenda) (Figure 11.23).

In the **male**, the pudendal artery supplies the ductus deferens and the ureter. Close to the receptaculum ductus deferens, the pudendal artery gives rise to a tuft of capillaries (corpus vasculare phalli). This vascular network is associated with nearby lymph chambers that participate in bringing about erection of the phallus (see Chapter 9 'Male genital organs').

In the **female**, the pudendal artery gives off the **caudal oviductal artery** (a. oviductalis caudalis) and the **vaginal artery** (a. vaginalis). In both sexes, the pudendal artery terminates as the **cloacal arteries** (aa. cloacales), branches of which supply the cloaca and bursa of Fabricius.

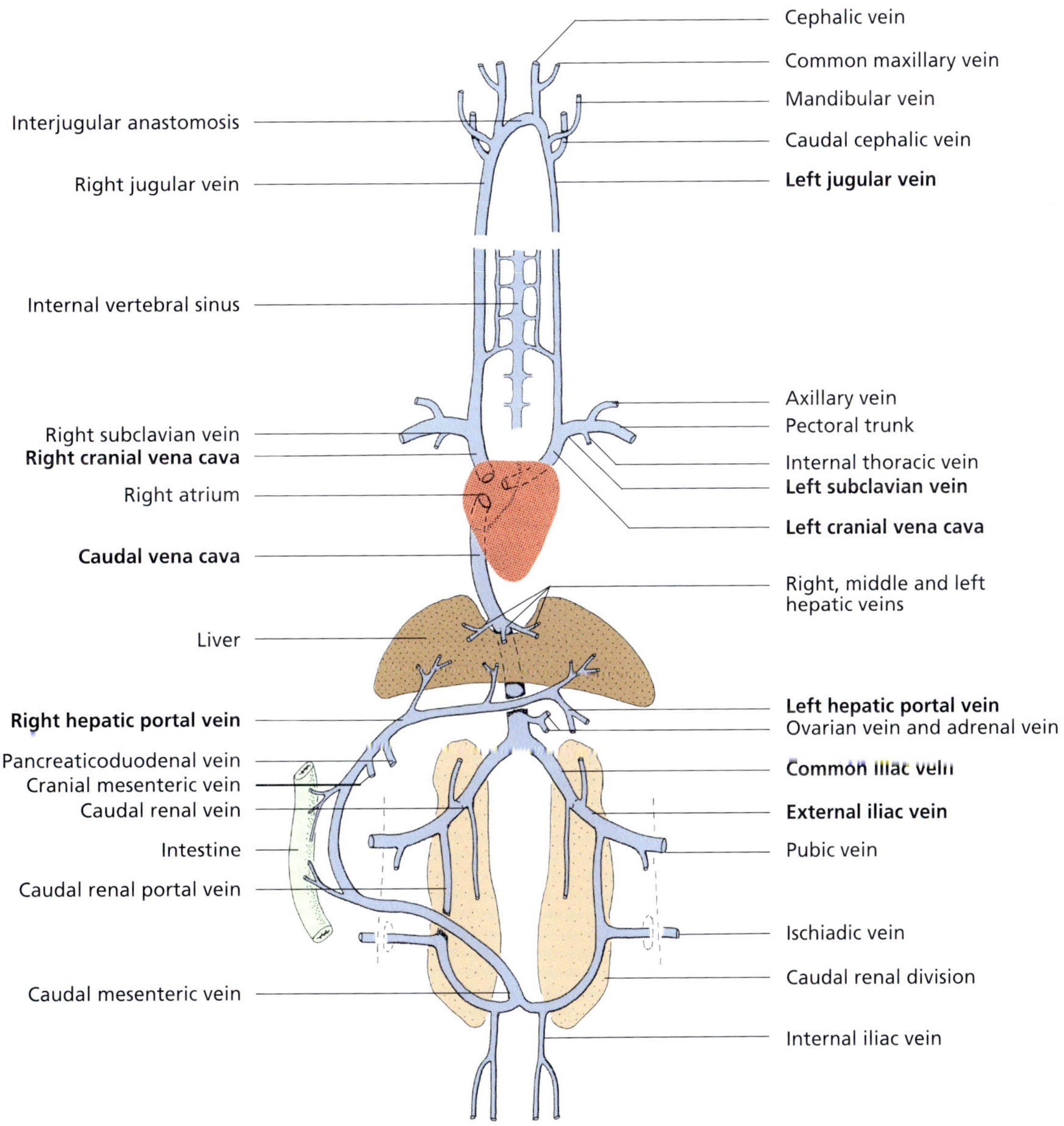

11.24 Large veins of the chicken (schematic), adapted from Ghetie, 1976.

Systemic veins

On both sides of the body, the **jugular vein** (v. jugularis) arises from the confluence of veins of the head (including those formed by the convergence of the retrobulbar veins with the veins of the ear and face). In most birds (not in pigeons), the **right jugular vein** is significantly larger than the left. The jugular vein also receives blood from the **dural venous sinuses** in the cranium. Caudoventral to the pharynx, the jugular veins are connected by a transverse **interjugular anastomosis** (anastomosis interjugularis) (Figure 11.24).

The subcutaneously positioned **jugular vein** passes caudally through the neck, parallel to the a. comes nervi vagi and the vagus nerve. **Jugular venipuncture** can be performed in the mid-cervical region, ideally on the right side (see Chapter 21 'Medication and blood collection techniques'). In juvenile birds, lobes of the thymus are fused with the jugular vein. The **jugular vein** and **subclavian vein** (v. subclavia) converge at the **venous angle** to form the **left** and **right cranial vena cava** (v. cava cranialis sinistrum and dextrum) (Figure 11.24).

Veins of the wing

The largest vein of the wing, the **ulnar vein** (v. ulnaris) is formed from the confluence of distal veins. Its more proximal continuation is referred to either as the **ulnar vein** or **v. basilica**. It crosses the middle of the humerus medially and is suitable for **venipuncture** (see Chapter 21 'Medication and blood collection techniques').

The veins of the wing are satellites of the arteries and are similarly named. Blood is received by the **axillary vein** (v. axillaris) from the cutaneous thoracoabdominal vein. The axillary vein then merges with the substantial **pectoral trunk** (truncus pectoralis) to form the **subclavian vein.**

Caudal vena cava (vena cava caudalis)

The caudal vena cava arises from the confluence of the **right** and **left common iliac veins** (v. iliaca communis dextra and sinistra). Both **common iliac veins** take up the veins of the pelvic limb – the **external iliac vein** (v. iliaca externa) and the **ischiadic vein** (v. ischiadica) – and the renal portal veins (Figure 11.24). In the proximal medial third of the femur, the more voluminous **external iliac vein** (also referred to as the femoral vein in this region) is joined with the smaller ischiadic vein by a **substantial anastomosis** (anastomosis ischiofemoralis).

The ischiadic vein is the proximal prolongation of the **popliteal vein** (v. poplitea), which is formed by the convergence of the **cranial** and **caudal tibial vein** (v. tibialis cranialis and v. tibialis caudalis). The caudal tibial vein represents the proximal continuation of the plantar superficial metatarsal vein (v. metatarsalis plantaris superficialis). Passing medially along the tarsometatarsus, the latter vessel is used for **venipuncture** (immediately distal to the intertarsal joint) in ducks, geese swans and ratites (see Chapter 21 'Medication and blood collection techniques').

At its caudal end, the caudal vena cava receives the **testicular veins** (vv. testiculares) in the male and the **ovarian vein** (v. ovaricae) in the female. The **adrenal vein** (v. adrenalis) flows into the caudal vena cava near the point of entry of the veins of the gonads.

The caudal vena cava then takes up the **left** and **right hepatic veins** (v. hepatica dextra and sinistra) and the **middle hepatic veins** (vv. hepaticae mediae) (Figure 11.24).

Hepatic portal system

A hepatic portal system is present in birds, as in mammals. Birds, however, have two **hepatic portal veins** (Figure 11.24).

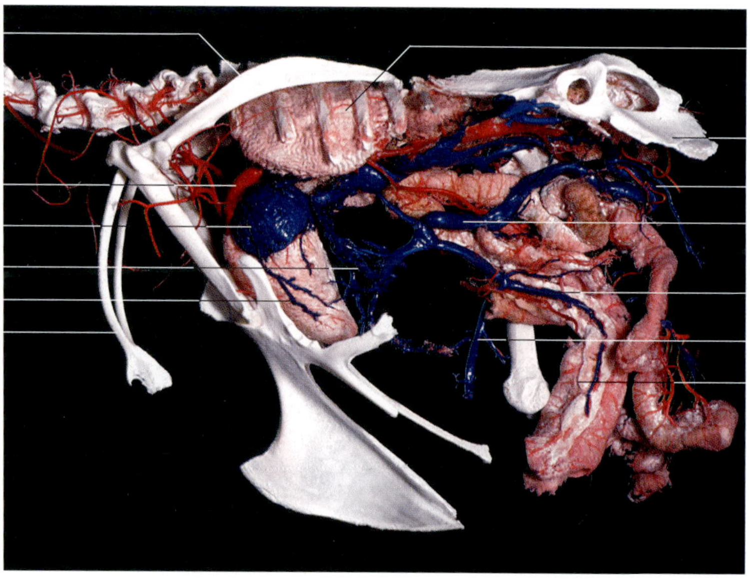

11.25 Blood vessels of the body cavity of a chicken (corrosion cast, left aspect).

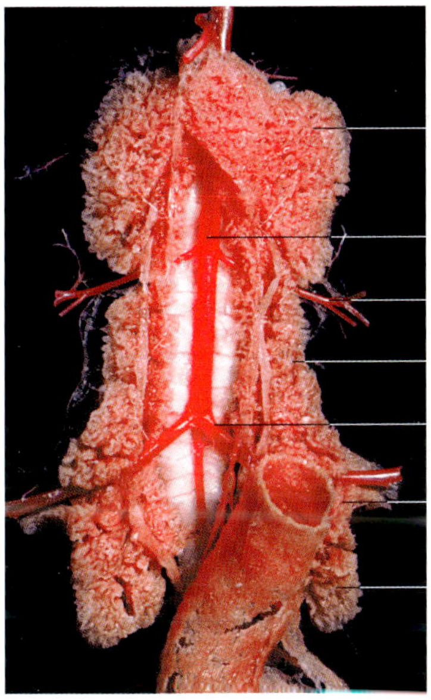

11.26 Renal arteries of a chicken (corrosion cast).

The smaller **left hepatic portal** vein (v. portae sinistrae hepatis) drains veins from the glandular and muscular stomachs. Venous blood from the small intestine, caeca, rectum, cranial cloaca, pancreas and spleen passes into the **right hepatic portal vein** (v. portae dextrae hepatis).

Blood drains from the liver via the previously described hepatic veins into the **caudal vena cava.**

Renal portal system

In birds, as in almost all vertebrates, a portal system is also present in the kidney. Only mammals lack this vascular feature. The renal portal system is located downstream of capillary beds in the pelvic limbs, pelvic region, tail, caudal segments of the intestine and the vertebral column (Figures 11.23 and 11.24). Most of the blood entering the renal portal system originates from the:

- external iliac vein (v. iliaca externa),
- ischiadic vein (v. ischiadica),
- internal iliac vein (v. iliaca interna).

These three vessels are connected at the lateral margin of the kidney, forming the **caudal renal portal vein** (v. portalis renalis caudalis). The **cranial renal portal vein** (v. portalis renalis cranialis) detaches from the common iliac vein and passes to the cranial renal division (Figure 11.24).

Branches of the renal portal veins enter the renal parenchyma where they combine with the peritubular capillary network of the renal lobules. Blood subsequently drains into the caudal renal vein (v. renalis caudalis). The **renal portal valve**, located in the common iliac vein (Figure 11.23), determines the resistance to direct venous return of blood, thus influencing the amount of blood that enters the renal portal system.

The left and right portal systems are connected by several routes: via the vv. intersegmentales trunci and **internal vertebral sinus** (see Figure 11.24), by an anastomosis between the left and right internal iliac veins, and via the caudal mesenteric (coccygeomesenteric) vein.

These vascular connections, together with the renal portal valve, function as an autonomically controlled mechanism for regulating blood flow to the peritubular capillary network of each kidney. For example, blood flow to the kidneys is increased at rest by closure or narrowing of the renal portal valve. This system is of considerable physiological significance with respect to water conservation and uric acid production.

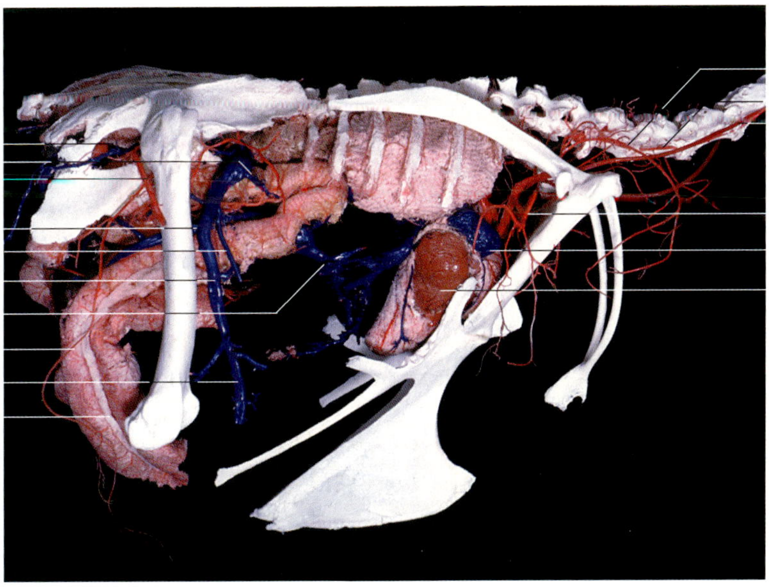

Figure11.27 Blood vessels of the body cavity of a chicken (corrosion cast, right aspect).

Clinical aspects

Proficiency in blood collection and intravenous drug administration is of fundamental importance in avian medicine. Techniques referred to in this chapter are addressed in greater detail in Chapter 21 'Medication and blood collection techniques'. These procedures are **indispensable** for **rapid diagnosis** and **efficient administration of pharmacological agents.** The latter is important in individual animal medicine for quickly achieving effective blood and tissue concentrations, as disease progression in birds can be very swift. Blood collection techniques are also essential for disease management in flocks (e.g., commercial poultry).

It is often stated in the anatomical literature that, due to the existence of the renal portal system, drugs **injected into the muscles of the thigh** may be partially excreted by the kidney before therapeutic levels are achieved in the target tissue. From a clinical perspective, however, the lateral thigh muscles represent a valuable alternative for intramuscular injection when the usual site (m. supracoracoideus) is not practicable (e.g., cachectic patients, intramuscular vaccination of day-old chicks).

Species-specific variation in vascular anatomy is infrequently described in the anatomical literature, yet can be of clinical significance. In the common buzzard (Buteo buteo), for example, the ulnar artery is surrounded by a **venous plexus** that can make arterial cannulation (e.g., for intra-arterial/direct blood pressure monitoring) more difficult.

In members of the class Aves, the autonomous conducting system of the heart is more extensive than in mammals. This is clinically relevant in terms of **anaesthetic monitoring** as an electrocardiogram or phonendoscope (augmented stethoscope) may continue to register cardiac electrical activity for some time post-mortem, potentially masking intraoperative complications or death. However, pathological changes in the ECG pattern and auscultatable arrhythmias usually arise in these situations, so timely detection of abnormalities is crucial. The aforementioned factors are also responsible for the rhythmic contractility that may be observed in an isolated heart when a bird is dissected shortly after death or slaughter.

Auscultation of the heart is an essential part of **routine anaesthetic monitoring**. With its dorsocranially directed base lying on the sternum, and its sides partially covered by the liver (see Figures 6.41 and 6.45), the relatively large avian heart can be auscultated (with some species variation) over a region bounded dorsally by the lungs and cranioventrally by the coracoid bone.

The **auscultatory field** is centred upon the cranial trisection point of a line connecting the **shoulder joint** and the **caudal margin of the sternum**. By placing the middle finger on the shoulder joint and the thumb on the caudal sternal margin, the site of auscultation can be localised with the index finger. The same approach can be used to identify the optimal site for cardiac puncture (see Chapter 21 'Medication and blood collection techniques' for further detail). Since the right lobe of the liver is larger in many birds, extending further into the auscultatory field, the heart is more accessible on the left side.

While monitoring of **heart rate** and **rhythm** is also important in **perianaesthetic period**, the resting heart rate of birds is usually too high to be measured using auscultation. Due to the reflex-driven behaviour of birds, and the highly developed fight or flight response, handling and restraint techniques often also influence the frequency and rhythm of the heart beat in conscious animals.

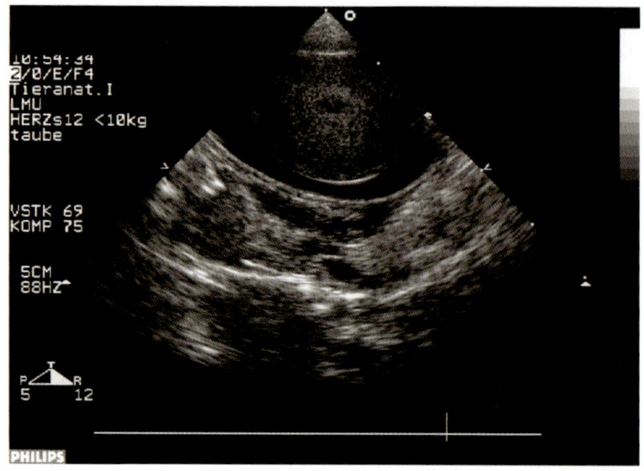

11.28 Sonogram (88Hz) of the heart of a racing pigeon (Columba livia).

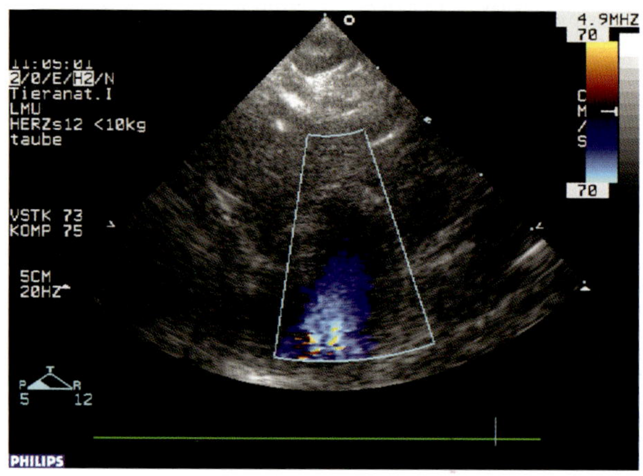

11.29 Sonogram (power Doppler, 20Hz) of the heart of a racing pigeon (Columba livia) showing the direction of blood flow.

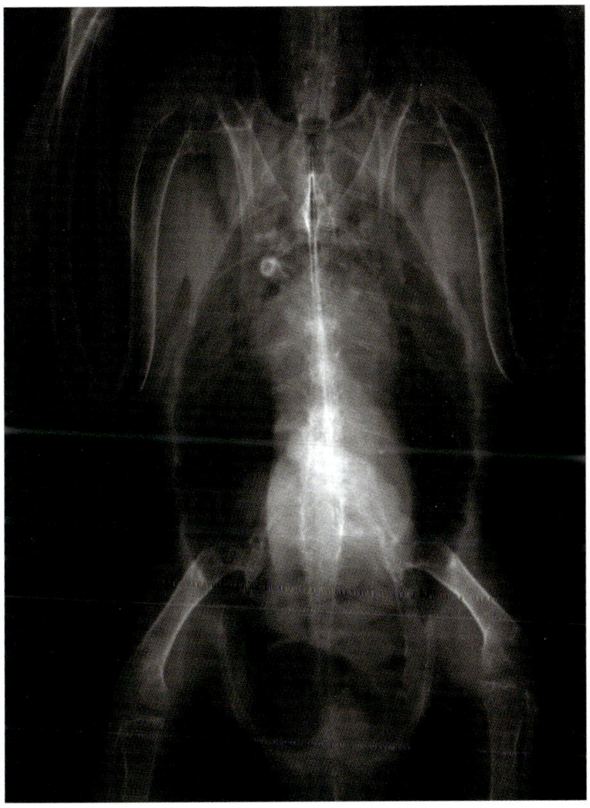

11.30 Radiograph (ventrodorsal view) of a gyrfalcon (Falco rusticolus). Note the relatively large heart.

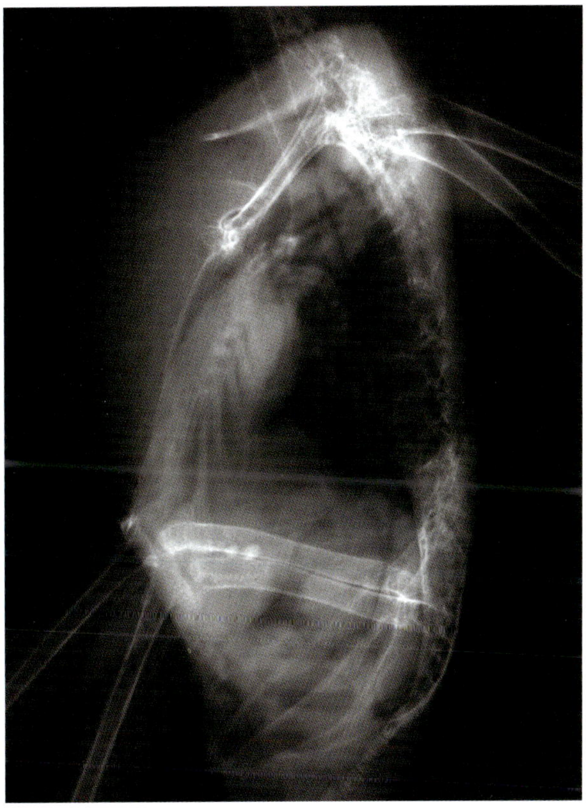

11.31 Radiograph (lateral view) of an African grey parrot with atheleroslerosis of the great vessels.

The following serves as an approximate guide for the **resting heart rate** in birds:

- racing pigeon (Columba livia): approx. 170–180 beats/min,
- common buzzard (Buteo buteo): approx. 280–320 beats/min,
- budgerigar (Melopsittacus undulatus): approx. 500–600 beats/min,
- canary (Serinus canariae): approx. 700–800 beats/min,
- various hummingbirds: up to 1,200 beats/min.

Tachycardia induced by handling and agitation increases the **risk of cardiovascular complications**. While this is particularly significant in compromised animals, it is also an important consideration in healthy birds, especially in some species (e.g., canaries [Serinus canariae], capercaillies [Tetrao urogallus] and falcons).

Appropriate restraint is essential for examining avian patients. The method of restraint should suit the circumstances, and the physical examination may need to be conducted in stages. Causes of excessive stress, such as ineffective capture techniques, should be avoided.

Routine monitoring of the cardiovascular system can also be conducted using **pulse oximetry**, a non-invasive method for determination of pulse rate and blood oxygen levels. Based on photometric principles, pulse oximetry is performed by passing light (red and infra-red) through well-vascularised tissue and measuring pulse-related changes in the absorption of each wavelength by erythrocytes in the blood. Oxygenated and deoxygenated blood have different absorption characteristics, allowing oxygen saturation to be calculated. In birds, this technique is performed by placing the light emitters and sensors on the lower pelvic limb (m. gastrocnemius). Depending on the size of the bird, use of a paediatric monitor is usually preferable. Cloacal pulse oximetry sensors have proven useful in birds for preventing artefacts caused by movement.

Relatively little research has been undertaken with respect to **specific diseases of the cardiovascular system of birds. Atherosclerotic** lesions, possibly related to diet, may be seen in the walls of the great vessels at the base of the heart in very old animals, and may occur due to unknown aetiologies in some bird species (Figure 11.31). Numerous infectious diseases can result in cardiac pathology. Avian influenza, for example, may be associated with petechial haemorrhages in the myocardium and in coronary fat. Degeneration of heart muscle fibres (myodegeneration cordis), appearing as pale stripes within the myocardium, can be seen within seconds of death/euthanasia. This must be taken into account when conducting a post-mortem examination or anatomical dissection.

Disruption of the microcirculation can lead to **wingtip oedema** in raptors, and also plays a part in other pathological processes such as fractures and infection. Increased perfusion of the microvasculature can be achieved by **warming** the affected area.

Microvascular derangements caused by staphylococcal infection can result in an ascending necrosis of the toes, particularly in finches. Other infectious aetiologies may also cause vascular pathology (e.g., petechial haemorrhages in bare areas of the pelvic limbs in cases of **avian influenza** and **Newcastle disease**).

Cardiac arrhythmias may be observed in birds of prey in association with prolonged inhalational isoflurane anaesthesia (around 50 minutes post-induction) and as a consequence of lead poisoning.

While clinical investigation of cardiovascular disease in birds is less developed than in mammals, **imaging techniques** such as radiography (Figures 11.30 and 11.31) and sonography (including power Doppler sonography, Figures 11.28 and 11.29) are nevertheless useful. Other modalities such as computed tomography (see Chapter 19 'Imaging techniques') have limited application in routine diagnosis, particularly in view of their cost.

Haematology is a valuable and efficient diagnostic technique in avian medicine. Haematocrit and total protein are routinely evaluated in pet, zoo and wild birds as well as raptors. At **45–55 per cent** the reference range for the **packed cell volume (haematocrit) of birds** is generally higher than in mammals. In commercial flocks, serology (measurement of antibody titres) plays an essential part in the diagnosis and management of numerous diseases (e.g., salmonellosis, avian influenza, tuberculosis).

Immune system and lymphatic organs (organa lymphopoetica)

12

K.-D. Budras, H. E. König and R. Korbel

The lymphatic system of the class Aves includes the following distinguishing features:

- lymphatic hearts (cor lymphaticum),
- mural lymphatic formations in the lymphatic vessels,
- the cloacal bursa (bursa of Fabricius, bursa Fabricii).

The primary function of the **lymphatic system** and its cells, particularly lymphocytes and macrophages, is to provide the protective response referred to as **immunity**. Anatomically, the lymphatic system includes the **lymphatic vessels** (including the **lymph hearts**) and **lymphatic tissue.** The main components of the lymphatic tissue are the **lymph nodes**, diffuse **lymphoreticular formations** (with or without a germinal centre) found predominantly in parenchymatous organs such as endocrine glands and the liver (**formationes lymphoreticulares parenchymatosae**), the **lymphatic organs** (thymus, cloacal bursa, spleen and bone marrow) and the **tonsils** (tonsillae pharyngis, oesophagi and caeci).

In **phylogenetic terms**, the avian lymphatic system can be considered a transitional stage between that of reptiles and mammals, exhibiting characteristics of both.

A multidirectional system, incorporating anastomosing lymphatic vessels and lymph hearts, is seen in amphibians and reptiles. In mammals, on the other hand, the flow of lymph through the lymphatic system has become exclusively unidirectional.

Embryonically, birds have both a **unidirectional** system, in which lymph drains into veins at the **paired venous angle**, and an 'alternative route', by which lymph passes through paired lymph hearts (located either side of the first caudal vertebrae) to enter the **dorsal pelvic veins**. The persistence of lymph hearts post-hatching varies according to species.

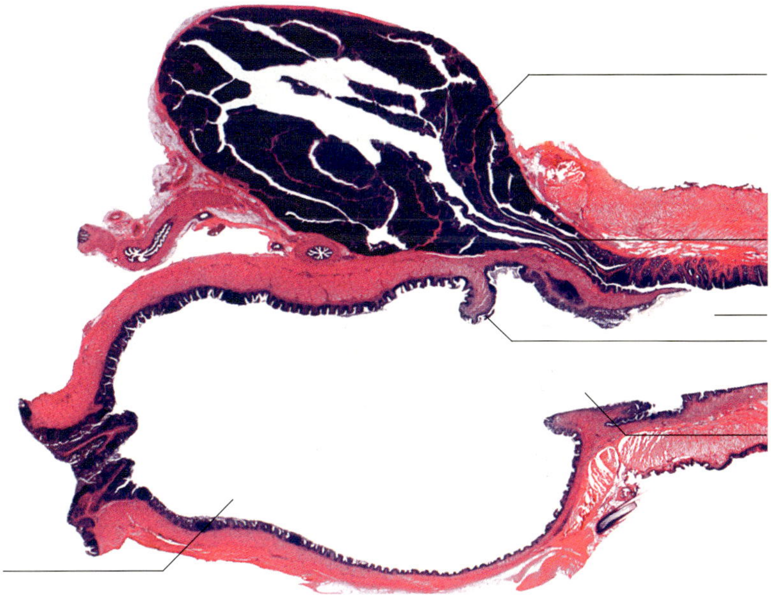

12.1 Histological section of the cloacal bursa (bursa Fabricii) of a chicken.

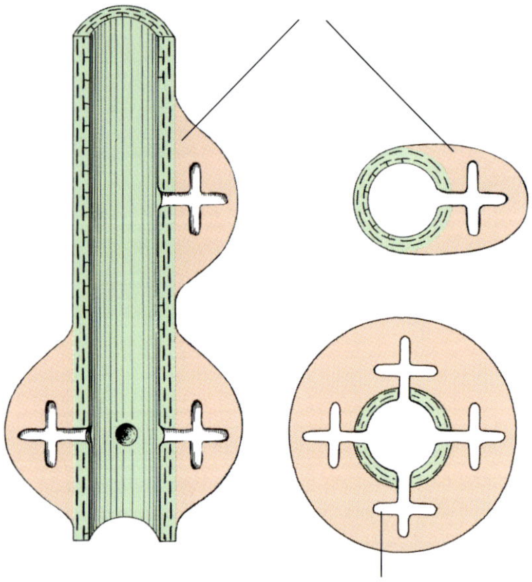

12.2 Wall of the initial segment of a collector vessel (vas lymphaticum myotypicum) with external lymphoreticular formations (schematic; left – longitudinal section, right – transverse section). Adapted from Berens von Rautenfeld et al. 1983.

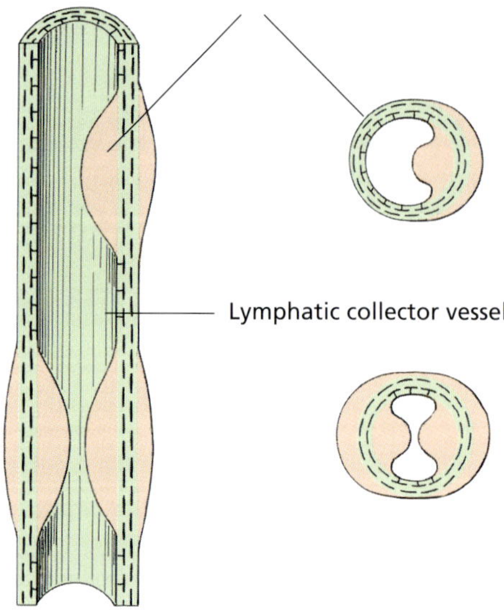

12.3 Wall of the terminal segment of a collector vessel (vas lymphaticum myotypicum) with internal mural reticular formations (schematic: left – longitudinal section; right – transverse section). Adapted from Berens von Rautenfeld et al. 1983.

Other features peculiar to the avian lymphatic system are the **cloacal bursa**, a primary lymphatic organ for the **differentiation of B-lymphocytes** (Figure 12.1), and a specialised system of lymphatic chambers in the **cloaca and phallus** that brings about erection (see Chapter 9 'Male genital organs').

Birds also appear to represent an intermediate phylogenetic stage with respect to the development of **lymph nodes**. These are present only in **water** and **marsh species**, as paired **cervicothoracic** and **lumbar nodes**. While avian lymph nodes can reach substantial lengths (up to 40mm in the goose), their structure differs significantly from those of mammals, bearing greater resemblance to **mural lymphoreticular formations** (Figures 12.2 and 12.3). Only just visible to the naked eye, these latter structures are typical of the lymphatic tissue found in reptiles, and are widely distributed in the lymphatic vessels of birds.

Lymphatic vessels (systema lymphovasculare)

The lymphatic vascular system drains fluid from most of the tissues of the body. As in mammals, lymphatic capillaries (and thus lymphatic drainage) are absent in epithelial tissue, cartilage, bone marrow, the thymus and in much of the central nervous system.

The **lymphatic capillaries** (rete lymphocapillare) are blind-ended. These structures are also referred to as 'initial lymph sinuses', a term that better reflects their variable calibre. Fluid travels through pre-formed fissures within the tissue towards the **capillaries** (**sinuses**). This phenom-

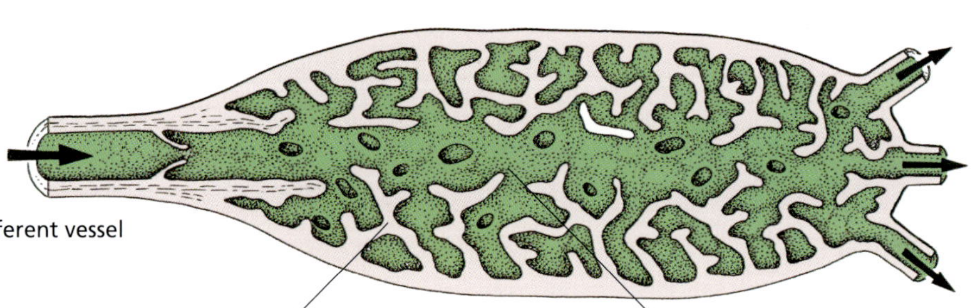

12.4 Lymph node of a duck (schematic), adapted from Berens von Rautenfeld and Budras, 1983.

enon, which can be observed particularly clearly in organs such as the testis, is the exclusive means of tissue fluid drainage in the central nervous system, where lymphatic capillaries are lacking.

As in mammals, fluid enters the lymphatic capillaries through 'open junctions' ('opening apparatuses') between the endothelial cells. Some of the water returns to the tissues and the intraluminal fluid is concentrated to become lymph. Specific lymph production occurs only in the phallus, as described in greater detail in Chapter 9 'Male genital organs'.

The open junctions are up to 5μm wide. Whole cells such as lymphocytes, macrophages and neoplastic cells can pass through the wall of the capillary, especially when the vessel is stretched or when pressure in the surrounding tissues is high. This has important implications for metastasis. **Hormones**, including testosterone and oestrogen, are also transported by the lymphatic route. Long-chain fatty acids in the form of chylomicrons are taken up by lymphatic capillaries within the digestive tract and distributed via the lymph.

The **post-capillary vessels** (vasa lymphatica fibrotypica = precollectorium) are similar in their basic structure to lymphatic capillaries, with open junctions present. In addition, they are furnished with pocket-shaped **valves** and a **subendothelial connective tissue collar** containing occasional smooth muscle cells. Their lumen is traversed and stabilised by connective tissue **trabeculae** (trabeculae fibroendotheliales) covered in endothelium. These 'pre-collector' vessels, like the capillaries (sinuses) that precede them, lie within the organ that they drain. In the limbs, they include short epifascial segments.

In the subsequent **'collector' vessels** (vasa lymphatica myotypica = collectorium), open junctions are no longer evident. Smooth muscle cell layers are present in the media (Figure 12.3).

Collector vessels transport lymph away from the organ of origin. In the limbs, they occur as relatively long subfascial segments, arranged in an **anastomosing** double array. The collector vessels are named according to the blood vessels they accompany (e.g., femoral lymphatic vessel). Lymph is conveyed from the collector vessels into **large lymphatic vessels** with three-layered walls (stratum internum, medium and externum).

The **large lymphatic vessels**, consisting of the paired **thoracoabdominal trunk** (truncus thoracoabdominalis) and the paired **jugular lymphatic vessel** (vas lymphaticum jugulare), transport most of the lymph to the **venous angle**, where the jugular vein and subclavian vein empty into the cranial vena cava (Figures 12.5 and 12.6). Lymph

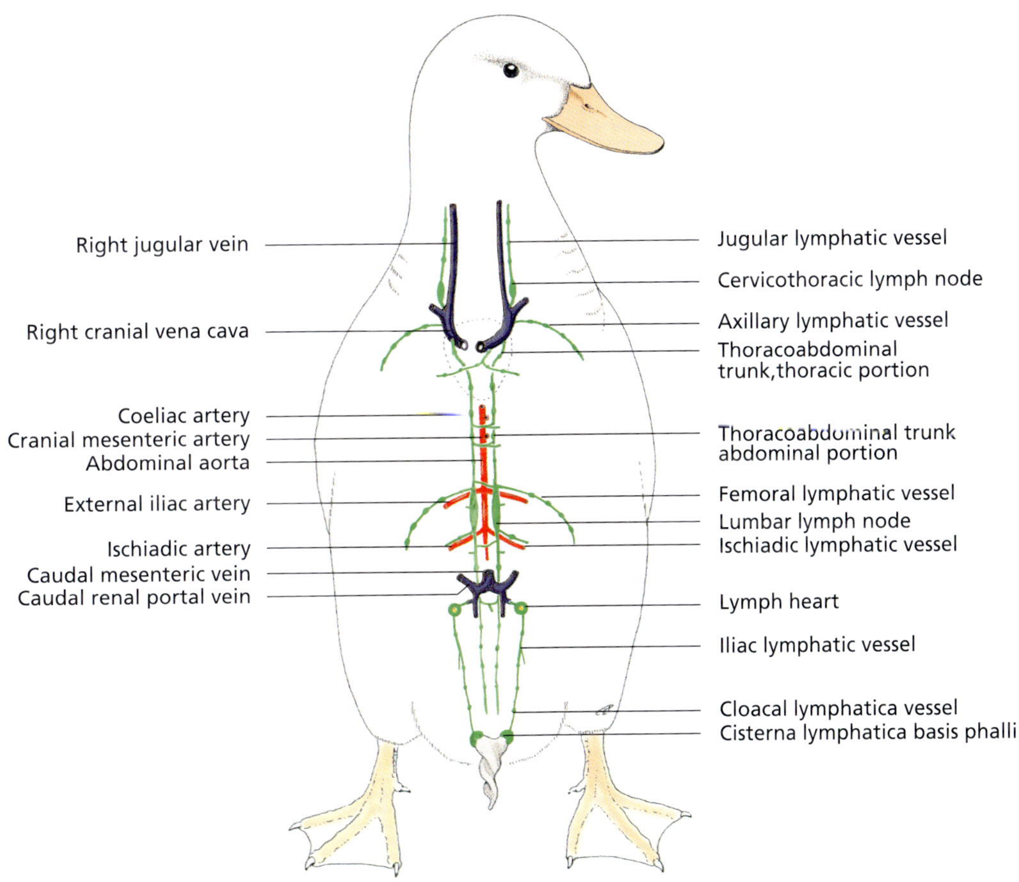

12.5 Large lymphatic vessels of the duck (schematic, not to scale).

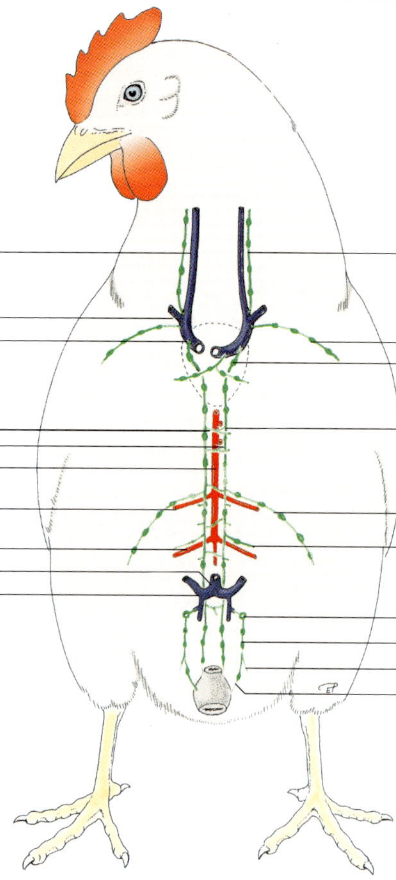

12.6 Large lymphatic vessels of the chicken (schematic, not to scale).

from the head and neck is delivered to the venous angle by the **jugular lymphatic vessel** either directly or indirectly via the thoracic portion of the thoracoabdominal trunk.

The paired **thoracic portion** (pars thoracica) of the **thoracoabdominal trunk** accompanies the caudal vena cava. It receives lymph from the gastrointestinal tract and the pelvic limbs via the external iliac lymphatic vessel (vas lymphaticum iliacum externum). Caudally the **abdominal portion** (pars abdominalis) of the thoracoabdominal trunk passes alongside the aorta and gathers lymph from the pelvic limbs, kidneys and gonads by way of the ischiadic lymphatic vessel (vas lymphaticum ischiadicum). Lymph flows either cranially towards the venous angle, or caudally to the lymph hearts (Figures 12.5 and 12.6).

The **paired lymph hearts** also receive lymph from the caudal body region, primarily from the pudendal and cloacal lymphatic vessels (vasa lymphatica pudendalia and cloacalia), which unite to form the internal iliac lymphatic vessel (vas lymphaticum iliacum internum).

Prior to reaching the lymph hearts, the internal iliac vessel becomes the **afferent vessel of the lymph heart**. While still within the pelvic cavity, this forms an extensively branching network that anastomoses with its counterpart on the other side of the body. It then passes between the synsacrum and the first free caudal vertebrae before opening into the lymph heart.

Lymph heart (cor lymphaticum)

The lymph heart consists, from exterior to interior, of the following components:

- adventitia,
- myocardium,
- endocardium.

The **lymph heart** is located outside the body cavity, at the caudal end of the synsacrum and dorsal to the transverse process of the first free caudal vertebra, adjacent to the ilium (Figures 12.5 and 12.6). In the foetus and newly hatched chick, the lateral half of the lymph heart lies subcutaneously under a fat pad. The medial half is covered by the m. levator caudae, which increases in size with age, eventually covering the entire lymph heart.

The **lymph heart** is an elongated dorsoventrally flattened organ. Fully formed, it reaches a length of 2mm in the chicken, 5mm in the duck, 15mm in the swan and approximately 25mm in ratites (ostrich, emu, nandu).

The wall of the **lymph heart** consists of an endothelium-lined **endocardium** (with a connective tissue sheath incorporating **smooth muscle**) and a myocardium. Its outer layer comprises an **adventitia** of loose connective tissue with multivacuolated white adipocytes that serve as pressure dampers. In ratites the myocardium is up to 2mm thick, consisting of **lymph heart muscle cells**, **myosatellite cells** and **conducting cells**.

The **contractile cells** of the **lymph heart** exhibit characteristics of striated skeletal muscle cells, striated cardiac muscle cells and smooth muscle cells, and are thus classified as a **fourth**, specialised and as yet relatively poorly understood **cell type**.

The crisscrossing **striated** lymph heart muscle fibres are arranged in **segments** and are bound with the **synsacrum** and first free **caudal vertebrae** by predominantly tendinous attachments. Other muscle fibres terminate without a skeletal attachment in the septa and trabeculae of the lymph heart. **Innervation** is supplied by myelinated and non-myelinated fibres.

The lymph heart is partially divided into a **ventrolateral inflow compartment** and **medial outflow compartment**, the latter being equipped with valves. A bicuspid valve marks the beginning of the inflow chamber. The lumen is traversed by incomplete septa and trabeculae.

Up to three efferent vessels, in which valves are present, leave the **medial outflow chamber**, although these diminish with age (especially in chickens and water birds). The **vas lymphaticum cordis efferens craniale** opens into a segmental vein that passes between the synsacrum and the first free caudal vertebra into the pelvic cavity and thence into the renal portal vein.

The **function of the lymph heart** is age- and species-dependent. In the **embryo** and **foetus** it serves exclusively as a pump for systemic lymph. Since the developing chick is essentially immobile within the egg, and is not subjected to the rhythmic compressions associated with breathing, the lymph heart provides an alternative mechanism for propelling lymph throughout the body. Later, this propulsion system is no longer needed.

As young birds mature, the bidirectional **flow of lymph in the caudal segment** of the thoracoabdominal trunk (towards the venous angle and the lymph hearts) gives way to a unidirectional **cranial flow**, with lymph passing exclusively towards the **venous angle**.

In species that have a well-developed copulatory apparatus and a protrusible phallus, such as ratites and water birds, the role of the lymph heart undergoes an **ontogenic transformation**, becoming wholly or predominantly incorporated into the **lymphatic apparatus of the copulatory organ** (including the cisterna lymphatica basis phalli, see Figures 12.5 and Chapter 9 'Male genital organs'). After erection, the lymph heart pumps a substantial amount of lymph (around 10ml in the drake) out of the phallus and directs it **towards the venous system**. In species with a less prominent, non-protrusible phallus, the lymph heart undergoes substantial regression.

Acting in an **ancillary capacity**, the lymph hearts also regulate blood pressure in the **internal vertebral sinus** and the **renal portal system**. Blood in the portal veins is required to pass through a second capillary network, necessitating an increased venous pressure.

Avian lymph nodes and mural lymphoreticular formations

Avian lymph nodes

The presence of lymph nodes is limited to **certain species of water bird**, and then only in the form of a:

- paired cervicothoracic lymph node (nodus lymphaticus cervicothoracicus),
- paired lumbar lymph node (nodus lymphaticus lumbalis).

The paired **cervicothoracic lymph node** (nodus lymphaticus cervicothoracicus) is a fusiform proliferation within the wall of the terminal segment of the **jugular lymphatic vessel**. It is located near the opening of the jugular vein into the cranial vena cava. In the goose and duck it assumes dimensions of up to 30mm in length and 5mm in thickness. Similarly, the paired **lumbar lymph node** (nodus lymphaticus lumbalis) is a modification of the wall of the **thoracoabdominal** trunk, slightly larger than the cervicothoracic node (up to 40mm by 5mm in the goose and duck). The lumbar lymph nodes flank the aorta between the external iliac and ischiadic arteries.

Avian lymph nodes are composed of a labyrinth of **lymph sinuses** interspersed with **lymphoreticular cords** (Figure 12.4). The sinuses branch from the afferent component of the lymphatic vessel. They are lined with endothelium and contain valves. At the efferent end of the lymph node, the sinuses drain into the lumen of the **same lymphatic vessel**. The system of sinuses is not subdivided into specific regions (e.g., marginal, intermediate, medullary), although a central sinus may be apparent (Figure 12.4).

The **lymphoreticular cords**, lying between the sinuses, are composed of diffuse collections of T-lymphocytes and areas (avian germinal centres) containing B-lymphocytes. Reticular fibres (type III collagen) extend to the discontinuous basal membrane of the lymph sinus but do not penetrate it. **Antigen processing** takes place between the lumina of the sinuses (containing numerous lymphocytes, macrophages and occasional red blood cells and granulocytes) and the lymphoreticular cords.

Lymphocytes **leave the lymph node** by the lymphovascular route, after passing through interendothelial openings in the lymph sinus, and by the haemovascular route, via postcapillary venules.

Mural lymphoreticular formations

Mural lymphoreticular formations are found in all of the larger lymphatic vessels in the chicken and are yet to be found lacking in any bird species. These spindle-shaped thickenings of the walls of the lymphatic vessels are just visible to the naked eye and can be considered as a modified and smaller version of the avian lymph node. Mural lymphoreticular formations occur as both internal variants within the intima of the lymphatic vessel, and as external variants in the thickened adventitia. Lymph sinuses, originating from the lymphatic vessel, are only found in external formations (Figure 12.2).

Lymphoreticular cords within mural formations contain areas of T- and B-lymphocytes as well as postcapillary venules for lymphocyte recirculation. Mural formations are stimulated by the presence of antigens within their drainage area, potentially becoming so enlarged that they are only distinguishable from lymph nodes by the presence of the media in the lymph vessel wall. As a whole, mural formations have **considerable immune potential** and fulfil the **role of lymph nodes**.

Lymphatic organs (thymus, cloacal bursa and spleen)

Thymus

The thymus is a lymphoepithelial organ. Its reticular epithelial cells are derived from the third and fourth pharyngeal pouches. The lymphoreticular tissue of the thymus is bounded externally by a basal membrane surrounded by a connective tissue capsule. During embryonic development, lymphocytic stem cells migrate to the thymus from the wall of the yolk sac and, later (in the foetal stages), from the bone marrow. Within the thymus, these cells mature into immunocompetent T-cells.

The **thymus** is divided into **lobes** that lie adjacent to the jugular vein from caudal to the third cervical vertebra to the cervicothoracic boundary (Figure 12.7). **Connective tissue septa** divide the lobes incompletely into **pseudolobuli** composed of a medulla and cortex.

The cortex and medulla consist of a **reticuloepithelial cell network** containing lymphocytes (**thymocytes**). These are particularly densely packed within the cortex. The medulla also contains macrophages, plasma cells and some granulocytes. Occurring exclusively in the medulla are **Hassall's corpuscles** (corpuscula thymi), composed of reticuloepithelial cells in a concentric 'onion skin' arrangement. The **lymphocytes in the cortex** are shielded from **antigen exposure** by a **blood-thymus-barrier**. This is absent in the medulla, where immunocompetent T-lymphocytes enter the bloodstream.

Involution of the thymus, mainly involving the cortex, is particularly pronounced from the fourth to the eighth week of life, but is less extensive than in mammals.

Cloacal bursa (bursa cloacalis, bursa Fabricii)

The cloacal bursa is the site of **B-lymphocyte** maturation. It is **unique to the class Aves**.

In most avian species, the cloacal bursa is a pedunculated **dorsal appendage of the proctodeum** (Figure 12.1). Ratites are an exception, their non-pedunculated bursa being integrated into the dorsal proctodeal wall. The bursa is subdivided by **longitudinal primary folds** arising from the cloaca. These are lined with epithelium and incorporate **secondary** and **tertiary** folds. The folds contain lymphatic nodules, or **lobuli lymphatici**, consisting of a central **pars lymphoepithelialis** and a peripheral **pars lymphoreticularis**. In ratites, this arrangement is reversed. Thus the terms 'cortex' and 'medulla' are confusing and inappropriate.

Lymphoid progenitor cells of the B-cell line mature within the cloacal bursa to reach **humoral immunocompetence**.

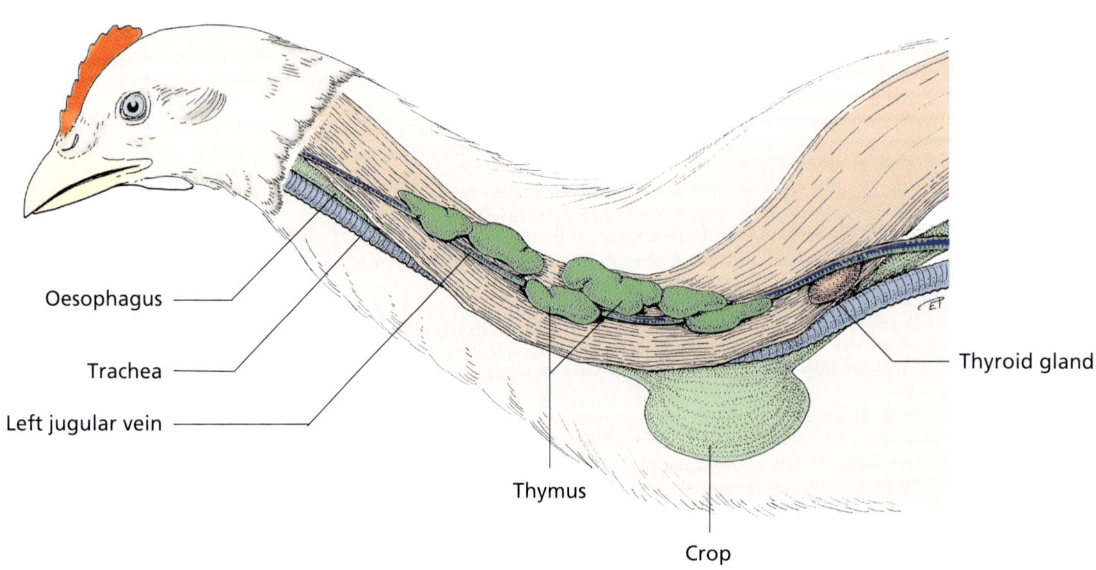

12.7 Thymus of the four-month old chicken (schematic, lateral view), adapted from Cotofan et al., 1971.

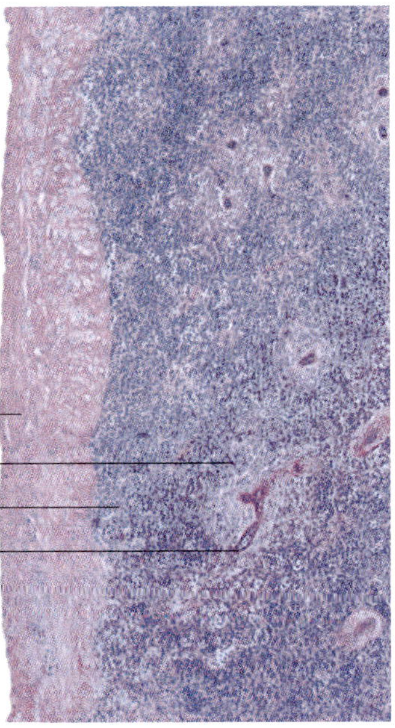

12.8 Histological section of the periphery of the spleen of a chicken.

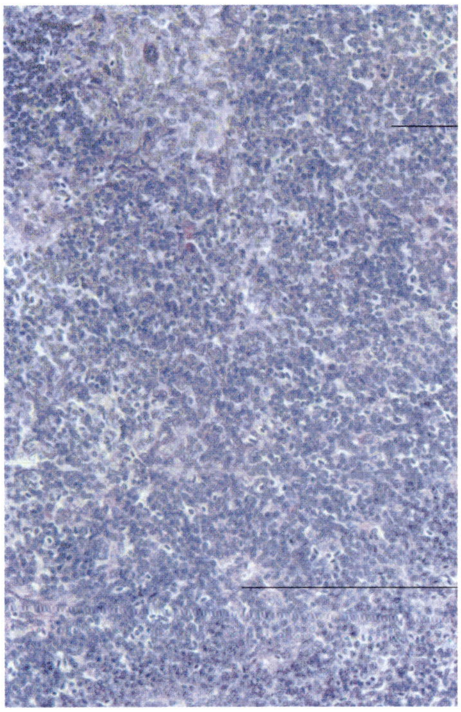

12.9 Histological section of the spleen of a chicken (detailed view).

At sexual maturity the **cloacal bursa undergoes involution**, which manifests as loss of the bursal folds and lobuli.

Spleen (lien, splen)

With its **red pulp** and indistinctly defined **white pulp**, the spleen belongs both to the blood vascular (red pulp) and lymphatic (white pulp) systems. It weighs approximately 1.4–4g in the chicken and duck, and around twice as much in the goose. Brown to cherry red in colour, the spleen is spherical in the chicken, tending towards a more flattened triangular shape in water birds. It lies medial to the junction between the glandular and muscular stomachs, near the visceral surface of the liver. The vessels entering and leaving the spleen divide in a similar manner to those in mammals.

The **basic structure** of the spleen comprises lymphoreticular tissue. Periarterial lymphatic sheaths (**T-cell zones**) and splenic nodules (**B-cell zones**) make up the white pulp (Figures 12.8 and 12.9). These are associated with the immune response, including lymphocyte proliferation and recirculation. Old red blood cells are broken down in the red pulp.

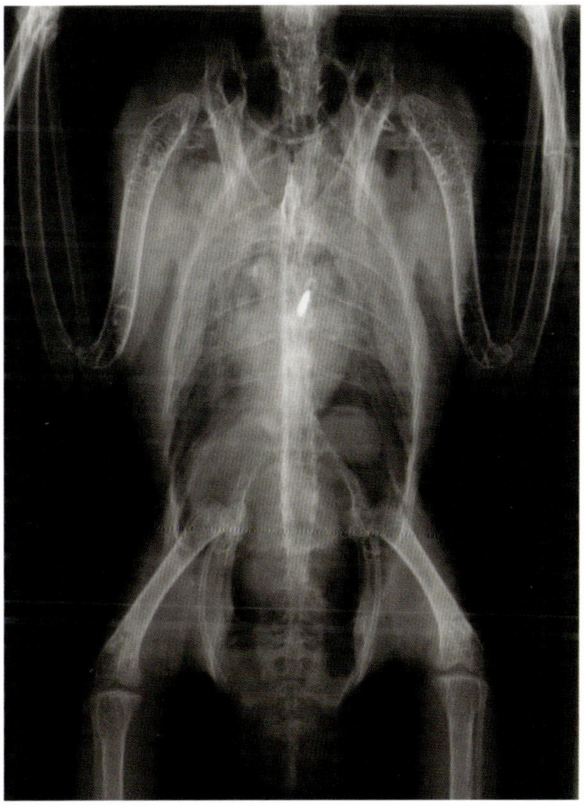

12.10 Radiograph (ventrodorsal view) of an African grey parrot (Psittacus erithacus) with splenomegaly. A radiographic marker is clearly visible in the cranial third of the sternum, in the m. supracoracoideus.

Clinical aspects

As most birds lack lymph nodes, the detection of a **bacteraemia** (e.g., in a blood smear) is not necessarily indicative of disease. In contrast to mammals, a small number of micro-organisms may also be observed in the liver of healthy birds, having entered via the enterohepatic circulation. In cases of bacterial infection, organisms may be present in the blood in large numbers (e.g., bacteraemia associated with Pasteurella multocida infection following a cat-bite injury). Conversely, the absence of a demonstrable bacteraemia in cases where the history is uncertain may be a consequence of previous antibiotic administration.

Infectious disease may be associated with clinically observable **splenomegaly**. In pet birds, splenic enlargement (Figures 12.10 and 12.11) is a characteristic finding in cases of psittacosis / ornithosis (Chlamydophila psittaci infection) and can sometimes be detected radiographically. Diagnosis and identification of the causative organism requires laboratory confirmation (e.g., conjunctival, choanal and cloacal swabs).

Occasionally a **fully developed thymus** may be encountered in adult birds. This is typically an indicator of chronic disease (e.g., mycobacteriosis / tuberculosis). In healthy birds, a non-involuted thymus is usually only present in juveniles.

In commercial poultry production, examination of the **cloacal bursa** is included in clinicopathological investigation of various infectious diseases, including infectious bursitis.

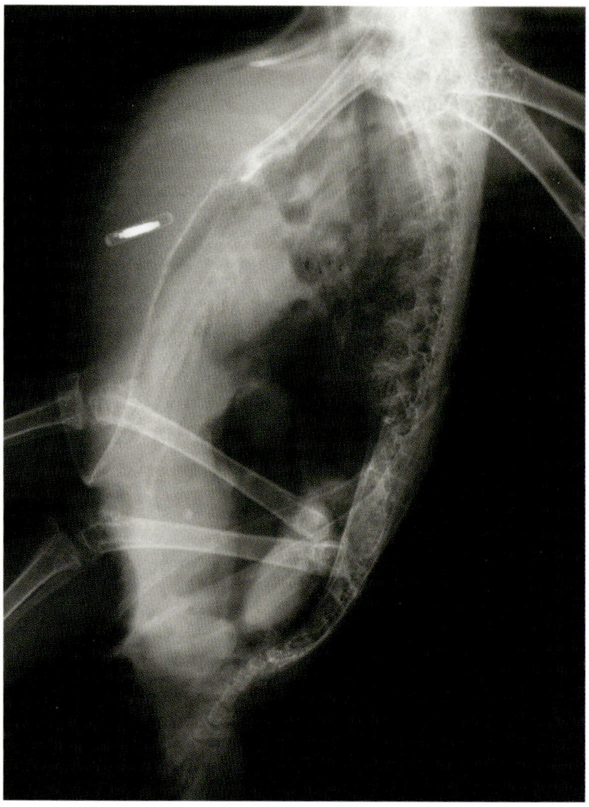

12.11 Radiograph (lateral view) of an African grey parrot (Psittacus erithacus) with splenomegaly and renomegaly.

13

Nervous system (systema nervosum)

H. E. König, I. Misek, H.-G. Liebich, R. Korbel and C. Klupiec

Central nervous system (systema nervosum centrale)

In birds, as in mammals, the central nervous system (CNS) consists of the **brain** and the **spinal cord** (Figure 13.1).

The following description assumes an understanding of the mammalian central nervous system, thus not all structures referred to in the text are illustrated in the accompanying images.

The avian brain is **lissencephalic** (smooth, lacking gyri) and, compared with mammals, is lower in weight relative to the spinal cord. Many functions that are regulated by the brain in mammals are localised to the **spinal cord in birds**, occurring as reflex arcs with minimal modification by higher centres. The approximate relationship between the weight of the CNS components in the chicken, dog and human is as follows:

	Spinal cord	Brain
• chicken	1	1
• dog	1	4
• human	1	25

As **birds are strongly visual animals** with large eyes, there is extensive development of CNS nuclei concerned with processing of optical signals.

Spinal cord (medulla spinalis)

Unlike its counterpart in mammals, the spinal cord of birds is the same length as the vertebral column, extending from the **foramen magnum** to the **last caudal vertebra**. The spinal nerves thus exit the intervertebral foramina in a lateral, rather than caudolateral, direction and there is no **cauda equina**. The **roots of the spinal nerves** emerge through **separate openings in the dura mater**. They merge within the intervertebral foramen, where the thus formed **spinal ganglion** (ganglion spinale) also resides.

The **cervical intumescence** (intumescentia cervicalis) is larger than the **lumbosacral intumescence** in flying birds, while in flightless birds the lumbosacral intumescence is greater in size. At the level of the lumbosacral intumescence, the left and right dorsal funiculi of the white substance and dorsal columns of the grey substance are separated by the **rhomboidal sinus** (sinus rhomboideus). This is occupied by the **gelatinous body** (corpus gelatinosum) (Figures 13.3 and 13.17).

The corpus gelatinosum is a richly vascularised structure of as yet undetermined function. It consists of **modified glycogen-rich glial cells** and is most substantial at the level of the third to sixth sacral segments of the spinal cord. The **central canal** passes ventral to the gelatinous body (Figure 13.3).

The ventral surface of the spinal cord bears a **ventral median fissure** (fissura mediana ventralis). Passing along the dorsal surface of the cord is the indistinct dorsal median sulcus (sulcus medianus dorsalis), from which a **glial dorsal median septum** (septum dorsale medianum) descends almost as far as the grey substance (Figures 13.2 and 13.3). Together, the fissure, sulcus and septum divide the entire spinal cord into symmetrical halves.

As is the case in mammals, the internally located **grey substance** (substantia grisea) consists of a **dorsal column** (columna dorsalis) and a **ventral column** (columna ventralis). In transverse section, these manifest as the **dorsal horn** (cornu dorsale) and **ventral horn** (cornu ventrale) (Figures 13.2 and 13.3). Based on architectural and functional criteria, the cells of the grey substance are subdivided into **layers**, or **laminae**. The laminar grey substance represents the 'relay system' of the spinal cord.

The **ventral horn** contains **motor neurons**. In the thoracic and lumbar regions, primary sympathetic preganglionic neurons form the **column of Terni**, dorsal and lateral to the central canal. **Neurons** present in the laminae of the **dorsal horn** are associated with sensory function. The left dorsal and ventral horns are connected with those on the right by the **grey commissure** (commissura grisea).

Along the length of the spinal cord, a group of nerve cells located near the lateral surface of the white substance form the **marginal nucleus** (nucleus marginalis). These neurons are considered to have a motor function. At the

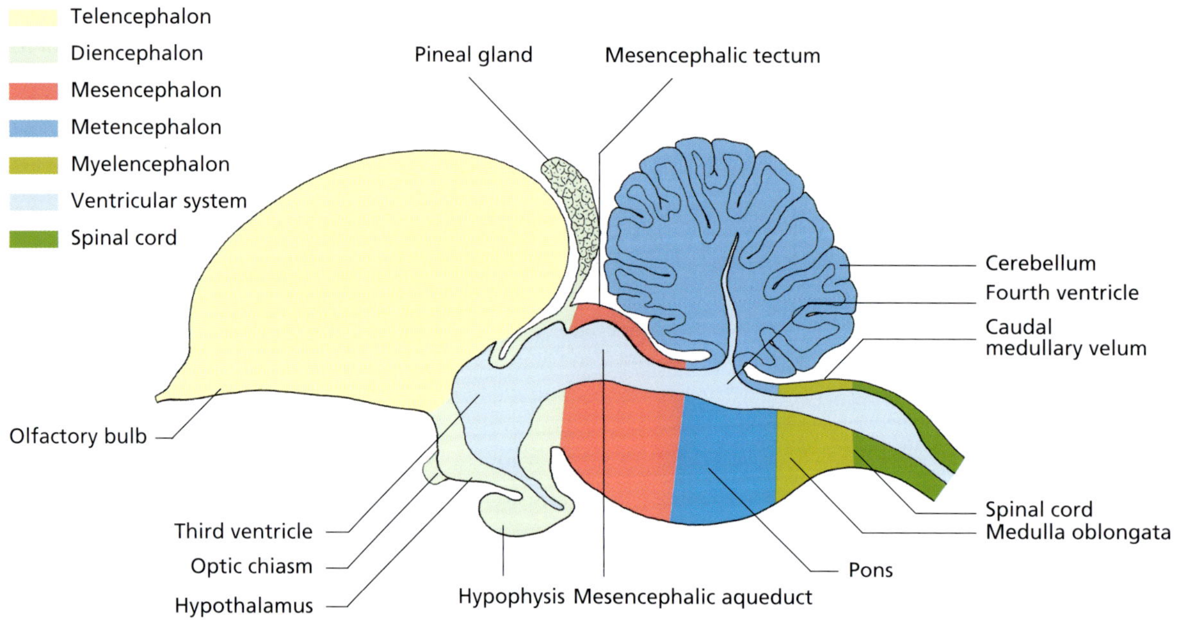

13.1 Brain of the chicken (schematic; median section), adapted from Romer, 1966.

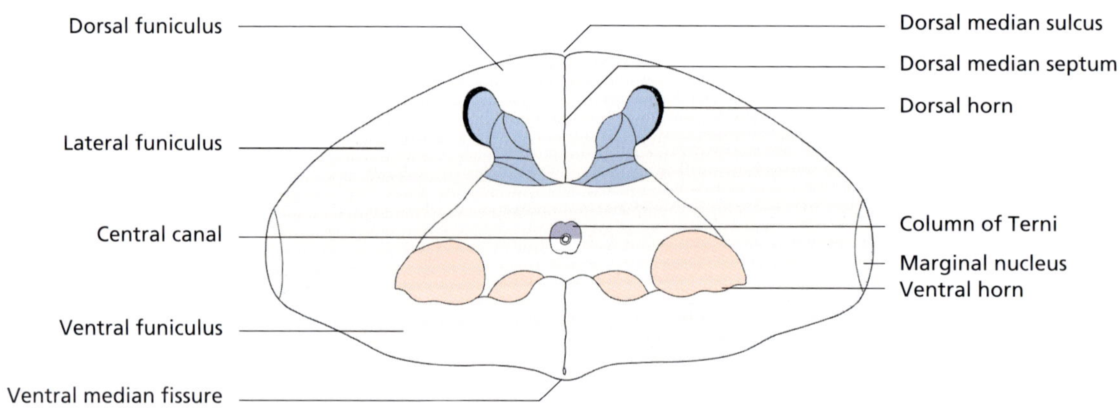

13.2 Cervicothoracic spinal cord of the chicken (schematic; transverse section), adapted from Breazile and Kuenzel, 1993.

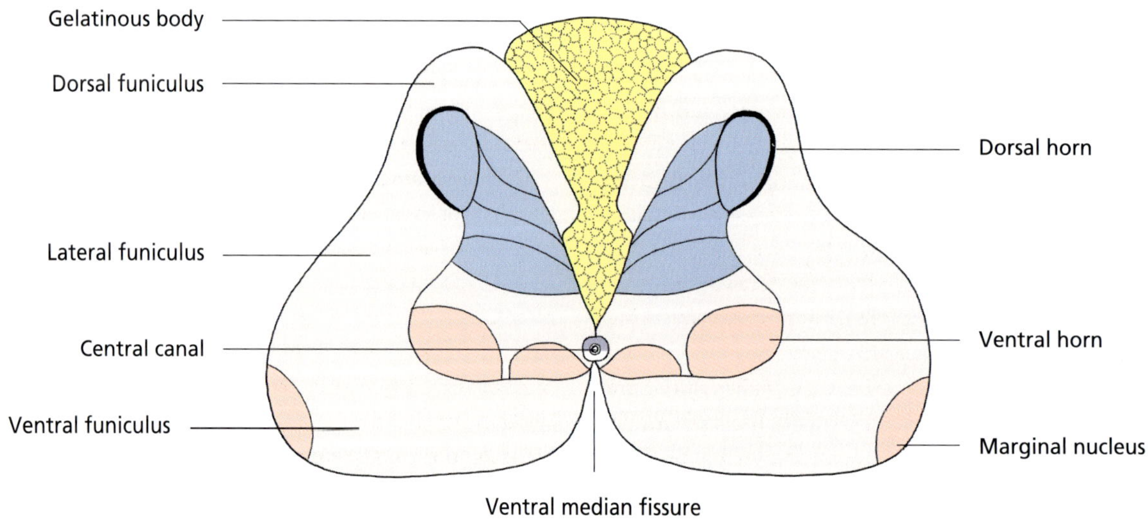

13.3 Lumbar spinal cord of the chicken (schematic; transverse section), adapted from Breazile and Kuenzel, 1993.

level of the lumbosacral intumescence, the nucleus protrudes slightly from the cord.

The **white substance** (substantia alba) surrounds the grey substance. It is incompletely divided by the columns of the grey substance into the paired (Figures 13.2 and 13.3):

- dorsal funiculus (funiculus dorsalis),
- lateral funiculus (funiculus lateralis),
- ventral funiculus (funiculus ventralis).

The **propriospinal (intraspinal association) fibre systems** (**fasciculi proprii**), lying adjacent to the grey substance, are more pronounced in birds than in mammals, while the long spinal tracts passing to and from the brain are relatively poorly developed. As such, segmentation of the spinal cord is more clearly defined in birds than in mammals.

Neither the propriospinal nor the spinal tracts are visible as discrete structures within the spinal cord. Thus they are not represented in Figures 13.2 and 13.3.

The ascending spinal tracts include the:

- spinothalamic tract (tractus spinothalamicus),
- spinotectal tract (tractus spinotectalis),
- spinoreticular tract (tractus spinoreticularis),
- dorsal spinocerebellar tract (tractus spinocerebellaris dorsalis) and
- ventral spinocerebellar tract (tractus spinocerebellaris ventralis).

Descending tracts described in birds include the:

- rubrospinal tract (tractus rubrospinalis),
- lateral vestibulospinal tract (tractus vestibulospinalis lateralis),
- medial longitudinal fasciculus (fasciculus longitudinalis medialis) and
- hypothalamospinal tract (tractus hypothalamospinalis).

Named according to their course, **the ascending tracts** are predominantly composed of **primary afferent nerve fibres** originating from the **spinal ganglia**. The first three tracts indicated above transmit pain, pressure, touch and temperature signals. The remainder form the key connections between the spinal cord and the cerebellum and are responsible for transmission of sensory information relating to coordination of muscular contraction and balance.

As in mammals, the (descending) **rubrospinal tract** (tractus rubropsinalis) originates in the **red nucleus** (nucleus ruber) within the **mesencephalon**. It is one of the most important **motor tracts**. The **lateral vestibulospinal tract** descends in the ventral funiculus and transmits impulses regulating muscle tone and postural adjustments.

The **medial longitudinal fasciculus** is formed by fibres originating from the **tectum** (tectospinale), the **vestibular nuclear complex** (vestibulospinale) and from the **reticular formation** (reticulospinale). These are involved in modulating the movement of the eyes and head.

The **hypothalamospinal tract** extends throughout the length of the spinal cord and is responsible for regulating autonomic functions.

In the spinal cord of birds, the ascending pathways are located in the dorsal funiculi, the descending tracts are within the ventral funiculi, and both ascending and descending pathways are represented in the lateral funiculi. Avian spinal cord tracts have received relatively little attention with most currently available information coming from studies conducted in the pigeon.

Brain (encephalon)

The avian brain essentially consists of the same components as that of mammals (Figures 13.4 and 13.5) and is divided into the following regions:

- **forebrain (prosencephalon)**:
  - telencephalon,
  - diencephalon;
- **midbrain (mesencephalon)**,
- **hindbrain (rhombencephalon)**:
  - metencephalon and
  - myelencephalon.

Features of the avian brain that differ from those of mammals include:

- the lack of a clear boundary between the medulla oblongata and the pons,
- the absence of discernible pyramids,
- the lack of an externally visible trapezoid body.

The **medulla oblongata** (myelencephalon) continues the spinal cord cranially. As in mammals, the boundary between the spinal cord and medulla oblongata is conventionally considered to be represented by the plane passing between the last pair of cranial nerves and the first pair of cervical spinal nerves.

At the **foramen magnum**, the spinal cord undertakes a tight dorsoconvex bend before joining the medulla oblongata. The **medulla oblongata** continues, without an obvious demarcation, as the **pons** (Figure 13.1). Separated rostrally from the mesencephalon by an obvious transverse furrow, the pons (part of the metencephalon) presents as a band of transversely oriented fibres.

At the level of the medulla oblongata, the **central canal** expands to form the **fourth ventricle** (ventriculus quartus). This is covered by the caudal and rostral medullary vela (velum medullare caudale and rostrale), which in turn are overlain by the **cerebellum** (Figure 13.1).

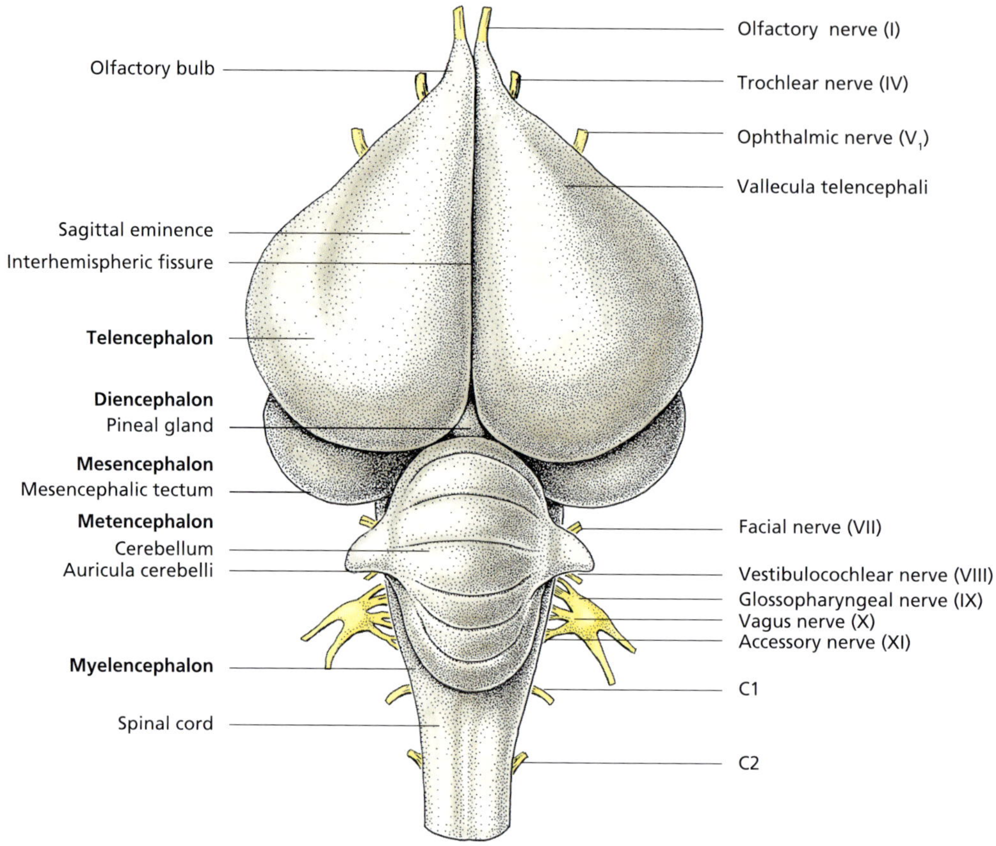

13.4 Brain and origin of the cranial nerves (schematic; dorsal view), adapted from Ghetie, 1976.

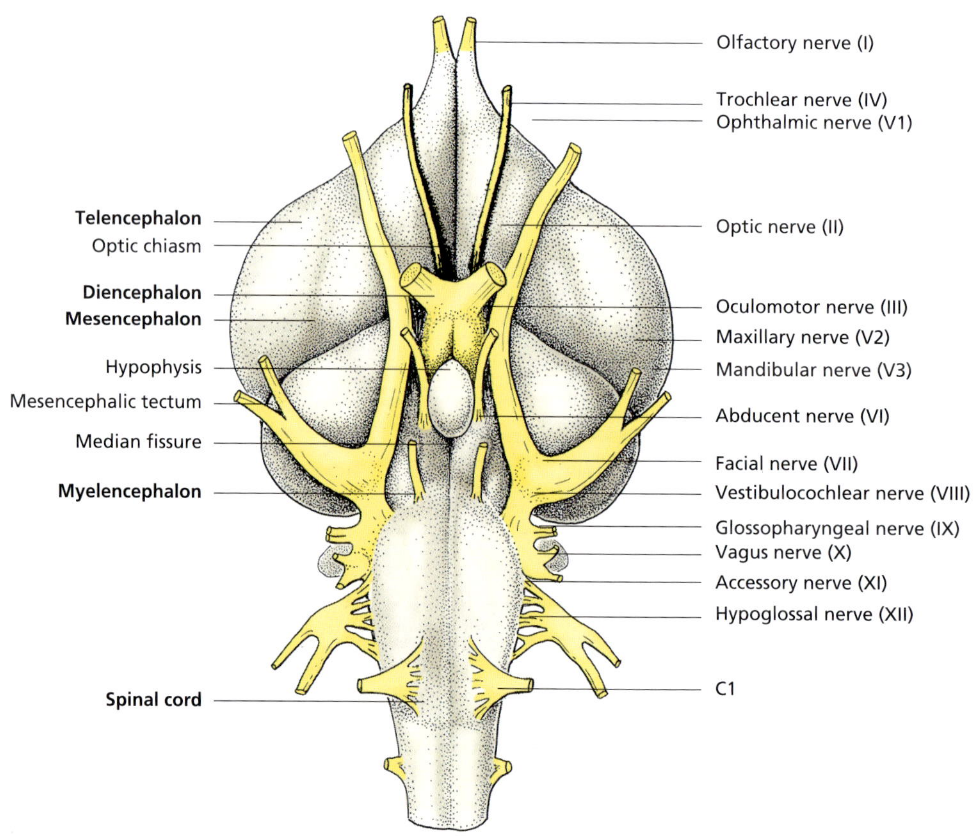

13.5 Brain and origin of cranial nerves of the chicken (schematic; ventral view), adapted from Ghetie, 1976.

The **median fissure**, a continuation of the ventral median fissure of the spinal cord, extends along the ventral surface of the medulla oblongata and the pons (Figure 13.5).

The **pyramids** and **pyramidal decussation**, as seen in mammals, are absent. The abducent nerve (VI) and hypoglossal nerve (XII) arise on either side of the median fissure (rostrally and caudally, respectively), while the trigeminal (V), facial (VII), vestibulocochlear (VIII), glossopharyngeal (IX), vagus (X) and accessory (XI) nerves arise from a more lateral position (Figures 13.4 and 13.5).

A discrete trapezoid body is not distinguishable externally, although equivalent auditory tract fibres decussate under the surface in this region.

Nuclei of the medulla oblongata and pons

The following nuclei are located within the medulla oblongata and pons:

- nuclei of cranial nerves V to XII,
- olivary nuclei,
- pontine nuclei,
- nucleus gracilis and cuneate nucleus,
- reticular formation nuclei.

Motor fibres of the trigeminal nerve (V) arise from a single nucleus consisting of lateral, median and ventral sections. The sensory **trigeminal nucleus** begins in the mesencephalon and ends in the cervical spinal cord, where it takes up connections to the dorsal columns. Its rostral mesencephalic portion serves as the origin of the **quintofrontal tract** (tractus quintofrontalis), which projects to the frontal lobe of the telencephalon.

The **nucleus of the abducent nerve** is thought to consist of principal and accessory subnuclei. Also subdivided (into dorsal, middle and ventral portions) is the **motor nucleus of the facial nerve**, which lies adjacent to the trigeminal motor nucleus.

The **vestibular nuclear complex**, responsible for regulation of balance, consists of six nuclei. As in mammals, fibres pass between these nuclei and those of the cranial nerves that supply the muscles of the eye (III, IV and VI). The auditory system is represented by **three cochlear nuclei**. Their fibres decussate at a location equivalent to the site of the trapezoid body of mammals and form the lemniscus lateralis.

The **vagus nerve** has a dorsal and a ventral motor nucleus. Nerve fibres extend between the ventral nucleus and the **glossopharyngeal nerve**. The **motor nucleus of the accessory nerve** extends from the medulla oblongata into the spinal cord as far as C4. Two nuclei associated with the **hypoglossal nerve** represent the continuation of the ventral columns of the spinal cord.

The substantial **caudal olivary nuclei** (complexus olivaris caudalis) are largely responsible for the girth of the medulla oblongata. Fibres pass from the olivary nuclei to the cerebellum, higher motor centres, the red nucleus and the telencephalon.

Fibres of the small **pontine nuclei** extend to the cerebellum. Unlike those of mammals, the avian pontine nuclei do not communicate with the corticospinal pyramidal tracts, as the pyramids are not present in birds.

The dorsal funiculi, which are relatively small in birds, terminate at the **nucleus gracilis** and **cuneate nucleus**. They continue rostrally as the modestly developed **lemniscus medialis**, from which only relatively few fibres extend as far as the thalamus. Groups of axons forming the **external arcuate fibres** (fibrae arcuatae externae) project to the cerebellum within the spinocerebellar tract.

The **reticular formation** is particularly prominent in the medulla oblongata and pons. It consists of groups of neurons with extensively branching dendrites that form connections with the ascending and descending tracts. Nuclei of the reticular formation appear to play a role in important visceral processes such as breathing and cardiovascular function.

Metencephalon

The metencephalon incorporates the pons (described above) and cerebellum. The **cerebellum** lies over the medulla oblongata, pons and the mid-section of the mesencephalon (Figures 13.1, 13.4 and 13.6 to 13.8). It is connected bilaterally to the brainstem by **rostral** and **caudal cerebellar peduncles** (pedunculi cerebellares rostrales and caudales). Discrete **pedunculi cerebellares medii (or pontocerebellares)** are not externally distinguishable in birds. Additional connections with the mesencephalon and medulla oblongata are established via the medullary vela. The cerebellum forms the roof and lateral walls of the **fourth ventricle,** and encloses the **ventriculus cerebelli**.

The cerebellum is composed of the large unpaired median **corpus cerebelli**, equivalent to the vermis in mammals, and the small paired cerebellar hemispheres, the main components of which (**flocculus** and **partes dorsalis et ventralis** of the **paraflocculus**) form the **auricula cerebelli** (Figure 13.6).

The body is divided by primary and secondary fissures into the lobi rostralis, caudalis and flocculonodularis. **Transverse grooves** (sulci cerebelli) further divide the cerebellum into ten **primary lobules** (lobuli cerebelli). The pedunculus flocculi connects the **flocculus** with the lobulus noduli (lobule X).

The internal **medullary body** (corpus medullare) consists of white substance and contains intracerebellar connecting fibres as well as afferent and efferent fibres. Three **nuclei cerebellares** are also embedded within the medulla.

The functional organisation of the avian cerebellum is similar to that of mammals. Fibres pass between the

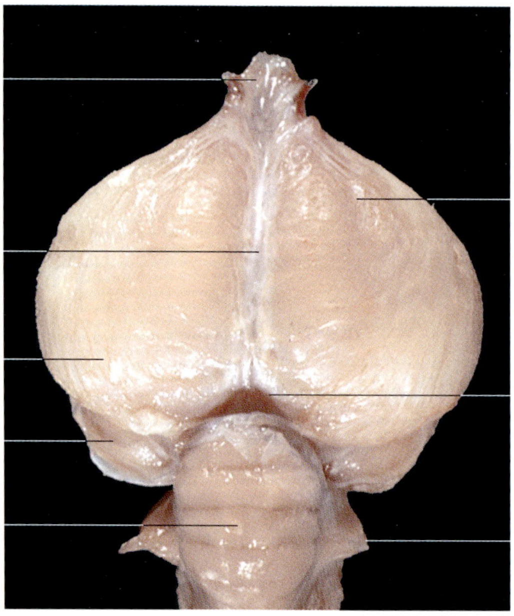

13.6 Brain of a chicken (dorsal view).

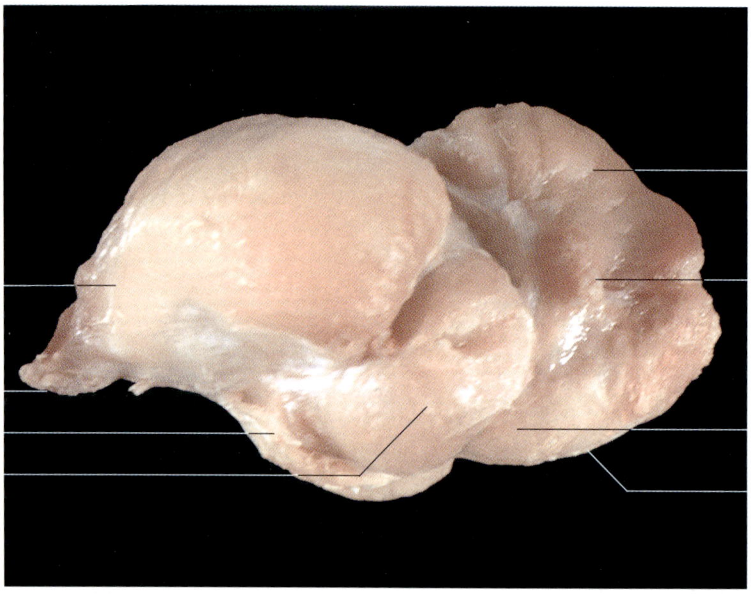

13.7 Brain of a chicken (lateral view).

cerebellum and the olivary and vestibular nuclei and to higher motor centres of the brain. Visual and auditory stimuli are also routed through the cerebellum. It is likely that the cerebellum also receives fibres from the trigeminal nucleus. **Ascending fibres** transmit proprioceptive signals. **Efferent tracts** pass to the reticular formation, the red nucleus and the vestibular nuclei, and are thence incorporated into the somatic efferent system.

The cerebellum is responsible for coordination of **motor function** associated with **posture** and **locomotion**. Damage to the cerebellum results in loss of coordination and a marked increase in muscle tone.

Mesencephalon

The midbrain adjoins the rostral rhombencephalon. Ventrally, the boundary between these sections is delineated by a distinct transverse furrow. The bulk of the roof of the midbrain is formed by the **mesencephalic tectum** (tectum mesencephali) (Figure 13.4). It projects laterally and ventrally, coming to lie lateral to the more ventrally located tegmentum.

The tectum receives fibres from the optic tract and bears functional similarity with the mammalian rostral colliculi. Although structures resembling the caudal colliculi of mammals are not visible on the exterior of the tectum in birds, the internal **nucleus mesencephalicus lateralis** participates in auditory and vestibular processing. A significant portion of the central auditory pathway (**lemniscus lateralis**) terminates in the nucleus mesencephalicus lateralis and in the nucleus lemnisci lateralis.

The axons of the **nucleus mesencephalicus lateralis** project rostrally to the ovoid nucleus (nucleus ovoideus) of the diencephalon (homologous with the medial geniculate body of mammals). After giving off the bulbotectal tract into the roof of the midbrain, the **lemniscus medialis** ends in the diencephalon.

The **isthmo-optic nucleus** (nucleus isthmo-opticus) of the complexus isthmi (comprising four nuclear regions) sends efferent fibres to the retina via the isthmo-optic tract.

The mesencephalic tegmentum houses the nuclei of cranial nerves III (oculomotor) and IV (trochlear). In addition, the trigeminal nucleus and the reticular formation extend into this region.

The **red nucleus** (nucleus ruber) is located in the ventromedial tegmentum. It receives afferent fibres from the dentate nucleus (cerebellum) and the dorsal telencephalon. The main descending tract emerging from the red nucleus is the **rubrospinal tract**. Components of the limbic system are combined with the motor nuclei of the tegmentum by the **tractus habenulointerpeduncularis**. The **nucleus ectomamillaris (or nucleus basalis tractus optici)** receives impulses of optical origin and transfers them to the reflex centres of the brain stem.

As indicated by these neural connections, the avian midbrain coordinates optical, auditory and vestibular stimuli and acts as an integration centre.

The **oculomotor nerve** arises from the ventromedial mesencephalon, in line with the origins of the abducent (VI) and hypoglossal (XII) nerves (Figure 13.5). Cranial nerve IV, the **trochlear nerve,** arises dorsolaterally between the cerebellum and the tectum. Dorsally, the midbrain is joined to the cerebellum by the rostral cerebellar peduncles.

A relatively large channel, the **mesencephalic aqueduct** (aqueductus mesencephali), passes through the midbrain (connecting the third and fourth ventricles). This passage is surrounded by **periaqueductal grey matter** (substantia grisea centralis). Lateral expansions of the aqueduct into the mesencephalic colliculi give rise to the ventriculus tecti mesencephali.

A cerebral peduncle, as seen in mammals, is not recognised in birds.

Diencephalon

The components of the diencephalon (Figure 13.5) are the:

- epithalamus,
- thalamus,
- hypothalamus.

The diencephalon forms the rostral continuation of the mesencephalon and represents the rostral limit of the brain stem. Dorsolaterally the diencephalon is completely covered by the **cerebral hemispheres.** The epithalamus, thalamus and hypothalamus surround the third ventricle, which takes the form of a vertical cleft.

At its base, the **third ventricle** extends into the neurohypophysis as the **recessus neurohypophysialis**, and into the mamillary body (corpus mamillare) as the **recessus inframamillaris**. The third ventricle communicates

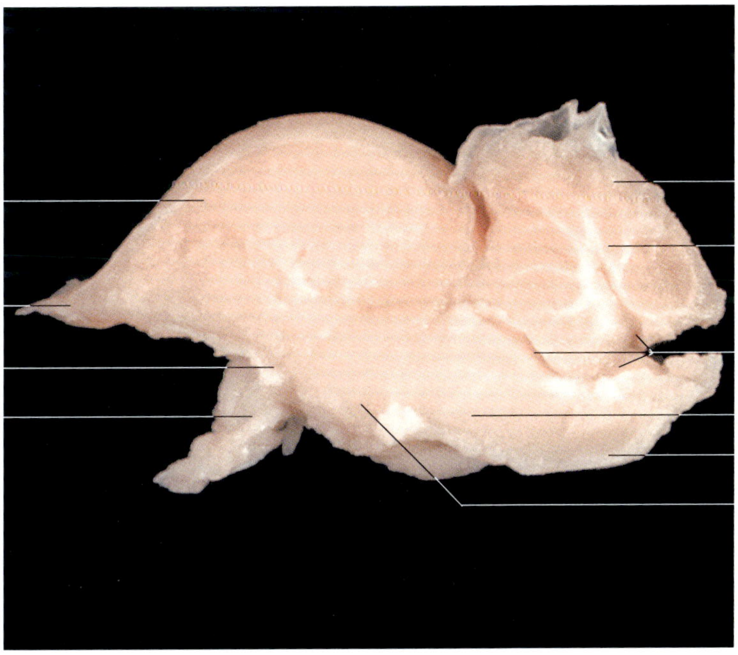

13.8 Brain of a chicken (median section).

rostrally with each of the lateral ventricles via an interventricular foramen. Caudally, it is continuous with the mesencephalic aqueduct.

The dorsal section of the diencephalon, the **epithalamus**, consists of the habenular nuclei and fibres, and the **pineal gland** (glandula pinealis), situated in the triangular space between the cerebral hemispheres and the cerebellum (Figures 13.1 and 13.4).

In the chicken, the pineal gland is approximately 3.5mm long and 2mm wide. It projects dorsally into the cerebral transverse fissure (fissura transversa encephali). The pineal gland is connected to the diencephalon by two bands that arise from the prominent longitudinal **habenular stria** (stria habenularis) and merge dorsally in the habenular commissure (commissura habenularis). Nerve tracts pass through the habenular stria from the basal olfactory region, the hypothalamus and the archistriatum (arcopallium) to the habenular nucleus. The pineal gland is highly responsive to light stimuli and contains cells that are structurally similar to photoreceptors. Together with the hypothalamus, the pineal gland contributes to regulation of reproduction.

The **epithalamus** forms the roof of the third ventricle and contains its choroid plexus.

The **thalamus** constitutes the largest component of the diencephalon. Its two halves lie lateral to the third ventricle. There is typically no interthalamic adhesion in birds. As in mammals, the thalamus is the last relay centre before information from afferent pathways ascends into the cerebral hemispheres.

There are relatively few fibres in the **spinothalamic tract**, thus the lemniscus medialis primarily contains fibres of the **tractus quintofrontalis**, that arises from the sensory nucleus of the trigeminal nerve.

In birds, the dorsal section of the thalamus is more developed than the ventral portion. The **dorsal thalamus** is the optical centre in which several groups of nuclei combine to form the nucleus (complexus) opticus principalis (homologue of the mammalian lateral geniculate body). The **ventral** component of the **thalamus**, termed the area ventralis thalami, is relatively small.

Several components of the central visual pathway are also integrated in the thalamus. The nucleus rotundus receives afferent fibres from the tectum mesencephali. Other regions of the thalamus receive fibres directly from the retina. Also located in the thalamus is the ovoid nucleus, a major component of the auditory pathway.

As in mammals, the **hypothalamus** forms the ventral part of the diencephalon. It contains the paired nucleus praeopticus, nucleus paraventricularis, nucleus supraopticus and nucleus infundibularis. Of these, all except the nucleus praeopticus are connected with the **neurohypophysis**. They are neurosecretory and form part of the **hypothalamo-hypophyseal system.** Fibres from caudally positioned hypothalamic nuclei project into the reticular formation of the mesencephalon. Afferent tracts pass from the hypothalamus, within the fasciculus medialis prosencephali, to the basal olfactory region and the telencephalon.

Through these connections the hypothalamus regulates higher-order autonomic functions, particularly thermoregulation, breathing, blood circulation, appetite and reproduction, as well as aggression and defensive reactions.

On the ventral hypothalamus, the **tuber cinereum** is separated from the **infundibulum of the neurohypophysis** by the sulcus tuberoinfundibularis. The tuber cinereum contains fibres originating from nuclei that pass within the infundibular tract to the interface between the adeno- and neurohypophysis.

Through its close association with the **hypophysis**, the hypothalamus serves as a link between the nervous and endocrine regulatory systems (Figure 13.5) (see also Chapter 14 'Endocrine glands').

The **optic chiasm** (chiasma opticum) is located rostral to the hypophysis (Figure 13.5). Within this structure, fibres of the optic nerve decussate before extending to the tectum of the midbrain and the thalamic nuclei.

Telencephalon

Distinguishing features of the avian telencephalon (Figures 13.4 to 13.8) include the following:

- the external surface is smooth,
- the corpus callosum is absent.

The two **cerebral hemispheres** (hemispherium cerebri) are separated by the **interhemispheric fissure** (fissura interhemispherica). A shallow, slightly caudolaterally oriented groove, the **vallecula telencephali**, is present on the dorsal surface of each hemisphere (Figure 13.9). The protuberance between the interhemispheric fissure and the vallecula is the **sagittal eminence** (eminentia sagittalis). A small rostrally tapering **olfactory bulb** (bulbus olfactorius) is situated at the rostral pole of the hemispheres (Figures 13.4, 13.6 and 13.9). A recess on the lateral surface of the hemispheres, the **fovea limbica**, is formed by the relatively large eyes.

The **hippocampus** is located on the medial surface of the hemispheres. Caudally and laterally, the hemispheres overlie the mesencephalic tectum. The **lateral ventricle** (ventriculus lateralis) extends far into the periphery, reaching to **just beneath** the dorsolateral surface of each hemisphere, such that the overlying cortex is only around 1mm thick. The caudomedial portion of this chamber contains the **choroid plexus of the lateral ventricle** (plexus choroideus ventriculi lateralis) (Figure 13.11).

The **cerebral cortex** (pallium) includes **olfactory** and **limbic** components, the latter incorporating the hippocampus, which forms a major portion of the dorsomedial

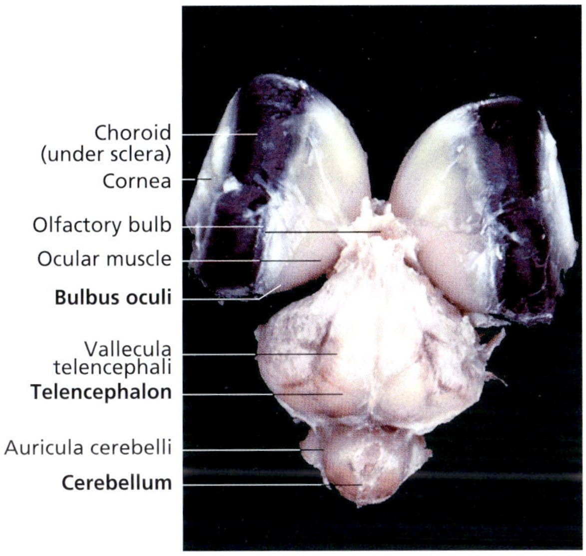

13.9 Brain of a wild turkey with eyeballs in situ (dorsal view).

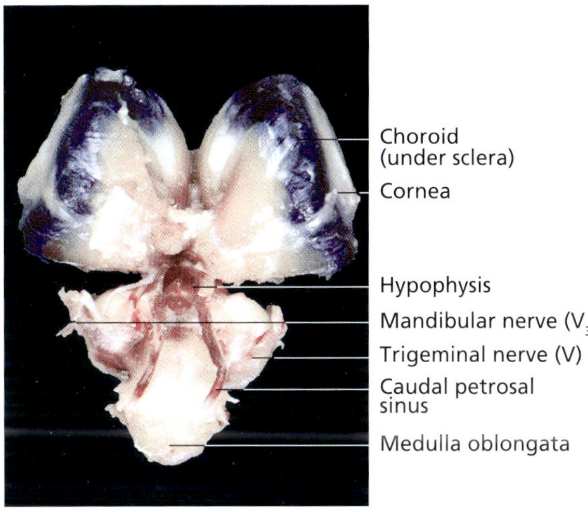

13.10 Brain of a wild turkey with eyeballs in situ (ventral view).

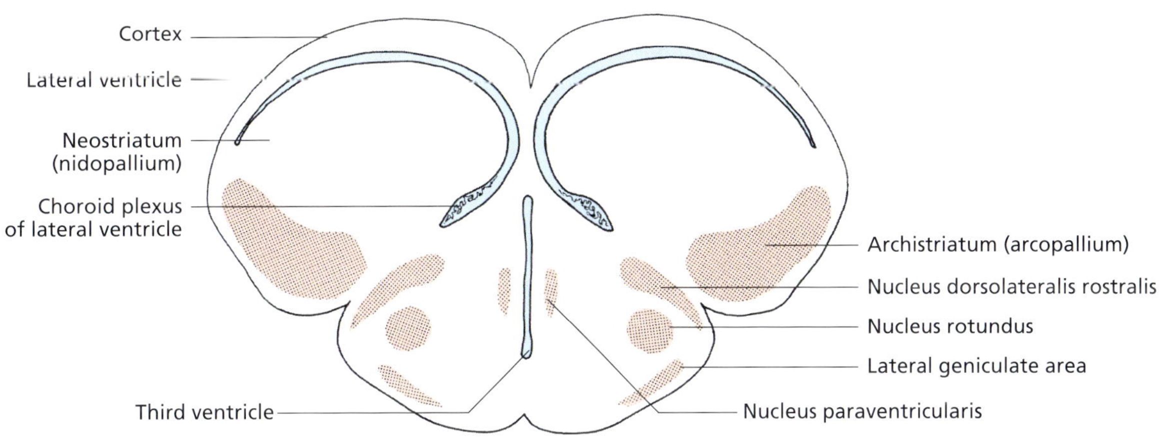

13.11 Transverse section of the telencephalon and diencephalon of the chicken (schematic), adapted from King and McLelland, 1978.

cortex. Non-cortical structures such as the **septum** and the **amygdaloid body** (corpus amygdaloideum) also contribute to the limbic system.

The nomenclature pertaining to the avian telencephalon has undergone considerable review in recent years. In particular, the telencephalon is no longer considered to consist largely of greatly enlarged basal nuclei with a rudimentary cortex. Rather, the pallial region, while differing in its architecture to that of mammals, appears to be well-developed in birds. For clarity, the following overview includes previously used terms as well as new terms in italics.

The subpallial region lateral to the ventricles, including the palaeostriatum (*lateral striatum* and *globus pallidus*), contains homologues of the basal nuclei of mammals. The overlying neostriatum (*nidopallium*), hyperstriatum (*hyperpallium* and *mesopallium*) and archistriatum (*arcopallium*) are pallial in nature. These lack the layered structure ('stria') of mammals, consisting instead of a continuum of nuclei.

The hyperstriatum, which forms the sagittal eminence, receives fibres of the optical pathway from the thalamus. Connections between the hemispheres are represented by the commissura rostralis and commissura pallii. An **internal capsule** and a **corpus callosum are lacking**.

The olfactory components of the avian telencephalon are poorly developed, reflecting the greater reliance of birds (other than the kiwi) on vision than olfaction.

Ventricles of the brain (ventriculi cerebri)

The central canal of the spinal cord and ventricles of the brain are filled with cerebrospinal fluid produced by choroid plexuses in the ventricles. At the level of the rhombencephalon, the **central canal** of the spinal cord opens into the voluminous **fourth ventricle** (ventriculus quartus). Ventrally, the fourth ventricle contains the

fossa rhomboidea. The roof of the ventricle is formed by the:

- rostral medullary velum (velum medullare rostrale),
- cerebellum,
- caudal medullary velum (velum medullare caudale).

Dorsally, the fourth ventricle extends into the cerebellum, giving rise to the ventriculus cerebelli. Caudolaterally, the **choroid plexus of the fourth ventricle** (plexus choroideus ventriculi quarti), protrudes deep into the subarachnoid cavity to form the recessus lateralis and recessus caudalis. To date, openings allowing communication between the fourth ventricle and the subarachnoid space, as seen in mammals, have not been demonstrated in birds.

The fourth ventricle continues rostrally as the relatively broad **mesencephalic aqueduct**, from which lateral expansions (ventriculi tecti mesencephali) extend deep into the mesencephalic tectum.

Rostral to the mesencephalic aqueduct, the **third ventricle** (ventriculus tertius) manifests as a narrow median cleft (Figure 13.11). Its roof is formed by the plexus choroideus ventriculi tertii. The ventricle expands dorsocaudally as the **recessus suprapinealis** and ventrally, into the infundibulum, as the **recessus infundibuli**. A connection between the third and lateral ventricles is formed by small **foramina interventricularia**.

The large **lateral ventricles** are located in the medial and occipital regions of the cerebral hemispheres (Figure 13.11). Rostrally the lateral ventricles project into the **olfactory bulb**. The **choroid plexus** of the lateral ventricles is small. It lies at the level of the foramen interventriculare, where it is continuous with the plexus of the third ventricle.

Meninges and meningeal blood vessels

As in mammals, the avian meninges consist of the:

- **dura mater** (pachymeninx or dura mater encephali et spinalis),
- **leptomeninges** (leptomeninx), comprising the:
  - arachnoid membrane (arachnoidea encephali et spinalis) and
  - pia mater (pia mater encephali et spinalis).

The dura mater is the strong outer meningeal layer enclosing the brain (dura mater encephali) and spinal cord (dura mater spinalis). Most of the **dura mater encephali** is fused with the periosteum of the cranium. Between the two hemispheres, the dura mater forms a shallow longitudinal partition. A more substantial transverse partition, the **tentorium cerebelli** (or plica tentorialis) extends between the cerebral hemispheres and the cerebellum (dorsally) and the mesencephalon (laterally). The **diaphragm sellae** is a ring-shaped projection of the dura mater that surrounds the **hypophyseal stalk** (infundibulum). Embedded within the diaphragma sellae is the small venous **sinus cavernosus**.

The olfactory bulb is flanked bilaterally by an **olfactory sinus** (sinus olfactorius). Both olfactory sinuses commmunicate ventrally with the unpaired olfactory sagittal sinus (sinus sagittalis olfactorius) and dorsally with the dorsal sagittal sinus (sinus sagittalis dorsalis). The latter is continuous caudally with the **transverse sinus** (sinus transversus) that runs along the tentorium cerebelli.

The dorsal sagittal and transverse sinuses communicate caudally with the large, median **sinus occipitalis** that passes over the surface of the cerebellum. Blood travels from the sinus occipitalis into the jugular vein, the vertebral vein and the **internal vertebral sinus** (sinus venosus vertebralis internus).

In mobile regions of the vertebral column, the **dura mater spinalis** contains large numbers of elastic fibres. An **epidural space** (cavum epidurale) is present, except in the synsacral region where the dura fuses with the **endorhachis** (periosteal layer of the vertebral canal). The epidural space contains a gelatinous substance that protects the spinal cord from excessive mechanical impact.

The **leptomeninges** are composed of the arachnoid membrane (arachnoidea encephali et spinalis) and the delicate pia mater (pia mater encephali et spinalis).

The **arachnoidea encephali** is closely apposed to the internal surface of the dura mater. Fine fibres connect the arachnoidea encephali with the pia mater. At the convex surfaces of the brain, the pia mater lies against the arachnoid. Elsewhere the **subarachnoid space** (cavum subarachnoidale) between the arachnoid and the pia mater is filled with **cerebrospinal fluid** (liquor cerebrospinalis) and accommodates the larger vessels of the brain. The subarachnoid space extends into crevices between the ventral diencephalon and mesencephalon, and between the cerebellum and medulla oblongata, to form **subarachnoid cisterns** (cisternae subarachnoidae).

In the cervical and thoracic segments of the spinal cord the **arachnoidea spinalis** comes to lie very close to the pia mater and the subarachnoid space is very narrow. Dorsal to the lumbosacral intumescence, the subarachnoid space widens into a large, cerebrospinal fluid-filled cistern. As mentioned in the previous section, the means by which cerebrospinal fluid passes from within the ventricular system to the subarachnoid space has not yet been established.

The **pia mater**, comprising the pia mater encephali and pia mater spinalis, is intimately attached to the entire surface of the brain and spinal cord. It forms a thin lining over the vessel bundles of the choroid plexuses. Between the spinal nerves, the pia mater gives off lateral extensions to the dura mater forming the serrate **ligamentum denticulatum**, which is particularly well-developed in the cervical region and at the level of the gelatinous body.

The unpaired **internal vertebral sinus** passes dorsal to the spinal cord within the **epidural space.** At the **foramen magnum** it is continuous with the **sinus occipitalis**. Caudally it is briefly interrupted at the level of the gelatinous body. The internal vertebral sinus receives the veins of the spinal cord and is connected by segmental branches with the jugular veins, vertebral veins and rami of the renal portal system.

Peripheral nervous system (systema nervosum periphericum)

The peripheral nervous system (PNS) connects the central nervous system with the tissues of the body. Based on morphological and functional criteria, the peripheral nervous system can be divided into **two broad categories**:

- **somatic nerves** and **ganglia**:
  - cranial nerves (nn. craniales),
  - spinal nerves (nn. spinales);
- **autonomic nerves** and **ganglia** (autonomic or vegetative nervous system).

Cranial nerves (nervi craniales)

Like mammals, birds have 12 pairs of cranial nerves (Figures 13.12 and 13.13; see also 'Parasympathetic system' below), although these exhibit considerable species variation.

Olfactory nerve (I)

The exclusively sensory olfactory nerve (n. olfactorius) (Figures 13.12 and 13.13) arises from the convergence of unmyelinated axons of bipolar nerve cells in the olfactory mucosa. This nerve fibre bundle passes caudally along the medial wall of the orbit and enters the cranial cavity, through the **foramen nervi olfactorii**, before dividing into numerous filaments, or **fila olfactoria**, that extend to the **olfactory bulb**. Near the foramen nervi olfactorii, the olfactory nerve detaches dorsal and ventral rami. The dorsal branch supplies the roof of the nasal cavity and the nasal septum, while the ventral branch innervates the ventral surface of the caudal nasal septum.

Optic nerve (II)

The optic nerve (n. opticus) (Figures 13.12 and 13.13) transmits nerve impulses from the retina to the diencephalon. It consists of the initially unmyelinated axons of the multipolar ganglion cells of the retina. Upon penetrating the sclera, the axons acquire a myelin sheath and the nerve is surrounded by meninges. The combined cross-sectional area of the substantial optic nerves exceeds that of the cervical spinal cord, highlighting the functional significance of the avian eye. After entering the cranium through the foramen opticum, the optic nerves decussate completely (in most birds) at the **optic chiasm**, rostral to the hypophysis. This is significant with respect to ophthalmic examination

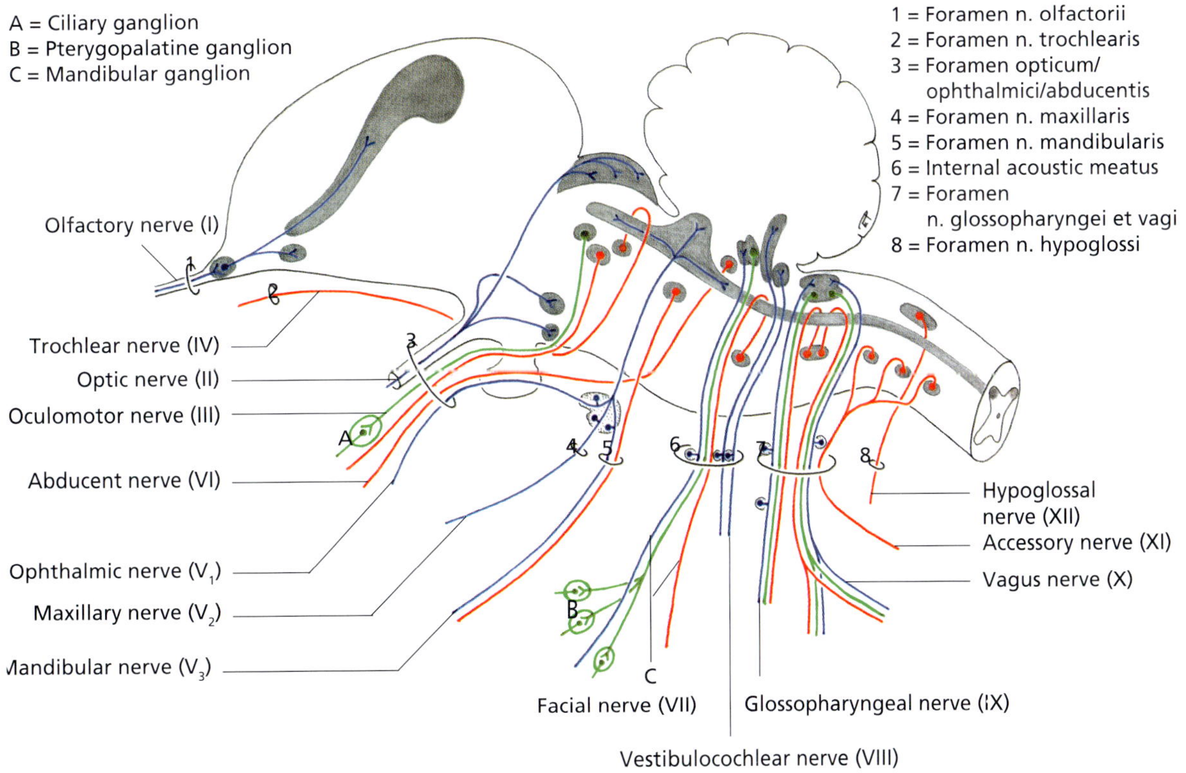

13.12 Origin of the cranial nerves of the chicken (schematic), adapted from Goller, 1972. Blue – sensory fibres, red – motor fibres, green – parasympathetic fibres.

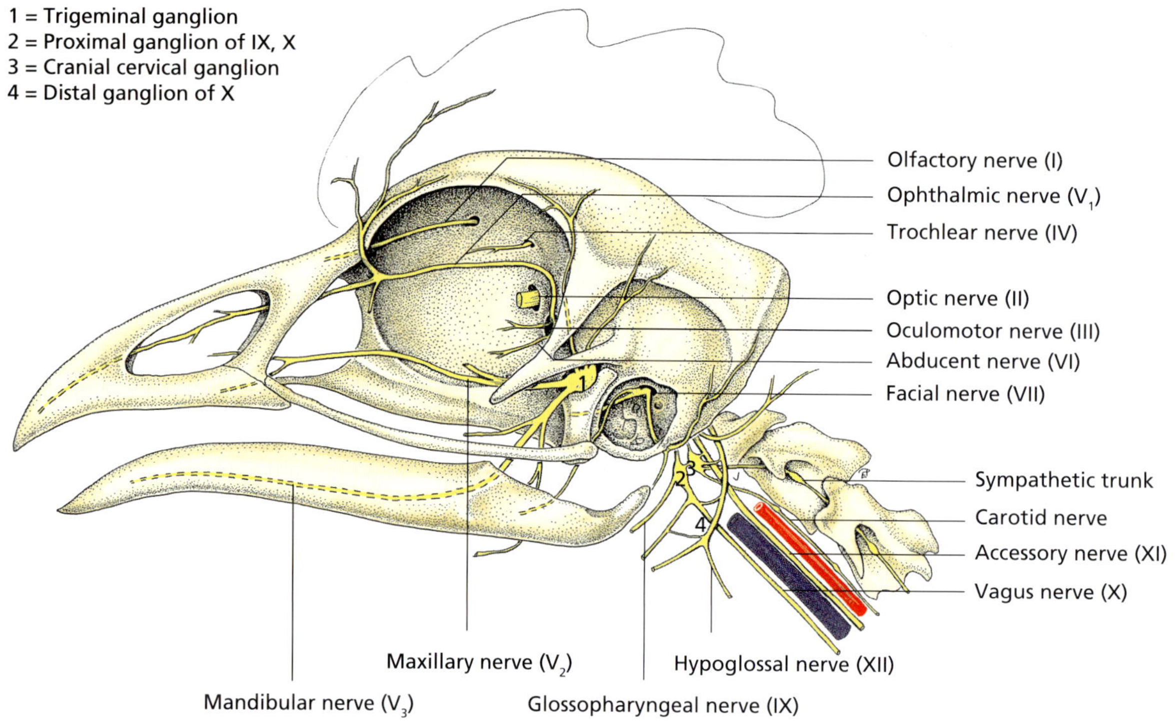

13.13 Major branches of the cranial nerves of the chicken (schematic), adapted from King and McLelland, 1978.

as, in contrast to mammals, a true consensual pupillary light reflex does not occur in birds (see Chapter 18 'Clinical examination').

Oculomotor nerve (III)

The third cranial nerve (Figures 13.12 and 13.13) arises from the mesencephalon, immediately caudal to the hypophysis. Ventrolateral to the foramen opticum, the oculomotor nerve traverses the **foramen nervi ophthalmici** to reach the orbit. It gives off a parasympathetic branch to the **ciliary ganglion** (ganglion ciliare), which lies between the optic nerve and the oculomotor nerve. The continuation of the oculomotor nerve contains **motor fibres** that innervate the m. rectus medialis, m. rectus ventralis, m. obliquus ventralis, m. pyramidalis and the muscles of both eyelids.

Postganglionic **parasympathetic fibres** pass within the **nn. ciliares breves** (nn. choroidales) to innervate the gland of the third eyelid, the choroid and the pecten, and within the **n. iridociliaris** (n. ciliaris longus) to supply the muscles of the iris and ciliary body.

Trochlear nerve (IV)

The trochlear nerve arises dorsolaterally from the mesencephalon. Before leaving the brainstem, its fibres cross over in the **rostral medullary velum**. The trochlear nerve enters the orbit through the foramen nervi trochlearis, just dorsolateral to the foramen opticum. As in mammals, the sole structure innervated by the trochlear nerve is the m. obliquus dorsalis (Figures 13.12 and 13.13).

Trigeminal nerve (V)

Immediately caudal to the mesencephalic tectum, the trigeminal nerve (n. trigeminus) leaves the brain stem as motor and sensory roots (radix motoria and radix sensoria). Located in the sensory root, close to its origin, is the large **trigeminal ganglion** (ganglion trigeminale). The sensory root then divides into two branches, the **ophthalmic nerve** and the common trunk of the **maxillary** and **mandibular nerves.** Fibres of the motor root pass lateral to the trigeminal ganglion before joining the mandibular branch of the sensory root to form the mixed mandibular nerve (Figures 13.12 and 13.13).

The sensory **ophthalmic nerve (V1)** enters the orbit together with the oculomotor nerve through the foramen nervi ophthalmici. It gives off the n. ciliaris longus (n. iridociliaris) that innervates the choroid. In the rostral orbit, the ophthalmic nerve divides into a dorsal and ventral branch. The dorsal branch innervates the upper eyelid and the skin of the forehead and comb. The nasal cavity and, in the chicken, the bulk of the upper beak are innervated by the ventral branch.

The sensory **maxillary nerve (V2)** passes into the orbit through the foramen nervi maxillare (often together with the mandibular nerve through a common opening).

It gives off branches to the upper and lower eyelids, palate, upper beak, nasal cavity, infraorbital sinus and the lacrimal gland. In the duck and goose, the maxillary nerve is the predominant source of sensory innervation of the upper beak. The maxillary nerve receives parasympathetic fibres from the pterygopalatine ganglion.

The sensory and motor **mandibular nerve (V3)** innervates the masticatory muscles, the lower beak, tongue and the wattle. Its lingual branch is smaller than that of the glossopharyngeal nerve (IX). The chorda tympani of the facial nerve contributes parasympathetic fibres to the mandibular nerve. At the angle of the beak, these fibres innervate the mucosa and glands as far as the lateral palatine region, and the salivary glands located on the floor of the oropharyngeal cavity, lateral to the tongue.

Abducent nerve (VI)

The somatomotor abducent nerve (n. abducens) (Figures 13.12 and 13.13) arises ventrally from the rostral end of the medulla oblongata.

It traverses a dedicated bony canal to enter the orbit, ventrolateral to the oculomotor nerve and medial to the ophthalmic nerve, through the foramen nervi abducentis. The abducent nerve innervates the m. rectus lateralis and the striated muscles of the third eyelid, the m. quadratus and m. pyramidalis.

Facial nerve (VII)

The somatosensory and motor facial nerve (n. facialis) (Figures 13.12 and 13.13) emerges ventrolaterally from the medulla oblongata. Within the gently curved facial canal, it carries the very small (sensory) geniculate ganglion (ggl. geniculi). The major petrosal nerve and chorda tympani, arising at the level of the geniculate ganglion, establish connections with the maxillary nerve (pterygopalatine ganglion, ggl. pterygopalatinum) and the mandibular nerve (mandibular ganglion, ggl. mandibulare). Upon exiting the facial canal, the facial nerve receives sympathetic fibres from the cranial cervical ganglion (ggl. cervicale craniale). It continues rostrally to innervate the hyoid muscles, the cervical cutaneous muscles and the large depressors of the mandible.

Vestibulocochlear nerve (VIII)

The sensory vestibulocochlear nerve (Figures 13.12 and 13.13) arises laterally from the medulla oblongata, caudal to the facial nerve. Located within the vestibular part of the nerve, the vestibular ganglion (ggl. vestibulare) receives fibres from the semi-circular ducts, the macula utriculi, the macula sacculi and the crista neglecta. The cochlear component is formed by the convergence of the **n. cochlearis** and the **n. lagenaris**, each of which has a separate ganglion. In birds, the cochlea and spiral ganglion are **not helical**. The ganglion lagenare is connected to the macula lagenae and is located near the tip of the cochlea.

Glossopharyngeal nerve (IX)

The cranial nerves IX, X and XI arise close together as a group of small rootlets on the ventrolateral margin of the medulla oblongata (Figures 13.12 and 13.13). Proximal ganglia of the glossopharyngeal and vagus nerves lie close together in a small excavation in the bone.

The **glossopharyngeal** (n. glossopharyngeus) and **vagus** (n. vagus) nerves exit the cranium through separate openings. After traversing the foramen nervi glossopharyngeum, the glossopharyngeal nerve comes to lie – together with the vagus nerve – medial to the cerebral carotid artery, the external ophthalmic artery and the facial nerve. Enclosed between the glossopharyngeal and vagus nerves is the **cranial cervical ganglion**, from which both nerves receive sympathetic fibres. A small, barely visible distal ganglion is located within the glossopharyngeal nerve. Near this ganglion, a fibre bundle connects the glossopharyngeal and vagus nerves.

The **glossopharyngeal nerve** subsequently divides into three terminal branches, the **ramus lingualis**, **ramus pharyngeus rostralis** and the **ramus pharyngeus caudalis**. The ramus lingualis primarily innervates the tongue and the larynx. Fibres of the ramus pharyngeus rostralis pass to the pharynx and larynx. The ramus pharyngeus caudalis forms the continuation of the glossopharyngeal nerve and passes caudally in the neck, parallel to the jugular vein, providing innervation to the oesophagus and trachea.

Vagus nerve (X)

The vagus nerve (Figures 13.12 and 13.13) leaves the cranium distal to the proximal ganglion through the foramen nervi vagi. It appears that connections between the vagus and glossopharyngeal nerves contain vagal fibres that come to join the ramus pharyngeus rostralis in innervating the larynx, pharynx and cranial oesophagus.

Lying parallel to the jugular vein, the vagus nerve extends to the base of the neck. The only branches detached in the cervical region consist of small rami that supply the thymus. Caudal to the thyroid gland, the vagus nerve carries the distal ganglion, from which fibres are given off to the thyroid gland, parathyroid glands, ultimobranchial body and carotid body.

Upon entering the body cavity, caudal to the distal ganglion, the vagus nerve gives off the **nn. cardiaci craniales**, the **ramus pulmonalis** and, at the level of the pulmonary trunk, the **recurrent nerve** (n. recurrens).

The **left recurrent nerve** curves around the ligamentum arteriosum, while the right winds around the **aortic arch**. Both pass towards the head and supply the oesophagus, trachea and the crop. Additional branches of the vagus nerve include the rr. pulmonales and the nn. cardiaci caudales. The left and right vagus nerves approach one another on the ventral oesophagus, forming a trunk that continues caudally. This gives off branches to the stomach, liver, spleen and pancreas. Ultimately the vagal branches join the sympathetic intestinal nerve.

Accessory nerve (XI)

In addition to the aforementioned rootlets from the medulla oblongata, the accessory nerve (n. accessorius) (Figures 13.12 and 13.13) receives branches from the first and second cervical segments. These enter the cranium through the foramen magnum and combine with the cranial roots. After pursuing a brief (5mm) common course, the accessory nerve separates from the vagus nerve and continues to innervate the superficial cervical muscles.

Hypoglossal nerve (XII)

Several small rootlets arise from the ventral surface of the medulla oblongata and merge to form **two nerve trunks** that exit the cranium through separate foramina nervi hypoglossi. Fibres from the first two cervical spinal nerves join the caudal trunk, after which the two trunks combine. The hypoglossal nerve (n. hypoglossus) takes a rostral course and crosses the vagus and accessory nerves, exchanging fibres with both nerves (Figures 13.12 and 13.13).

The hypoglossal nerve subsequently gives off a **ramus trachealis** that, together with the **ramus pharyngeus of the glossopharyngeal nerve**, passes caudally along the neck and innervates the muscles of the trachea. Near the larynx, the hypoglossal nerve divides into the **ramus lingualis**, supplying the muscles of the tongue, and the **ramus laryngeus**, extending alongside the trachea to innervate the syrinx and tracheal muscles.

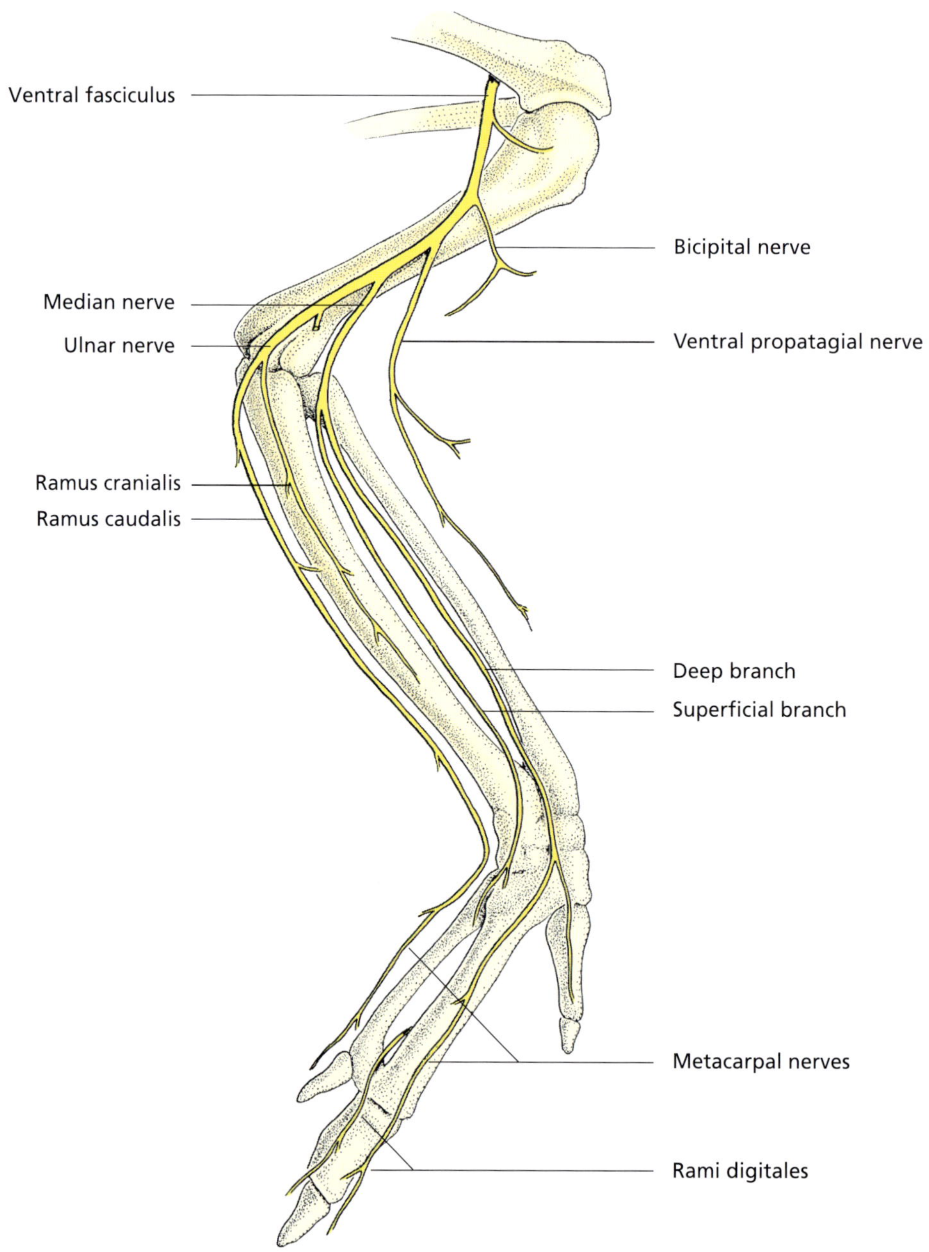

13.14 Innervation of the wing of the chicken (schematic; ventral view), adapted from Salomon, 1992.

Spinal nerves (nervi spinales)

As in mammals, the spinal nerves are formed by the union of a larger, sensory **dorsal root** (radix dorsalis) and a smaller **ventral root** (radix ventralis). The first cervical nerve has only a ventral root. From the third cervical nerve onwards, the dorsal root includes a **spinal (dorsal root) ganglion** (ggl. spinale).

In the cranial half of the vertebral column, the spinal ganglion is situated within the intervertebral foramen. More caudally, it lies just outside this opening. In the chicken, the thoracic spinal ganglia are fused with the sympathetic ganglia of the sympathetic trunk. The spinal nerves are numbered according to their corresponding vertebra, thus the number of spinal nerves varies with the number of individual vertebrae in each species. In the chicken there are 41 pairs of spinal nerves comprising:

- 15 cervical nerves (nn. cervicales),
- seven thoracic nerves (nn. thoracici),
- 14 synsacral nerves (nn. synsacrales) and
- five caudal nerves (nn. caudales).

Outside the intervertebral foramen, each spinal nerve divides into a dorsal and ventral branch. The somewhat thinner **dorsal branches** supply the intrinsic spinal muscles and overlying skin while the **ventral branches** innervate the lateral and ventral regions of the body including the limbs.

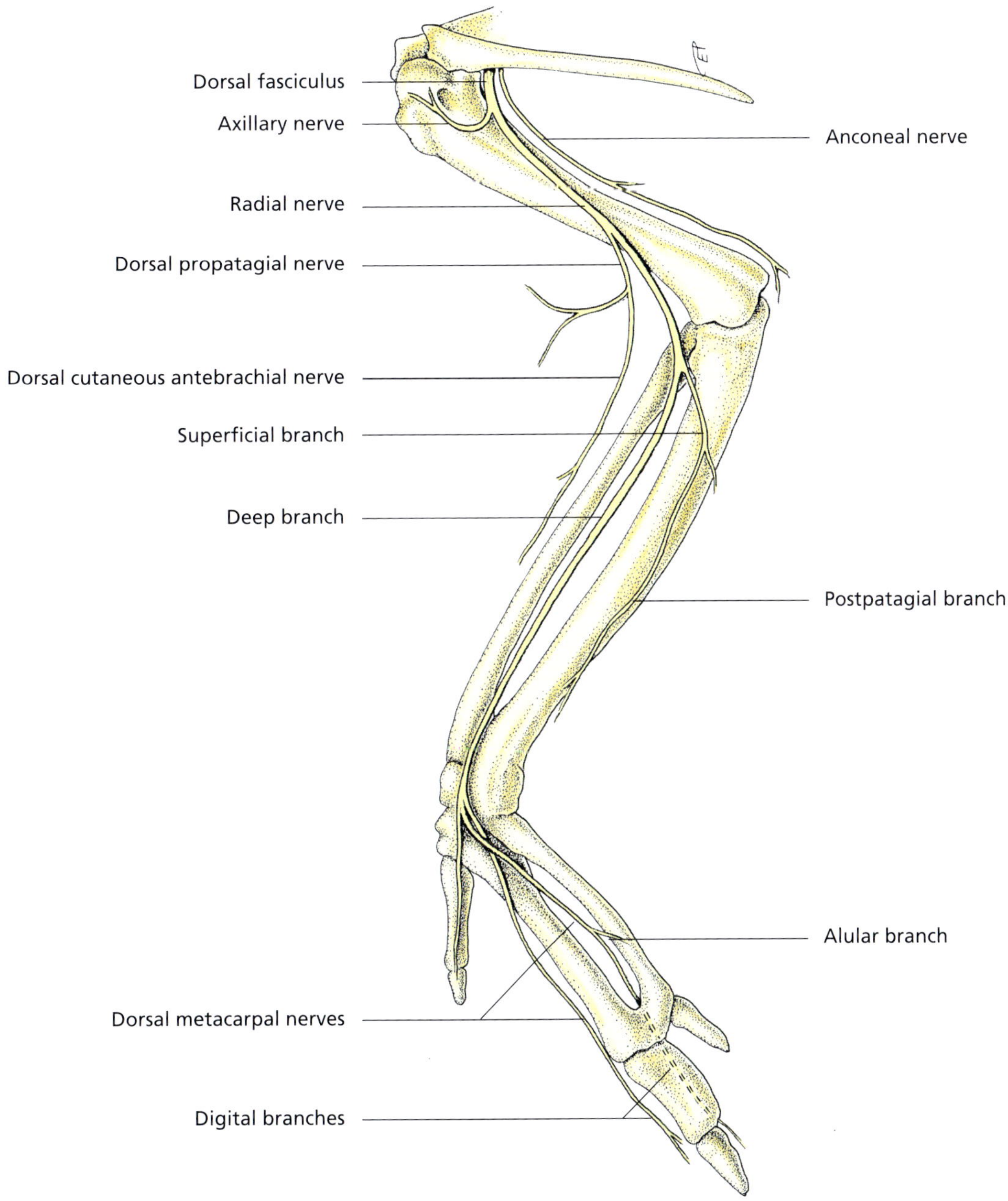

13.15 Innervation of the wing of the chicken (schematic; dorsal view), adapted from Salomon, 1992.

In the cervical and thoracic region, the **ganglia of the sympathetic trunk** lie closely apposed to the ventral surface of the ventral branches. Near the synsacrum and tail, short connecting rami (rami communicantes) pass between the ventral branches and the ganglia of the sympathetic trunk.

Brachial plexus (plexus brachialis)

The **brachial plexus** innervates the muscles of the shoulder girdle and the wing. It is usually formed from 4–5 ventral branches of the last cervical and first thoracic spinal nerves. In chickens, these are the spinal nerves 13–17. The brachial plexus is divided into dorsal and ventral components. The **dorsal section** gives off nerves that innervate the muscles associated with the scapula. Extensive branching of the cranial nerve roots gives rise to a small accessory plexus.

The **ventral portion** of the brachial plexus is distributed to the muscles of flight. Arising ventrocaudally, the pectoral nerves innervate the **m. pectoralis** (depressor of the wing), while the **supracoracoid nerve**, originating ventrocranially, supplies the **m. supracoracoideus** (elevator of the wing).

The **muscles of the wing** are innervated by the **dorsal and ventral fasciculus** (fasciculus dorsalis and ventralis), which arise from the plexus either side of the axillary artery.

The **dorsal fasciculus** innervates the dorsal surface of the wing, the elevators of the humerus, and the extensors of the elbow, carpal and digital joints. Its direct continuation, the **radial nerve** (n. radialis), passes from the caudal aspect of the wing onto the dorsal surface of the humerus and continues on the dorsocranial antebrachium. At the level of the elbow joint the radial nerve gives off the **dorsal propatagial nerve** (n. propatagialis dorsalis) that innervates the propatagium.

The large **axillary nerve** (n. axillaris) arises from the proximal dorsal fasciculus. It supplies the shoulder joint and the mm. deltoidei. The **anconeal nerve** (n. anconealis), which supplies the m. triceps brachii (Figure 13.15), also originates proximally.

On the antebrachium the **n. radialis** gives off the **dorsal cutaneous antebrachial nerve** (n. cutaneus antebrachialis dorsalis) before dividing, distal to the elbow joint, into **superficial and deep branches** (rami superficialis et profundus). The smaller superficial branch terminates at the

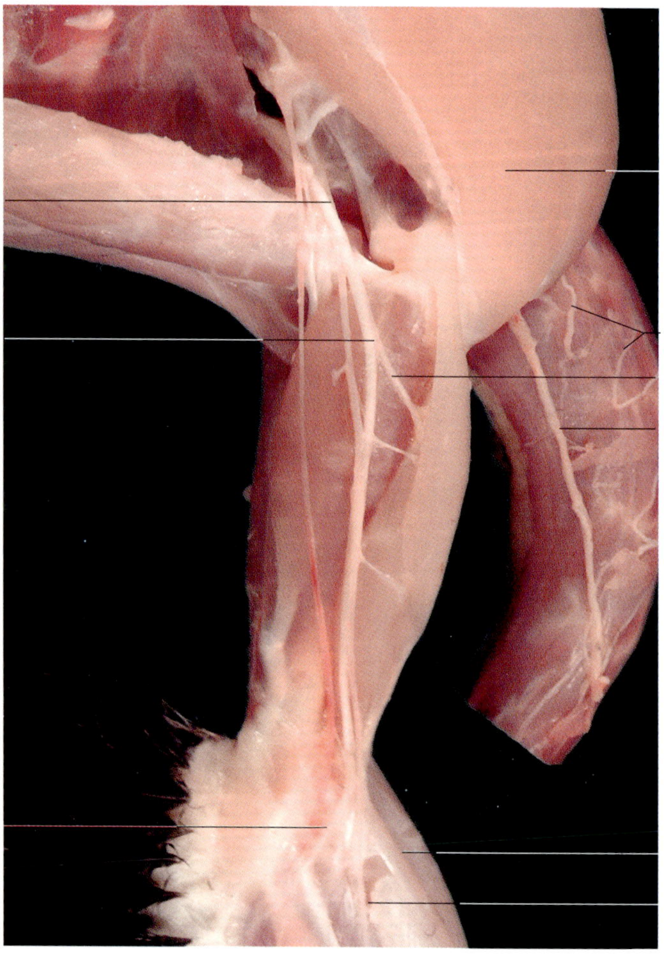

13.16 Innervation of the left wing of a chicken (ventral view).

level of the carpus and detaches **postpatagial branches** (rami postpatagiales). The more substantial deep branch gives rise to the **dorsal metacarpal nerves** (nn. metacarpales dorsales) and **alular branches** (rami alulares) that supply the extensors of the carpus and major digit as well as the skin between the quills of the primary remiges (Figure 13.15).

The **ventral fasciculus** innervates the skin on the ventral surface of the wing as well as the flexors of the elbow, carpus and digital joints. It gives off the **bicipital nerve** (n. bicipitalis) for the m. biceps brachii before dividing into the **ulnar nerve** (n. ulnaris) and the **median nerve** (n. medianus) that supply the aforementioned flexors. Further branches are depicted in Figure 13.14.

Lumbosacral plexus (plexus lumbosacralis)

In most birds, the lumbosacral plexus is formed by the ventral branches of the **synsacral nerves 2–4** (**plexus lumbalis**) and **4–9** (**plexus sacralis**). Due to the intimate associations around the fourth synsacral nerve, these plexuses are combined into the **lumbosacral plexus** (plexus lumbosacralis). The nerves of the lumbosacral plexus supply the pelvis, hindlimbs and the tail (Figures 13.17 to 13.19).

Lumbar plexus (plexus lumbalis)

The lumbar plexus (Figure 13.17) is located close to the kidneys. Nerves arising from the plexus innervate the skin of the thigh and crus as well as the skin and muscles of the ventral body wall. These are the:

- lateral cutaneous femoral nerve (n. cutaneus femoralis lateralis),
- medial cutaneous femoral nerve (n. cutaneus femoralis medialis),
- pubic nerve (n. pubicus or n. ilioinguinalis),
- femoral nerve (n. femoralis) and
- obturator nerve (n. obturatorius).

The **lateral cutaneous femoral nerve** (n. cutaneus femoralis lateralis) innervates the muscles and skin of the craniolateral thigh. Its medial counterpart, the **n. cutaneus femoralis medialis**, supplies the skin and muscle of the proximal, medial region of the thigh. More caudally, the lumbar plexus gives off the **pubic nerve** (n. pubicus or n. ilioinguinalis), which innervates the muscles of the abdomen.

For innervation of the muscles of the pelvic limb, particularly the extensors of the knee, the lumbar plexus detaches the **femoral nerve** (n. femoralis). The **obturator nerve** (n. obturatorius), sheathed in serosa, passes to the obturator foramen and supplies the m. obturatorius lateralis and the m. obturatorius medialis.

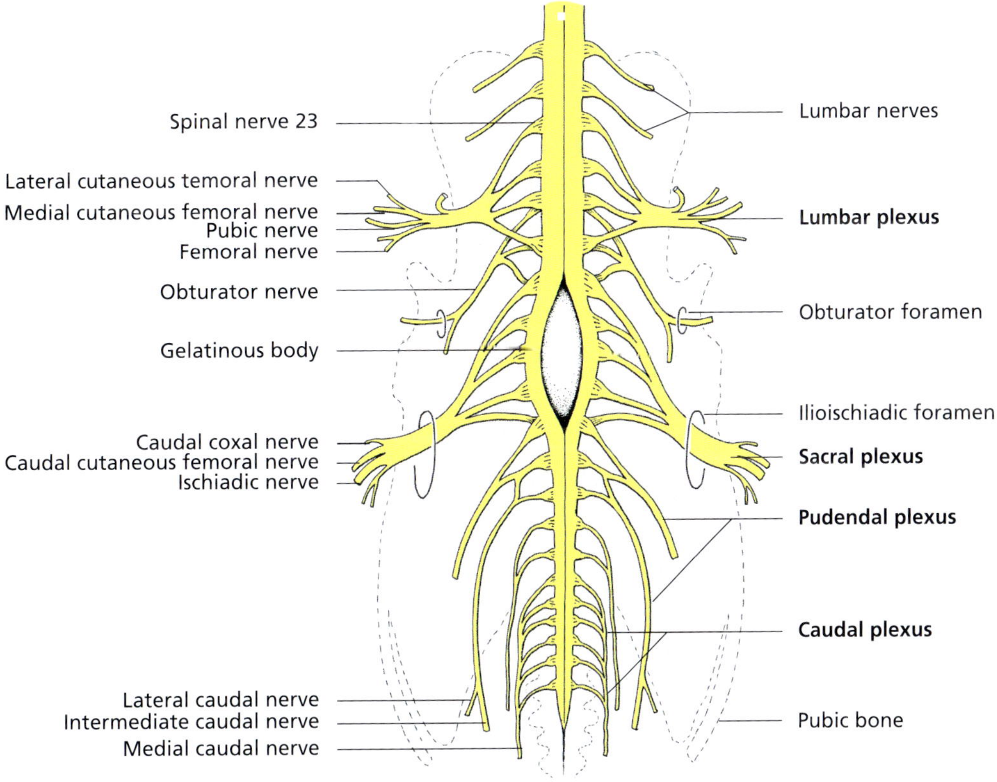

13.17 Lumbar, sacral, pudendal and caudal plexuses of the chicken, adapted from Ghetie, 1976 (in this case four nerves are shown to contribute to the lumbar plexus).

Sacral plexus (plexus sacralis)

The sacral plexus gives rise to the following nerves (Figure 13.17 to 19):

- caudal coxal nerve (n. coxalis caudalis),
- caudal cutaneous femoral nerve (n. cutaneus femoralis caudalis),
- ischiadic nerve (n. ischiadicus):
  - tibial nerve (n. tibialis) and
  - fibular (or peroneal) nerve (n. fibularis or peroneus).

The **caudal coxal nerve** (n. coxalis caudalis) innervates the m. caudo(ilio)femoralis and the mm. flexor cruris medialis and lateralis. It extends into the middle of the thigh, giving off nerve fibres to the muscle and skin.

The **ischiadic nerve** (n. ischiadicus) is the largest peripheral nerve in the bird. Lying caudomedial to the femur and approximately parallel to the ischiadic artery, it courses towards the popliteal region. Proximal to the knee joint, it divides into the **tibial nerve** (n. tibialis) and the **fibular nerve** (n. fibularis).

The larger of the two, the **tibial nerve,** innervates the extensors of the intertarsal joint and the flexors of the joints of the digits. It gives rise to the medial and lateral plantar nerves, the latter passing along the metatarsus to the fourth digit as the **plantar metatarsal nerve** (n. metatarseus plantaris).

The **fibular nerve** passes to the craniolateral aspect of the crus and divides into the **superficial and deep fibular nerves** (n. fibularis superficialis and profundus) that supply the flexors of the intertarsal joint and the extensors of the

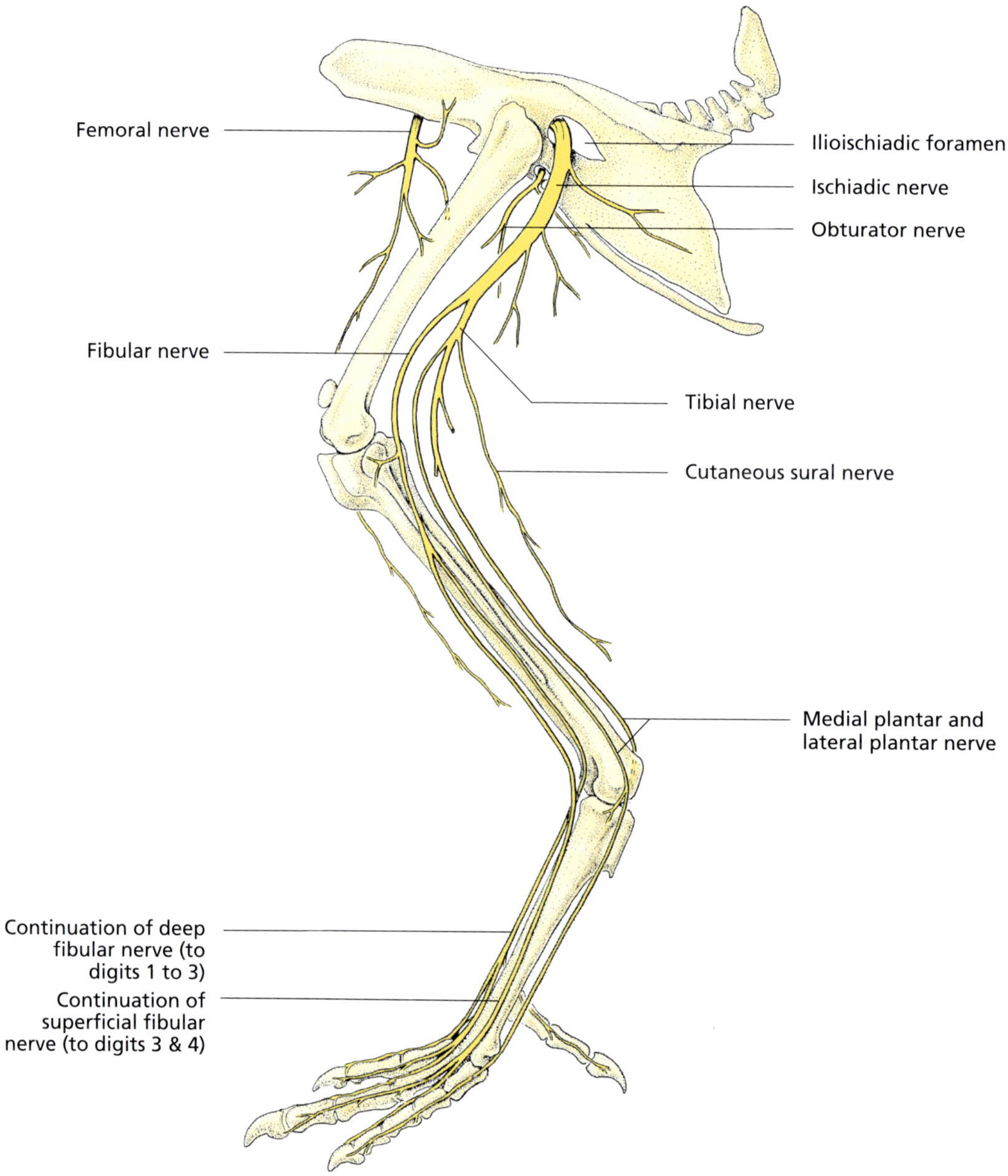

13.18 Innervation of the pelvic limb of the chicken, adapted from King and McLelland, 1978.

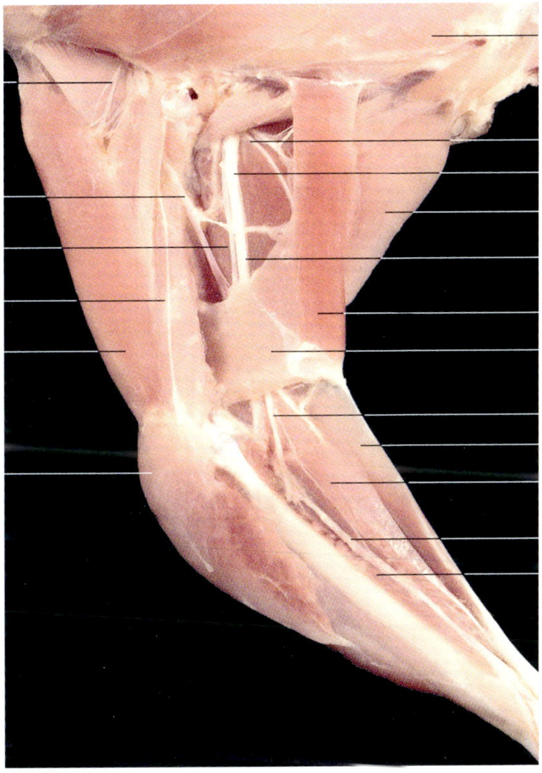

13.19 Innervation of the pelvic limb of a chicken (medial view, m. iliofibularis removed and m. gastrocnemius, pars medialis partially resected). Courtesy of Dr Annette Kaiser, Munich.

digits. These branches continue along the metatarsus to supply the third and fourth digits (superficial fibular nerve) and the first to third digits (deep fibular nerve).

Pudendal plexus (plexus pudendus)

The major nerve emerging from the pudendal plexus (Figure 13.17) is the **pudendal nerve** (n. pudendus) that passes towards the cloaca, accompanying the pudendal artery. Structures innervated by the pudendal plexus include the ventral and lateral tail muscles, the muscles of the cloaca and the surrounding skin. The pudendal nerve terminates in the cloaca at the end of the oviduct or deferent duct.

The **lateral caudal nerve** (n. lateralis caudae) supplies the ventrolateral muscles and skin of the tail and abdomen. Motor fibres pass from the **intermediate caudal nerve** (n. intermedius caudae) to the cloacal sphincter and continue to the vent and ventral region of the tail.

Caudal plexus (plexus caudae)

Some authors describe a separate caudal plexus arising from the ventral branches of the last spinal nerves. These branches form the **medial caudal nerve** (n. medialis caudae) that primarily supplies the skin and muscles of the tail and the uropygial gland (Figure 13.17).

Autonomic nervous system (systema nervosum autonomicum)

The avian autonomic nervous system, like that of mammals, consists of **sympathetic** and **parasympathetic components**. In birds, the preganglionic cells of the parasympathetic system are located in the brain stem and the sacral segments of the spinal cord. Their counterparts in the sympathetic system are situated in the thoracolumbar spinal cord (Figure 13.20).

The autonomic nervous system includes all nerves that supply the glands and vessels, the thoracic and abdominal organs and the internal muscles of the eye. Generally all structures are innervated by **both sympathetic and parasympathetic fibres**. Both afferent and efferent pathways play an important part in regulating and coordinating organ function.

The autonomic system also includes ganglia located in the walls of hollow organs. Together these form the **intramural nervous system**.

Sympathetic system

The paired **sympathetic trunk** (truncus paravertebralis) (Figure 13.20) extends from the base of the skull to the pygostyle. Each trunk contains **paravertebral ganglia** (ganglia paravertebralia) that correspond with the segments of the spinal cord. The preganglionic nerve cells are located dorsolateral – and immediately adjacent – to the central canal of the spinal cord. Preganglionic nerve fibres pass into the ventral roots of the spinal cord, from which they detach as the **rami communicantes** to join the paravertebral ganglia. Postganglionic fibres join the spinal nerves and continue to the skin or, after synapsing in **prevertebral ganglia**, to their target organs.

The sympathetic trunk is divided into the following parts that correspond with the sections of the vertebral column:

- cervical (pars cervicalis),
- thoracic (pars thoracica),
- synsacral (pars synsacralis) and
- caudal (pars caudalis).

The **cervical part** commences at the **cranial cervical ganglion**, from whence its main component, the **cervical sympathetic trunk** (truncus paravertebralis cervicalis), accompanies the vertebral artery through the transverse foramina of the vertebrae. **A paravertebral ganglion** is present in each segment.

The **cranial cervical ganglion** is the **largest sympathetic ganglion** in the bird. It is located medially, adjacent to the point at which the glossopharyngeal and vagus nerves exit the skull. Postganglionic fibres pass from the cranial cervical ganglion to cranial nerves V, VII, IX, X and XII. In addition, fibres are distributed to glands within the orbit and the pterygopalatine ganglion. Others form periarterial networks in effector organs of the head.

A further paired cervical component of the sympathetic nervous system emerges from the cranial cervical ganglion to run alongside the internal carotid artery. Termed the **subvertebral trunk** (truncus subvertebralis), it is smaller than the cervical sympathetic trunk, yet also contains segmentally organised ganglia that are connected to the cervical spinal nerves by thin rami communicantes. The **subvertebral trunk** forms a neural network known as the **plexus subvertebralis**.

In the **thoracic part** of the sympathetic trunk, the fibres passing between the ganglia diverge around the heads of the ribs. Rami communicantes are barely discernible in the thoracic and synsacral regions of the sympathetic trunk.

The first thoracic ganglion gives rise to the **cardiac nerve** (n. cardiacus) that innervates the heart and lungs. This detaches the rami pulmonales that form the **plexus pulmonalis**. The **nn. splanchnici thoracici** arise from the thoracic ganglia and converge on the **coeliac ganglion** (ggl. coeliacum) and **cranial mesenteric ganglion** (ggl. mesentericum craniale) to form the **plexus subvertebralis thoracicus** around the base of the cranial mesenteric and coeliac arteries.

Postganglionic fibres from the coeliac and cranial mesenteric ganglia give rise to the **plexus coeliacus et mesentericus cranialis**, components of which supply the spleen (**plexus splenicus**), the liver (**plexus hepaticus**), the pancreas and small intestine (**plexus pancreaticoduodenalis**), and the stomach (**plexus proventricularis et gastricus**). These fibres reach their effector organs directly or by joining the intestinal nerve (n. intestinalis).

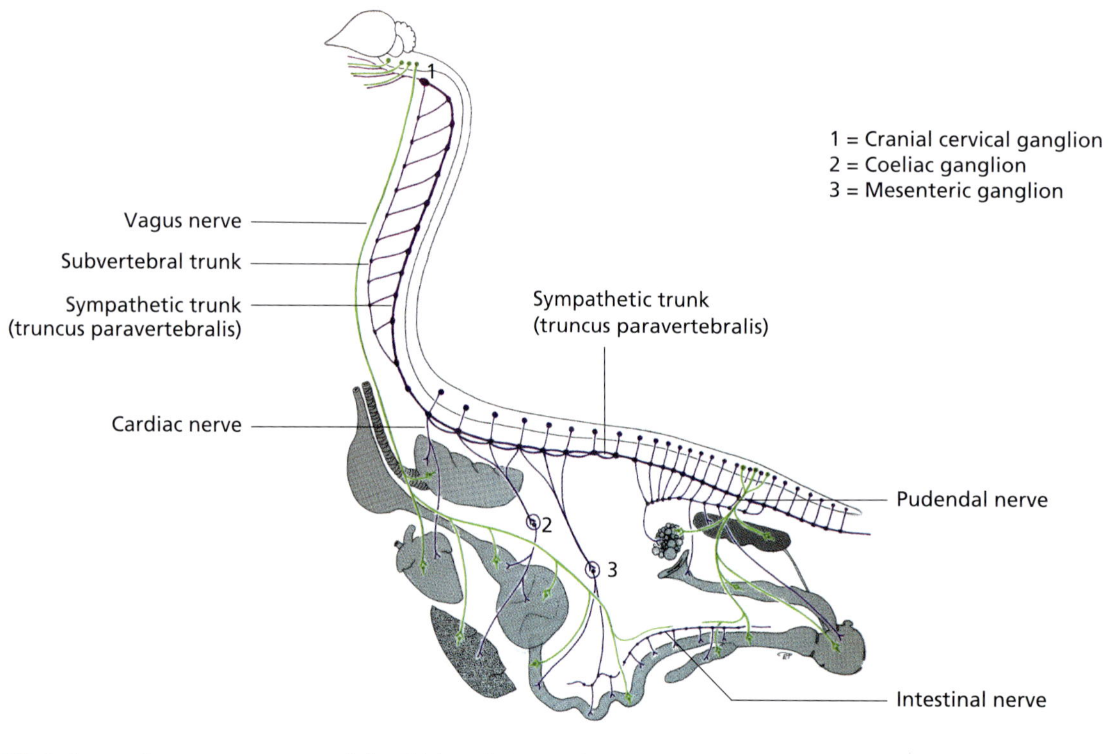

13.20 Autonomic nervous system of the chicken (schematic overview), adapted from King and McLelland, 1978.

The **intestinal nerve**, also known as the **nerve of Remak**, arises from the plexus mesentericus cranialis et caudalis and the plexus aorticus and runs within the mesentery close to its attachment to the jejunum, ileum and large intestine. Parasympathetic fibres are usually also present within the intestinal nerve at its cranial and caudal ends. The nerve contains microscopic ganglia.

On its ventral aspect, the **synsacral part** of the sympathetic trunk is largely covered by the kidneys. Ventral to the caudal vertebrae, the right and left trunks merge to form the **caudal part** of the sympathetic trunk, which contains small unpaired ganglia along its course to the pygostyle. The segmentally arising **nn. splanchnici synsacrales** form the **plexus subvertebralis synsacralis**, terminating in ganglia that lie along the aorta. These include the **adrenal ganglia** that form part of the **aortic plexus**. Postganglionic fibres are distributed to the organs of the urogenital tract, the adrenal glands and the cloacal bursa.

Parasympathetic system

Based on the location of preganglionic cells, the **parasympathetic system** is divided into **cranial** (cell bodies located in the mesencephalon and rhombencephalon) and **synsacral** (cell bodies located in the synsacral spinal cord) components. Essentially, all visceral nerves that do not arise from the sympathetic trunk belong to the parasympathetic nervous system (Figure 13.20).

The **preganglionic parasympathetic fibres** emerge from the brain in conjunction with cranial nerves III, VII, IX and X. **Parasympathetic fibres** of the oculomotor nerve (III) pass to the **ciliary ganglion,** from which fibres extend to the ophthalmic and abducent nerves. Postganglionic fibres pass within the n. iridociliaris (*syn*. ciliarus longus) to the m. sphincter pupillae and parts of the m. ciliaris.

Parasympathetic fibres of the **facial nerve (VII)** project to the dorsal and ventral pterygopalatine ganglia (Figure 13.12). Postganglionic fibres innervate glands of the orbit and nasal cavity, and most of the glands in the oropharyngeal cavity.

The **chorda tympani** arises from the facial nerve (VII) and passes to the mandibular ganglion. Postganglionic fibres innervate glands at the base of the oropharyngeal cavity. Parasympathetic fibres detaching from the **glossopharyngeal nerve (IX)** also supply glands in this region, as well as those near the larynx.

The largest bundle of parasympathetic fibres leaving the brain as part of the cranial parasympathetic system is located within the **vagus nerve**. This supplies the organs situated between the cranial and sacral components of the autonomic system. It exchanges fibres with the cranial cervical ganglion and the glossopharyngeal nerve. Adjacent to the jugular vein, the vagus nerve gives off rami laryngeales, pharyngeales, tracheales and thymici. Postganglionic rami glandulares innervate the thyroid gland, parathyroid gland and the ultimobranchial body.

A further branch arising from the vagus is the **recurrent nerve**. Its branches include the rami bronchiales, oesophageales, ascendens and descendens. The descending ramus of the recurrent nerve joins with fibres of the coeliac plexus, while branches of the ascending ramus supply the oesophagus, crop, trachea, bronchi and the heart.

The two **vagus nerves** merge in the vicinity of the proventriculus. Additional branches innervate the stomach, jejunum, pancreas and the liver. These are named according to the organs they supply. Vagal fibres extend as far as the intestinal nerve.

Preganglionic fibres of the synsacral part of the parasympathetic system emerge from the spinal cord together with the ventral roots of spinal nerves 30–33, giving rise to the **pudendal nerve** (n. pudendus). This accompanies the ureter to reach the dorsal wall of the cloaca. Together with fibres from the cloacal ganglia, it innervates the ureter, the receptaculum (males) or oviduct (females) and the cloaca. Parasympathetic pudendal nerve fibres combine with sympathetic fibres of the intestinal nerve in forming the cloacal plexus.

Clinical aspects

Due to the highly developed avian autonomic flight response, eliciting and evaluating reflexes as part of a clinical neurological examination is challenging in birds, compared with mammals. Assessment of reflexes is nevertheless an important component of anaesthetic monitoring.

Presenting clinical signs such as opisthotonus (hyperextension of the neck), torticollis (twisting of the neck, Figure 13.21), functional deficits or abnormal positioning of body parts and the absence of the pupillary light reflex may be observed in association with numerous infectious or non-infectious disorders of the central nervous system. Several of these infectious diseases are potential **zoonoses** and are therefore also of clinical significance for the owner of the bird. These include **salmonellosis**, **Newcastle disease (atypical avian influenza)** and **avian influenza.**

Avian paramyxovirus 1 (aPMV 1) infection, first described in the early 1980s, is particularly important in pigeon flocks, where outbreaks can have considerable economic consequences. In addition to the typical neurological signs such as torticollis, opisthotonus and lameness or ataxia, presenting signs of paramyxovirus infection include polyuria. Prophylaxis is achieved through vaccination.

Proventricular dilatation disease (PDD, neuropathic gastric dilatation disease of psittacines) is an important disease of parrots. Its aetiology is incompletely understood, but it has been associated with paramyxovirus and, most recently, with bornavirus infection. The disease is characterised by neurological abnormalities and excretion of

13.21 Budgerigar (Melopsittacus undulatus) with torticollis due to lead intoxication.

undigested seed, resulting from pathological changes in intramural ganglia in the proventriculus, ventriculus, crop and duodenum, and in the spinal cord. Diagnostic testing includes radiography (see Chapter 6 'Digestive system'), which may reveal marked dilatation of the proventriculus, ventriculus and cranial duodenum. Definitive identification of the aetiological agent is difficult.

Avian encephalomyelitis in commercial poultry (chickens) is a typical example of neurotrophic virus (pircornavirus) infection. The virus is transported along the optic nerve causing neurological and ocular lesions. In young birds, the disease typically manifests as 'epidemic tremor', characterised by high-frequency tonic-clinic convulsions induced by central nervous system pathology. Cataracts are seen in up to 60 per cent of adult survivors.

As well as ocular abnormalities and tumour formation, neurological deficits such as paralysis and crop distension (pendulous crop) are typical findings in herpesvirus-induced Marek's disease. These result from lesions in the lumbosacral plexus and the ischiadic and vagus nerves. Pathological changes include loss of striation, yellow discolouration and enlargement of the lumbosacral plexus and vagus nerve.

No specific treatment is available for avian encephalomyelitis or Marek's disease. Prophylaxis involves vaccination of young birds (avian encephalomyelitis) or day-old chicks (Marek's disease) (see Chapter 20 'Handling, restraint and anaesthesia'). When administered by the intramuscular route, Marek's disease vaccine is injected (using a semi-automatic applicator) into the muscles of the lateral thigh, as the breast muscles are insufficiently developed (see Chapter 20 'Handling, restraint and anaesthesia').

In addition to those infectious aetiologies in which neurological signs are strongly indicative of – if not pathognomonic for – the underlying disease, a number of other infectious diseases (including chlamydial psittacosis/ornithosis) can involve the nervous system.

Particularly in pigeons, **chronic salmonellosis** must be considered alongside paramyxovirus as an important differential diagnosis in the presence of central nervous system deficits. Additional signs observed in longstanding cases of salmonella infection include swollen joints and ocular lesions (keratitis and iritis, cataracts secondary to uveitis). Racing pigeons are vaccinated against salmonella from around four weeks of age and prior to breeding.

Physiological (saccadic) eye oscillations in birds (see also Chapter 15 'The eye') should not be confused with pathological nystagmus. These rhythmic oscillatory 1–2 degree eye movements correspond with rhythmic (10–20/minute) vibrations that can be felt through the animal's skull.

Among non-infectious disorders of the nervous system, **head trauma** and **toxicity** are of particular significance. The former may be associated with few externally visible signs and is often diagnosed by identification of haemorrhage in the ear opening or the vitreous body (see Chapter 15 'The eye').

Lead poisoning is one of the most clinically relevant toxicities in avian medicine. Exposure is through the oral route, typically by ingestion of lead present in household sources such as curtain weights, lead-light windows and Tiffany lamps. In advanced cases, haemorrhagic diarrhoea is accompanied by central nervous signs (Figure 13.21). Severe lead intoxication can also occur in wild birds as a result of gunshot wounds through absorption of lead from bullet fragments (see Chapter 19 'Imaging techniques').

The incidence of lead poisoning may exhibit seasonal fluctuations, increasing around Christmas and New Year when birds are exposed to decorations and trimmings containing lead, as well as wine and champagne bottle foil.

Identical clinical signs are seen in cases of **zinc toxicity**. This is also known as **'new wire disease'**, as it often results from the inappropriate use of galvanised wire in the construction of aviaries and cages.

Acute onset of central nervous system dysfunction followed by sudden death is observed in association with **Teflon® poisoning** (polytetrafluoroethane toxicity). Intoxication results from inhalation of vapour (odourless to humans) produced by dry-heated Teflon®-coated non-stick cooking utensils or low-quality heating elements.

Prophylactic use of insecticidal strips and sprays in the management of air sac mites is also a potential cause of fatal central nervous system disease. Seen particularly in breeding finches and canaries, this can also affect individual birds kept as pets. Remnants of feather calami and skin scales, which play an important role in waterproofing the plumage, may be incorrectly identified by owners as mites (see Chapter 17 'Common integument'), leading to inappropriate treatment.

Botulism is another important toxin-induced disorder of the nervous system. In water birds, mass fatalities due to botulism may be seen during the summer months, in association with high water temperatures. Raptors may also be affected. Clinical signs include central nervous system deficits and paralysis of smooth muscle associated with swallowing and respiration.

Central nervous system morbidity and mass death of water birds around lakes and dams during the summer months in Germany has also been attributed to infection with *Aeromonas hydrophila*. Such outbreaks are often confused with botulism. Increased faecal contamination of water due to excessive bird numbers results in massive microbial overgrowth during the warm summer months when water temperature rises. This is exacerbated by the human population through feeding of wild birds, and by secondary botulism due to elaboration of botulinum toxin in birds that have succumbed to aeromonas infection.

Traumatic injury of the spinal cord at the structurally vulnerable junction between the notarium and the synsacrum (see Chapter 2 'Head and trunk') is characterised by paraplegia, abnormal tail posture and abnormal excretion.

Particularly in African grey parrots (Psittacus erithacus) and cockatoos (Cacatua spp.), acute tonic-clonic convulsions, with severe central nervous system deficits may be the result of **hypocalcaemic syndrome**. The aetiology of this condition is believed to involve abnormal function of the parathyroid glands with derangement of blood calcium and glucose concentrations.

Cerebral tumours with associated central nervous system deficits are sometimes seen in budgerigars (Melopsittacus undulates).

Anaesthesia, which in birds typically involves isoflurane-based inhalational anaesthesia, can in some respects be considered a form of **reversible intoxication** of the nervous system. The plane of anaesthesia is assessed using instrument-based monitoring of various physiological parameters and by the evaluation of up to 13 different reflexes. Optimal surgical anaesthesia is achieved when **all but three of these reflexes are lost**. The remaining reflexes are a sluggish, yet complete corneal reflex (relatively slow but complete protrusion of the third eyelid in response to touching the cornea at the temporal angle of the eye), partial pupillary dilation and a reduced pupillary light reflex.

13.22 Eurasian sparrowhawk (Accipiter nisus) with posttraumatic epistaxis and concussion following an in-flight collision. Note the subtle lesions in the region of the supraorbital ridge – these are suggestive of internal haemorrhage, which may only be detectable using ophthalmoscopy.

Evaluation of the pupillary light reflex is unreliable in birds since, in contrast to mammals, the predominantly striated internal ocular musculature (and its corresponding innervation) permits voluntary control over the pupillary response. Auditory and tactile stimuli can also elicit changes in pupillary diameter.

Apparently paradoxical responses (e.g., dilation of the pupil in response to illumination) may be detected during anaesthesia. For example, mydriasis may be observed if light enters the eye when the anaesthetised patient's eyelids are opened during surgery. This is not considered abnormal.

Due to the complete decussation of optic nerve fibres at the optic chiasm in most birds, a true **consensual pupillary light reflex** does not occur in birds (see Chapter 15 'The eye').

14

Endocrine glands (glandulae endocrinae)

H. E. König, G. Weissengruber and R. Korbel

The endocrine organs of birds, like those of mammals, are characterised by a glandular structure, the **absence of excretory ducts** (glandulae sine ductibus) and a rich **vascular supply**. Their biologically active products (hormones), which they secrete directly into the blood vascular system, regulate and coordinate a range of functions including metabolism, growth, development and reproduction.

The endocrine organs interact closely with the central nervous system (CNS), particularly the hypothalamus, and the autonomic nervous system. **Feedback mechanisms** exist among the endocrine glands, and between the endocrine organs and the central nervous system. Increased hormone release into the bloodstream results in reduced hormone synthesis within the endocrine organ of origin. The resulting decrease in circulating hormone concentrations in turn stimulates increased hormone production.

Endocrine organs (Figure 14.1) may be classified according to their **embryonic origins**:

- **pharynx**:
  - adenohypophysis (from Rathke's pouch),
  - thyroid gland,
  - parathyroid gland and
  - ultimobranchial body;
- **neurectoderm**:
  - neurohypophysis and
  - epiphysis;
- **mesoderm**:
  - gonads and
  - adrenal cortex;
- **neural crest cells (neurectoderm)**:

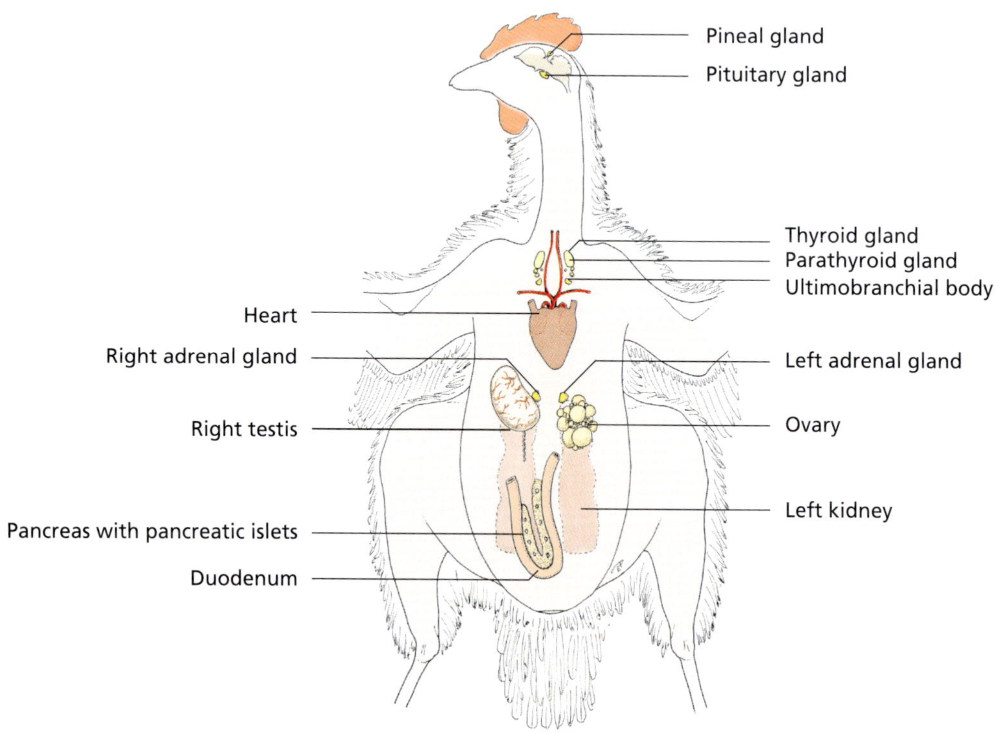

14.1 Endocrine glands of the chicken (schematic, ventral view).

- adrenal medulla and
- paraganglia;
- **pancreatic buds**:
  - pancreatic islets.

Hypophysis, pituitary gland (glandula pituitaria)

With respect to both morphology and function, the hypophysis (Figure 14.2) is closely associated with the hypothalamus. It is unpaired and lies immediately caudal to the **optic chiasm**. In relative terms, it is situated further from the brain in birds than in mammals.

The hypophysis consists of two clearly distinguishable parts, each of which is subdivided as follows:

- **adenohypophysis**:
  - pars tuberalis,
  - pars distalis;
- **neurohypophysis**:
  - median eminence (eminentia mediana),
  - infundibulum and
  - neural lobe (lobus nervosus).

The **adenohypophysis**, also referred to as the **anterior lobe** in view of its position, is functionally connected with the hypothalamus by a system of veins. Referred to as the **hypophyseal portal system,** this venous network delivers hormones produced by the nerve cells of the **nucleus infundibularis** and other nuclei within the **tuber cinereum** to the **adenohypophysis**.

These 'releasing' and 'inhibiting' hormones direct the function of various types of secretory cells in the adenohypophysis. The structure of these cells varies according to the hormone they produce, namely:

- ACTH (adrenocorticotropic hormone),
- TSH (thyroid stimulating hormone),
- STH (somatotropin, growth hormone),
- MSH (melanocyte stimulating hormone),
- FSH (follicle stimulating hormone),
- LH (luteinising hormone) and
- PRL (prolactin).

The **hormones produced by the adenohypophysis** regulate the function of other endocrine glands. **Adrenocorticotropic hormone** (**ACTH**) acts upon the cortex of the adrenal gland. The target organ for **thyroid stimulating hormone** (**TSH**) is the thyroid gland, where it controls not only the synthesis of thyroxine but also its release from thyroid follicles. **Follicle stimulating hormone** (**FSH**) induces the growth of ovarian follicles as well as follicular secretion of oestrogen. In males, FSH influences the growth of seminiferous tubules and spermatogenesis.

Luteinising hormone (**LH**) participates in the induction of ovulation. In males, LH stimulates androgen

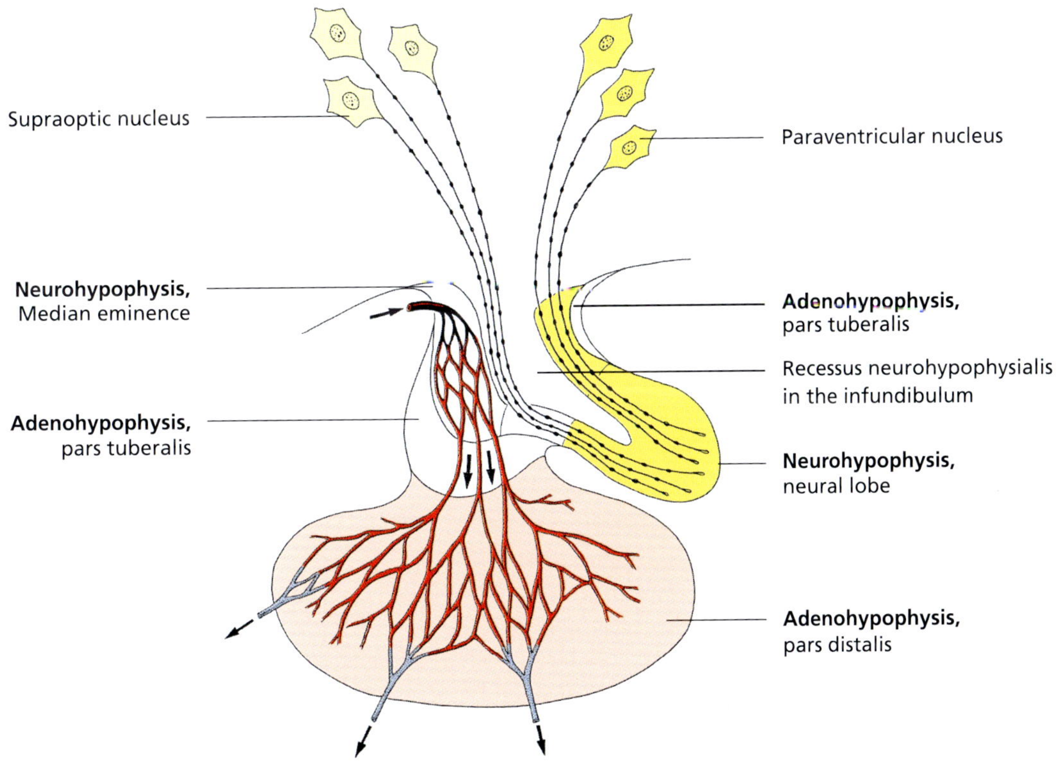

14.2 Hypophysis of the chicken (schematic). Arrows indicate the flow of blood in the hypophyseal portal system.

production by Leydig cells in the interstitium of the testes. **Prolactin (PRL)** is responsible for initiating and maintaining brooding behaviour. It promotes the development of the brood patch and secretion of crop milk. In birds, prolactin synthesis appears to be controlled directly by the hypothalamus.

Somatotropin (STH) directs skeletal growth as well as other metabolic processes. **Melanocyte stimulating hormone (MSH)** regulates melanin (pigment) synthesis by melanocytes, melanocyte expansion and distribution of pigment.

The **neurohypophysis** is not a site of hormone synthesis. Rather, it stores the hormones **vasopressin** (antidiuretic hormone; increases blood pressure) and **oxytocin** (stimulates smooth muscle in the oviduct) produced by neuronal cell bodies located in hypothalamic nuclei (see below).

The **median eminence** (eminentia mediana) of the neurohypophysis forms the rostroventral portion of the floor of the third ventricle. It merges without a distinct boundary with the tuber cinereum.

Together, the **infundibulum** of the neurohypophysis and the pars tuberalis of the adenohypophysis form the **hypophyseal stalk**. The infundibulum serves as the thoroughfare for the nerve fibres of the hypothalamo-hypophyseal tract. An extension of the third ventricle projects into the infundibulum forming the **recessus neurohypophysialis**.

The **lobus nervosus** joins the caudal end of the infundibulum. The perikarya of the axons reaching the lobus nervosus are situated within the hypothalamus, where they form the **supraoptic nucleus** (nucleus supraopticus) and **paraventricular nucleus** (nucleus paraventricularis). These are the sites of vasopressin and oxytocin synthesis.

Neurosecretory granules containing these hormones give rise to teardrop or bulb-shaped dilatations (**Herring bodies**) in the axons. The granules accumulate near the capillaries of the neural lobe.

Epiphysis, pineal gland (glandula pinealis)

The epiphysis (Figure 14.1) develops as an unpaired evagination of the caudal segment of the roof of the diencephalon. It is bulb-like in shape and is attached to the **roof of the diencephalon** by a thin stalk. Communication with the third ventricle is lost post-hatching, with small vesicles forming the remnants of the internal cavity. The epiphysis is particularly richly supplied with sympathetic nerve fibres.

With increasing age in chickens, lymphocytes accumulate in the **epiphyseal stalk** and **calcified concretions** appear within the **parenchyma**.

As in mammals, the pineal gland produces the antigonadotropic hormone **melatonin**, which is involved in the regulation of circadian and circannual photoperiod driven rhythms, particularly seasonal breeding. Melatonin also exerts a photoperiod dependent influence on chromatophores.

Thyroid gland (glandula thyroidea)

During embryonic development, the paired thyroid gland migrates further caudally in birds than in mammals. Each ovoid gland is located at the **thoracic inlet** at the **base of the neck**, in the angle between the subclavian and common carotid arteries. Its lateral border lies adjacent to the jugular vein (Figures 14.1, 14.3 and 14.4).

The thyroid gland exhibits seasonal variation in size and function. Its hormones, **triiodothyronine** and **thyroxin**, influence metabolic rate and participate in thermoregulation. Other functions include regulation of moulting and the development and function of the gonads.

Parathyroid gland (glandula parathyroidea)

The parathyroid glands are present in varying numbers of pairs (three in the chicken, two in other domestic species) and are partly fused to the thyroid gland. They are irregularly spherical and are consistently darker than the thyroid gland (Figures 14.1 and 14.3).

The parathyroid glands produce **parathyroid hormone (parathormone)**, which raises the calcium concentration in the blood by increasing calcium absorption from the gut and releasing calcium from the skeleton. This is of particular significance during the laying period when large amounts of calcium are mobilised from the medullary bone for incorporation into the eggshell.

Ultimobranchial body (glandula ultimobranchialis)

The ultimobranchial body is a paired, translucent organ. It lies caudad of the caudal pole of the thyroid gland, near the parathyroid glands, the **carotid paraganglion** and the distal **ganglion of the vagus nerve** (Figure 14.3). The right ultimobranchial body lies further caudal than the left, and is frequently located near the aortic arch. Fusion of the ultimobranchial body with the thymus or the parathyroid glands is frequently observed. Identification of the ultimobranchial body is often difficult, due to its variable shape and transparent appearance.

The ultimobranchial body is responsible for synthesising the hormone **calcitonin**, which prevents resorption of calcium from the bone.

Adrenal gland (glandula adrenalis)

The avian adrenal glands lie caudal to the lungs on the medial edge of the cranial pole of the kidney, either side of the descending aorta (Figure 14.1). Ventrally they are related to the testes in the male and, on the left side, to the ovary in the female. The adrenal glands are generally

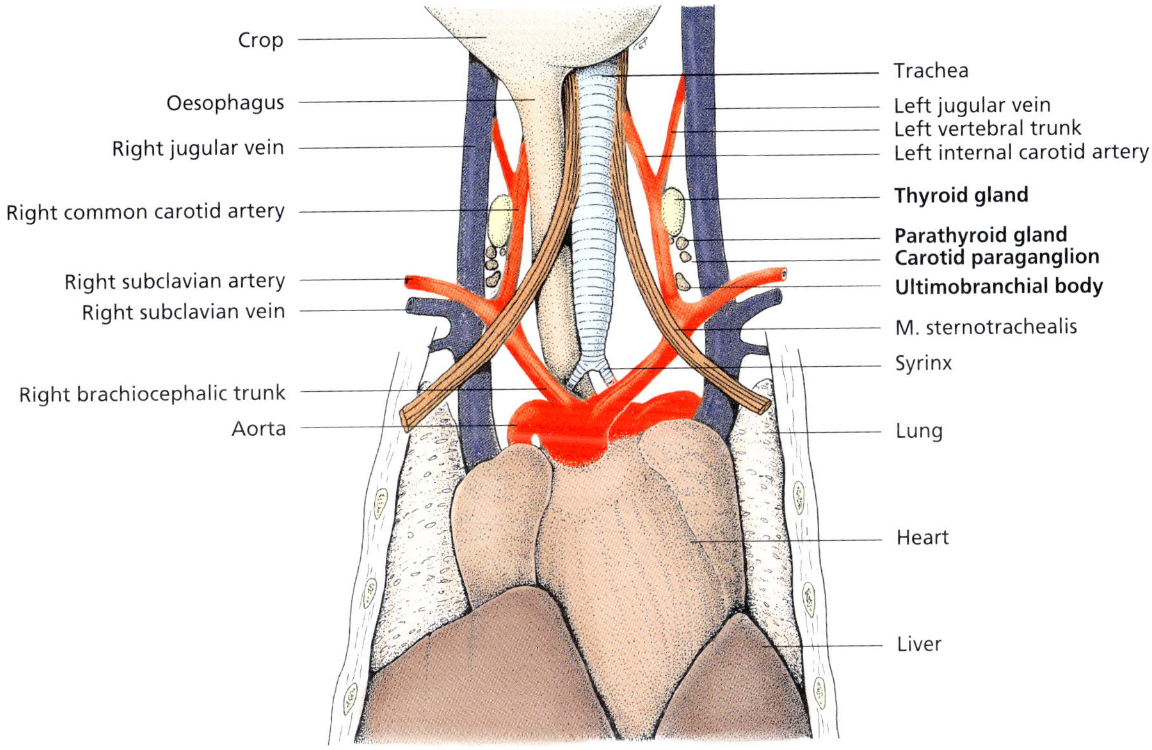

14.3 Thyroid gland, parathyroid gland and ultimobranchial body in the chicken (schematic), adapted from Stöger, 1996.

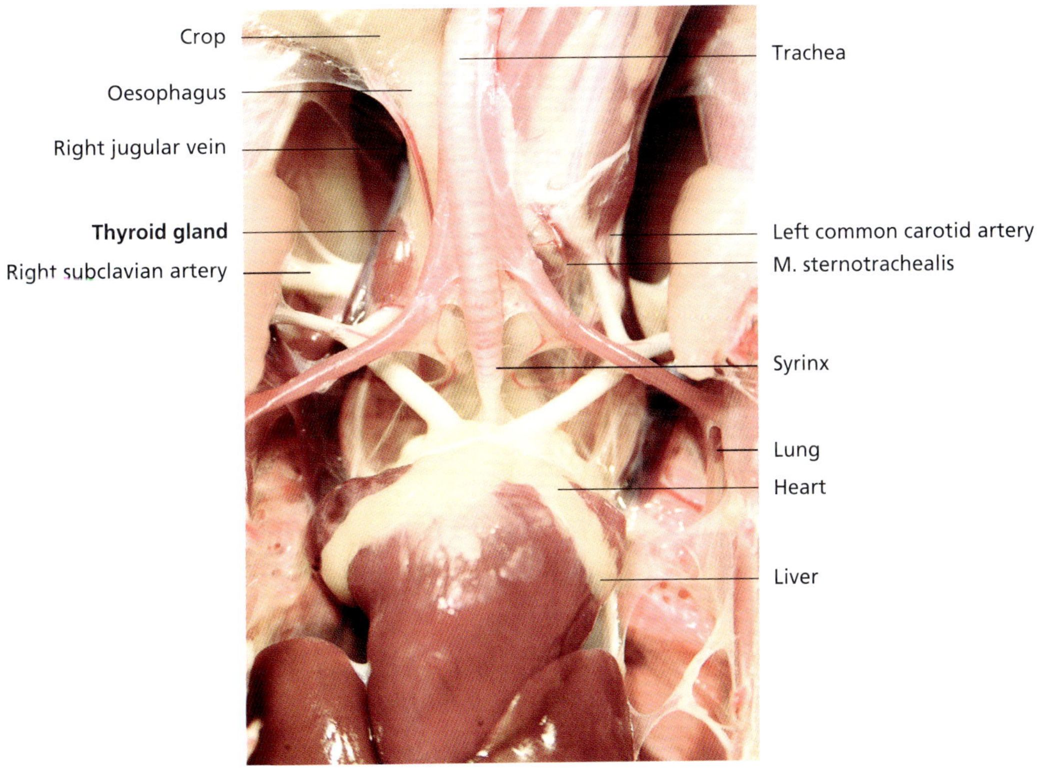

14.4 Thyroid gland of a chicken.

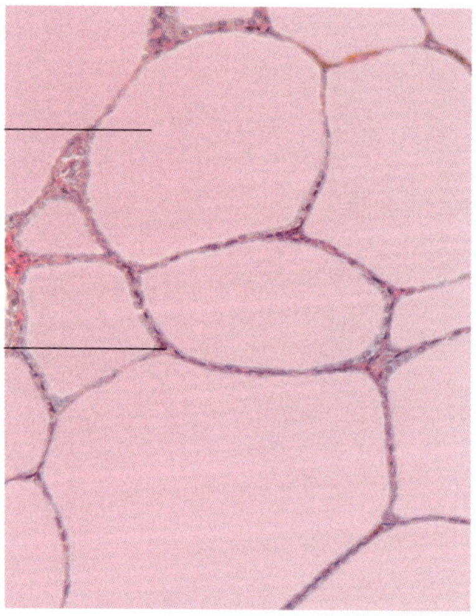

14.5 Histological section of the thyroid gland of a chicken.

shaped like a triangular pyramid, varying in colour from greyish- to orange- or reddish-yellow.

In contrast to mammals, in which the adrenal glands are divided into a medulla and cortex, catecholaminergic adrenal cells (corresponding to the mammalian adrenal medulla) and interrenal cells (equivalent to the renal cortex) are intermingled. Consequently, clumps of adrenal cells and cords of interrenal cells are in direct contact with one another (Figures 14.6 and 14.7).

Adrenaline and **noradrenaline** are synthesised by the adrenal cells, while the **interrenal cells** are responsible for production of **corticosterone** and **aldosterone**. The latter play an important role in the metabolism of carbohydrates and lipids (mainly corticosterone) and electrolytes (mainly aldosterone).

The connective tissue capsule surrounding the adrenal glands contains numerous **large autonomic nerve trunks and ganglion cells**. A sympathetic ganglion is located at the cranial and caudal pole of each gland.

Clusters of catecholaminergic cells, referred to as paraganglia, occur in various locations. The **carotid paraganglion** (paraganglion caroticum), or **carotid body**, is a paired ovoid structure situated, together with the thyroid gland, parathyroid gland and ultimobranchial body, in the angle between the subclavian and common carotid arteries (Figure 14.3). **Accessory paraganglia** are frequently present along the arteries that supply the carotid paraganglion.

Intravagal paraganglia within the vagus nerve or distal vagal ganglion have also been described. Additional paraganglia are located in the vicinity of the adrenal glands and the abdominal portion of the sympathetic trunk, and also in the walls of the large veins within the body cavity.

Pancreatic islets (insulae pancreaticae)

In contrast to mammals, **more than one type** of pancreatic islet exists in birds. Under light microscopy, a distinction can be made between **dark A islets** (mainly A cells), **light B islets** (mainly B cells) and **mixed type islets**. The islets of the chicken are considered to be of the mixed type. A greater concentration of islets is seen in the splenic lobe than in other parts of the pancreas (Figures 14.1 and 14.8).

The hormone products of the pancreatic islets include **glucagon** (A cells), **insulin** (B cells) and **somatostatin** (D cells). Glucagon appears to be of greater importance than insulin in the regulation of lipid and glucose metabolism in birds. The insulin content of the pancreas of birds is only around one-tenth of that observed in mammals.

Gonads (testis, ovarium)

In the testis, androgen-producing **Leydig cells** are found as individual cells or in small groups between the seminiferous tubules. Additional Leydig cells can be recruited by

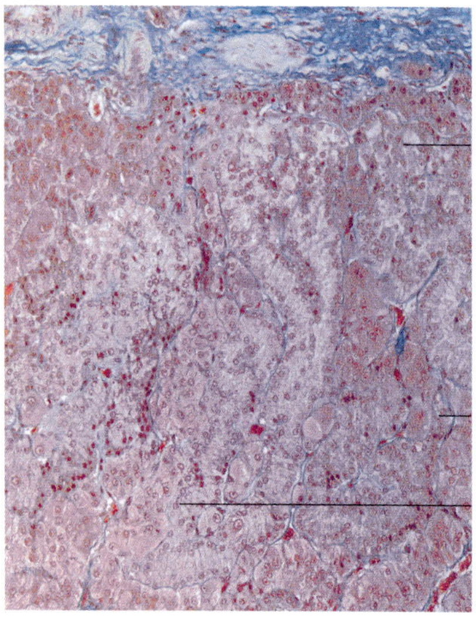

14.6 Histological section of the adrenal gland of a chicken, subcapsular region.

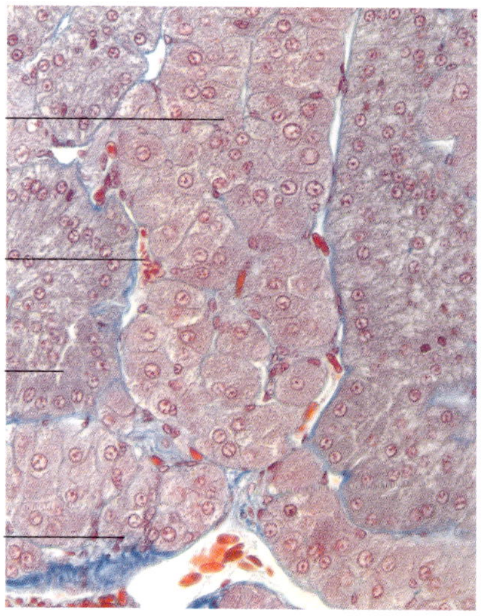

14.7 Histological section of the adrenal gland of a chicken.

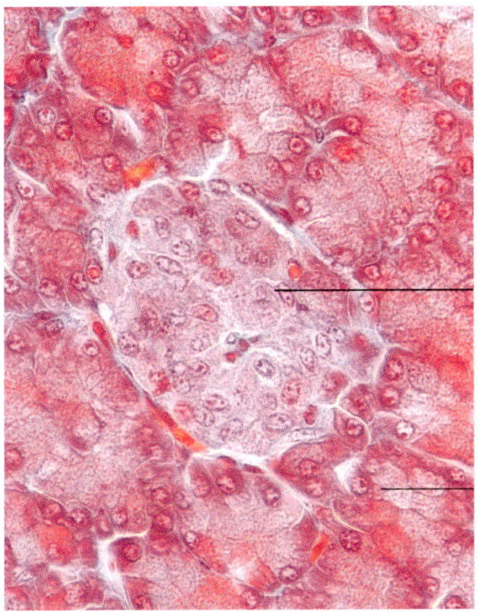

14.8 Histological section of an islet in the pancreas of a chicken.

differentiation of cells in the testes, epididymis and adrenal gland.

In the ovary, the cells of the **theca folliculi interna** produce oestrogens and progesterone. Together these cells are referred to as the 'thecal gland'.

Clinical aspects

Disorders of the endocrine glands occur in all clinically encountered avian cohorts, including commercial flocks and pet birds. However, in terms of clinical investigation and treatment of individual animals, endocrine disorders are of greatest significance in companion bird species.

In commercial flocks, the **epiphysis** is significant in terms of animal welfare and productivity. This relates to the role of the pineal gland in regulating circadian rhythms, and the impact of **artificial lighting regimens** used in intensive poultry production on the function of the gland. The use of precisely controlled light–dark cycles to regulate the duration, intensity and spectrum of light to which the birds are exposed can influence important production parameters, including weight gain in broilers and egg-laying in hens. It is sometimes also used to induce a forced moult to maintain productivity in hens kept in production for extended periods.

An example of a **24-hour lighting regimen** used in broiler production consists of a repeated cycle of four hours of darkness followed by eight hours of light. The use of such lighting schemes, some of which incorporate very short dark phases and low light intensities, has important welfare implications. Also noteworthy in this regard are differences between birds and humans, in terms of the characteristics of the visual apparatus (see Chapter 15 'The eye'). Many birds exhibit **pentachromatic perception** of light (including ultraviolet wavelengths and iridescent hues). Thus their interpretation of light intensity in a barn illuminated with fluorescent lighting, for example, differs significantly from that of humans.

Thyroid disease is commonly seen in budgerigars (Melopsittacus undulatus). Enlargement of the thyroid gland due to **iodine deficiency** can result in pressure on the adjacent trachea and syrinx. Associated clinical signs include stenotic inspiratory sounds. Treatment involves administration of iodine in the food or drinking water.

Hypocalcaemia caused by parathyroid hormone derangement is frequently observed in African grey parrots (Psittacus erithacus), cockatoos (Cacatua spp.) and other large parrots. Disease manifestation is dramatic, involving convulsive tonic-clonic muscle contractions. Prompt administration of calcium and glucose leads to rapid resolution of clinical signs.

In budgerigars (Melopsittacus undulatus), **abnormal pancreatic function** and associated **protein maldigestion** results in excretion of pathognomonic light-ochre coloured faeces that rapidly dry to a mass resembling solid foam ('popcorn faeces'). As a target organ in avian influenza, the pancreas is of diagnostic value in identifying this disease. Pathological changes include extensive petechial haemorrhages, bruising, inflammation and necrosis.

Hormonally active **Sertoli cell tumours** are common in male budgerigars (Melopsittacus undulatus) and may be associated with external manifestations (change in the colour of the cere from blue to brown).

The eye (organum visus)

15

S. Reese, R. Korbel and H.-G. Liebich

Birds are heavily reliant on their **visual sense**, thus the eyes are of considerable importance (Figure 15.1). While the basic structure of the avian and mammalian eye is similar, numerous differences are also apparent.

The visual organ consists of:

- the eyeball or bulb (bulbus oculi), that lies sheltered within the orbit (orbita), and
- the adnexa or accessory organs of the eye (organa oculi accessoria) including the:
  - ocular muscles (mm. bulbi),
  - eyelids (palpebrae) and
  - lacrimal apparatus (apparatus lacrimalis).

Neural impulses generated when light enters the eyeball are transmitted by peripheral efferent nerves and central nerve tracts (**optic nerves** and **optic tract**) to the visual cortex of the brain for processing.

15.1 In contrast to other birds, the upper eyelid of owls (great horned owl, Bubo virginianus, pictured here) is larger and more mobile than the lower eyelid.

Orbit (orbita)

The **bony orbit** is formed by components of the frontal, squamosal, laterosphenoid, lacrimal and ectethmoid bones. Ventrally, the orbit is mostly open (Figure 15.2).

The dorsal and caudal osseous boundary of the orbit is delineated by the **supraorbital margin** (margo supraorbitalis). Many parrots also have a bony **ventral orbital margin** (margo infraorbitalis) in the form of a **suborbital arch** (arcus suborbitalis) (Figure 15.3). In most other species, the orbit is closed ventrally by a fascial band, **the suborbital ligament** (ligamentum suborbitale).

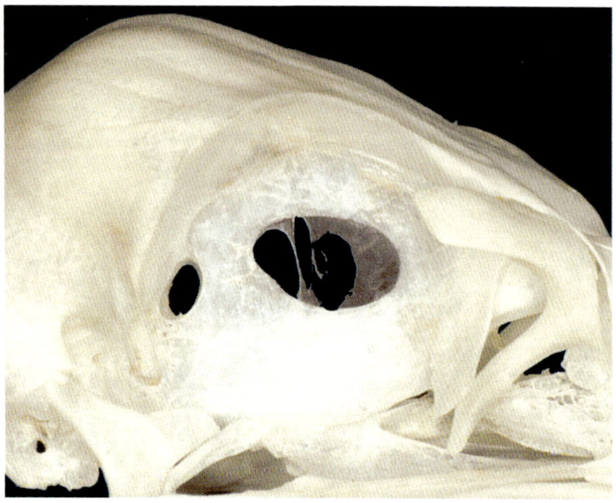

15.2 Skull of a Eurasian sparrowhawk (Accipiter nisus) with open ventral orbit. The interorbital septum is completed centrally by soft tissue.

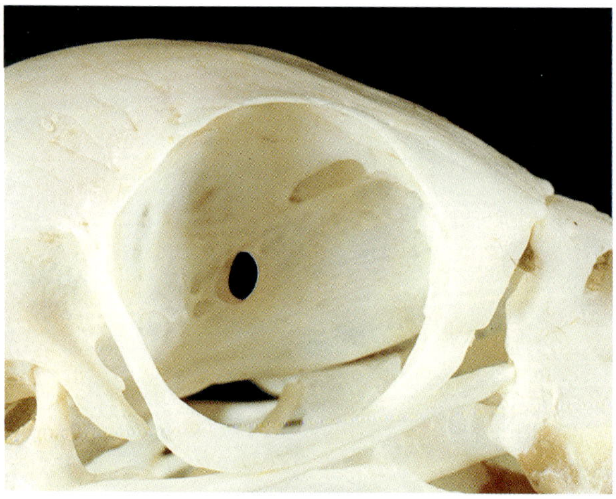

15.3 Skull of a blue fronted Amazon (Amazona aestiva) with orbit completed by an osseous suborbital arch.

The left and right orbits are very close together, separated only by the osseous **interorbital septum** (septum interorbitale). In all but crepuscular (active at dusk) species, the septum is very thin. Particularly in young raptors and corvids, it may have a membranous centre (Figure 15.2). In radiographs, the eyeballs of these birds may appear to be in contact with one another (Figure 15.60).

The large **eyeball** (bulbus oculi) of diurnal birds resides almost completely within the **orbit**, which provides protection against external trauma. In diurnal birds of prey, the long **supraorbital process** (processus supraorbitalis) projects caudolaterally from the **lacrimal bone**, forming a roof over the eyeball. This is clearly visible under the skin. In contrast, the orbit of many crepuscular species, such as owls, is shallow and offers the eyeball little protection. Instead, this role is performed by the bony **scleral ring** (anulus ossicularis sclerae) (Figures 15.8 and 15.9).

Unlike its mammalian counterpart, the eyeball of birds is not embedded in the **corpus adiposum orbitae** (fat body). Instead, it lies on the infraorbital sinus. Rather than being bounded by bone, the paranasal sinus is a soft tissue sac akin to an air cushion, upon which the eyeball rests.

Eyeball, bulb (bulbus oculi)

Size, shape and position

The size and weight of the avian eyeball is correlated with its function. In most bird species, the relative weight of the eyeball is greater than in mammals, constituting 7–8.5 per cent of the weight of the head in chickens, 17–21.5 per cent in pigeons and birds of prey, and up to 22–32 per cent in owls. This is considerably greater than the equivalent figure in humans (1 per cent).

The avian eyeball is not spherical. Its **anterior segment** (bulbus oculi anterior), bounded by the cornea, has a smaller radius of curvature (thus bulges more prominently) than the **posterior segment** (bulbus oculi posterior), which incorporates the shallow, dish-like fundus (fundus oculi) (Figure 15.5). The anterior and posterior segments are connected by a concave annular section, which is supported by the scleral ring (Figures 15.8 and 15.9).

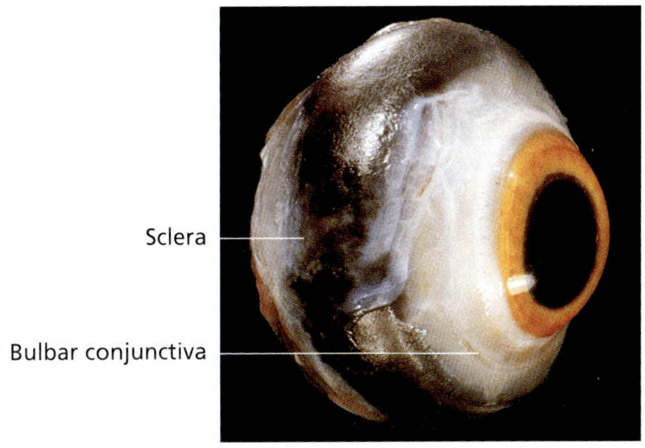

15.4 Eyeball of a chicken (craniomedial view).

The shape of the eyeball can be classified as:

- flat (e.g., pigeons),
- globose (e.g., diurnal raptors),
- tubular (owls) and
- flat-globose mixed type (e.g., ducks).

In all avian species, the eyeball is slightly **asymmetrical**. The scleral ossicles are narrower at the nasal aspect of the eye (Figure 15.9), resulting in nasotemporal asymmetry of the axis bulbi. This is thought to play a role in binocular vision (Figure 15.4).

In most birds, the bulbs are positioned laterally on the head, resulting in a wide, predominantly monocular visual field (up to 360 degrees in woodcocks and African penguins). Owls possess the most rostrally directed eyeballs (Figure 15.1). To compensate for the associated restriction of the visual field, the cervical spine of owls is highly mobile. Jackdaws are able to rotate their eyeballs caudally to some extent, giving them a panoramic view without the need for tell-tale head movements.

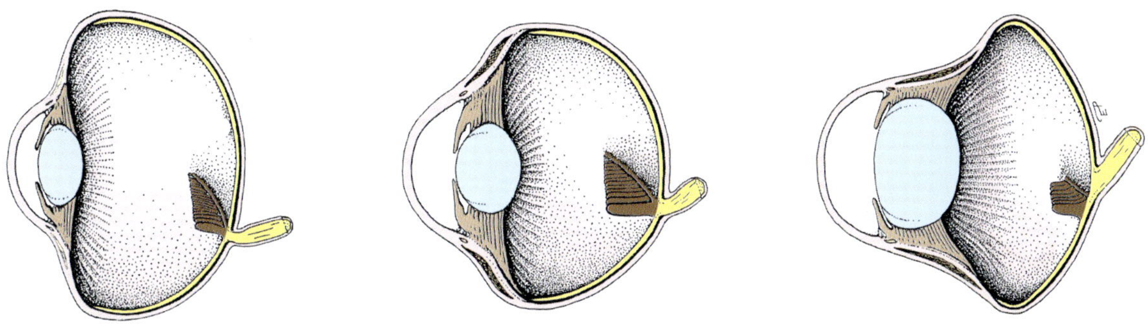
15.5 Flat (left), globose (middle) and tubular (right) eyeball shapes in meridional section (schematic), adapted from Walls, 1942.

Structure of the eyeball

The wall of the eyeball (Figure 15.7) consists of:

- a fibrous outer layer (tunica fibrosa bulbi),
- a middle vascular layer (tunica vasculosa bulbi, uvea) and
- an inner layer (tunica interna bulbi, retina).

The bulbar wall encloses the chambers of the eye (Figure 15.7):

- the anterior and posterior chamber (camera anterior and posterior bulbi)
- containing the aqueous humour (humor aquosus) and
- the vitreous chamber (camera vitrea bulbi) containing the vitreous body (corpus vitreum).

Outer fibrous layer (tunica fibrosa or externa bulbi)

The fibrous outer layer of the eye maintains the stability and shape of the eyeball, resists internal ocular pressure and protects the delicate internal structures from external insult. It consists of the posterior, opaque (white) sclera and the anterior, transparent cornea (Figures 15.6 and 15.7).

SCLERA

Embedded within the connective tissue of the sclera is a hyaline **cartilaginous lamina** (lamina cartilaginea sclerae) and an **osseous scleral ring** (anulus ossicularis sclerae). The cartilaginous lamina (Figures 15.10 and 15.11) reinforces the posterior wall of the eyeball. It may become ossified near the point of entry of the optic nerve, forming the horseshoe-shaped **os nervi optici**.

The **osseous scleral ring** confers mechanical stability upon the concave annular portion of the eyeball. It also serves as a buttress during accommodation that, in contrast to mammals, involves active compression of the lens. The scleral ring consists of 10–18 (usually 15) **individual ossicles** (ossicula sclerae) that overlap in a manner resembling fish scales (Figures 15.8 and 15.9).

The scleral ring exhibits nasotemporal asymmetry (Figure 15.9), resulting in medial convergence of the axis bulbi. Ventro-temporally, the scleral ring bears a groove

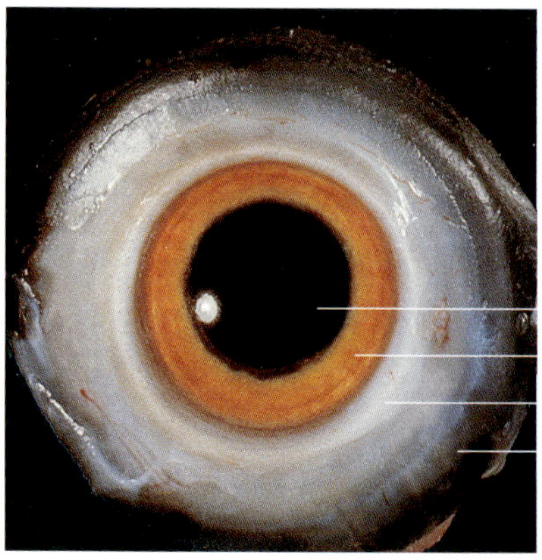

15.6 Eyeball of a chicken (frontal view).

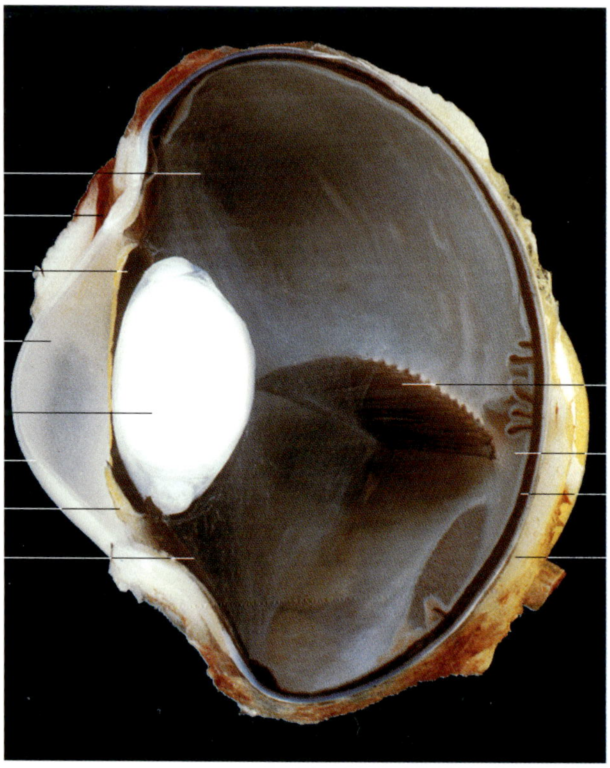

15.7 Eyeball of a chicken (meridional section).

that guides the tendon of the **m. pyramidalis**, which is responsible for movement of the third eyelid.

CORNEA

The cornea is composed of five layers:

- anterior corneal epithelium (epithelium corneae externum),
- Bowman's membrane (lamina limitans superficialis),
- connective tissue stroma (substantia propria corneae),
- Descemet's membrane (lamina limitans profunda) and
- posterior corneal epithelium (epithelium corneae internum).

Except in water birds and several diurnal birds of prey, the avian cornea is typically relatively thin. Compared with mammals, its relative diameter is usually also small. The radius of curvature, on the other hand, exhibits considerable species variation. In water birds, the cornea is comparatively flat, whereas in owls it is strongly curved with a correspondingly deep anterior chamber. The transitional zone between the sclera and the cornea (**junctura corneoscleralis**) is marked by an annular depression. At the outer edge of the cornea (**limbus corneae**), the corneoscleral junction contains pigment deposits that can be differentiated gonioscopically into an inner and outer pigment band (**annulus corneae**) (Figures 15.12 and 15.13).

The cornea is invested with a large number of **sensory nerve fibres** and is therefore extremely sensitive.

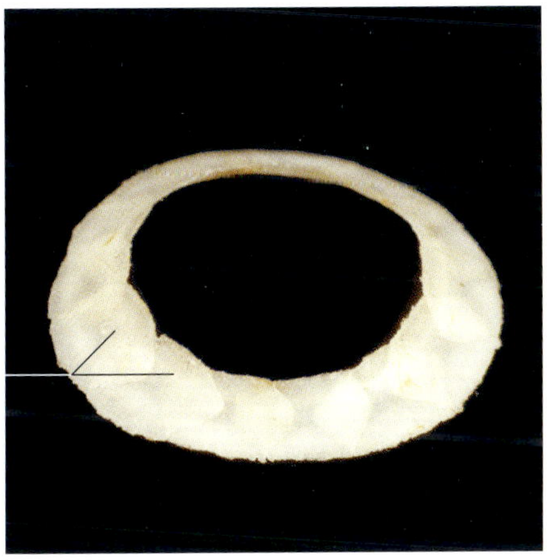

15.8 Flattened scleral ring of the right eye of a chicken.

15.9 Tubular scleral ring of the left eye of a Eurasian eagle-owl (Bubo bubo). Temporonasal asymmetry is evident in the height of the scleral ossicles.

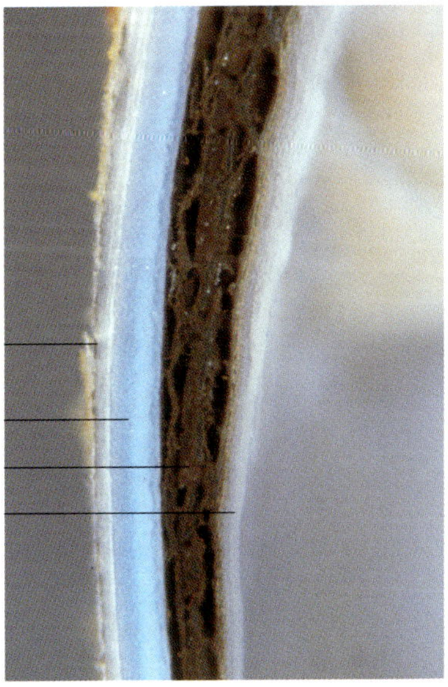

15.10 Histological section of the fundus oculi of a chicken (meridional section).

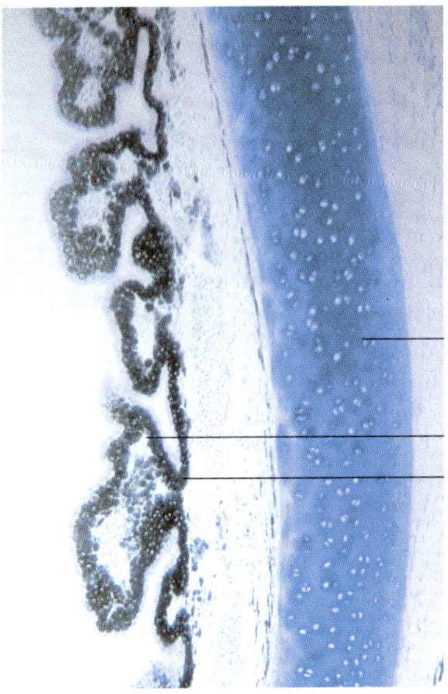

15.11 Histological section of the pars plicata ciliaris of a chicken (meridional section).

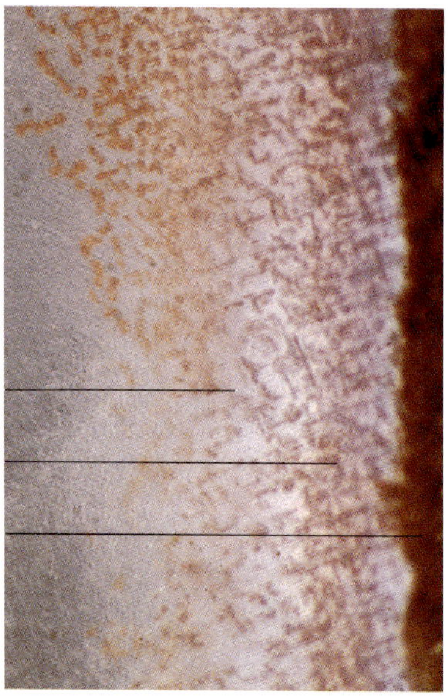

15.12 Pigment band at the corneoscleral junction, viewed from the anterior chamber, in a tawny owl (Strix aluco).

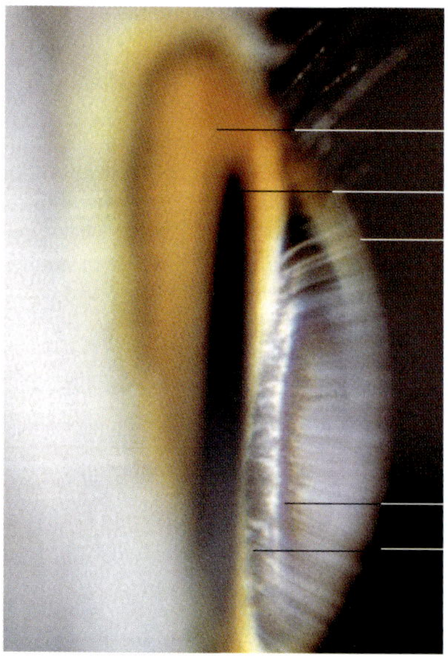

15.13 Gonioscopic view (nasotemporal) of the iridocorneal angle of a Eurasian sparrowhawk (Accipiter nisus).

In contrast, blood vessels – that would compromise the transparency of the cornea – are absent. The cornea is nourished instead by diffusion of nutrients from the aqueous humour and the pre-corneal tear film.

Bowman's membrane is relatively thick, compared with mammals, and makes a substantial contribution to the structural integrity of the cornea. **Descemet's membrane** is comparatively thin and is not present in all bird species. The corneal stroma is composed of collagen fibrils and plentiful hydrophilic chondroitin sulphate-rich ground substance. Corneal transparency results from the arrangement of the collagen fibrils and regulation of water content in the stroma.

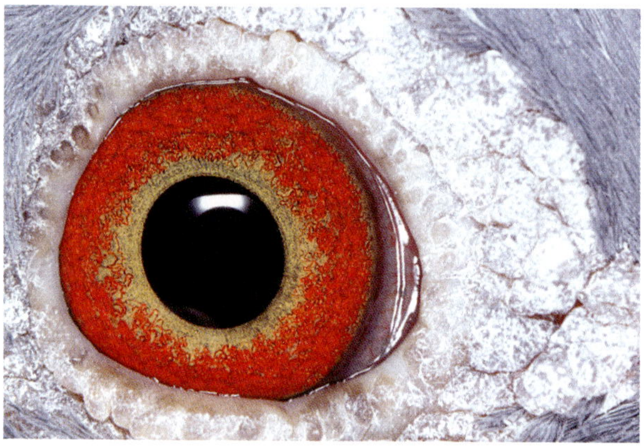

15.14 Right eye of a domestic pigeon with red iris and so-called 'circle of correlation' at the pupillary margin of the iris.

Middle vascular layer (tunica vasculosa or media bulbi, uvea)

The middle layer consists, from anterior to posterior, of the:

- iris,
- ciliary body (corpus ciliare),
- choroid (choroidea).

IRIS

The iris forms an 'aperture ring' that surrounds the usually round, occasionally transversely ovoid **pupil** (pupilla) (Figures 15.6 and 15.14ff.). It separates the anterior chamber (between the cornea and the iris) from the shallow posterior chamber (between the iris and the lens) (Figures 15.7, 15.18 and 15.19).

The **colour of the iris** (Figure 15.14 to 15.17) varies with species and, in some cases, with gender. In some monomorphic species, such as cockatoos, iris colour may

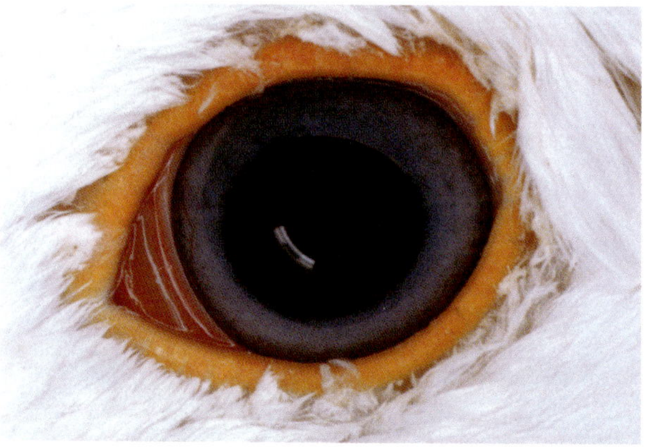

15.15 Left eye of a domestic goose (Anser anser) with blue iris.

15.16 Bright yellow iris and iris coloboma at '2 o'clock' in a juvenile Eurasian sparrowhawk (Accipiter nisus).

15.17 Orange iris in a Eurasian sparrowhawk (Accipiter nisus), frequently observed in older birds.

thus be utilised for gender determination (see Chapter 18 'Clinical examination'). Nutritional and seasonal factors can also influence the colour of the iris.

The iris is lined by an anterior epithelium and a darkly pigmented posterior epithelium (stratum pigmentum iridis). Between these layers are the pigmented stroma and muscles of the iris.

A feature unique to pigeons is the presence of a **tapetum lucidum iridis** consisting of reflective **iridocytes**. Also particular to pigeons is the **anulus iridis**, a non-pigmented region with few blood vessels. This region appears dark because of the underlying stratum pigmentum iridis (Figures 15.14, 15.18 and 15.19). Referred to by pigeon breeders as the 'circle of correlation', it has no significance with respect to vision. In blue-footed boobies, the portion of the iris adjacent to the pupil contains dark pigment deposits. The resulting contrast with the yellow of the remainder of the iris makes the pupil appear deceptively large.

The **width of the iris**, and thus the diameter of the pupil, is controlled by the **mm. sphincter** and **dilatator pupillae**. In the class Aves, these muscles are predominantly striated, enabling birds to adjust more quickly than mammals to **changes in light exposure**.

The **striated pupillary muscle** of birds is under voluntary control and, in contrast to mammals, is not responsive to commonly used ophthalmic drugs. Thus, while parasympatholytics are routinely used in mammalian patients to achieve mydriasis for ophthalmoscopic examination of the fundus, this practice is ineffective in birds. In a number of diving birds, including penguins and cormorants, the m. sphincter pupillae is particularly well developed and the lens protrudes through the pupil during accommodation (see below). This increases the refractive capacity of the lens, compensating for reduced corneal refraction under water.

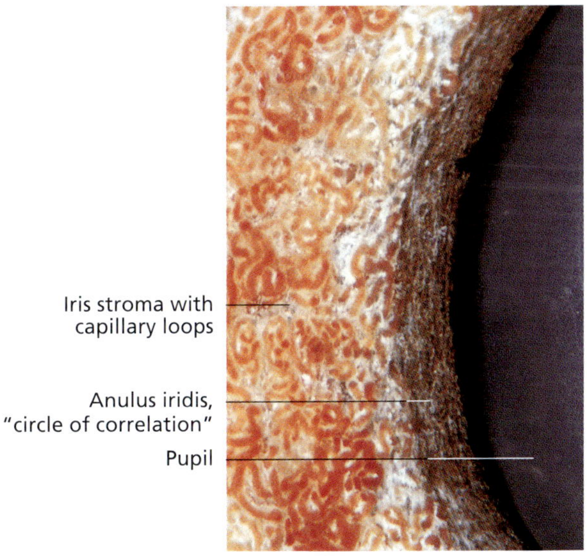

15.18 Anterior surface of the iris in a domestic pigeon.

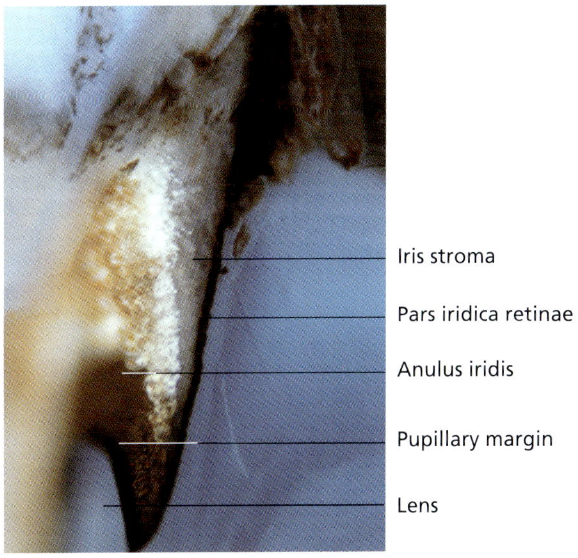

15.19 Iris of a domestic pigeon (meridional section).

CILIARY BODY (CORPUS CILIARE)

The ciliary body has several important functions:

- it fixes the lens within the eyeball,
- its muscles alter the shape of the lens and cornea,
- it is a central component of the mechanism of accommodation.

The ciliary body is also the site of production (by the ciliary processes) and drainage (via the cilioscleral sinus) of aqueous humour.

The ciliary body presents as a ring-shaped thickening between the **base of the iris** (margo ciliaris iridis) and the **choroid**. It is anchored externally to the **scleral ring**. The internal surface of the ciliary body is covered in numerous meridionally oriented folds (**plicae ciliares**) forming the **pars plicata** (*syn.* corona ciliaris). These folds diminish on the posterior internal surface giving rise to the flat **pars plana** (*syn.* orbiculus ciliaris) (Figure 15.20).

Numerous **ciliary processes** extend from the ciliary folds. In contrast to mammals, the ciliary processes attach directly to the periphery of the lens, their tips fusing with the lens capsule (Figures 15.20, 15.21 and 15.23).

The internal surface of the ciliary body is lined with a **double layer** of **epithelium** of neuroepithelial origin (**pars ciliaris retinae**) consisting of an outer (towards the exterior of the eye) pigmented layer and an inner (towards the interior of the eye) layer of columnar epithelial cells (Figure 15.25). In most bird species, as in mammals, the inner epithelial layer is non-pigmented. **Nocturnal owls** are an exception. In these birds, the cells of the inner layer contain large granules filled with **lipofuscin** (rather than melanin) (Figure 15.26). This additional pigment aids in the absorption of scattered light by the long, tubular ciliary body and prevents peripheral illumination of the sclera, which is incompletely covered by the orbit.

The **zonular fibres** (fibrae zonulares) arise from the basal membrane of the inner epithelial layer and pass to the lens capsule. Their involvement in the attachment of the ciliary body to the lens is considerably less significant than in mammals (Figure 15.20).

In contrast to mammals, the **ciliary muscle** (m. ciliaris) (Figure 15.24) is **striated**. It is embedded within the stroma of the ciliary body. The ciliary muscle consists of two parts:

- m. ciliaris anterior (Crampton's muscle),
- m. ciliaris posterior (Brücke's muscle).

A portion of muscle previously described separately as Müller's muscle is now referred to as the fibrae radialis and is considered by some authors to be part of the m. ciliaris posterior.

The **m. ciliaris anterior** and **posterior** both arise from the sclera deep to the scleral ring. While the former passes in an anterior direction to attach to the corneal stroma, the latter extends in the opposite direction to join the base

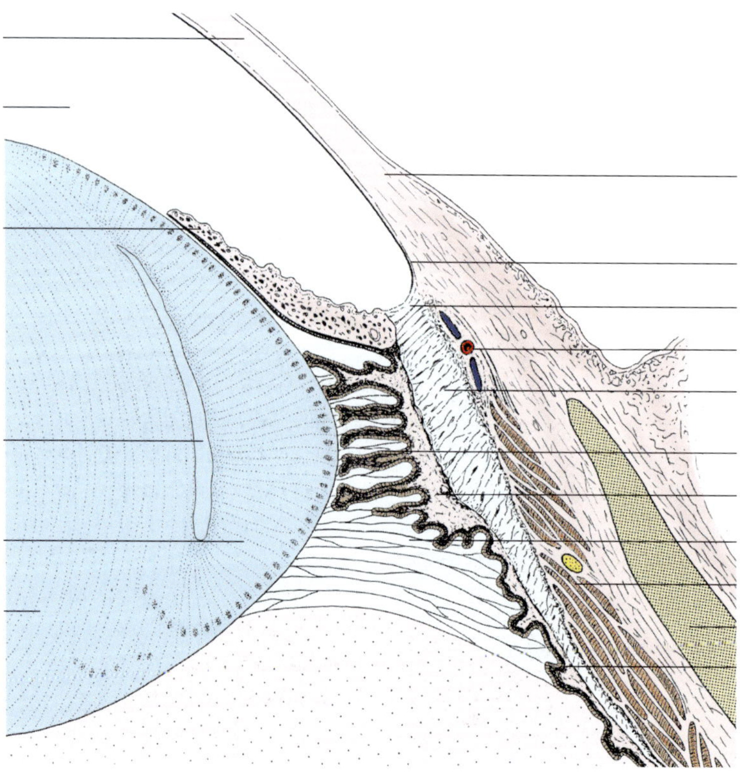

15.20 Ciliary body and iridocorneal angle of the domestic pigeon (meridional section).

15.21 Cilioscleral sinus, pectinate ligament and trabecular reticulum in a chicken.

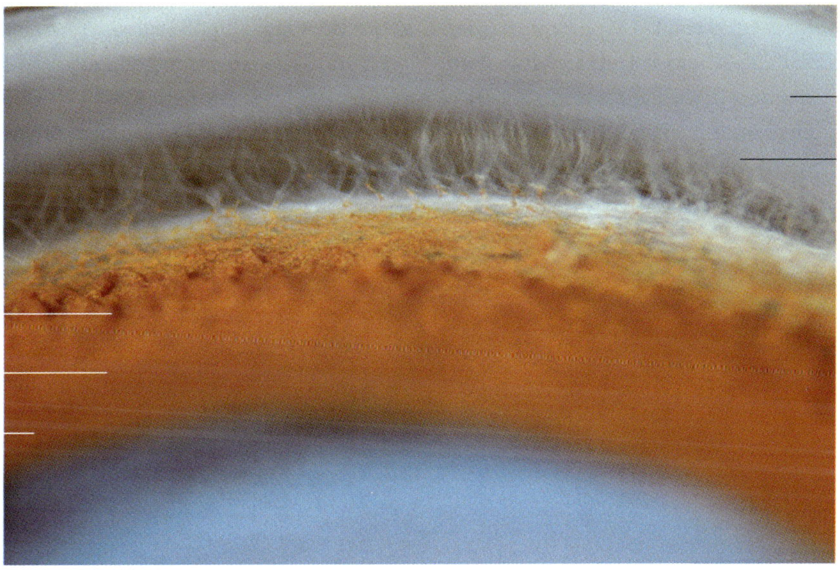

15.22 Iridocorneal angle and pectinate ligament in a ring necked pheasant (Phasianus colchicus).

of the ciliary body. These muscles are involved in the process of accommodation. The main action of the **m. ciliaris anterior** is to alter the curvature of the cornea (**corneal accommodation**) (Figure 15.24). Contraction of the **m. ciliaris posterior** draws the ring-shaped ciliary body forwards, thus also reducing its diameter as the scleral ring narrows anteriorly. Constriction of the ciliary body alters the shape of the lens, increasing its convexity (**lenticular accommodation**) (Figure 15.24).

In nocturnally active birds, such as owls, accommodation is predominantly corneal. The m. ciliaris anterior is well-developed in these birds, while the m. ciliaris

15.23 Posterior surface of the iris and ciliary processes of a chicken with the lens removed. Transparent remnants of the lens capsule can be seen adhering to the ciliary processes and lying flat against the iris. In the posterior portion, the vitreous membrane appears as a cloudy layer over the ciliary folds.

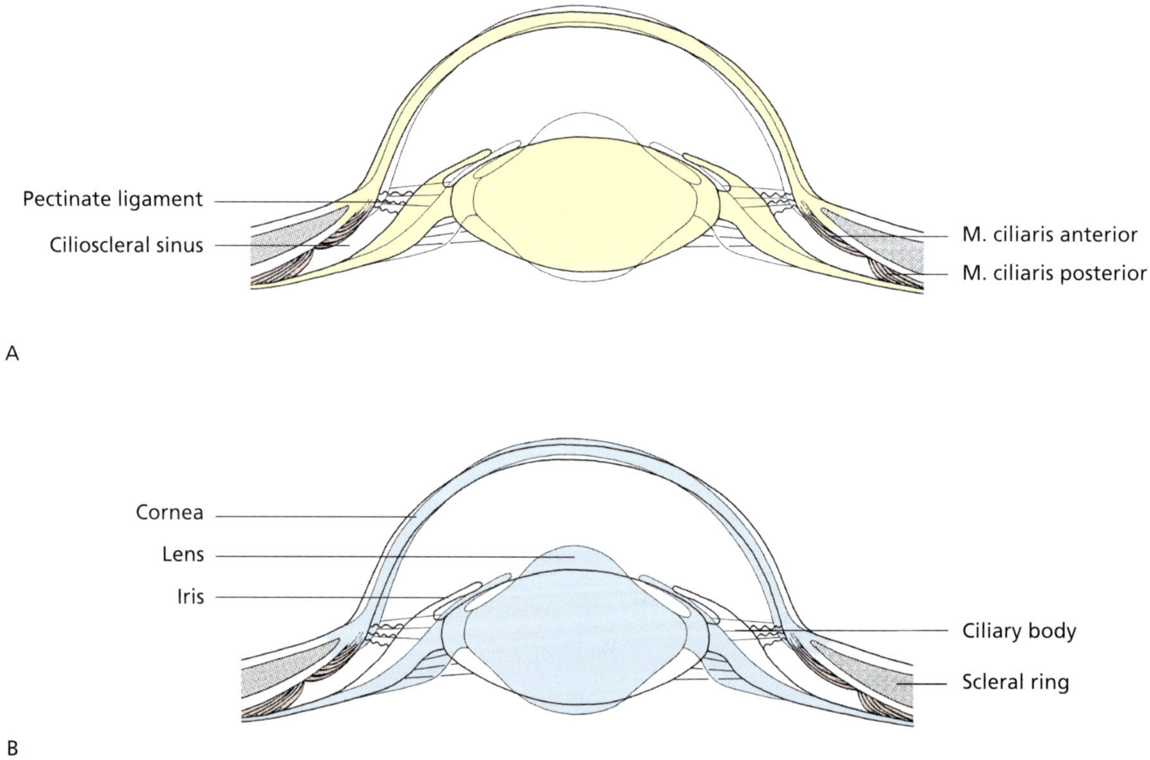

15.24 Corneal and lenticular accommodation (schematic). The upper yellow image represents the relaxed state associated with distance vision (A). The lower blue image illustrates the configuration of the ciliary body and the shape of the lens resulting from contraction of the m. ciliaris for focusing on near objects (B).

posterior is a rudimentary structure. In diving birds, which rely almost exclusively on lenticular accommodation, the degree of muscle development is reversed.

The **cilioscleral sinus** (sinus cilioscleralis) divides the ciliary body into inner and outer sections. These are connected at the posterior ciliary body, in the region of the pars plana. At the level of the base of the iris, the cilioscleral sinus communicates with the anterior chamber at the iridocorneal angle (angulus iridocornealis) (Figures 15.20 and 15.21).

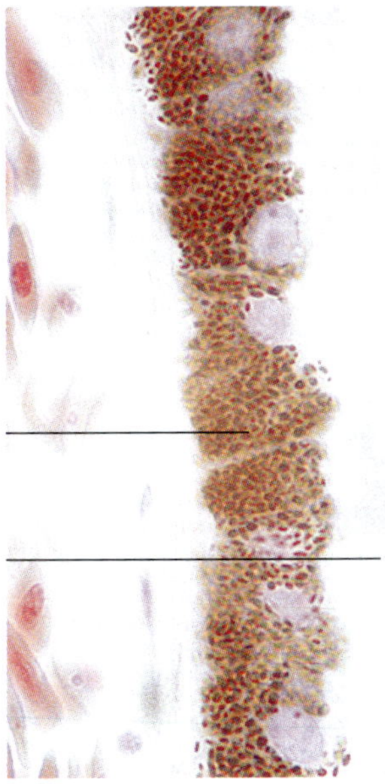

15.25 Pars ciliaris retinae of a common buzzard (Buteo buteo).

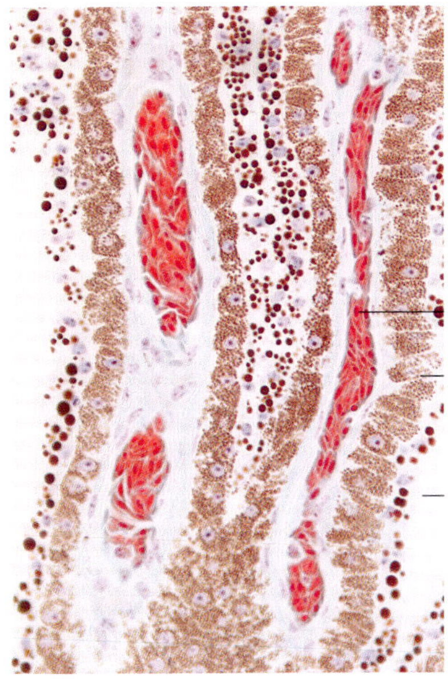

15.26 Pars ciliaris retinae of a tawny owl (Strix aluco) with pigment deposits in the inner epithelial layer.

The cilioscleral sinus is traversed by a mesh of delicate connective tissue fibres, or **trabecular reticulum** (reticulum trabeculare), enclosing the **spaces of Fontana** (spatia anguli iridocornealis) (Figures 15.20 and 15.21). Aqueous humour drains through these spaces from the anterior chamber into the **scleral venous sinus** (sinus venosus sclerae).

The transition to the iridocorneal angle is demarcated by strong, straight fibre bundles that anchor the base of the iris to the sclera. These are referred to as the **pectinate ligament** (lig. pectinatum) (Figure 15.21, 15.24 and 15.38).

The **cilioscleral** sinus permits the inner section of the ciliary body and its associated folds to become displaced, thus serving to underpin the avian mechanism of lenticular accommodation. For this reason, the cilioscleral sinus of diving birds (in which extreme compression of the lens results in a particularly wide range of accommodation) is relatively deep.

CHOROID (CHOROIDEA)

The choroid arises from the posterior boundary of the ciliary body at the **ora serrata** (the anterior limit of the sensory retina, see below). Its outermost layer, the **lamina suprachoroidea (fusca)**, is attached only loosely to the internal surface of the sclera. In contrast, the innermost **lamina basalis** forms a relatively firm attachment with the retinal pigment epithelium. The choroid is relatively thick and highly vascular, its primary function being the nourishment (by diffusion) of the avascular retina (Figure 15.10). Choroidal vessels are arranged in two layers, situated between the lamina suprachoroidea and the lamina basalis. The outer **lamina vasculosa** consists predominantly of arterioles that supply the capillaries of the inner **lamina choroidocapillaris**.

Diurnal birds have a heavily pigmented choroid, whereas little if any pigment is present in crepuscular species (Figures 15.28 and 15.29). A **tapetum lucidum choroideae**, present in several species of crepuscular mammal, has not been observed in birds. A white reflective area (tapetum lucidum retinae) is present in the dorsal fundus of the European nightjar, but this is associated with the retinal pigment epithelium and is not related to the choroidal tapetum of mammals.

Exclusively in woodpeckers, sinuses containing mucous substances are present within the choroid. These act as shock absorbers to dampen the impact of pecking.

INNER LAYER (TUNICA INTERNA BULBI, RETINA)

Developmentally, the inner layer of the eyeball commences as a vesicular evagination of the brain. Through subsequent invagination, this becomes a double walled 'cup-shaped' structure that lies against the deep surface of the middle vascular layer, with its free border situated at the **pupillary margin** (margo pupillaris). Its outer wall consists of a continuous layer of simple **pigmented epithelium** (stratum pigmentosum retinae). The inner leaf, or **stratum nervosum retinae**, incorporates the sensory, light-sensitive retina.

According to its structure and function the inner layer of the eyeball is divided into two sections:

- the posterior pars optica retinae (sensory)
- the anterior pars caeca retinae (non-sensory).

The boundary between these sections is the ora serrata, which also demarcates the transition from the choroid to the ciliary body. The **pars caeca retinae** (Figure 15.25 and 15.26) is further divided into:

- the pars ciliaris retinae and
- the pars iridica retinae.

The **retinal pigment epithelium** is firmly attached to the choroid. In diurnal birds, it contains large quantities of **melanin granules**. These absorb light that has passed through the light-sensitive retina, thus preventing interference by scattered light. In parrots, pigeons and diurnal raptors, the **fundus** is particularly heavily pigmented, making ophthalmoscopic examination more difficult (Figure 15.29). The degree of pigmentation varies considerably, however, both between individuals and species, and is correlated with plumage colour.

In crepuscular and nocturnal avian species, the retinal pigment epithelium is less developed. As a result, the underlying choroidal vessels give the fundus a striated, so-called 'tigroid' appearance on ophthalmoscopic examination (Figure 15.28).

The retinal pigment epithelium adheres relatively loosely to the stratum nervosum retinae. In the pars optica retinae, the stratum nervosum is held in place largely by the vitreous body. Changes in internal ocular pressure can result in **retinal detachment** (ablatio retinae), whereby the stratum nervosum retinae becomes separated from the retinal pigment epithelium.

The **pars optica retinae** consists of layers of dense neuronal networks (Figure 15.27). While this arrangement is similar to the retina of other vertebrate species, the retina of birds has specific features that contribute to the exceptional visual capacity of many avian species.

The avian retina contains a large number of neuronal connections that, in mammals, are established in higher

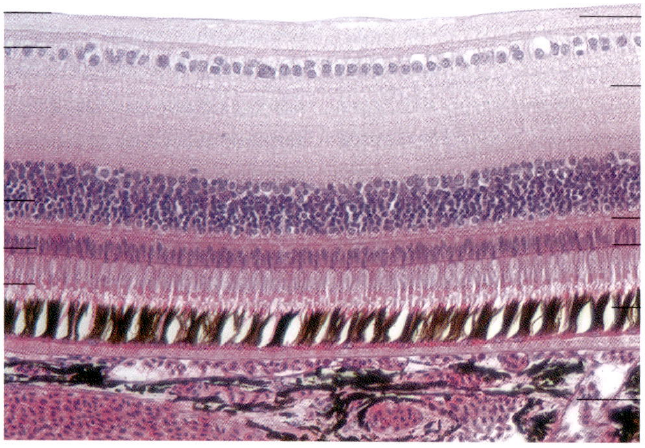

15.27 Pars optica retinae of a chicken.

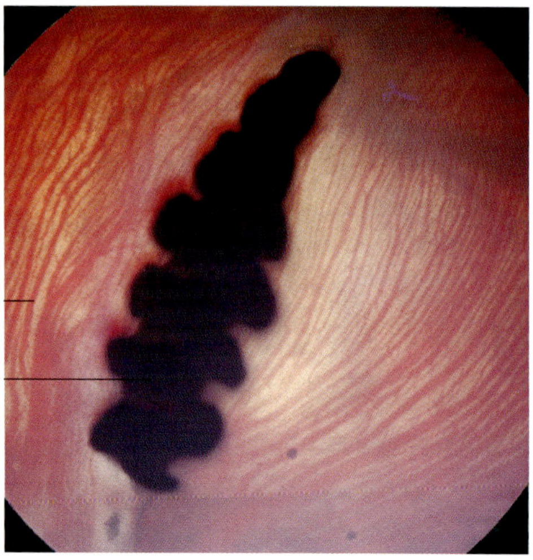

15.28 Sparsely pigmented fundus of a nocturnal bird (tawny owl, Strix aluco).

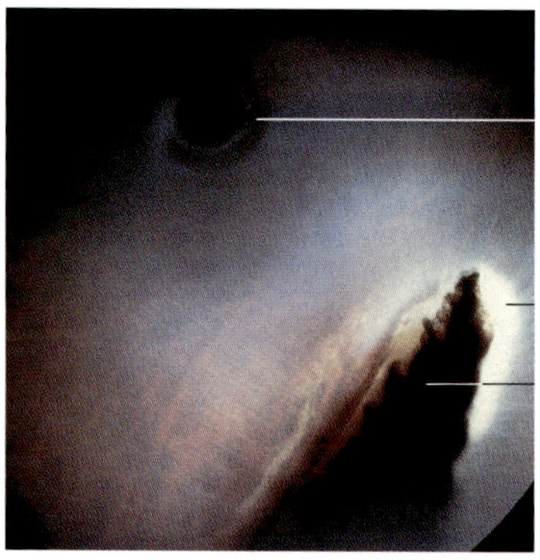

15.29 Heavily pigmented fundus of a diurnal bird (short-toed snake eagle, Circaetus gallicus).

neural centres. Thus, the pars optica retina of birds is comparatively thick. The retina of the pigeon, for example, with its well-developed inner nuclear and plexiform layers, is twice as thick as that of humans. The **photoreceptor layer** contains:

- cone cells for daytime and colour vision (photopic vision) and
- rods for crepuscular and nocturnal vision (scotopic vision).

In birds, the visible light spectrum ranges from 320 to 680 nm. A particular feature of avian cone cells is the presence of oil droplets containing **carotenoid** pigments. At least five types of pigment have been identified, each with a different absorption spectrum. They are believed to act as intraocular chromatic filters. A sixth, colourless type of droplet has been linked with perception of **ultraviolet light**. Rods are considerably more light-sensitive and responsible for non-colour-dependent vision at low light intensity. They contain **rhodopsin**, a pigment with an absorption maximum in the range of 490 to 506nm.

Cone cells predominate in the retina of diurnal birds, **rods** being far fewer in number and restricted to the periphery. In many diurnal birds of prey, the density of cones is greater than in humans. Moreover, there are only very few cone cells per efferent nerve cell (**low convergence**), resulting in a high degree of visual resolution. In contrast, the high light sensitivity of the retina of owls results from a large number of densely packed rods (representing up to 90 per cent of photoreceptors) and the connection of more than 1,000 rods with each bipolar nerve cell (**high convergence**).

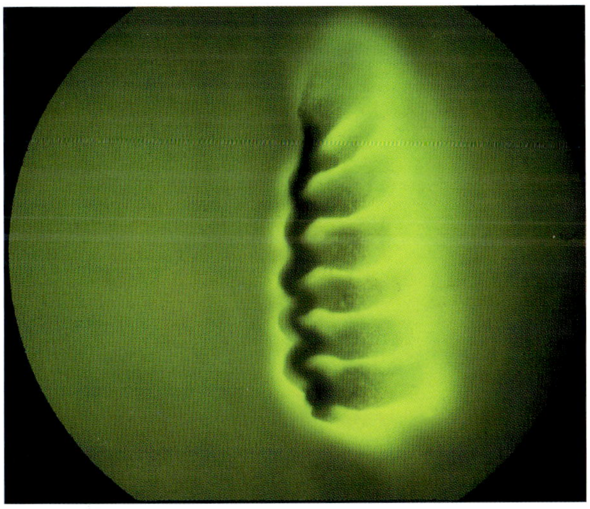

15.30 Fluorescein angiography (pecten phase) in a great horned owl (Bubo virginianus). Fluorescent dye can be seen diffusing from the pecten into the vitreous body.

The term **area retinae** is used to describe well-defined slightly thickened regions of the retina in which the density of cone cells is especially high, and that are thus capable of particularly high resolution. In these regions, the ratio of photoreceptors to efferent neurons can reach the optimal value of 1:1. Visual resolution associated with these areas is further enhanced by the peripheral displacement of overlying neurons, resulting in a central depression known as the **fovea retinae**.

The position and shape of the **areas** and **foveae** varies considerably with species. Most bird species have a central round **area centralis rotunda** with a fovea centralis (absent in the chicken), representing the site of highest monocular visual resolution (Figure 15.29). In many water birds, the central area is linear (termed the **area centralis horizontalis**), and contains a fovea centralis.

Many diurnal raptors also possess an area temporalis with a **fovea temporalis**, which contributes to binocular, stereoscopic vision. Owls, on the other hand, have only an area temporalis.

In contrast to the retina of most mammals, the avian retina is **avascular**. It receives its nutrition by diffusion from the capillary network of the lamina choriocapillaris and from the richly vascularised **pecten** (pecten oculi) (Figures 15.28 to 15.32).

OPTIC NERVE (NERVUS OPTICUS)

The point at which the optic nerve leaves the retina (the equivalent of the optic papilla or optic disc in mammals) is oval in birds and is largely covered by the **pecten**. It is therefore only visible ophthalmoscopically as a narrow whitish-yellow margin at the base of the pecten (Figure 15.29).

In most birds, **all of the nerve fibres** within each optic nerve decussate at the **optic chiasm**. Thus, a true **consensual pupillary light response** does not occur. Throughout its extracranial course, the optic nerve is sheathed in pia and dura mater. It incorporates very little slack in most avian species, and none at all in birds with minimal eye movement (e.g., owls).

A further peculiarity of the avian optic nerve is the presence of efferent nerve fibres originating from the isthmo-optic nucleus in the mesencephalon. These are responsible for further enhancing visual acuity.

PECTEN (PECTEN OCULI)

The pigmented, highly vascular pecten arises from the retina, at the ovoid exit point of the optic nerve, and protrudes into the vitreous body. It is unique to the eye of birds. The only known structure that bears similarity to the pecten is the conus papillaris observed in many species of reptile.

- pleated (pecten plicatus oculi),
- vaned (pecten vanellus oculi) and
- conical (pecten conicus oculi).

Structurally, three types of pecten have been described:

The **pleated** type is characterised by closely apposed vertical folds that are joined at their tip by a bridge, or **pons pectinis**. This type is typical of carinate birds (Figures 15.31 and 15.33).

In the **vaned** form, 25–30 vertical vanes are connected to a central lamina (Figure 15.32). This type is found in ratites, such as the ostrich, emu and nandu. Only the kiwi is known to possess a **conical** pecten, an undivided structure devoid of folds or laminae, resembling the conus papillaris of reptiles.

According to ontogenic studies, the pecten is **derived from the retina**. Its connective tissue scaffold, enclosing a dense capillary network (Figure 15.34), extends from the optic nerve.

Despite extensive investigation, the function of the pecten remains unclear. The manifold hypothesised roles include protection of certain regions of the retina from glare, reduction of scattered light, immunocompetence, participation in motion perception and sensing of magnetic fields for the purpose of orientation. It is generally agreed that the pecten has a **nutritional role**, supplying the vitreous body and avascular retina, and that it contributes to intraocular **pressure** and **temperature regulation**.

This is supported by the characteristics of the endothelium of the capillaries in the pecten, which are suggestive of constant, active transepithelial transport. The passage of substances from the capillaries to the vitreous body has been demonstrated using fluorescein angiography (Figure 15.30). Distribution of these substances throughout the vitreous body is facilitated by rhythmic contractions of the extrinsic muscle of the eye, which result in oscillatory eye movements and corresponding passive movement of the pecten.

15.31 Pleated pecten of a chicken. The height of the pecten decreases from ventronasal (image left) to dorsotemporal (image right).

15.32 Nasoventral view of a vaned pecten in a nandu (Rhea americana).

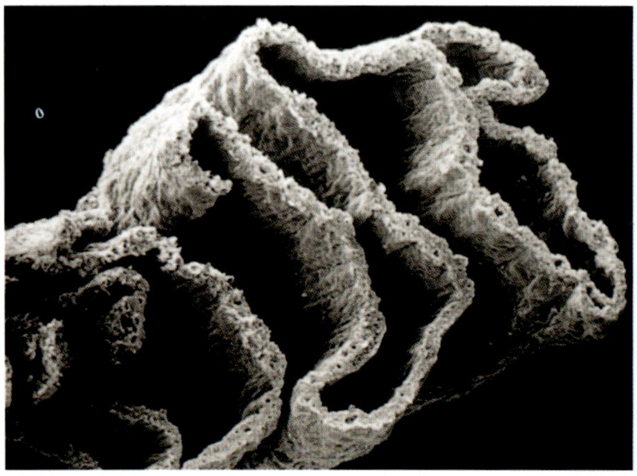

15.33 Scanning electron micrograph of the folds of the pecten of a common buzzard (Buteo buteo) in horizontal section.

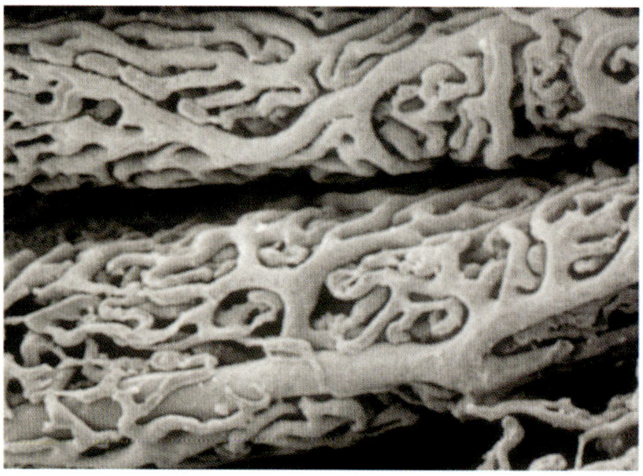

15.34 Scanning electron micrograph of the capillary network in the pleated pecten of a domestic pigeon (corrosion cast).

Internal structures of the eye

LENS

The lens is a transparent, biconvex structure of epithelial origin, positioned between the iris and the vitreous body. In altricial birds, the lens is opaque during the nestling stage, only becoming transparent after fledging (Figure 15.37). The lens of diurnal birds is relatively flat, while in water birds and nocturnal species it is spheroid in shape (Figure 15.35).

The lens is surrounded by a **capsule** (capsula lentis). Beneath the capsule is a layer of simple epithelium comprising cells that become elongated towards the **lens equator** (equator lentis). These give rise to radially oriented hexagonal **prisms** that combine to form the equatorial **annular pad** (pulvinus anularis lentis) that surrounds the central core of the lens (Figure 15.36). The annular pad is a characteristic feature of the avian eye. It is marked by a row of indentations that correspond in number with the ciliary processes to which they are attached (Figure 15.36).

The function of the annular pad is incompletely understood, although it is not considered to be part of the optical system. Rather, it is presumed to play an important role in the speed of **lenticular accommodation**, since the pad is particularly thick in fast flyers (especially diurnal birds of prey and pigeons) and less developed in diving birds. It is notably narrow in psittacines.

The annular pad may also have a **nutritional function**, as its cells secrete fluid, the **aqua vesiculae lentis**, into the cleft-shaped **lens vesicle** (vesicula lentis) (Figure 15.20) between the pad and the **central core of the lens** (corpus centrale lentis). This fluid may be absorbed by the lens core. Lens sutures, as seen in mammals, are not present in birds.

The **lens capsule** is derived from the basement membrane and forms a semi-permeable barrier through which nutrients can diffuse from the aqueous humour. It also forms a permanent barrier separating the protein within the lens from the immune system. Protein released as a consequence of lens trauma may be recognised as foreign and can thus result in phacogenic uveitis (endophthalmitis phacoanaphylactica).

In contrast to mammals, the avian **lens equator** (equator lentis) is closely associated with the ciliary processes, the tips of the processes being tightly fused with the capsule. Fixation of the lens is supplemented by the **ciliary zonule** (zonula ciliaris), composed of the **zonular fibres** (fibrae zonulares).

This intimate association with the ciliary body, and the relative pliability of the avian lens, facilitate the process of active **lenticular accommodation.** Accordingly, the lens is relatively flexible in diving birds, in which lenticular accommodation predominates, and relatively hard in owls, which rely almost exclusively on corneal accommodation. Age-related loss of lens pliability, seen in humans, dogs and cats, is not observed in birds.

ANTERIOR AND POSTERIOR CHAMBERS (CAMERA ANTERIOR AND POSTERIOR BULBI) AND AQUEOUS HUMOUR (HUMOR AQUOSUS)

The anterior and posterior chambers contain aqueous humour, which is secreted continuously by the inner epithelial layer of the pars caeca retinae into the **posterior**

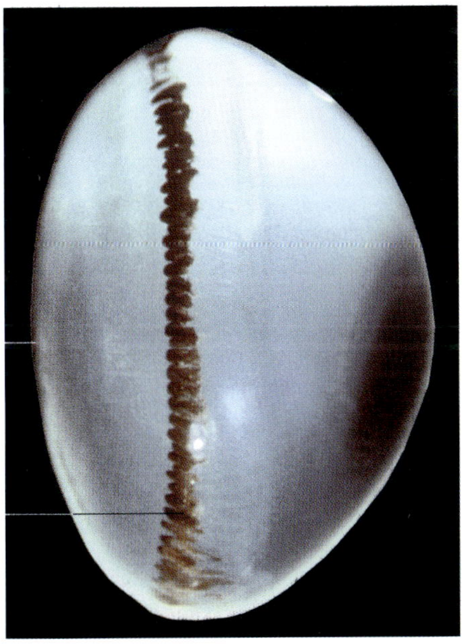

15.35 Lens of a Mallard duck (Anas platyrhynchos) with remnants of torn ciliary processes at the lens equator.

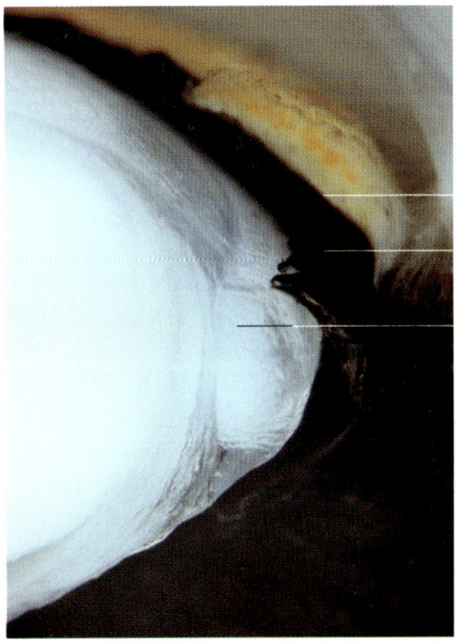

15.36 Annular pad (pulvinus anularis lentis) of a chicken (meridional section).

15.37 In altricial species the lenses of nestlings, branchlings and birds up to six weeks of age are immature and may be opaque. Shown here is a five-week-old tawny owl (Strix aluco).

chamber (camera posterior bulbi). Aqueous flows through the pupil into the **anterior chamber** (camera anterior bulbi) from whence it passes, at the iridocorneal angle, through the **pectinate ligament** into the **spaces of Fontana** of the **cilioscleral sinus** (Figure 15.38).

From there it diffuses into the wide, usually bipartite **scleral venous sinus** (Figure 15.39), which is located more superficially in birds than in domestic mammals. In owls, the aqueous humour contains mucous substances secreted by the posterior corneal epithelium, making anterior chamber paracentesis more difficult in these birds.

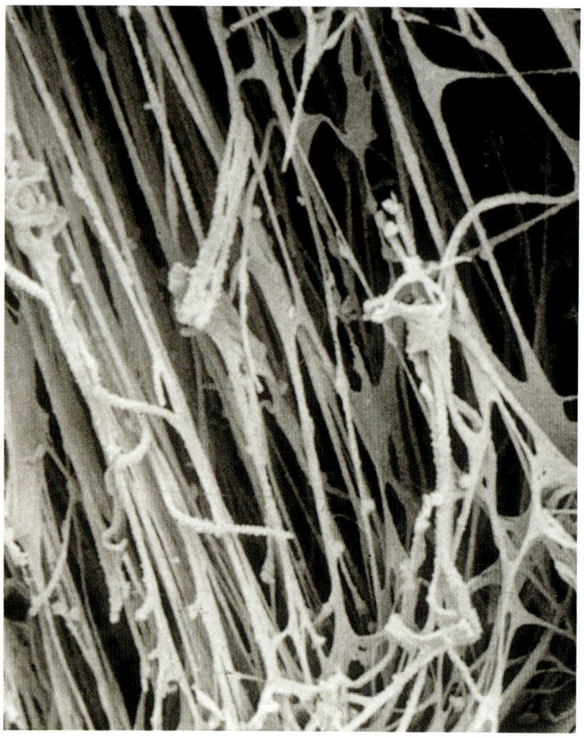

15.38 Taught fibres of the pectinate ligament in a yellow-crowned Amazon (Amazona ochrocephala).

VITREOUS BODY (CORPUS VITREUM; CAMERA VITREA BULBI)

The gelatinous vitreous body fills the **camera vitrea bulbi**, the intraocular space that lies posterior to the lens (Figure 15.7). It is avascular and is composed of viscous, transparent, extracellular substance comprising around 99 per cent water (**humor vitreus**). The basic structure of the gel comprises hydrophilic glycosaminoglycans embedded in a delicate framework of collagen fibrils, the **stroma vitreum**. Interweaving of superficial fibrils gives rise to the **vitreous membrane** (membrana vitrea), which attaches the vitreous to the base and bridge of the pecten, and to the ciliary body.

The **intraocular pressure** created by the vitreous body holds the retina in position. Displacement or disease of the vitreous body has the potential to affect the retina and may lead to partial or total **retinal detachment**. Compared with many mammalian species, the consistency of the avian vitreous body is relatively thin. This aids the diffusion of nutrients from the pecten to the retina and permits the oscillatory movements of the pecten that facilitate nutrient distribution.

Adnexa of the eye (organa oculi accessoria)

Extrinsic muscles of the eyeball

The configuration of the extrinsic muscles of the eyeball is similar to that in mammals (Figure 15.40), comprising: The mobility of the avian eye ranges from extremely pronounced to profoundly limited (as little as 2 degrees). Owls are an example of the latter, having only rudimentary extraocular muscle development. The restricted movement of the eyeball in these species is countered by

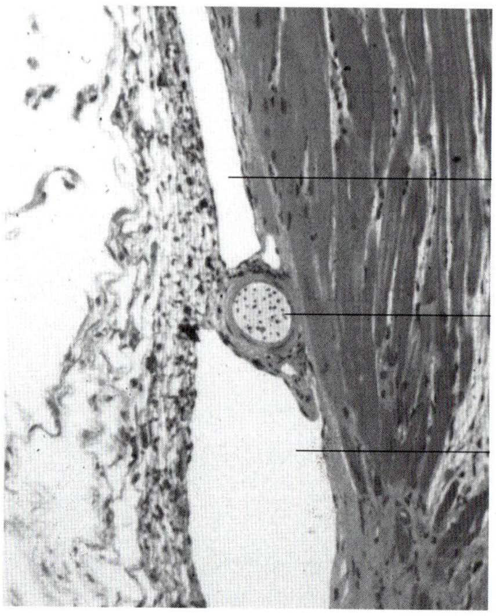

15.39 Bipartite scleral venous sinus in a common buzzard (Buteo buteo).

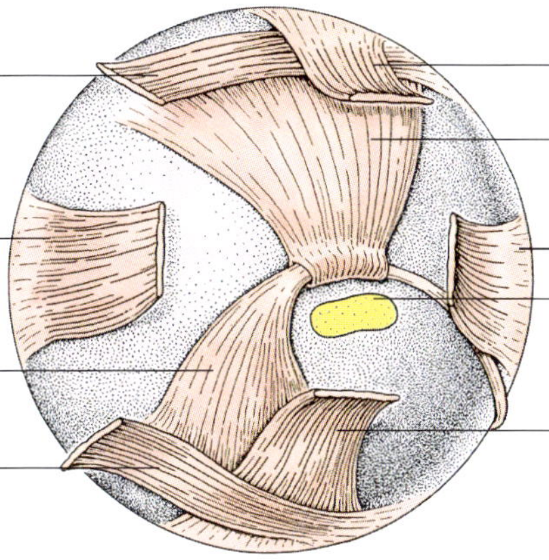

15.40 Posterior pole of the bulbus oculi with extrinsic muscles. The tendon of the m. pyramidalis is redirected by a trochlea formed by the m. quadratus, adapted from Frewein and Sinowatz, 2004.

the high degree of mobility of the head and neck. The m. retractor bulbi of mammals is absent in birds. In its place, birds possess the well-developed **mm. quadratus et pyramidalis membranae nictitantes** that lie against the posterior surface of the bulb. Their action is to draw the third eyelid across the cornea.

- **four straight muscles**:
  - m. rectus dorsalis,
  - m. rectus ventralis,
  - m. rectus temporalis,
  - m. rectus nasalis and
- **two oblique muscles**:
  - m. obliquus dorsalis and
  - m. obliquus ventralis.

Dorsal to the optic nerve, the **m. quadratus** forms a **tendinous loop** (trochlea) for the passage of the tendon of the **m. pyramidalis** (Figure 15.40). This tendon, guided by a ventrotemporal sulcus in the **scleral ring**, radiates into the ventral free edge of the nictitating membrane.

Eyelids (palpebrae)

The eyelids (Figures 15.41 and 15.42) serve to protect the cornea. In most diurnal birds the **lower eyelid** (palpebra ventralis) is larger and more mobile than the **upper eyelid** (palpebra dorsalis). When closed, the lower lid almost completely covers the cornea. It is supported by a fibrous **plate** (tarsus palpebralis), which may be cartilaginous in birds of prey.

In nocturnal species, such as owls, the upper eyelid is larger and more mobile (Figure 15.1). Also in parrots, the upper eyelid exhibits considerable mobility. Three muscles are responsible for movement of the upper and lower eyelids:

- m. levator palpebrae dorsalis,
- m. depressor palpebrae ventralis and
- m. orbicularis oculi.

The **internal surface** of the eyelids, the **facies conjunctivalis**, is lined with **palpebral conjunctiva** (tunica conjunctiva palpebrarum). At the base (fornix) of the **conjunctival sac** (saccus conjunctivae), the palpebral conjunctiva reflects

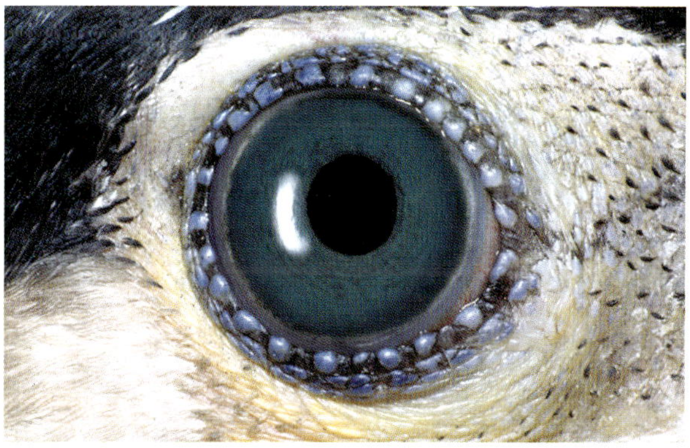

15.41 Right eye of a cormorant (Phalacocorax carbo). The lid margins are non-feathered and coloured blue.

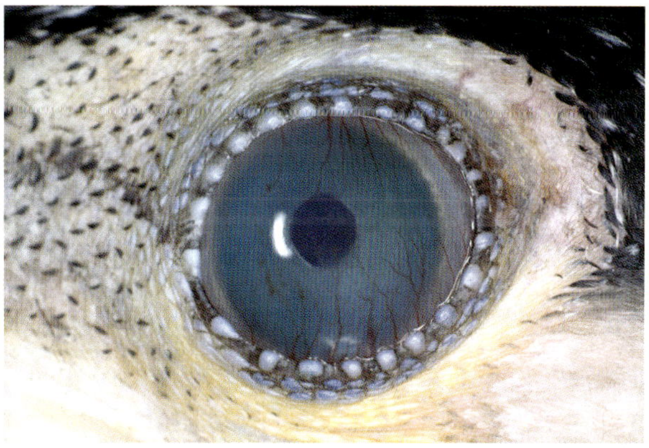

15.42 Left eye of a cormorant (Phalacocorax carbo) with nictitating membrane drawn across the cornea. The translucent and vascular nictitating membrane should not be confused with corneal vascularisation resulting from chronic corneal inflammation (keratitis).

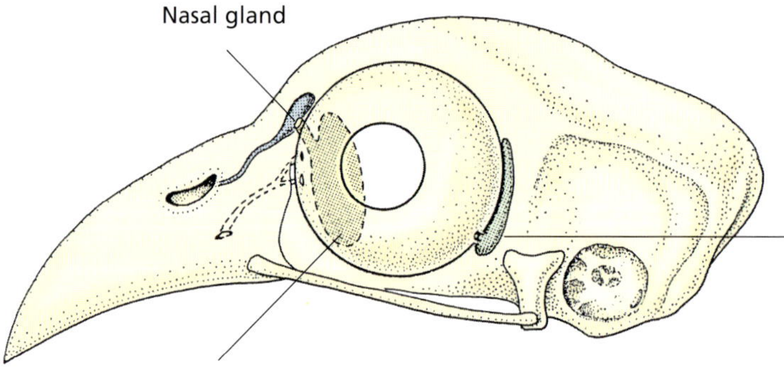

15.43 Lacrimal apparatus of the chicken, adapted from Baumel, 1993.

onto the eyeball to become the **bulbar conjunctiva** (tunica conjunctiva bulbi).

A pit-like depression in the upper eyelid of domestic poultry is a common site of infestation with ectoparasites such as lice and fleas. In pigeons, cormorants and other species, the **lid margin** (limbus palpebralis) may be completely bare (Figure 15.14, 15.41 and 15.42). It is sparsely feathered in chickens, while in parrots, raptors and ostriches it is lined with **'hair feathers'** (cilia palpebralia), characterised by the absence of vanes. Meibomian glands, located in the eyelid margin of mammals, are **absent in birds**. In many bird species, the lid margins, and sometimes also the external surface and surroundings of the eyelids, are accentuated with bright colours.

The **third eyelid** (palpebra tertia, membrana nictitans, nictitating membrane) is a thin, highly specialised fold of conjunctiva. It **protects the cornea** from desiccation during flight and distributes the complex multilayered **precorneal tear film**.

In contrast to mammals, the **third eyelid** extends over the cornea from the dorsonasal to the ventrotemporal quadrant of the eye (Figure 15.47). Its free, often pigmented margin is lined with featherlike epithelial processes that sweep the surface of the cornea clean. The third eyelid is translucent in some birds and white in owls. Most water birds, especially diving species, have a transparent third eyelid that serves as an additional refractive medium, akin to a dive mask, when the animal is underwater (Figure 15.42).

Lacrimal apparatus (apparatus lacrimalis)

The avian lacrimal apparatus (Figure 15.43) consists of:

- the lacrimal gland (glandula lacrimalis),
- the gland of the third eyelid (glandula membranae nictitantis),
- the nasal gland (glandula nasalis) and
- the lacrimal ducts.

In domestic birds, the relatively small **lacrimal gland** is located at the temporal angle of the eye, between the periorbita and the palpebral conjunctiva. It empties by a narrow duct (or ducts) into the **conjunctival sac** near the conjunctival fornix. Owls, which produce only a small volume of tears, do not have a lacrimal gland.

In most bird species the **gland of the third eyelid**, previously referred to as the Harderian gland, is more than double the size of the lacrimal gland. It lies ventrally on the caudomedial aspect of the bulb at the nasal angle of the eye. As well as producing mucoid tears, the gland of the third eyelid has an important role in cell mediated immunity (secretion of immunoglobulin A, aggregation of lymphocytes and plasma cells).

The **nasal gland** is a modified lacrimal gland located dorsonasally within the orbit. In certain seabirds, it functions alongside the kidney as an additional means of salt excretion. Thus the increased lacrimation often observed in seagulls represents a physiological mechanism for ridding the body of salt (e.g., after ingestion of sea water).

Tear drainage occurs at the nasal angle via either one (e.g., penguins) or more commonly two openings, the **ostia canaliculi lacrimalis**. These are located on the internal surface of the upper and lower eyelids, clearly separated from the lid margin. Short lacrimal canaliculi (canaliculi lacrimales), only a few millimetres long, convey the tears into the **nasolacrimal duct** (ductus nasolacrimalis) that opens into the nasal cavity beneath the nasal conchae. From there, the tears flow through the choana directly into the oral cavity.

Innervation of the eye

The **optic nerve** is responsible for sensory innervation of the retina of the bulbus oculi. In contrast to mammals, the optic nerves cross over completely at the optic chiasm in most bird species, thus a true **consensual pupillary reflex is not observed**. Sensory, motor and autonomic innervation of the remaining structures of the eye is supplied, with species variation in configuration and extent, by the:

- oculomotor nerve (n. oculomotorius, III),
- trochlear nerve (n. trochlearis, IV),
- ophthalmic branch of the trigeminal nerve (n. trigeminus, V_1),
- abducent nerve (n. abducens, VI) and
- palatine branch of the facial nerve (n. facialis, VII).

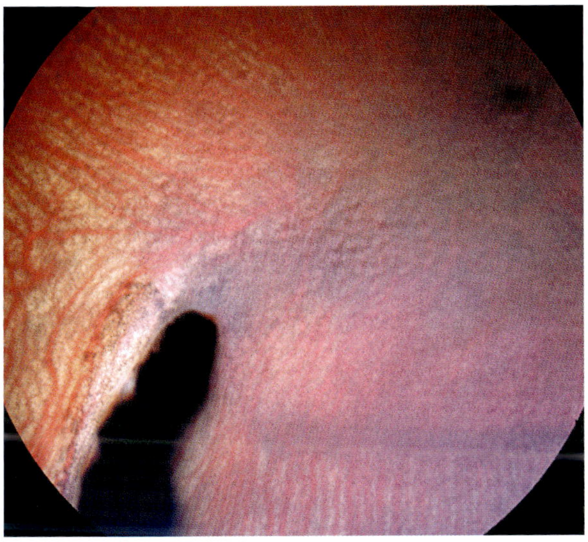

15.44 Central region of the fundus of a great horned owl (Bubo virginianus). The horizontally oriented a. ciliaris posterior longa nasalis gives off multiple aa. parallelae choroideae that form dorsally and ventrally coursing vascular cascades. The fovea (top right) and pecten (bottom left) are visible.

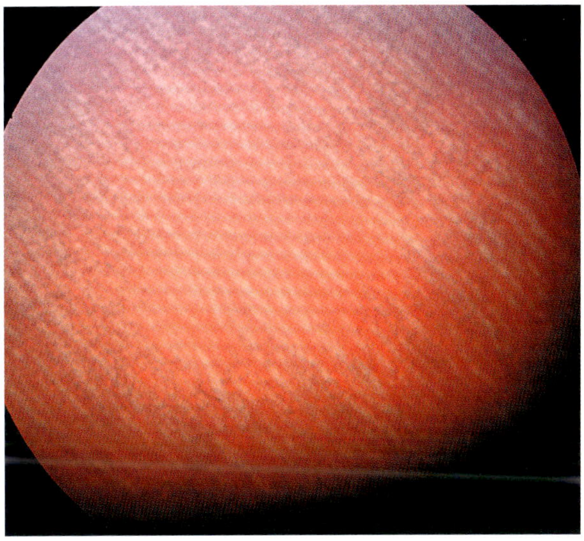

15.45 Nasal region of the fundus of a great horned owl (Bubo virginianus) with aa. parallelae choroideae branching off in a fan-like arrangement from the a. ciliaris posterior longa.

Blood vessels of the eye

The main vessel supplying the eyeball and adnexa is the **external ophthalmic artery** (a. ophthalmica externa), that ramifies ventrotemporally within the orbit into the **rete mirabile ophthalmicum**. This is of clinical significance as it can be a source of extensive intraoperative bleeding. The **ophthalmotemporal artery** (a. ophthalmotemporalis) is the principal artery leaving the rete mirabile, providing blood to the bulbus oculi. Three branches of the ophthalmotemporal artery supply the lamina choriocapillaris of the choroid:

- the a. ciliaris posterior longa nasalis,
- the a. ciliaris posterior longa temporalis and
- the aa. ciliares posteriores breves.

The **a. ciliaris posterior longa nasalis** is the continuation of the ophthalmotemporal artery and supplies the medial eyeball (Figures 15.44 and 15.45). The small **a. ciliaris posterior longa temporalis** arises immediately distal to the rete ophthalmicum and supplies a small lateral portion of the bulb.

The **aa. ciliares posteriores breves** penetrate the bulbus oculi around the optic nerve and directly supply the choroid. In contrast, the aa. ciliares posteriores longae divide into a multitude of **aa. parallelae choroideae**, that pass dorsally and ventrally towards the ciliary body throughout the vascular layer of the choroid. Over their course, they give off branches into the lamina choriocapillaris. The pecten is supplied by a dedicated branch of the ophthalmotemporal artery, the **a. pectinis oculi** (Figure 15.55).

Venous drainage of the choroid occurs via **veins** that mostly run parallel to the aforementioned arteries and are thus accordingly named.

Anterior ciliary arteries (aa. ciliares anteriores), which anastomose with the aa. ciliares posteriores, arise as branches of the vessels supplying the eyelids and provide blood to the ciliary body.

The eyelids are supplied by small vessels that originate from the **supraorbital artery** (a. supraorbitalis) and give off fine branches as they pass parallel to the lid margin. Transection of these vessels can lead to significant haemorrhage. This is an important consideration when performing a lateral canthotomy (widening of the palpebral fissure by a skin incision at the temporal angle of the eye), for example during surgical removal of the eyeball in owls.

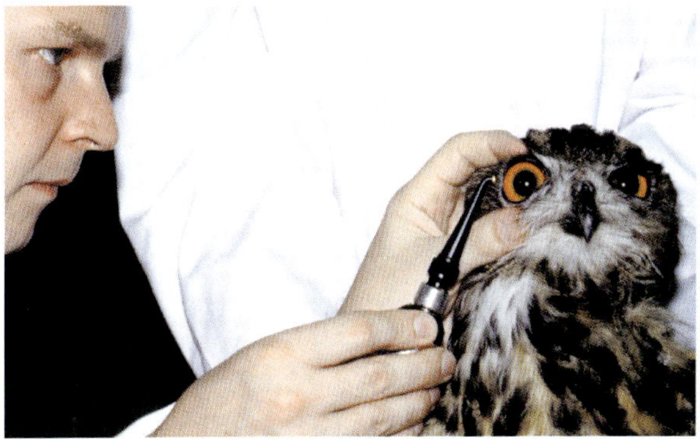

15.46 Inspection of the anterior segment of a Eurasian eagle owl (Bubo bubo) using lateral illumination of the lens-iris diaphragm during routine ophthalmic examination.

Clinical aspects

Ophthalmic examination

The eye is the most important sense organ in birds. A fully functional visual apparatus is essential for activities such as flight and acquisition of food or prey, the remaining senses having only a very limited capacity to compensate for visual impairment. Even minor or partial loss of visual function typically has significant, prognostically relevant consequences.

15.47 Elicitation of the corneal reflex in a long-eared owl (Asio otus) by touching the peripheral cornea with a sterile moistened cotton bud (the central cornea is avoided to minimise the risk of potentially vision-impairing corneal epithelial trauma).

When performing an ocular assessment, it is particularly important for the examiner to be aware of the considerable species-specific variation in the normal anatomy of the avian eye. This typically requires a degree of experience in ophthalmic examination.

The incidence of ophthalmic disease in avian medicine is approximately 7 per cent. Not infrequently, relevant infectious aetiologies are **zoonoses** and therefore also have implications for the owner. Many infectious causes of ocular disease also present with classic, if not pathognomonic clinical signs. These include:

- salmonellosis,
- mycobacteriosis (tuberculosis),
- chlamydiosis (psittacosis / ornithosis) or
- paramyxovirus infection (Newcastle disease).

The majority of ophthalmic abnormalities in birds represent ocular manifestations of systemic disease. In this way, the eye serves as 'diagnostic window' into the functional status of the internal organs. Avian ophthalmology is therefore less a specialised discipline than an integral component of routine examination of avian patients. While it may be argued that the same applies in other species, it is particularly relevant in birds.

In all bird species, ocular trauma is involved in a relatively large proportion of case presentations. Lesions are observed in more than 32 per cent of trauma cases, usually involving the posterior segment of the eye (predominantly intravitreal haemorrhage originating from the pecten or the choroid). In these patients, the anterior segment may appear completely normal. Thus, in order to preserve the animal's most valuable sense, ophthalmoscopic examination of trauma patients is crucial for prompt establishment

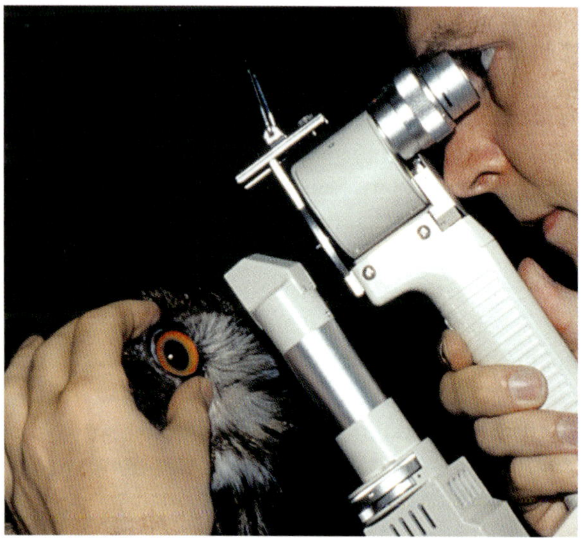

15.48 Examination of the anterior segment of a Eurasian eagle owl (Bubo bubo) using a slit lamp biomicroscope.

15.49 Tonometry of the left eye in a blue-fronted amazon (Amazona aestiva), illustrating the importance of using appropriately sized instruments in avian ophthalmology.

of an accurate diagnosis and implementation of appropriate treatment. This also highlights the importance of a thorough and systematic approach to routine **ophthalmic examination**.

History

Changes suggestive of a disturbance of vision include:

- an obvious reluctance to fly,
- inability to locate perches and food,
- altered socialisation and
- absence of the flight response when approached.

Observation and vision testing

Examination commences with observation of the unrestrained patient in an unobtrusive manner, in an environment with minimal auditory stimuli. Factors that can be assessed in this setting include:

- general behaviour,
- the ability of the bird to orient itself within the cage,
- feeding behaviour,
- social behaviour (towards companion birds, owners) and vocalisation, as well as
- flight characteristics (where possible).

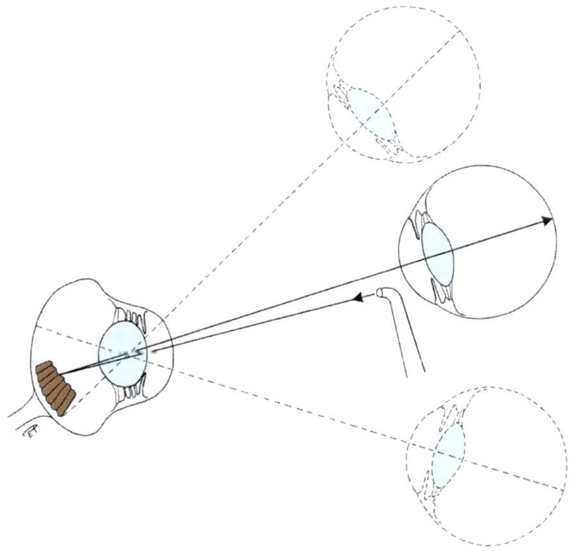

15.50 Principles of ophthalmoscopic examination. Ophthalmoscopy is defined as examination of the segment of the eye located posterior to the lens by directing a focused beam of light through the pupil and observing the light reflected from the fundus. The optical basis for this technique is the coaxial alignment of the incident and reflected light with the visual axis of the examiner (through the dilated pupil of the patient) such that the reflected beam is projected onto the examiner's retina.

The transition to examination of the restrained patient begins with assessment of vision, including menace testing and evaluation of the response to subtle tactile stimuli. Sudden excessive flight responses are suggestive of visual impairment or blindness.

General ophthalmic examination

Examination of the restrained patient is conducted under dimmed natural or artificial light. It may be necessary to undertake the examination in stages to minimise stress and prevent shock.

The ophthalmic examination is based largely on the use of a focused light source (otoscope light, preferably a transilluminator) to illuminate the eye from different directions and at varying angles of incidence.

Individual components of the examination include:

- **Comparison of both eyes from in front and from above**: permits evaluation of the surroundings, size, position and shape of the bulbs, as well as the size and appearance of the palpebral fissure.
- **Examination of the ear openings**: haemorrhage into the ear openings (see Chapter 16, Figures 16.2 and 16.3; ear openings in a long-eared owl) is suggestive of intraocular bleeding, usually originating from the pecten. In owls, the temporal wall of the bulb and the external temporal periphery of the fundus (see Chapter 16, Figure 16.2) can also be inspected.
- **Frontal inspection with frontal illumination**: provides an overview of the periocular region and adnexal structures (e.g., eyelids), the cornea and the anterior chamber as well as the shape and position of the lens and pupillary opening and the pupillary light reflex.
- **Frontal examination with lateral transillumination (light source directed towards the temporal limbus):** permits inspection of the structures of the anterior chamber as far as the lens.
- **Frontal examination with illumination from various angles**: enables examination of the transparent media of the dioptric (light refractive) apparatus as far as the anterior region of the vitreous body. Also permits evaluation of the pupillary light reflex. Note that the pupillary light reflex can be unreliable as the striated pupillary muscle enables the patient to exert voluntary control over the pupillary response. The location of opacities within the anterior refractive apparatus can be assessed by observing the position, shape and migration of the so-called Purkinje-Sanson images while swinging the light source from side to side.
- **Lateral examination with lateral illumination**: the light source is directed at the temporal lim-

bus (Figure 15.46) and the examiner's line of sight is oriented parallel to the light beam (i.e., along the lens–iris diaphragm). This permits evaluation of the shape, depth and contents of the anterior chamber, examination of the iris and detection of phacodonesis (abnormal vibration of the lens during spontaneous oscillatory eye movements, caused by varying degrees of lens luxation [disruption of the suspensory apparatus of the lens]).

- **Examination with retroillumination under strongly dimmed light or in total darkness**: this technique is used to identify opacities in the refractive apparatus, including the vitreous body, by passing light through the pupil and observing the light reflected by the fundus (fundus reflex). Note that transmitted and reflected light must be aligned along the same axis.

Specialised ophthalmic examination

Specialised procedures in avian ophthalmic examination include:

- slit lamp biomicroscopy,
- Schirmer tear test (STT) or phenol red test (PRT),
- tonometry,
- gonioscopy,
- ophthalmoscopy,
- electroretinography,
- fluorescein angiography,
- imaging (radiography, sonography [inc. 3D], optical coherence tomography [OCT]),
- scanning digital ophthalmoscopy [SDO]).

Slit lamp biomicroscopy (Figure 15.48) is a technique in which light focused into a sheet (slit) is combined with a source of magnification. It permits examination of the **anterior refractive apparatus** (cornea, anterior chamber and lens) in 'optical section'.

Using the slit lamp biomicroscope, the **Purkinje-Sanson images** (reflections on the cornea and the anterior and posterior lens capsule) can be used to determine the position of foreign bodies and intraocular opacities. With optical modifications, the slit lamp biomicroscope can also be used to examine the iridocorneal angle, the vitreous body and the fundus.

The **Schirmer tear test (STT)** is used to evaluate the function of the **tear-producing glands**. It is performed by placing the tip of a standardised strip of **absorbent paper** (2mm wide for small birds and owls, 3mm wide for psittacines, 5mm wide for raptors; may incorporate a colour indicator and scale) into the temporal third of the lower conjunctival sac. The distance over which the paper becomes moistened is read after exactly one minute.

In birds a distinction is made between:

- STT I: performed without prior anaesthesia of the cornea; assesses the response to the presence of a foreign body (paper),
- STT II: performed after application of topical anaesthesia; assesses baseline tear production and
- STT III: performed under general anaesthesia (to avoid excessive stress without use of topical anaesthesia).

Depending on species, normal values range from <1mm to 12mm. The Phenol Red Thread test is an alternative to the STT, offering certain advantages in birds.

Tonometry (Figure 15.49) is used to identify pathological changes in **intraocular pressure (IOP)** (as seen, for example, with glaucoma). It is usually performed using a fast-acting electronic applanation tonometer (Tonopen XL®). The main advantage of the Tonopen® is that it can be used independently of head position, thus IOP can be determined quickly and with a minimum of stress. Particularly useful in birds are the TonoVet® or TonoLab®, as these are suitable for very small corneal diameters. The Schiotz tonometer, which functions on the principle of indentation (versus applanation), is a considerably cheaper device, but is only suitable for use in the very largest birds. Normal intraocular pressure in birds ranges from 9mmHg to 22mmHg, according to species.

Gonioscopy (Gk. gonía = knee, angle; skopein = to look) is utilised to examine the **iridocorneal angle**. It is performed using either the **direct** technique (without a contact lens, particularly in birds with a strongly curved cornea, e.g., raptors) or the **indirect** technique (with a contact lens) (Figure 15.13). Gonioscopy permits assessment of the size of the filtration angle and the morphology of

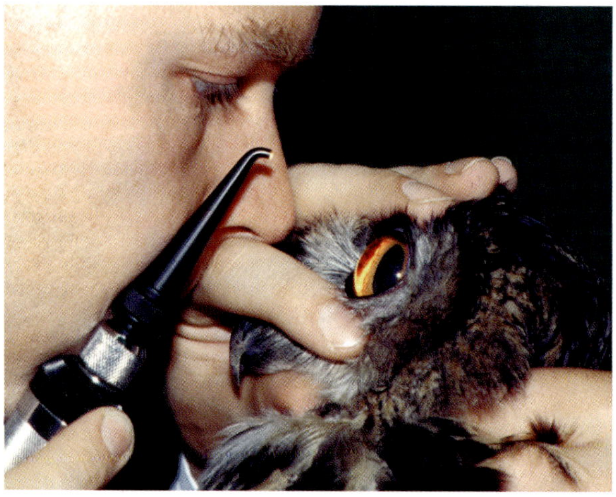

15.51 Examination of the posterior segment of the eye of a Eurasian eagle-owl (Bubo bubo) using monocular direct ophthalmoscopy.

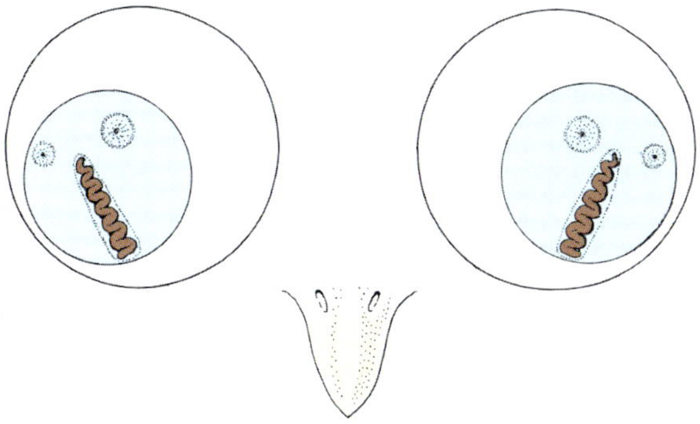

15.52 Field of view (schematic) of the avian fundus using monocular direct ophthalmoscopy (blue circles). Includes the pecten and the central and temporal foveae.

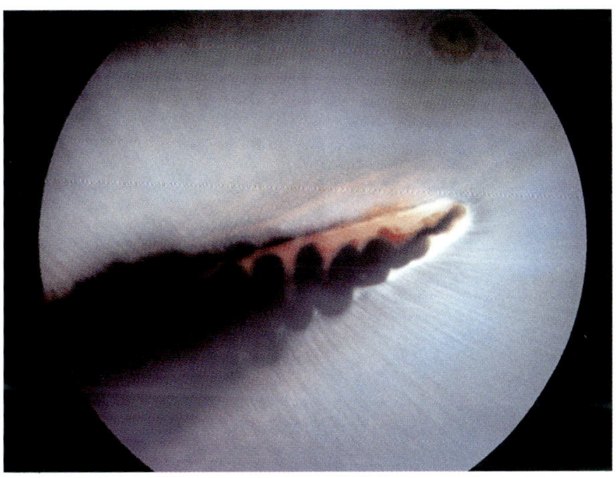

15.53 Typical ophthalmoscopic view of the fundus of a diurnal bird (in this case a bald eagle, Haliaeetus leucocephalus), including fovea centralis.

the pectinate ligament between the base of the iris and the cornea. This can be used to establish the aetiological basis of glaucoma, whereas tonometry merely identifies one of the clinical manifestations of this disease (increased intraocular pressure). Since the iridocorneal angle of birds is very open, primary glaucoma is rare.

Ophthalmoscopy (Figures 15.50 to 15.53) (Gk. ophthalmos = eye; skopein = to look) permits examination of the fundus (internal ocular structures located posterior to the lens). The fundus is illuminated with a beam of light and the reflected light is viewed by the examiner (Figure 15.50).

Types of ophthalmoscopy used in birds include:

- direct monocular ophthalmoscopy,
- indirect monocular ophthalmoscopy,
- indirect binocular ophthalmoscopy.

Direct monocular ophthalmoscopy requires less complex equipment than indirect ophthalmoscopy and permits very detailed examination of the fundus. The image of the fundus is magnified five to seven times by the patient lens, and is upright and true-sided (not reversed).

A disadvantage of direct ophthalmoscopy is that it offers a relatively narrow field of view. This makes it difficult to examine the periphery of the fundus, increasingly so with decreasing pupillary diameter (Figure 15.52). A wider field of view is achieved using an indirect ophthalmoscope. The greater the refractive index of the ophthalmoscopic lens, the greater the field of view. Thus, as the pupillary diameter decreases, the required power of the lens increases.

The direct procedure can be performed using a **focused source of light** such as a transilluminator or **otoscope light** (depending on the size of the bird) (Figure 15.51). In order to align the light reflected from the fundus with the examiner's line of sight, the light source is positioned directly under the axis passing between the pupils of the examiner and the patient.

Beginning approximately 40cm from the patient, the transilluminator is positioned such that the examiner can look just over the top of the instrument. After locating the fundus reflex (reflection from the fundus), the examiner gradually brings their eye nearer to the patient (to a distance of around 3cm) without accommodating. As the distance between the examiner and patient decreases, the visible region of the fundus becomes larger.

Technical proficiency is particularly important when performing this procedure in birds, because of the relatively small diameter of the pupil. Ideally, a pupillary of diameter of at least 9mm is required. Depending on operator experience, the lower limit can be as little as 5mm. To avoid bite injuries (due to the proximity of the examiner's nose to the beak), it is advisable for the examiner to use their right eye to inspect the patient's right eye, and their left eye for the patient's left eye.

Direct ophthalmoscopy can also be performed using a monocular direct ophthalmoscope. This instrument

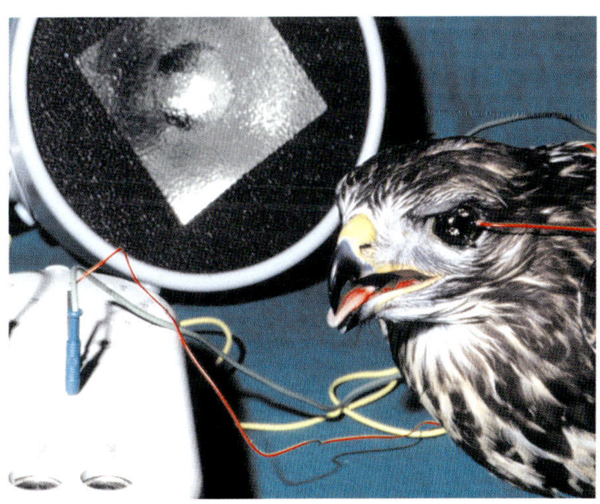

15.54 Electroretinography in a common buzzard (Buteo buteo): headbox, light source and patient with corneal electrodes in place.

combines a focal light source with a viewing system, such that the examiner's eye is positioned correctly to view the reflected light. However, as this technique is associated with a particularly small viewing angle (approximately 5 degrees), the instrument must be repeatedly repositioned, in very small increments, in order to view the entire fundus. For this reason, it is preferable to use either a transilluminator, as described above, or indirect ophthalmoscopy.

Indirect binocular ophthalmoscopy is performed using a head-mounted ophthalmoscope and a hand-held lens (30, 40, 60, 78 or 90 diopter). It is suitable for all avian species, including those with a small pupillary diameter (to c. 2mm), and is therefore the ophthalmoscopic method of choice in birds. Other advantages of indirect ophthalmoscopy include superior visualisation of the periphery of the fundus (as far as the ora serrata in some species, with partial visibility of the ciliary body in birds with tubular eyes), and the ability to share the image with other observers using a mirror or a video camera. The drawbacks of indirect ophthalmoscopy include the complexity of the equipment and the level of skill required to perform the procedure with competence.

Lens selection for indirect ophthalmoscopy is based on the size of the bird's eye and the diameter of the pupil:

- raptors: 30–40 diopter,
- Amazon parrots: 78 diopter,
- budgerigars: 90 diopter.

Lenses with high refractive power increase the angular field, permitting visualisation of a larger area of the fundus. The observed image is inverted, reversed and reduced in scale. For this reason, and because of the need to avoid the formation of reflections on the lens, a considerable amount of practice is required to become skilled in this technique.

Some of these disadvantages can be overcome by the use of a relatively expensive **indirect monocular ophthalmoscope**, which has an inbuilt system of mirrors and prisms that produces an upright non-reversed image. This is particularly useful for examining birds with small pupils (e.g., parrots).

Fundus cameras, which are optically analogous to indirect ophthalmoscopy, can be used to **document the findings of a fundus examination**.

Mydriasis (dilation of the pupil) is a prerequisite for technically competent ophthalmoscopy. However, as the internal ocular musculature of birds is predominantly striated, mydriatics routinely used in mammalian ophthalmology (e.g., the parasympatholytic atropine) are ineffective.

For diagnostic purposes (ophthalmoscopy), mydriasis can be achieved using anaesthesia. Inhalation anaesthesia using isoflurane is suitable for short examinations, whereas air sac perfusion anaesthesia (APA) is preferable for longer procedures. In owls and some raptors, transient stress-induced mydriasis following capture can be utilised for ophthalmoscopic examination.

An alternative means of inducing **mydriasis is the introduction of curare derivatives** into the anterior chamber. This results in paralysis of the intraocular musculature for up to 24 hours (depending on species). In diurnal species, mydriasis may commence within 15–20 seconds of administration. This interval is considerably longer in owls due to the relatively high viscosity of the aqueous in these birds, resulting in slower distribution of the injectate.

The drug of choice for inducing prolonged mydriasis is a **0.3 per cent solution of d-tubocurarine**. This may need to be compounded as required, as the benzalkonium chloride in some commercial preparations can induce phacogenic uveitis. The d-tubocurarine solution (0.01 to 0.03ml) is introduced directly into the anterior chamber by intracameral injection, performed at the temporal limbus at the transition between the cornea and the pigment band (Figures 15.7, 15.12 and 15.20) using a 27–30 gauge needle.

Risks associated with this **invasive procedure** include intraocular trauma and increased intraocular pressure (IOP), which can lead to glaucoma in predisposed patients. It may also result in the introduction of conjunctival micro-organisms into the anterior chamber. Excessive doses can

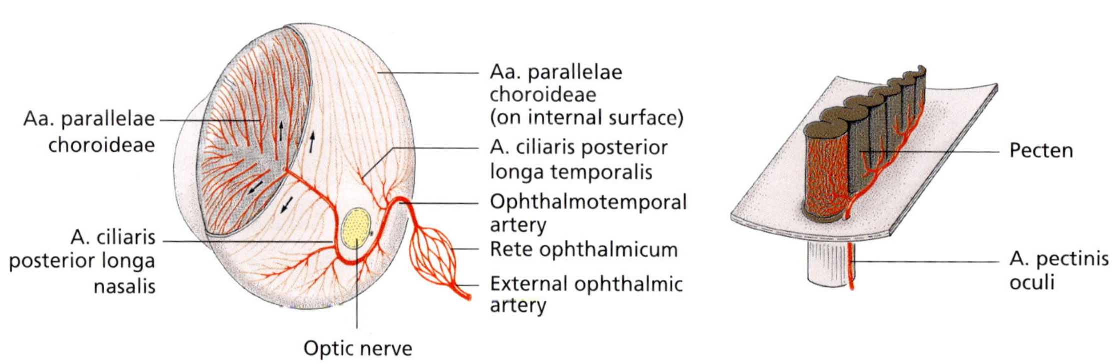

15.55 Arterial supply (schematic) of the right eye (left) and the pecten (right). Arrows indicate the direction of blood flow in the aa. parallelae choroideae.

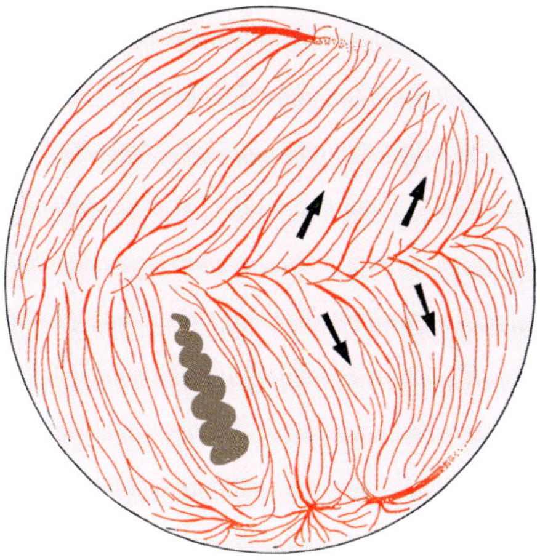

15.56 Blood flow in the avian fundus (schematic; arrows indicate direction of flow). In contrast to mammals, the aa. parallelae choroideae extend dorsally and ventrally into the fundus from a horizontal (equatorial) origin.

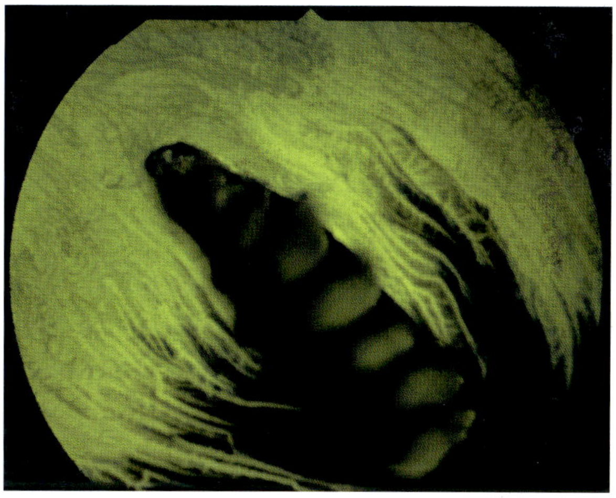

15.57 Fluorescein angiogram of part of the choroidal vasculature of a great horned owl (Bubo virginanus) 14 seconds after intravenous administration of sodium fluorescein (40mg/kg) under isoflurane inhalation anaesthesia.

15.58 Jet-like ejection of sodium fluorescein by the pecten into the vitreous chamber (schematic). This phenomenon is only observable in conscious animals.

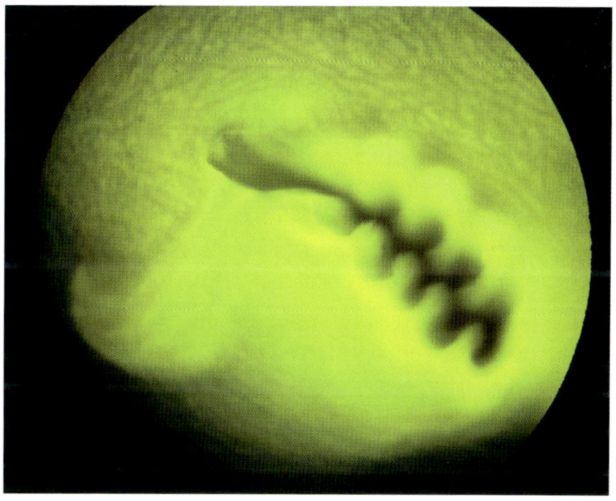

15.59 Dorsotemporally directed diffusion of sodium fluorescein following jet-like ejection of dye from the pecten into the vitreous body in a tawny owl (Strix aluco).

cause paralysis of the respiratory musculature which, in the worst case, can result in death.

The importance of these risks, particularly if the procedure is carried out inexpertly, cannot be overstated. Intracameral injection of curare derivatives should therefore not be used in conjunction with diagnostic procedures. Rather, it should be reserved for situations in which protracted mydriasis is required for prophylactic or therapeutic reasons (e.g., relief of pain and prevention of anterior and posterior synechiae as a result of uveitis-induced muscular spasm; patient monitoring after fluorescein angiography).

Electroretinography (ERG) (Figure 15.54) is an electrophysiological technique used to record electrical potentials (waveforms) generated by the retina in response to a standardised stimulus (light flash). An electroretinogram only assesses the **functional integrity of the retina** and cannot be used to assess the patient's vision. Electroretinography is typically indicated for evaluation of retinal function when opacities in the refractive media interfere with visual inspection of the fundus (e.g., prior to cataract surgery).

Fluorescein angiography (FAG) (Figures 15.57 to 15.59) involves the use of a fundus camera to document (approximately one image per second) the distribution of **intravenously administered fluorescent dye** (sodium fluorescein, 40mg/kg) in the vasculature of the fundus. A choroidal and pecten phase can be distinguished.

In the **choroidal phase** the vessels in the dorsal fundus fill in a ventrodorsal direction, while those in the ventral

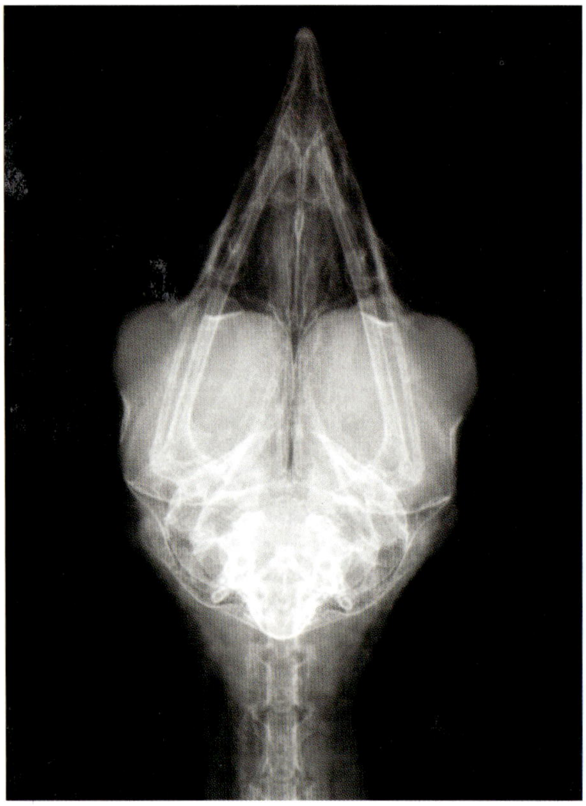

15.60 Radiograph (ventrodorsal view) of the skull and eyes of a common buzzard (Buteo buteo). Notable features include the relative diameter of the cornea and fundus, the size of the bulbi (almost in direct contact in the midline, separated only by a thin interorbital septum) and the osseous scleral rings.

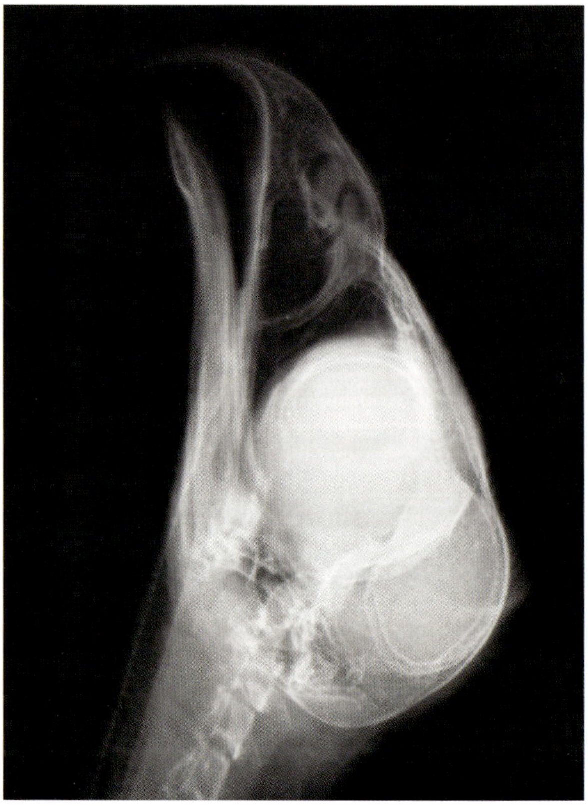

15.61 Radiograph (lateral view) of the skull and eyes of a common buzzard (Buteo buteo).

fundus are filled from dorsal to ventral. This contrasts with the centripetal filling of vessels in the mammalian fundus, and reflects the projection of the aa. parallelae choroideae into the dorsal and ventral fundus from the horizontally positioned a. ciliaris posterior longa nasalis (arising from the ophthalmotemporal artery) (Figures 15.55 and 15.56).

In the **pecten phase**, the dye diffuses from the pecten oculi into the vitreous body (Figure 15.30). Video-fluorescein angiography in owls reveals an additional rhythmic (8–18/min) jet-like ejection of dye into the vitreous body (Figures 15.58 and 15.59). Barely appreciable oscillatory eye movements rapidly propel the dye to the periphery of the fundus. Observations made using FAG have contributed substantially to the understanding of the function of the pecten, particularly its role in nourishing the avascular retina. This technique is also useful in diagnosing the following ocular abnormalities:

- small choroidal haemorrhages,
- haemorrhage associated with the pecten,
- atrophy of vessels and the retinal pigment epithelium,
- neoplastic lesions in the fundus,
- retinal detachment and other pathology of the fundus.

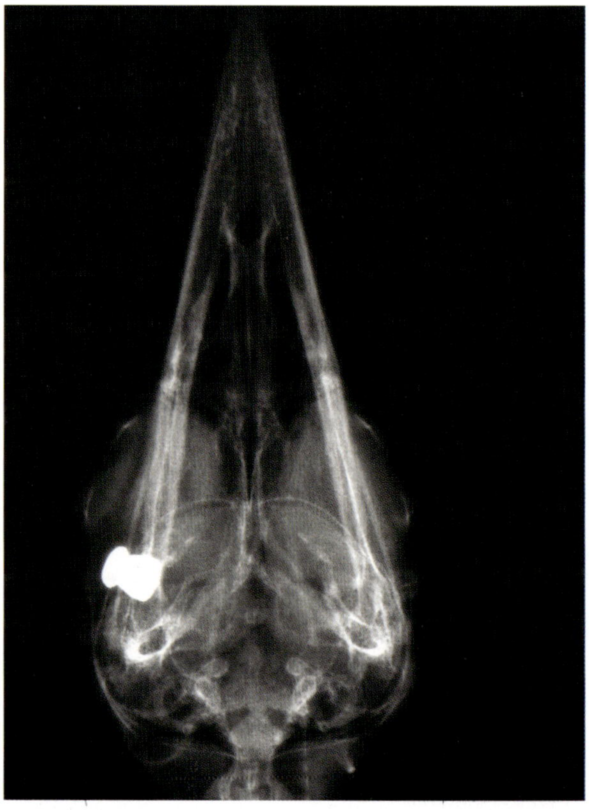

15.62 Radiograph (ventrodorsal view) of a carrion crow (Corvus corone) with a gunshot injury. An air gun pellet is visible in the temporal region of the orbit.

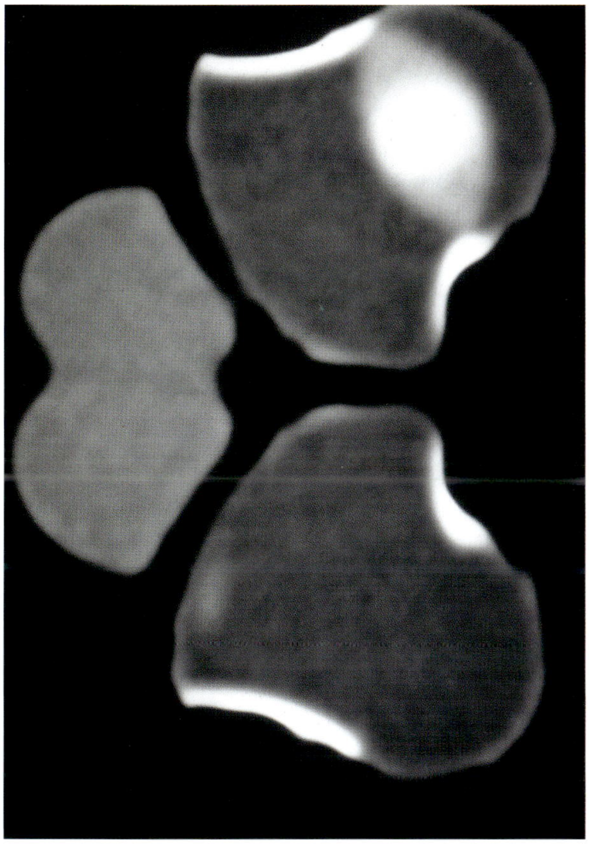

15.63 Computed tomographic image of a barn owl (Tyto alba) with aphakia (absence of the lens) in the right eye due to phacolysis (lens resorption induced by inflammation).

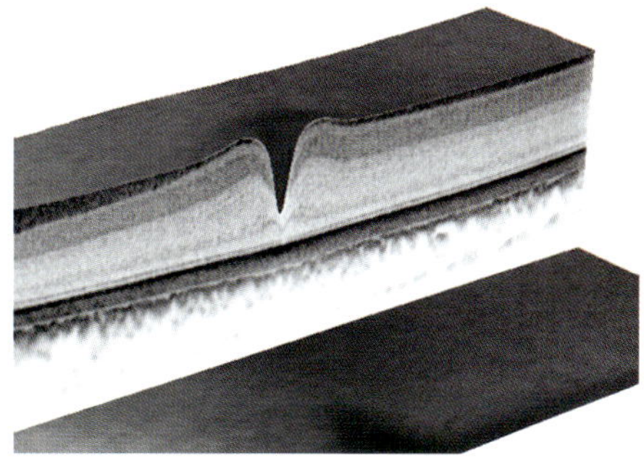

15.65 3D Optical Coherence Tomography (OCT) reconstruction of the retina and fovea centralis in a common buzzard (Buteo buteo, left eye) illustrating the width and depth of the central fovea under normal conditions.

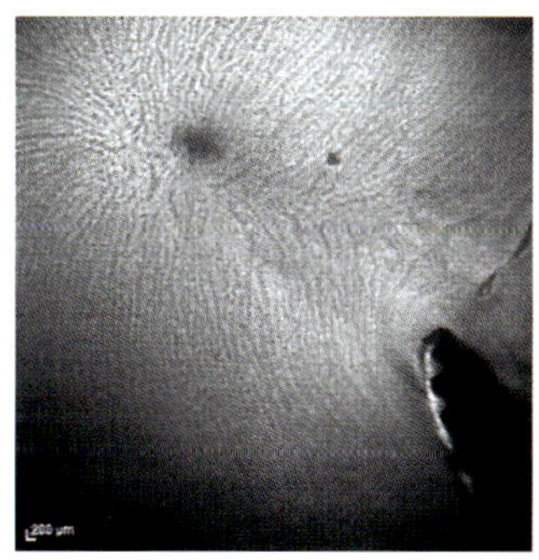

15.66 Optical coherence tomograph of the fundus oculi of the common buzzard (Buteo buteo, right eye) with the upper and lower vessel cascade, fovea centralis, pecten oculi (5 o'clock position) and retinal scar formation (central and 3 o'clock position).

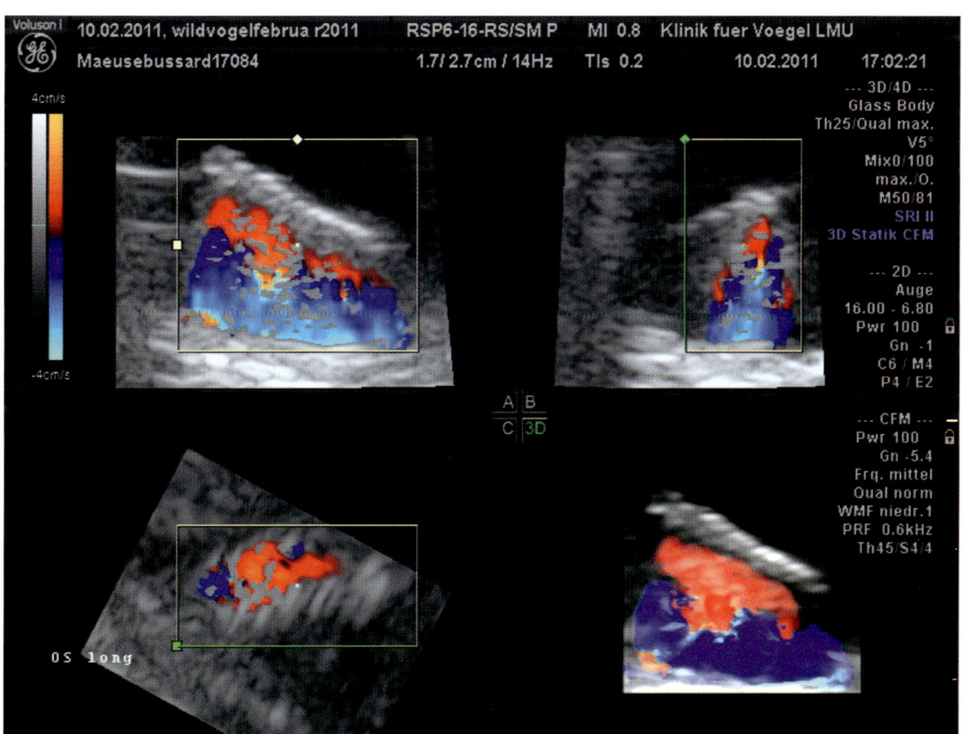

15.64 3D Doppler flow ultrasonography of the pecten oculi of the common buzzard (buteo buteo, right eye).

Imaging is also used in ophthalmic examination. Radiography (Figures 15.60 to 15.63) and, in particular, sonography are valuable diagnostic tools (see Chapter 19 'Diagnostic imaging'). Radiographic findings specific to birds (and reptiles) include fractures of the osseous scleral ring.

The most recent major advances in imaging in avian ophthalmology include Scanning Digital Ophthalmoscopy (SDO), 3D ultrasonography (Figure 15.64) and Optical Coherence Tomography (OCT) (Figures 15.65 and 15.66). The latter is a laser light based technique allowing visualisation of ocular structures of the anterior and posterior segment of the eye at almost histological resolution in living birds.

The ear (organum vestibulocochleare)

16

H. E. König, G. Weissengruber, I. Walter and R. Korbel

Birds are endowed with highly developed senses of balance and hearing. While the former assists with flying and swimming, the latter plays an important role in various functions, such as courtship, recognition of individual birds and conspecifics, and communication with young.

As in mammals, the vestibulocochlear organ, or ear, consists of the following divisions:

- external ear (auris externa),
- middle ear (auris media),
- internal ear (auris interna).

The vestibular organ, responsible for balance, is located entirely within the internal ear. The osseous components of the vestibulocochlear organ are depicted in Figure 16.1.

External ear (auris externa)

The external ear of birds comprises the external acoustic meatus and the tympanic membrane. There is no pinna.

In the chicken, the **auditory aperture** (apertura auris externae), or external ear opening, is 4–5mm in diameter and is situated above a **red** or **white ear lobe**. The ear opening is covered in most birds by modified contour feathers (auricular feathers; pennae auriculares) arranged in concentric rows on an annular fold (Figures 16.4 and 16.5). In the long-eared owl (Asio otis), the ears are covered by a rostral **skin fold** (operculum auris) (Figures 16.2 and 16.3) containing striated muscle that enables the fold to be moved to assist with localisation of sound. Examination of the auditory aperture is an important component of ophthalmic examination in birds (see Chapter 18 'Clinical examination').

The **external acoustic meatus** (meatus acusticus externus) lies caudal to the quadrate bone. It is approximately 4–7mm long. A small mound in the ventral wall contains an opening for the underlying **auricular glands** (glandulae auriculares). The opening of the duct draining these glands can be visualised under magnification.

The **tympanic membrane** (membrana tympanica) bulges slightly into the external ear canal. In the chicken it is 0.012mm thick with a surface area of approximately 25mm^2.

The tympanic membrane is attached to the surrounding bone, its margins reinforced by elastic fibres.

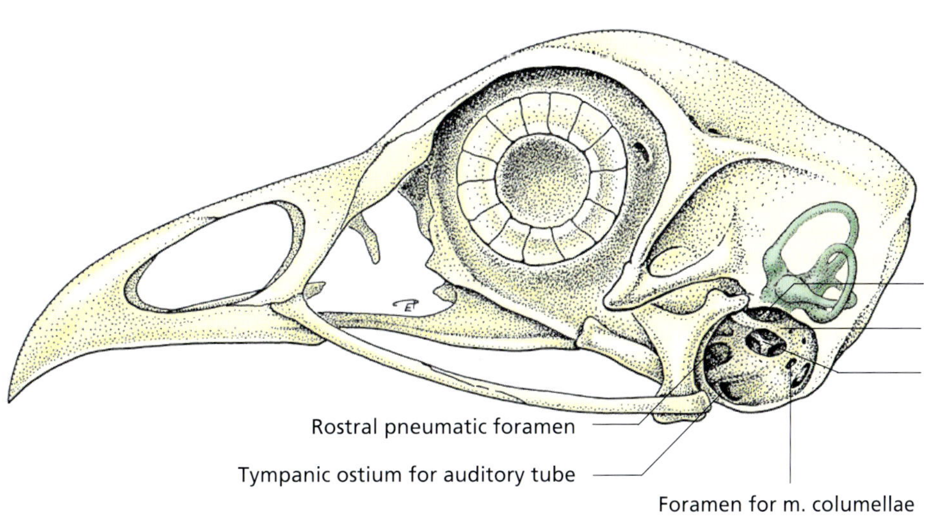

16.1 Osseous components of the vestibulocochlear organ in the chicken (schematic).

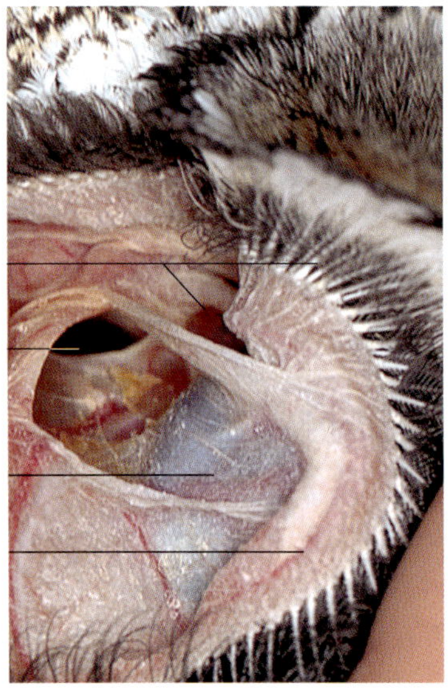

16.2 Ear opening of a long eared owl (Asio otus).

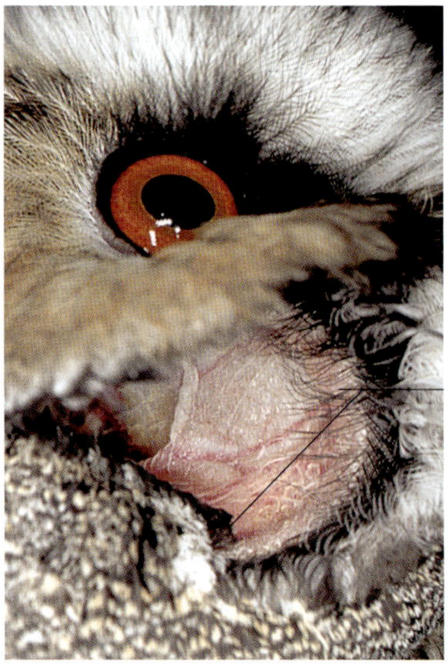

16.3 Ear opening of a long eared owl (Asio otus).

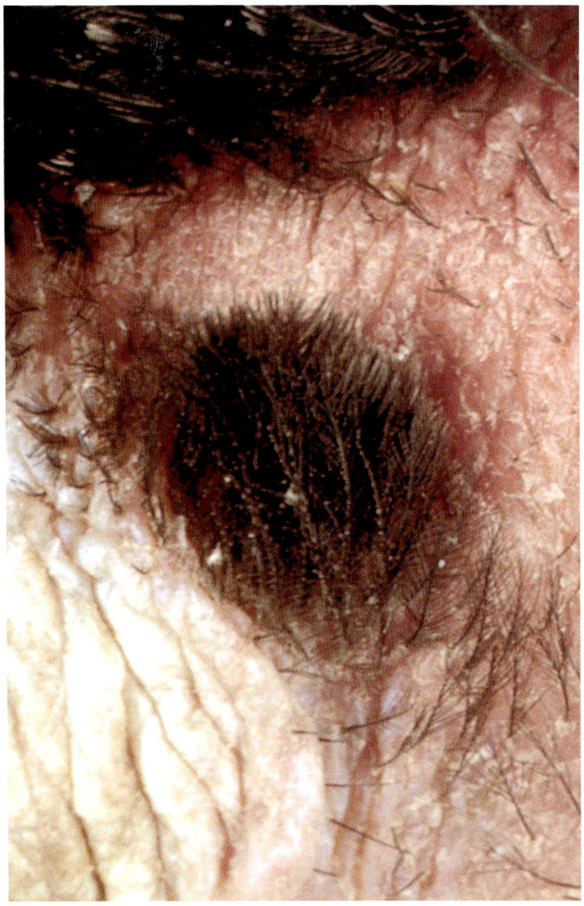

16.4 Ear opening of a chicken (closed), Courtesy of PD Dr S. Reese, Munich.

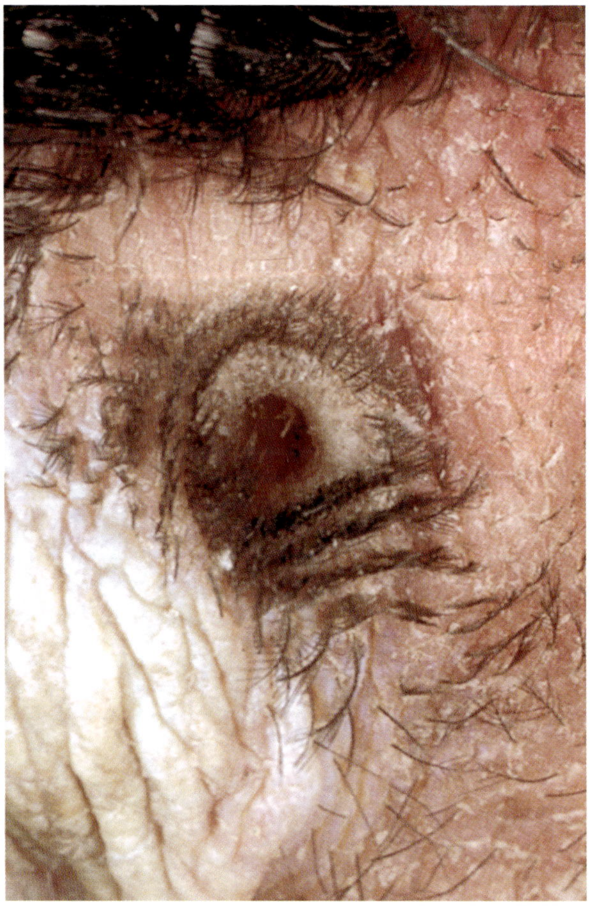

16.5 Ear opening of a chicken (open), Courtesy of PD Dr S. Reese, Munich.

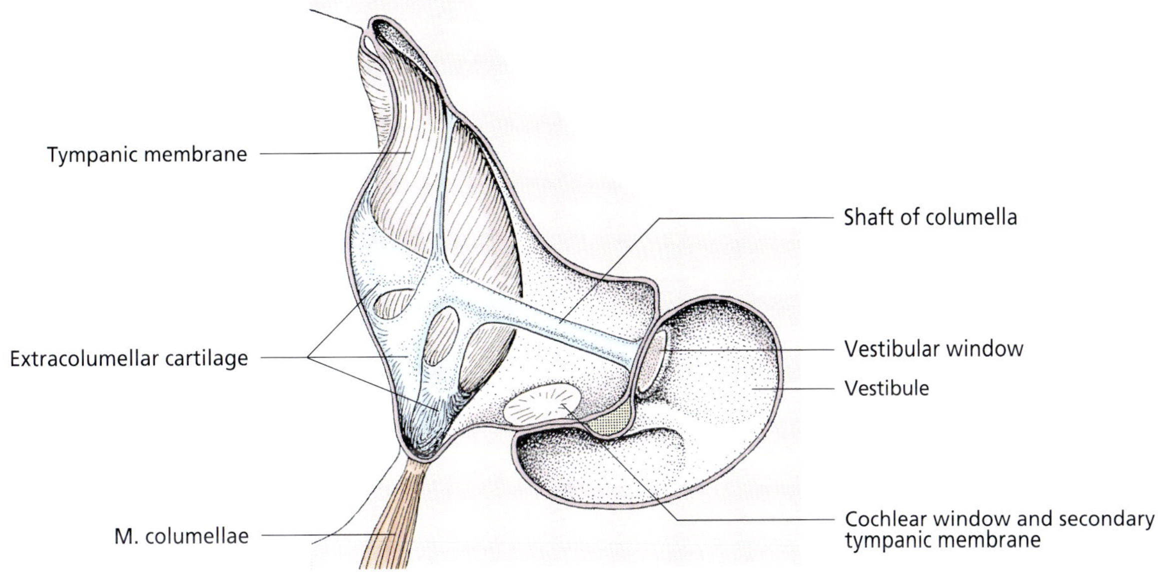

16.6 Section of the left middle ear of a chicken (schematic), adapted from Schwarze and Schröder, 1979.

Rostroventrally it incorporates a small air-filled cavity, the **sinus pneumaticus marginalis**. The tympanic membrane is tensed by the **m. columellae**. This muscle is equivalent to the m. stapedius in mammals and is innervated by the facial nerve. The internal surface of the tympanic membrane is attached to the cartilaginous processes of the **columella** (Figure 16.6).

Middle ear (auris media)

The middle ear contains the air-filled **tympanic cavity** (cavitas tympanica), which is connected to the oropharynx by the **auditory tube** (tuba auditiva *syn.* pharyngotympanica), via the infundibular cleft.

The **sole auditory ossicle** in the bird, the **columella**, spans the tympanic cavity. Three movably interconnected **cartilaginous extracolumellar processes** project from the columella to the tympanic membrane as the cartilago extracolumellaris (Figure 16.6). Resembling a tripod, the cartilago extracolumellaris is positioned caudally, ventrally and rostrally against the tympanic membrane. The **shaft (scapus) of the columella** expands proximally to form the **basis columellae**, which closes the vestibular window.

Lateral to the opening of the auditory tube, the tympanic cavity houses a small curved vesicular structure, the **organum paratympanicum**. Lying parallel to the longitudinal axis of the head, the organum paratympanicum contains ciliated sensory cells and is presumed to register **changes in air pressure**. The left and right tympanic cavities are connected by air-filled spaces.

In birds, the **vestibular window** (fenestra vestibularis) is located very close to the **cochlear window** (fenestra cochlearis) (Figure 16.6). Pressure from sound waves striking the tympanic membrane is transmitted by the base of the columella to the perilymph within the inner ear. Compression of the perilymph results in bulging of the **secondary tympanic membrane** (membrana tympanica secundaria) at the cochlear window.

Internal ear (auris interna)

The internal ear contains both the vestibular organ and the components of the cochlear organ responsible for the transduction of sound waves. As in mammals, the inner ear consists of the **osseous labyrinth** and, within it, the **membranous labyrinth**. The space between the membranous and osseous labyrinth is filled with **perilymph**, while the cavity within the membranous labyrinth contains the somewhat viscous **endolymph**.

The **osseous labyrinth** (labyrinthus osseus) is comprised of the central **vestibule** (vestibulum), the ventrally positioned osseous **cochlea** and the caudodorsally projecting **semi-circular canals** (canales semi-circulares ossei) (Figures 16.9 and 16.13). In contrast to the spiral cochlea of mammals, the avian cochlea is shaped like a blunt, slightly medially concave and rostrally convex club (Figures 16.6 and 16.7).

The **semi-circular canals** are arranged **perpendicular to one another**. According to their plane of orientation, they are termed the rostral vertical (anterior), caudal vertical (posterior) and lateral horizontal (lateral) canals. An **ampulla** is located at one end of each canal (Figures 16.9 and 16.10).

The **membranous labyrinth** (labyrinthus membranaceus) (Figure 16.7) is an extensively partitioned replica of the osseous labyrinth. As indicated above, it is surrounded by **perilymph** and filled with **endolymph**. The central section of the membranous labyrinth contains the larger, dorsally positioned tubular chamber known as the

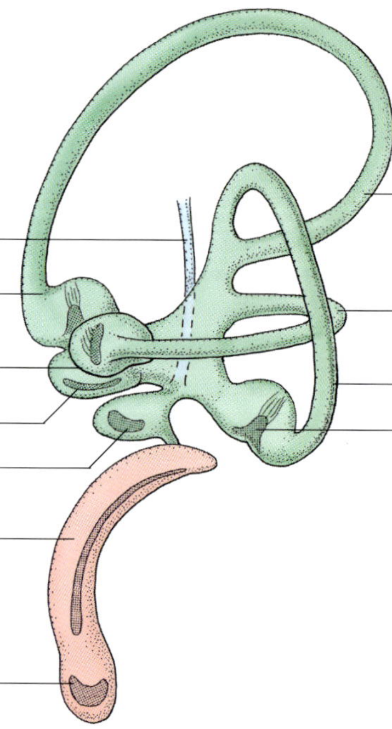

16.7 Membranous labyrinth of the internal ear of the chicken (schematic), adapted from Salomon, 1993.

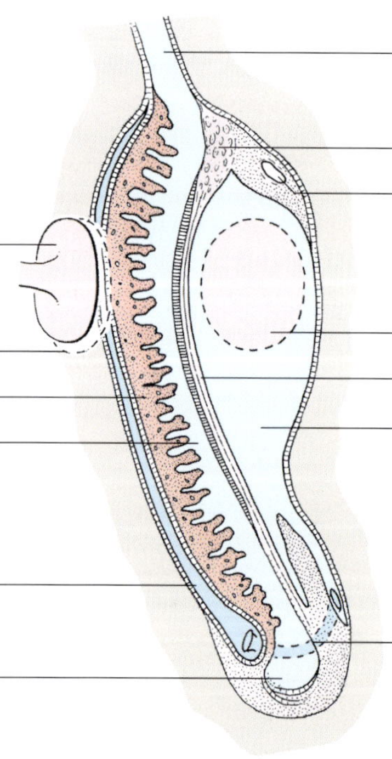

16.8 Internal ear of the chicken (schematic), adapted from Evans, 1982.

utriculus, as well as the smaller, ventrally situated **sacculus.** A small tube, the **ductus utriculosaccularis**, connects the two chambers. The narrow ductus endolymphaticus (Figure 16.7) projects from the sacculus, ending blindly between the meninges within the cranial cavity.

The **utriculus** gives rise to the semi-circular ducts, each bearing an expanded ampulla at one end. Considerably smaller in diameter than the semi-circular canals in which they lie, the semi-circular ducts (Figure 16.7) occupy a slightly eccentric position within the canals. They are attached to the periosteum of the canals by delicate fibres that pass through the perilymph.

The **sensory cells of the vestibular apparatus** are grouped to form the **cristae ampullares** within the ampullae, the **macula utriculus** and **crista neglecta** within the utriculus, and the **macula sacculi** in the sacculus.

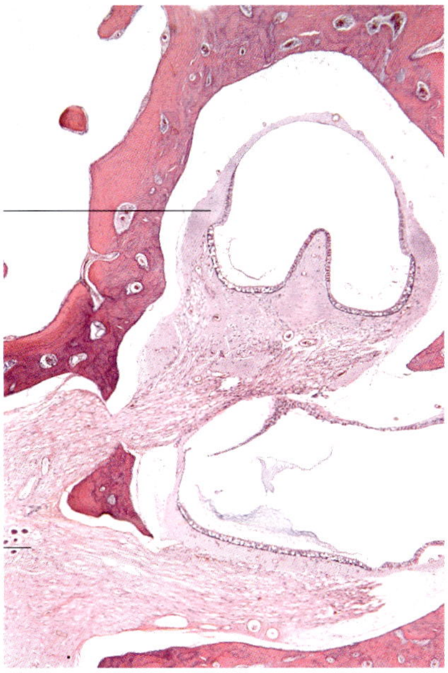

16.9 Histological section of the internal ear of a chicken with semicircular canal and duct.

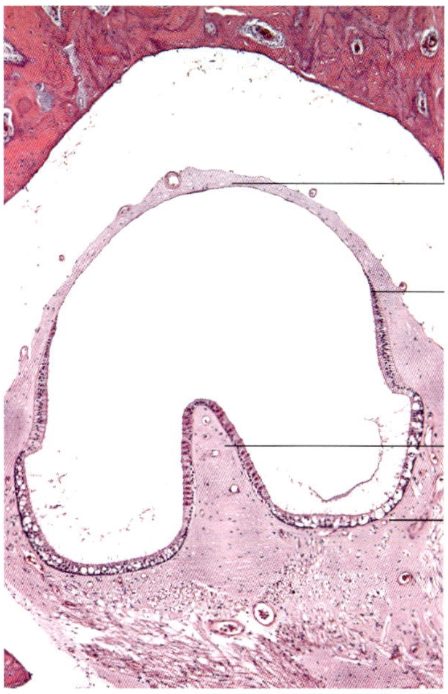

16.10 Histological section of a semicircular canal and duct in the internal ear of a chicken.

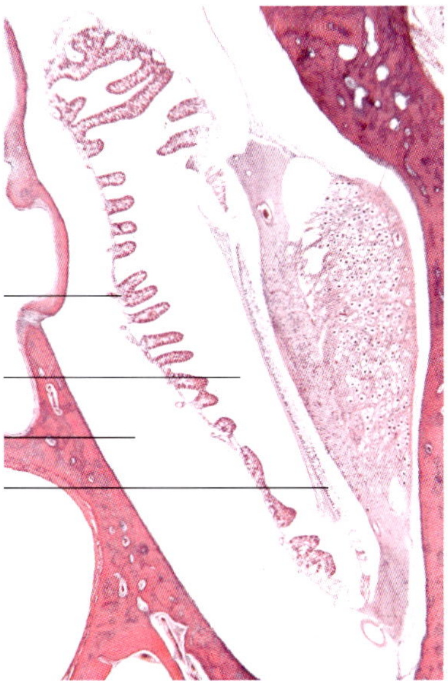

16.11 Histological section of the cochlea in the internal ear of a chicken.

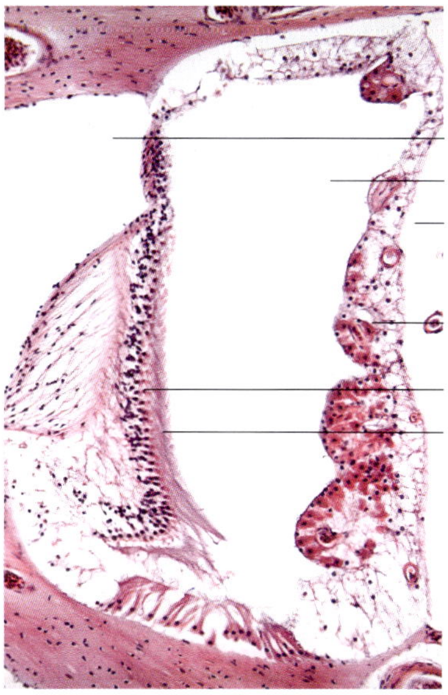

16.12 Histological section of the cochlea in the internal ear of a chicken.

The **hair cells** of the macula utriculi and the macula sacculi are in contact with **statoconia**. In contrast, the cristae ampullares and the crista neglecta are covered in a dome-shaped membrane **devoid of statoconia**. Movement of endolymph, induced by inertia, stimulates the hair cells, which are connected with fibres of the **vestibular ganglion** and the **vestibular nerve**.

The tubular **cochlear duct** (ductus cochlearis) (Figures 16.7, 16.8, 16.11 and 16.12) emerges ventrally from the sacculus. In the chicken it measures approximately 6mm in length. Its blind end, the **lagena**, reaches the tip of the cochlea.

Lying within the centre of the cochlea, the cochlear duct is framed by rostral and caudal cartilaginous ledges. This arrangement gives rise to two channels, the narrow **scala vestibuli** and the **scala tympani**. The two scala communicate at the apex of the cochlea via the **apical interscalar canal** (canalis interscalaris apicalis), the equivalent of the **helicotrema** in mammals.

The cochlear duct is separated from the scala vestibuli by the thick, folded **tegmentum vasculosum** (vestibular membrane or Reissner's membrane in mammals) (Figures 16.8, 16.11 and 16.12).

The **basilar membrane** (membrana basilaris), separates

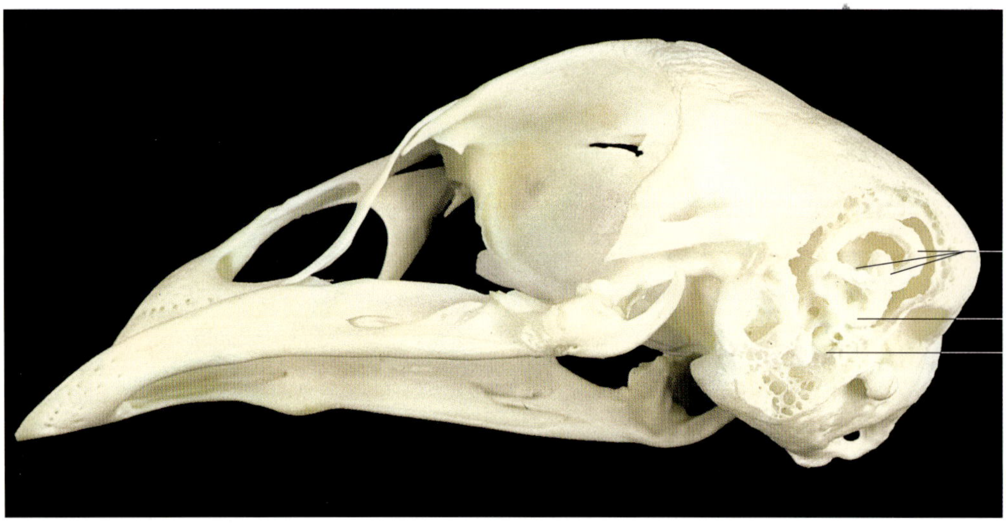

16.13 Osseous components of the vestibulocochlear organ of a chicken (surrounding spongy bone removed).

the cochlear duct and the scala tympani. Resting upon the basilar membrane, the **papilla basilaris** carries the **sensory cells** of the cochlea. These cells have stereocilia that are anchored in an overlying gelatinous layer, the tectorial membrane (membrana tectoria). Bipolar neurones of the cochlear ganglion form a delicate network of nerve fibres at the base of the sensory cells.

A further group of **sensory cells** is located within the **lagena**, forming the **macula lagenae**. Sensory hairs projecting from these cells are in contact with **otoconia**. The function of the macula lagenae has not been established.

Auditory sensitivity in birds is greatest between 1,000Hz and 6,000Hz.

Clinical aspects

Diseases of the ear tend to be inconspicuous in birds, largely because – in contrast to mammals – diurnal birds typically rely primarily on the visual sense for orientation. Furthermore, the use of hearing in birds has received relatively little attention, compared with the role of vision.

When examining the ears, it is important to be aware that the ear opening can be more difficult to locate in some birds than others due to species-related anatomical variation (Figures 16.2, 16.3 and 16.14). In owls, for example, the periauricular feathers serve as a mobile funnel that can be directed towards the source of sound through fine positional adjustments, brought about by underlying muscles.

Routine examination of the auditory aperture may reveal intraocular bleeding resulting from traumatic injury, for which there may be little external evidence (see Chapter 15 'The eye'). The ear openings are also common sites of infestation by parasites such as mites and ticks. Inflammatory aural discharge may be the result of pseudomonas infection. Ear diseases **affecting the vestibular organ** are frequently associated with **central nervous system deficits** (see Chapter 13 'Nervous system').

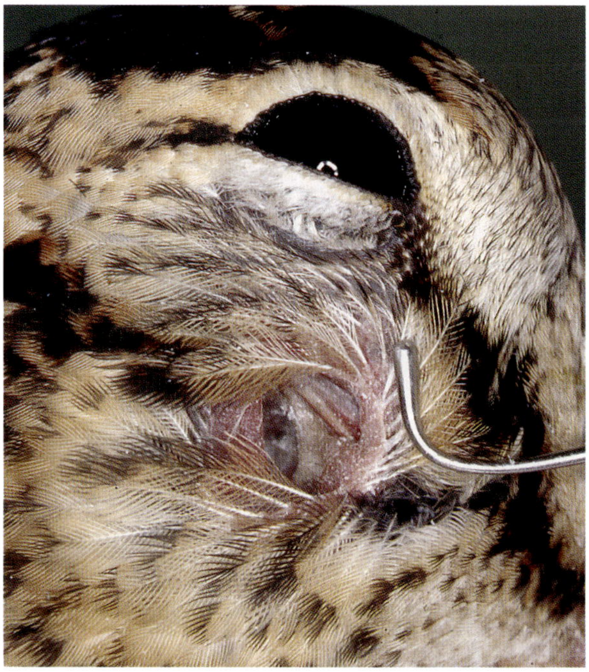

16.14 Ear opening of a Eurasian woodcock (Scolopax rusticola), located cranioventral to the eye.

17

Common integument (integumentum commune)

H. E. König, S. Reese, C. Mülling and R. Korbel

The majority of the avian body is covered with **feathers** (pennae). Originating from the epidermis, these structures are a characteristic and unique feature of the class Aves (Figure 17.1). The avian feather coat performs many of the functions for which hairy skin is responsible in mammals, including:

- provision of a barrier against irradiation and mechanical, thermal, chemical and biological influences,
- thermoregulation and
- communication.

In addition, the feathers are fundamental in enabling flight.

Where **feathers are present**, the **skin** is relatively thin. Feathered skin consists of an epidermis and dermis (corium) underlain by the subcutis (hypodermis) (Figure 17.2).

While the structure of the avian **epidermis** is homologous with that of mammals (Figure 17.3), differences are apparent in the layers of the **dermis**. In birds, the dermis consists of:

- a superficial layer (stratum superficiale) and
- a deep layer (stratum profundum) divided into:
  - the stratum compactum and
  - the stratum laxum.

The **superficial layer** (stratum superficiale) is composed of loose connective tissue. Discrete dermal papillae are only present in association with feather follicles. The **stratum compactum** of the deep layer contains dense connective tissue that gives the dermis its mechanical strength. Smooth muscle cells in the **stratum laxum** serve to tense the skin and connect the feather follicles, thus contributing to the movement of the feathers.

The **subcutis** serves as a mobile layer of tissue connecting the skin to the underlying structures. It contains adipose tissue concentrated into localised **fat bodies** (corpora adiposa). In chickens and turkeys, a **bursa** (bursa sterni) is located within the dermis over the cranial aspect

17.1 Black crowned crane (Balearica pavonina) with a crown of ornamentally modified coverts.

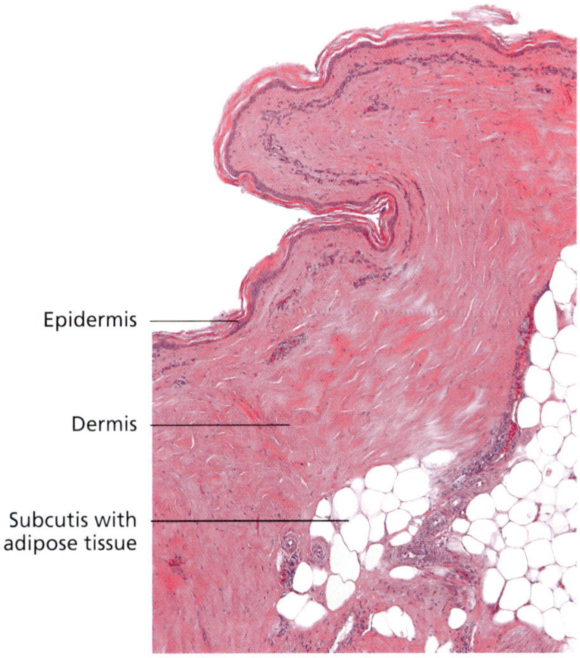

17.2 Histological section of the skin of the thigh of a chicken.

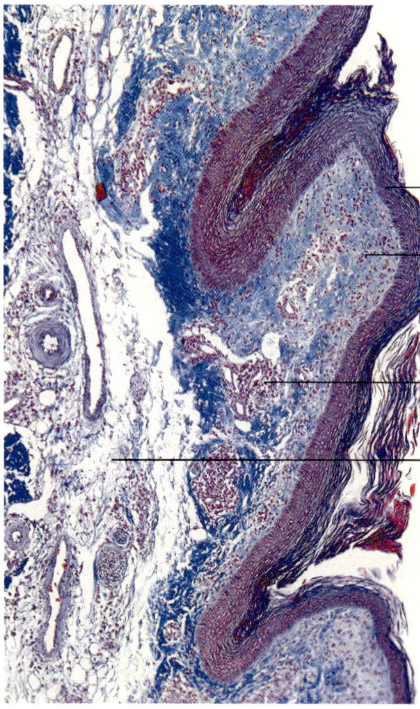

17.3 Histological section of the skin of the lateral trunk of a chicken.

of the sternum. This frequently becomes pathologically enlarged in broilers and laying hens housed under inappropriate conditions.

The lack of nerves in feathered skin, except in the vicinity of feather follicles, renders this tissue relatively insensitive.

During incubation, many bird species develop a median ventrosternal **brood patch** (area incubationis). Feathers are lost in this region and the vascularity of the dermis increases to facilitate the transfer of body heat to the eggs. Waterfowl, such as ducks and geese, do not develop brood patches. Instead, they warm their eggs with plucked down feathers.

Featherless body regions

The stratum corneum of the epidermis is thicker in featherless regions, corresponding with the mechanical forces to which these parts of the body are subjected. Various epidermal specialisations are also present, including:

- the horny beak (rhamphotheca) with
- the cere (cera),
- scales (scuta) and small scales (scutella),
- pads (pulvini),
- claws (ungues) and
- the spur (calcar metatarsale).

The horn (keratinised epidermis) is particularly hard at the edges of the beak, in the scales and spur, and on the dorsum of the claws. Soft horn is found at the cere, between the scales and on the plantar surface of the claws. Additional specialisations associated with the avian integument include:

- skin glands (glandulae cutis),
- accessory skin structures (appendices integumenti),
- the skin folds of the wings (patagia) and
- interdigital webs (telae interdigitales).

Skin glands

Birds **do not have sweat glands**. **Sebaceous glands** are found in only three locations:

- above the tail: the uropygial gland (glandula uropygialis),
- in the external acoustic meatus: the auricular glands (glandulae auriculares),
- in the cloaca: the vent glands (glandulae venti).

The **uropygial gland** is present consistently in chickens and water birds, but may be absent in psittacids and pigeons. It consists of two bilaterally symmetrical lobes, each with an excretory duct. The ducts open on the unpaired median **uropygial papilla** (papilla uropygialis), upon which a tuft of small down feathers (circulus uropygialis) is often present (Figure 17.6). The oily holocrine secretory product is used in preening to cover the feathers in a fatty waterproof film. It is thought that the uropygial gland may also play a role in the storage of Vitamin D, allowing this nutrient to be taken up by the beak during grooming.

The **glands of the external acoustic meatus** produce a waxy secretion containing numerous sloughed cells. A mucoid substance is secreted by the **glandulae venti** located on the labia of the cloaca.

Accessory cutaneous structures

Various species-specific **accessory cutaneous structures** (appendices integumenti) are observed on the head and neck of birds. In some cases, sexual dimorphism is evident in the degree to which these cutaneous modifications are developed. Those found in domestic birds include:

- the comb (crista carnosa),
- the wattle (palea),
- cheek- or earlobes (lobus auricularis),
- the snood or frontal process (processus frontalis),
- cutaneous caruncles (carunculae cutaneae),
- the crest or helmet (galea, crista ossea) and
- cutaneous papillae (papillae cutaneae).

The **comb, wattles and earlobes**, typical features of chickens, are variably developed in different breeds (Figures 17.4 and 17.5). In its basic form, the comb is erect and

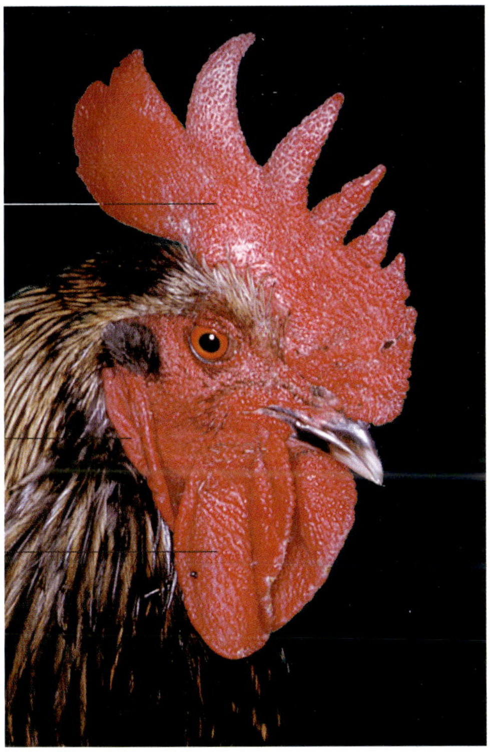

17.4 Accessory structures of the head of a male chicken.

17.5 Cutaneous appendages of the head of a hen.

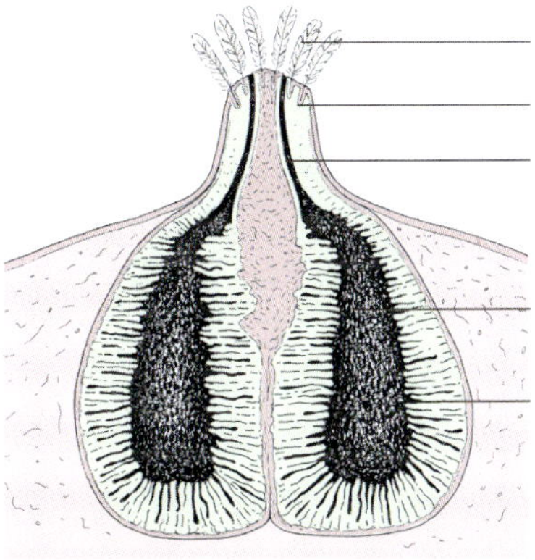

17.6 Cross section of the uropygial gland (glandula uropygialis) (schematic).

five-pointed. The wattle is a pendulous fold of skin, particularly elongated in older males, while the earlobe consists of a smaller featherless flap situated under the ear opening. An abundant deep dermal layer with embedded fibromucoid tissue is responsible for the tough elastic consistency of these structures. Their red colour is a consequence of sinusoidal capillary networks located in the superficial layer of the dermis (Figures 17.4 and 17.5).

In turkeys, the unpaired **snood** can reach considerable proportions in sexually aroused males. The head of turkeys is also covered in numerous wart-like cutaneous **caruncles** (Figures 17.7 and 17.8) that extend to the neck and breast. In fully grown turkeys, a beard comprising a tuft of hard, dark bristles is present on the neck.

The **crest** (helmet) of the guinea fowl consists of a cone-shaped peg of bone overlain by rough skin with a strongly keratinised epidermis. In females, the crest is blunt; in males it is distended caudally. The wattle of this species presents as a bulge of skin that may become engorged when the bird is aroused.

In Muscovy ducks the eyes and the base of the beak are surrounded by red **cutaneous papillae** (Figure 17.10).

Patagia

The flexor surfaces of the joints of the wings (alae) are spanned by membranous folds of skin referred to as **patagia**. Only at the shoulder joint are these skin folds present on both the flexor and extensor surfaces of the limb. Patagia associated with the cranial aspect of the wing are the:

- patagium cervicale (at the transition from the neck to the wing),
- propatagium (between the shoulder and carpus),
- patagium alulae (between the alula and digit 2).

Those located on the caudal aspect of the wing are the:

- metapatagium (extending from the breast to the elbow),
- postpatagium (between the elbow and the manus).

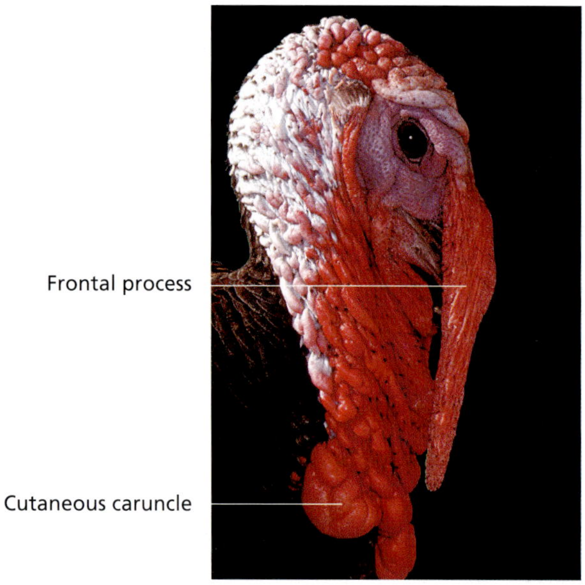

17.7 Accessory structures of the head and neck of a turkey tom.

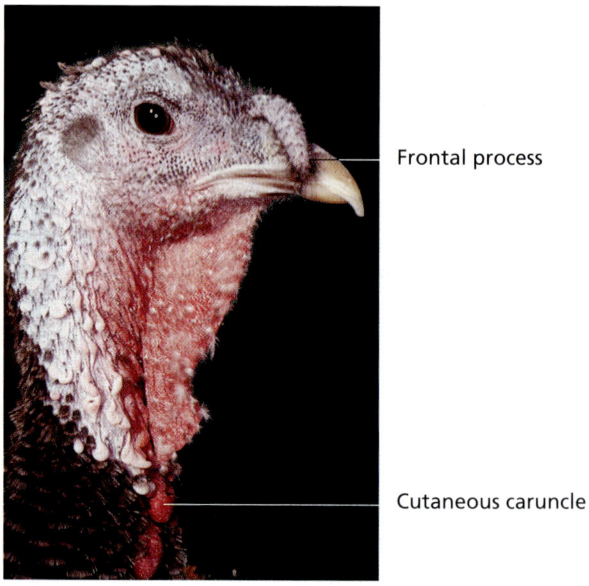

17.8 Accessory structures of the head and neck of a turkey hen.

17.9 Head of a female Muscovy duck with red cutaneous papillae (papillae cutaneae) at the base of the beak.

17.10 Head of a male Muscovy duck with red cutaneous papillae (papillae cutaneae) at the base of the beak.

Elastic membranes embedded in the patagia assist in holding the wings against the body without muscular effort.

Interdigital webs

The interdigital space between the second, third and fourth pedal digits is spanned by short folds of skin known as **interdigital webs** (telae interdigitales intermedia et lateralis) (Figures 17.11 and 17.22). In water birds such as geese and ducks, these are used for swimming and extend to the tips of the toes (Figure 17.13). Pelecaniformes possess an additional web between digits 1 and 2 (tela interdigitalis medialis) (Figure 17.14). Rails (Rallidae), such as the coot, lack interdigital webbing, possessing instead small flaps of skin that extend from the side of the second to fourth toes (Figure 17.12).

Rhamphotheca and cere

The epidermal covering of the beak forms a hard keratinised sheath, the **rhamphotheca**, which extends over the rostral portions of the maxilla and mandible. Its dermal layer is tightly attached to the periosteum. There is no subcutis. In functional terms, the beak takes the place of the lips and teeth of mammals.

The **shape of the rhamphotheca** varies according to feeding behaviour (see Chapter 6 'Digestive system'). In most species, the horn of the rhamphotheca is tough, particularly so in granivores. It is continually replaced and worn down through use. The rhamphotheca of ducks and geese is relatively soft and flexible. In these species, the edges of the beak are lined by sieve-like **lamellae**, used in filtering of food, and the horn of the bill tip is modified to form a hard plate, or **nail** (unguis maxillaris et mandibularis) (Figure 17.15).

17.11 Foot of a common buzzard (Buteo buteo) with small interdigital webs (telae interdigitales).

17.12 Foot of a Eurasian coot (Fulica atrata) with skin flaps extending from the side of the toes.

17.13 Foot of a Mallard duck (Anas platyrhynchos) with interdigital webs between digits 2–4.

17.14 Feet of a blue-footed booby (Sula nebouxii) with interdigital webs between digits 1–4. Courtesy of J. Eisner and Dr D. Schratter, Vienna.

17.15 Beak of a domestic goose (Anser anser) with nail (unguis maxillaris) at its tip.

17.16 Newly hatched chick with eye tooth on the dorsum of the beak tip.

In newly hatched chicks, a cone-shaped **egg tooth** is present on the rostral portion of the upper beak (Figure 17.16). Used by the chick to break open the egg shell during hatching, the egg tooth is subsequently shed.

The bill of ducks and geese is covered by a waxy skin. This is restricted to the base of the upper beak in chickens. In some species, the skin at the base of the upper beak is variably thickened and is referred to as the **cere** (cera). A prominent white cere is found in pigeons (Figure 17.17), while that of many diurnal raptors is bright yellow (Figure 17.18). The colour of the cere varies with gender in several types of budgerigar (blue in adult males, brown in mature females) and can therefore be used in sexing (Figures 17.19 and 17.20).

Aggregations of touch-sensitive receptors embedded in the rhamphotheca form the so-called **bill-tip organ**.

17.17 Domestic pigeon (Columba livia) with thick, white cere at the base of the upper beak.

17.18 Yellow cere at the base of the upper beak of a Eurasian sparrowhawk (Accipiter nisus).

17.19 Blue cere at the base of the upper beak of a male budgerigar (Melopsittacus undulatus).

17.20 Brown cere at the base of the upper beak of a female budgerigar (Melopsittacus undulatus).

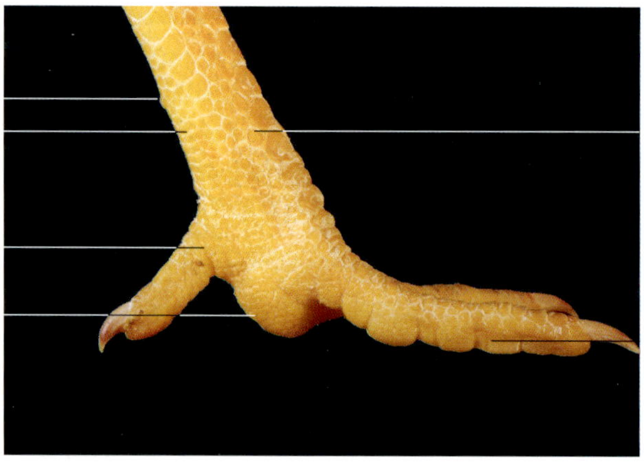

17.21 Foot of a female chicken with spur presenting as a small wart-like scale (medial view).

Particularly well developed in psittacines and water birds, this structure provides a high degree of sensitivity. The bill can thus be considered a tactile organ (see Chapter 6 'Digestive system').

Scales

The featherless skin of the tarsometatarsus and digits (podotheca) is covered in **scales.** On the dorsum these are relatively large (**scuta**), while those found on the plantar surface are smaller and typically hexagonal in shape (**scutella**) (Figure 17.21). In chickens the scales overlap, while those of ducks and geese butt against each other (Figure 17.13).

Pads

The joints of the pedal digits are underlain by **pads** (pulvini) (Figure 17.21). These include the **metatarsal pad**

17.22 Foot of an adult male chicken with well-developed spur (lateral view).

(pulvinus metatarsalis), which bears weight when the foot is placed on the ground, and the **digital pads** (pulvini digitales) that conform closely to the underlying surface during perching.

Claws

The **claws** (ungues) are located at the tips of the digits of the foot. Each claw forms a cone-shaped covering of the underlying bone. It consists of a **dorsal plate** (scutum dorsale) and a **plantar plate** (scutum plantare). In chickens, the claws are adapted for scratching and are thus shorter and relatively straight (Figure 17.21). In contrast, the claws of raptors are long, strongly curved and sharp (Figure 17.11). Muscovy ducks fold the claw of the second toe to the side, thus avoiding wear and preserving the sharp point for use in defence. **Claws on the wings** (ungues digiti manus) are an occasional atavistic finding in domestic chickens and geese.

Spurs

Well developed in male chickens and turkeys (Figure 17.22), the **spur** (calcar metatarsale) usually manifests only as a wart-like scale in females (Figure 17.21). The spur is located on the caudomedial tarsometatarsus where, in males, it is mounted on a large pyramidal bony protuberance. It grows approximately 1cm per year, up to a total of around 6cm, and can thus be used in estimating age.

Feathered body regions

Feathers

Related phylogenetically to the scales of reptiles, feathers are the distinguishing feature of all birds.

By virtue of their many specialised characteristics (e.g., lightweight construction, interlocking of the vanes, compliance and conformability), feathers confer upon birds the capacity for flight.

Particularly in males, feathers are often brightly and characteristically coloured.

Feather structure

The following description is based on the structure of a mature contour feather (Figure 17.23), consisting of the:

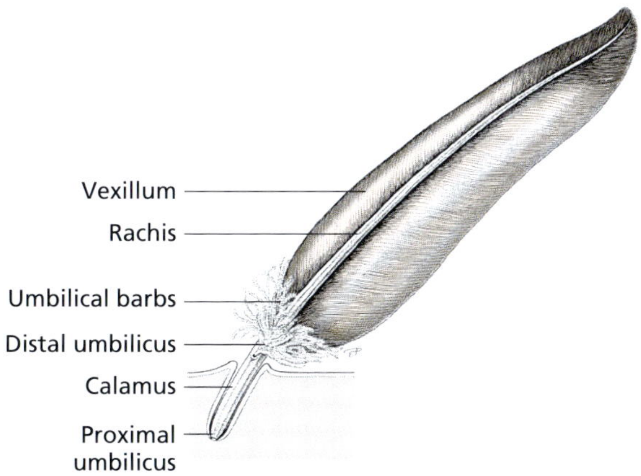

17.23 Structure (schematic) of a feather (penna) based on a flight feather (remex).

17.24 Scanning electron micrograph illustrating interlocking of hooklets and barbules.

- shaft (scapus),
- rachis (rhachis),
- calamus and
- internal and external vexillum.

The **shaft** is divided into the calamus and the rachis. A small opening, the **distal umbilicus** (umbilicus distalis), is located at the junction between these segments. The calamus is embedded in the skin, while the rachis forms the visible portion of the feather. Located at the tip of the calamus is a round opening, the **proximal umbilicus** (umbilicus proximalis), into which the dermal papilla projects. The papilla is covered by a layer of living epidermal cells, from which new feathers are formed after moulting.

The **calamus** is round in cross section and, in the mature feather, is predominantly hollow. During feather development, the calamus is filled with pulp consisting of a loose mesenchymal reticulum surrounding a central artery and vein. During **maturation**, the pulp recedes leaving behind a series of **air filled compartments**.

The **rachis** contains pith formed from epithelial cells. Its upper surface is convex, while the ventral side (facing the body) is marked by a groove (sulcus ventralis). The rachis bears two rows of slender, rigid **barbs** (rami). Arising from each barb are two rows of fine **barbules** (radii, barbulae). Barbulae of adjacent barbs cross one another at right angles. The **distally directed barbulae** (radii distales) possess tiny **hooklets** (hamuli) that interlock with the **proximally directed barbules** (radii proximales) (Figure 17.24). Disruption of this interlocking arrangement by external mechanical forces is corrected during preening, whereby the hooklets and barbules are reconnected ('zipping the feathers').

Together, the interconnected barbs form the **vane** (vexillum) of the feather. The vexillae of neighbouring feathers overlap in a shingle-like arrangement such that one is covered (**vexillum internum**) while the other is exposed (**vexillum externum**).

A small **afterfeather** (hypopenna) projects from the proximal umbilicus. This may consist of just a tuft of **umbilical barbs** (barbae umbilicales) or may – as is typical of covert feathers – have an **aftershaft** (hyporhachis) with two **hypovexillae**.

The **calamus** is set within a cylindrical cutaneous pit known as the **feather follicle**. A close fit is formed between the calamus and the follicle, resembling the relationship between the hair and hair follicle in mammals.

The **feather follicle** (Figure 17.25) has an inner epidermal and outer dermal wall. The **epidermis of the follicle** is continuous with the calamus at the proximal umbilicus. Here, the living epithelial cells of the follicle give way to the dead keratinised cells of the calamus. At the base of the feather follicle the small **dermal papilla** projects into the proximal umbilicus (Figure 17.26) and blends with the mesenchymal pulp within the calamus of the developing feather. In the young, growing feather, the dermal papilla is well vascularised to provide sufficient nutrition to the epidermis for the processes involved in feather development.

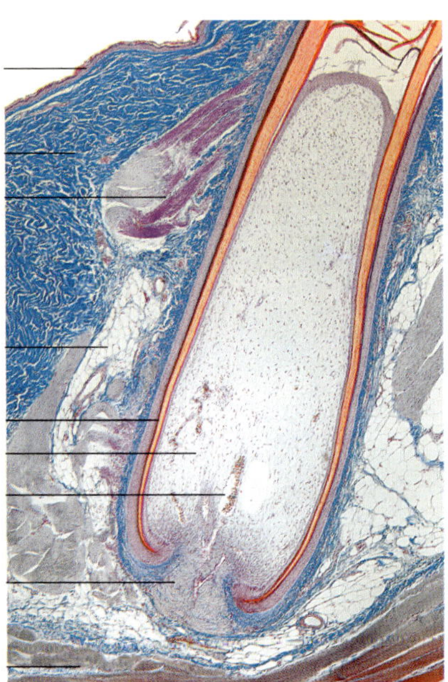

17.25 Histological section of a developing feather (penna) with surrounding skin and subcutis.

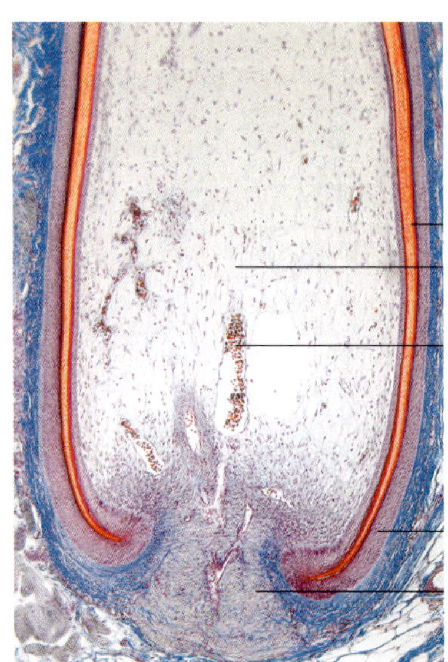

17.26 Histological section of the follicle of a developing feather (penna) (detailed view).

Types of feathers

Feathers fall into the following categories:

- Contour feathers (pennae contornae),
  - coverts (tectrices),
  - flight feathers of the wings (remiges),
  - tail feathers (rectrices),
- down feathers (plumae),
- semiplumes (semiplumae),
- powder down feathers (pluviplumae or plumae pulveraceae),
- filoplumes (filoplumae) and
- bristles (setae).

Coverts are the most numerous feathers, covering the bulk of the body of the bird. They are distinguished by the presence of an afterfeather with an aftershaft (Figures 17.28 and 17.29). The vexillae usually consist of a distal **pennaceous portion** (pars pennata) and a proximal **plumaceous portion** (pars plumacea) (Figures 17.27 to 17.31). In the latter the barbules are not interlocked. Apart from these characteristic features, coverts vary greatly in size, shape and colour, depending on their location and function. As **ornamental feathers** (Figures 17.1 and 17.32), for example, coverts make up the fan of the peacock, the sickle feathers of the rooster and the crest of the cockatoo. Coverts are distributed over the body in species-specific **tracts** (pterylae). Plumules (plumae) and semiplumes (semiplumae) may be present in areas of skin between the tracts (**apteria**).

The **remiges** (*sing.* remex) are located on the antebrachium, manus and the alula (Figure 17.34). Their number is consistent within species. Around 10–20 secondary remiges (remiges secundarii) arise from the **forearm,** while ten primary remiges (remiges primarii) originate from the **manus**. Up to four **alular remiges** (remiges alulares) are present. Each remex has accompanying coverts. There are no remiges in the **brachial region**, which is covered with coverts.

The **remiges are characterised** by a slightly curved shaft and asymmetrical vexillae (Figures 17.33 and 17.34). During the downstroke, the broad internal vexillae lie under their external counterparts, such that the surface of the wing is smooth and uninterrupted. When the wing is raised, the air stream pushes the internal vexillae downwards, rotating the feathers about their longitudinal axis and opening the wing like a venetian blind. To absorb the forces generated during flight, the calami are not only anchored to the skin, but are also attached by syndesmoses to the underlying bones.

Arising from the rump, the **rectrices** (*sing.* rectrix) form the basis of the tail. Their shaft is usually straight. Near the centre of the tail, the vexillae are generally symmetrical. Towards the periphery, there is an increasing tendency towards asymmetry of the vexillae (the medial vexillum becoming narrowed) and curvature of the shaft.

17.27 Covert of a chicken (Gallus gallus); the vane is divided into a pars pennata and a pars plumacea.

17.28 Covert of a chicken (Gallus gallus) with afterfeather containing an aftershaft (hyporhachis).

17.29 In the emu (Dromaius novaehollandiae), coverts and their afterfeathers are similarly proportioned.

17.30 Covert of a Eurasian jay (Garrulus glandarius).

17.31 Covert of a Japanese crested ibis (Nipponia nippon).

17.32 Ornamentally modified covert of a little egret (Egretta garzetta).

17.33 Primary remex (remex primarius) of a domestic pigeon (Columba livia).

17.34 Secondary remex (remex secundarius) of a common kestrel (Falco tinnunculus).

17.35 Middle rectrix of the tail of a domestic pigeon (Columba livia).

17.36 Down feather (plumule) of a Mallard duck (Anas platyrhynchos).

17.37 Semiplume (semipluma) of a chicken (Gallus gallus).

17.38 Transitional feather type exhibiting characteristics of a semiplume and covert (Gallus gallus).

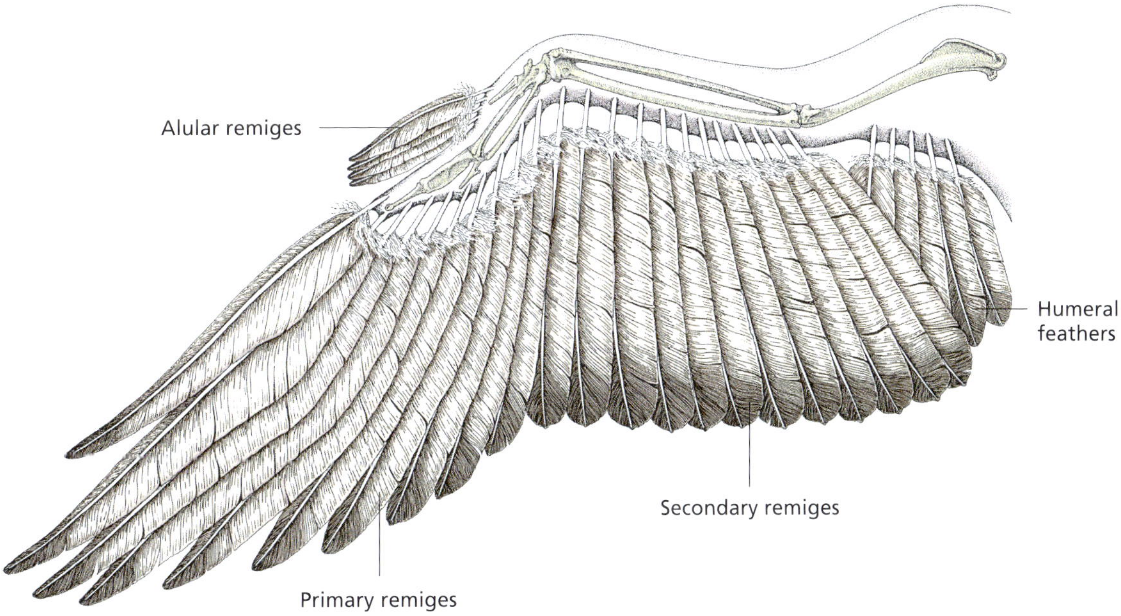

17.39 Right wing (schematic) of a peregrine falcon (Falco [Hierofalco] peregrinus) with skeleton and remiges (ventral view).

Plumules, or down feathers, consist only of a calamus with a large tuft of keratinised filaments at its tip (Figure 17.36). Definitive (adult) down feathers are particularly numerous in water birds.

Hardly any down feathers are found on chickens, their function being taken over in this species by **semiplumes** (Figures 17.37 and 17.38). Semiplumes are similar to down feathers, but have a clearly defined rachis. Their association with an afterfeather, including an aftershaft, indicates that semiplumes are actually a modified form of covert. Like down feathers, their role is to provide thermal insulation.

Powder down feathers are specialised down feathers that continually produce a powdery substance consisting of minute (approximately 1μm) keratin granules. They are especially plentiful in pigeons. It is presumed that the powder acts as a waterproof coating for the plumage.

Filoplumes are down-like feathers that lie immediately adjacent to the follicle of contour feathers. The follicles of filoplumes are rich in free nerve endings and **Herbst bodies** are found in the surrounding tissue.

Bristles are located on the head. They line the base of the beak and form the eyelashes. Bristles have a rigid shaft with few barbs and are associated with numerous mechanoreceptors.

Feather replacement and moulting

Most birds replace their plumage once a year, typically following the breeding season. This process, known as **moulting**, is regulated by the hormonal products of the thyroid gland. Chickens moult three times during the first six months of life. The first two moults are complete while the third is partial. Newly developing feathers are initially encased in a **feather sheath** (vagina pennae) that subsequently ruptures from the feather tip, allowing the vexillae to unfurl.

The plumage of hatchlings, known as **natal down or neoptile**, consists of down feathers. In precocious species, these are already fully developed at hatching, while in altricial birds they are formed in the first days of life. The yellow colouring of the feathers of chicks is derived from carotenoids in the yolk. Replacement of this first generation of feathers is accompanied by the development of contour feathers. From the second generation onwards the plumage is referred to as **teleoptile**.

Blood supply and innervation of the skin

Vascular networks are present in the subcutis, at the junction with the dermis and within the dermis itself. The dermal vascular system gives off **subepithelial capillaries** that also supply the dermal papillae of the feather follicles. **Arteriovenous anastomoses** are present in the toes, pads of the feet and interdigital webs. With the exception of the feather follicles, the skin is sparsely innervated. The beak is well vascularised and richly endowed with nerves that extend all the way to its tip.

Clinical aspects

Due to the marked anatomical variation among avian species, and a multitude of possible aetiologies, disorders of the skin and cutaneous appendages pose a considerable diagnostic and therapeutic challenge in avian medicine. A sound understanding of normal anatomy and a keen awareness of the possible role of species, breed, husbandry and season are indispensable, not least for avoiding serious errors in diagnosis and treatment.

For various reasons, a fully intact feather coat is essential for all types of birds, be they wild or domestic species. It is therefore a prerequisite for release of wild birds back into their natural environment. Even individual damaged or broken feathers can compromise important functions, such as hunting ability in falconry birds (see Chapter 24 'Falconry and raptor medicine') and performance in racing pigeons. More generalised feather loss can **interfere with thermoregulation** in pet and captive wild birds, potentially resulting in secondary disease (e.g., respiratory aspergillosis in parrots). In commercially raised poultry, bare patches of skin are a significant **stimulus for cannibalism**. Damaged plumage can sometimes be replaced or repaired using previously moulted feathers ('imping' of falconry birds, see also Chapter 24).

As a result of the considerable metabolic demands of moulting, birds are less immunocompetent – and thus more susceptible to infectious disease – during this period. Particularly in wild birds that are unaccustomed to humans, **exposure to stress** (e.g., due to inappropriate handling, see Chapter 18 'Clinical examination' and Chapter 20 'Handling, restraint and anaesthesia') can induce **spontaneous shedding** of large quantities of contour feathers. In the wild, this may serve as a means of escape from a predator. Simultaneous moulting of certain feather types can also be part of a **normal moulting strategy**. Loss of flight feathers during a wing moult may result in temporary partial or total flightlessness (e.g., goshawks [Accipiter gentilis] and Mallard ducks [Anas platyrhynchos]). In raptors, this occurs at different times in males and females, typically during the rearing of young. Consequently, the flightless adult taking care of the offspring is dependent on its partner for obtaining prey, and death of the flighted bird can result in the demise of its partner and young. Physiological moulting strategies should not be confused with those used to evade predation. Importantly, the latter may be accompanied by a rapidly fatal **Pasteurella infection**, resulting from associated bite and scratch wounds.

In captive wild bird species and pet birds, unilateral wing clipping is used as a minimally invasive means of **restricting flight** (some authors question the suitability of this practice in pet birds on animal welfare grounds). The procedure is performed by cutting 8–10 primary and secondary remiges (at the transition from the vane to the shaft) of one wing to bring about unbalanced flight. For cosmetic reasons, and to protect the alula, the first 3–4 primaries are left intact (such that clipping is not evident when the wing is rested against the body).

The **apteria** (regions of relatively bare skin between feather tracts) are useful sites for accessing superficial vessels for blood collection (see Chapter 21 'Medication and blood collection techniques'). Depending on **species, breed** and **season**, other areas may also be featherless and should not be confused with pathological changes. These include the bald patch under the crest in some cockatiels (Nymphicus hollandicus) and the hyperaemic brood patch

17.40 Gender determination based on sexual dimorphism in feather colouring: red undertail coverts (short red tail feathers) typical of a male African grey parrot (Psittacus erithacus).

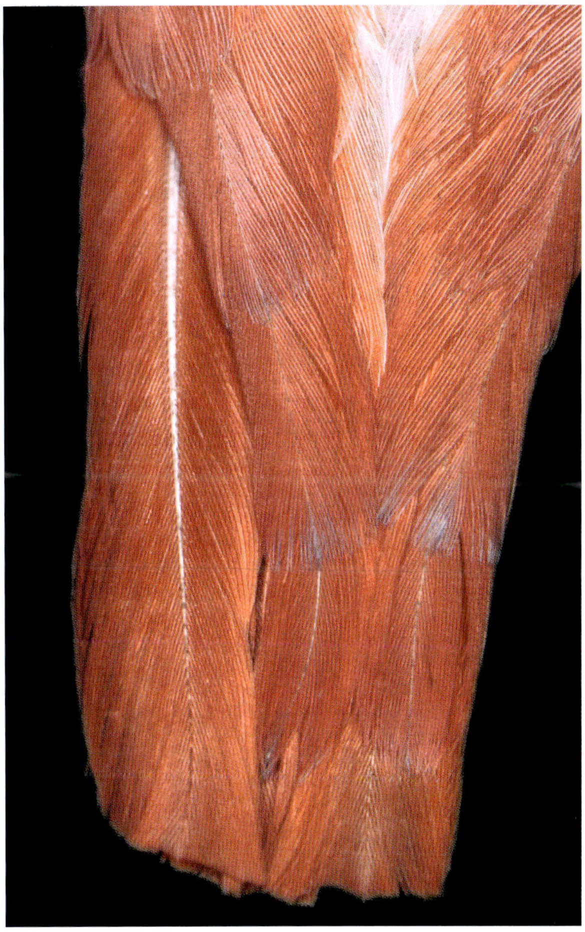

17.41 Gender determination based on sexual dimorphism in feather colouring: grey-black edged undertail coverts typical of a female African grey parrot (Psittacus erithacus).

that appears annually in the caudal sternal region of incubating females.

In wild water birds, bare areas in the lower cervical area or on the back of females may be the result of **excessive feather picking** by drakes and ganders during mating. Featherless patches in domestic water birds may indicate an inappropriate gender ratio, particularly an excess of males (ideally one male for 3–5 females). They may also occur when a male has a favourite female.

Subtle gender dependent differences in the plumage can be useful in sexing ('feather sexing') **parrots**, which typically do not exhibit obvious sexual dimorphism. However, this method is not as reliable as endoscopy (see Chapter 22 'Endoscopy') or PCR-based techniques. In **African grey parrots**, the presence or absence of dark edges on the vanes of undertail coverts can be used with a high degree of confidence to distinguish between males and females (Figures 17.40 and 17.41). Intense orange cheek patches and a completely yellow crest differentiate male **normal grey cockatiels** (Nymphicus hollandicus) from females, in which the cheek patches are duller and only the base of the crest is yellow (Figures 18.4 and 18.5). Female **opaline (pearl) cockatiels** can be identified by barring of the undertail coverts. One species that does exhibit marked sexual dimorphism is the **eclectus parrot** (Eclectus roratus), in which the plumage of the female is a vivid red, while that of the male is intensely green.

17.42 'Stress marks' or 'hunger traces' visible under backlighting as pale stripes, perpendicular to the shaft, on the remiges of a blackbird (Turdus merula). These result from abnormal feather development and typically signify chronic disease.

While the sex of many birds is indistinguishable to human eyes, the ability of birds to perceive **light in the ultraviolet spectrum** (see Chapter 15 'The eye') enables them to appreciate certain gender specific features (e.g., strongly UV-reflective feathers on the legs of canaries, UV-reflective patches around the beak of toucans). These characteristics can be verified using specialised UV photography, which thus serves as an alternative to endoscopy and PCR in **sexing**.

Of the many infectious and non-infectious (including husbandry- and behaviourally induced) **diseases of the skin and feathers**, 'stress marks' or 'hunger traces' are included here as a particularly conspicuous example (Figure 17.42). These are indicative of chronic disease. In birds kept on their own, behaviourally induced abnormalities of the plumage (feather picking) manifest as bare patches in body regions that the bird can access itself. Feather loss in inaccessible areas is suggestive of a systemic disease process.

As an example of the importance of distinguishing between anatomical variation and pathological processes, the air pockets under the skin of pelicans should not be confused with subcutaneous emphysema.

Clinical examination

R. Korbel, S. Reese and H. E. König

18

The following chapters are intended to provide an introduction to clinical examination and selected diagnostic techniques and therapeutic interventions in avian patients, with particular regard to underlying anatomical principles.

Clinical examination of birds (Figure 18.1) requires a systematic and meticulous approach. Careful recording of all findings is an important aspect of building a case for diagnosis. The extraordinary diversity within the class Aves, with its approximately 10,000 extant species and more than 20,000 subspecies, is reflected clinically in the many species-specific features that must be taken into account with respect to diagnosis and treatment.

Avian patients fall into various categories, including commercial birds, cage and aviary birds, pigeons, zoo birds and raptors (including falconry birds). For economic reasons, veterinary management of commercially reared birds, such as chickens, turkeys, geese, quail and pheasants, is concerned mainly with the health of the flock as a whole. Under these circumstances, examination of

18.1 Clinical examination of a budgerigar (Melopsittacus undulatus): palpation of the right wing.

18.3 Inappropriate housing of a budgerigar. The cage should be of sufficient size, rectangular, easy to disinfect and furnished with suitable equipment. It should be positioned approximately at the height of the owner's head.

18.2 Blackbird (Turdus merula) exhibiting typical signs of serious illness: ruffled feathers, lethargy and somnolence.

individual animals serves primarily as a means of evaluating the health status of the whole population.

In the veterinary care of **companion birds**, on the other hand, the focus is on the well-being of individual animals, many of which are extremely valuable (e.g. for breeding purposes or because they are particularly sought after).

In some populations, such as pigeons and zoo birds, intensive diagnostic evaluation may be justified on both a flock and individual basis. The latter applies particularly in the case of valuable racing pigeons and captive wild birds involved in international breeding programmes.

Clinical examination includes, in chronological order:

- history and signalment,
- visual appraisal,
- physical examination and
- additional examination procedures.

History and signalment

Comprehensive history taking should include **questioning of the owner** with respect to:

- whether the bird is kept as an individual or as part of a group,
- duration of illness,
- disease manifestations and
- the number of affected birds.

In **commercial birds**, additional information is required regarding:

- the nature of the enterprise, including husbandry conditions,
- flock performance and breeding data,
- vaccination protocols and
- any prior treatments.

Details relating to **patient signalment** should also be recorded. These include:

- breed,
- age,
- sex and
- intended purpose.

It should be noted in this context that **accurate estimation of age** in adult birds with a fully developed feather coat is virtually impossible.

In **production animals**, determination of gender is commercially significant. **Vent sexing** of day-old chicks is a challenging technique requiring considerable practice and is generally carried out by specially trained staff (see Chapter 9 'Male genital organs' and Chapter 10 'Female genital organs'). Feather characteristics can also be used to distinguish between males and females, as can gender specific differences in the development of accessory cutaneous structures in adult chickens (see Chapter 17 'Common integument').

The majority of **parrots** do not **exhibit phenotypic sexual dimorphism**, even as adults. Sexing of these birds can be achieved using:

- endoscopy,
- analysis of sex chromosomes,
- DNA analysis.

Chromosomal analysis is increasingly replacing endoscopy as a means of gender determination. This technique, which identifies the paired **Z chromosomes** of males and the **ZW chromosome pairs** of females, is typically performed using feather pulp. An alternative method involves the use of DNA probes to identify gender specific gene sequences. Both methods allow the risks of anaesthesia and surgical intervention to be avoided.

DNA analysis requires only small sample volumes, such as blood (EDTA) droplets or cells derived from feather pulp, and therefore has the additional advantage of being suitable for very small birds.

Pet bird species that do exhibit **sexually dimorphic characteristics** include the:

- **Cockatiel** (Nymphicus hollandicus): well demarcated orange cheek patch, non-barred undertail coverts and yellow crest in males, pale cheek patch, barring of undertail coverts and grey crest (specifically at the base) in females (Figures 18.4 and 18.5).
- **White, sulphur crested, red-tailed black and salmon crested cockatoo** (Cacatua spp.): black iris in males, brown to red iris in females. In the female galah (Eolophus roseicapillus) the iris may be light red.
- **Budgerigar** (Melopsittacus undulatus): cere is blue in males and brown in females. Accuracy of sexing based on cere colour is around 70 per cent in varieties with light coloured feathers (e.g., lutinos; yellow feather coat) and up to 95 per cent in opaline (blue) varieties (see Chapter 17, Figures 17.19 and 17.20).
- Many **diurnal raptors** (particularly falcons, also hawks and sparrowhawks): differences in size, males approximately one-third smaller than females. This is a possible etymological basis for the term 'tercel' (from the Latin, 'tertius') used to refer to male falcons. The male common kestrel is further distinguished by a grey cap and uniformly grey tail feathers. The tail of females is barred.
- **Canary** (Serinus canaria): song serves as a secondary

18.4 Male cockatiels are characterised by a brightly coloured, well demarcated cheek patch and yellow crest feathers.

18.5 In female cockatiels the cheek patch is paler and less clearly defined than in the male. The feathers of the crest are grey.

gender-specific characteristic; depending on reproductive status, males have a cone-shaped cloacal projection (not apparent in females).

- **African grey parrot** (Psittacus erithacus): uniformly red undertail coverts in males, grey-black tips (and sometimes vane edges) on undertail coverts in females (Figures 17.40 and 17.41).
- **Eclectus parrot** (Eclectus roratus): plumage bright green in males and a vivid red in females.
- **Lovebirds** (Agapornis spp.): vane of tail feathers is pointed in males and more rounded in females. Interpretation of this morphological characteristic requires considerable experience and, depending on the reproductive status of the bird, should be accompanied by examination of the distance between the pubic bones.

Observation

History-taking is followed by appraisal of the unrestrained bird with respect to:

- general condition and behaviour,
- condition of the feather coat.

It is important to be aware that clinical signs of disease are generally poorly expressed by birds. As a consequence, owners frequently only become alerted to the presence of disease at a relatively late stage. This can lead to a lower therapeutic success rate than might be expected in mammals, as well as a poorer prognosis. Obvious

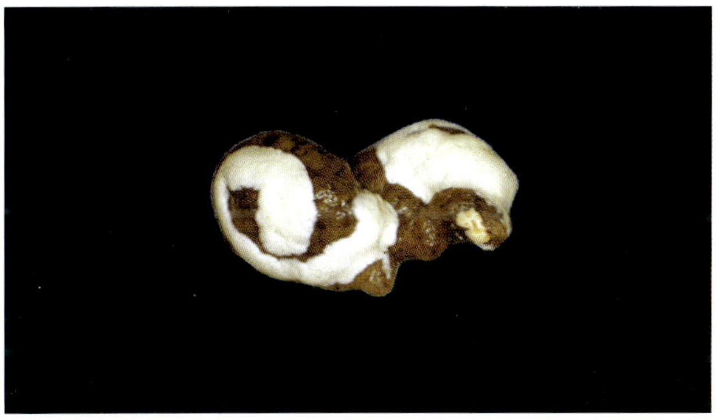

18.6 Normal excreta of a granivore such as a budgerigar (Melopsittacus undulatus) consisting of firm, easily distinguishable dark faecal and pasty white uric acid components.

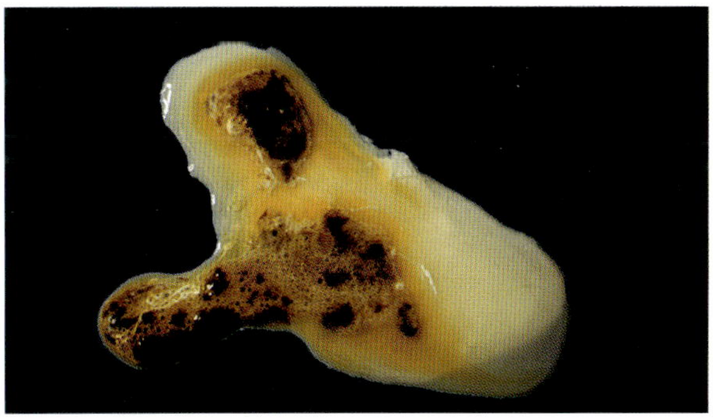

18.7 Normal excreta of a raptor such as a peregrine falcon (Falco peregrinus) consisting of viscous faeces (thin to thick consistency) mixed with urine.

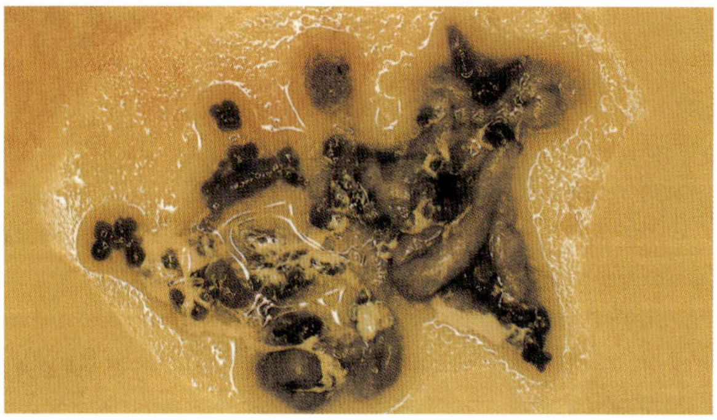

18.8 Normal excreta of 'soft feeders' (insectivores, frugivores) such as mynas (Gracula religiosa) consisting of thin viscous faeces mixed with urine. Colour and consistency (and the presence of partially digested food) varies with the specific ingredients of the diet.

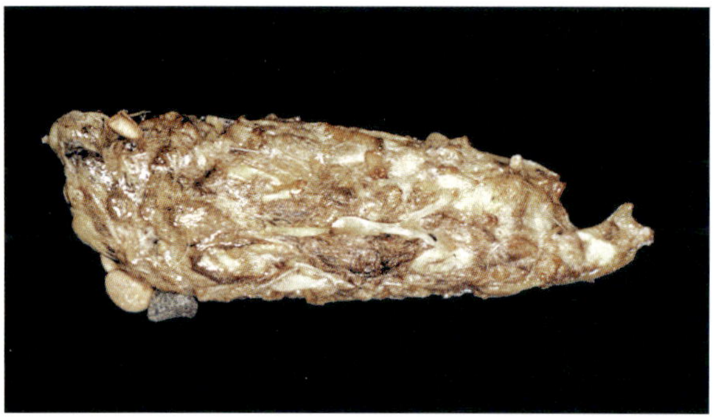

18.9 'Casting' of a great horned owl (Bubo virginianus). Castings are typically egested 12–14 hours after the last meal and are covered by a thin film of mucus that facilitates their passage.

indications of (advanced) disease necessitating urgent intervention include (Figures 18.2 and 18.10):

- ruffled feathers,
- lethargy and somnolence and
- absence of the flight response.

The absence of the **flight response** may be misinterpreted by owners as a sign of tameness, particularly if the bird was only recently acquired.

Other signs suggestive of disease include:

- loss of phonation (singing or talking),
- inappetence,
- increased food consumption (polyphagia),
- behavioural changes with an increased startle response,
- nervousness and hyperactivity, possibly extending to neurologic excitement.

Visual appraisal should include assessment of:

- general behaviour and posture,
- the feathers,
- the skin,
- accessory cutaneous structures,
- body openings (eye, ear, nose, cloaca) and
- respiratory function.

In addition, examination should include assessment of the excreta (quantity, colour, composition; Figures 18.6 to 18.9) in the cage, transport container, aviary or loft. It should be noted that owners frequently confuse **diarrhoea** with **polyuria,** as they are often unaware of the combined excretion of faeces and urine in birds.

18.10 Lethargy, somnolence and feather ruffling in the absence of specific clinical signs, indicative of advanced disease in a canary (Serinus canaria).

18.11 Marked torticollis in a Gouldian finch (Erythrura [Chloebia] gouldiae) resulting from paramyxovirus infection.

Polyuria/polydipsia presents as an increased liquid component surrounding a more or less firm faecal mass. In contrast, 'true' diarrhoea is typified by a loose to liquid faecal component (taking into account the normal, species-related variation in faecal consistency, e.g., soft food eaters and raptors, see Figures 18.7 and 18.8).

Signs suggestive of a **neurological abnormality** include:

- inability to fly,
- drooping of the wing(s),
- leg weakness,
- swaying,
- inability to perch and
- abnormal postures.

Abnormal body position (Figure 18.11) may be provoked or exaggerated by elicitation of the **menace response**.

The flight response should also be evaluated in the unrestrained patient by approaching the bird and observing its reaction. In caged birds, this is done by extending a hand towards the bird or applying subtle tactile stimuli. This is followed by physical examination of the patient using appropriate methods of restraint.

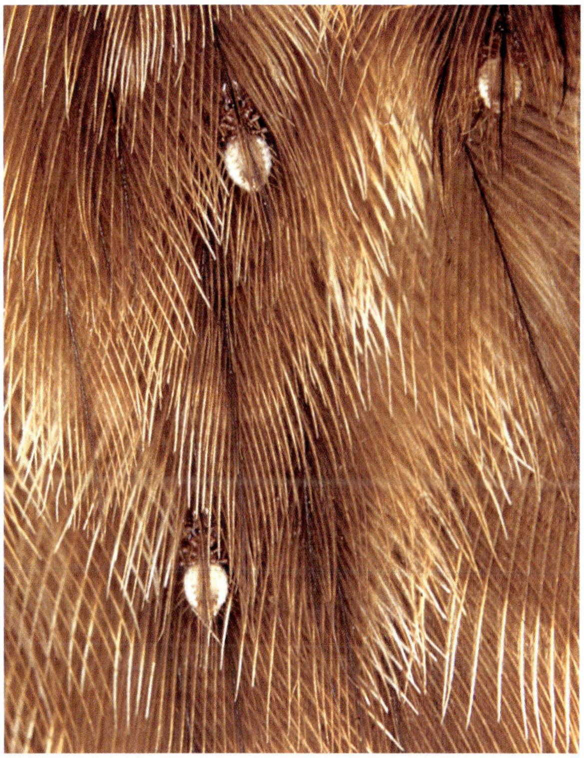

18.12 Bird lice (Mallophaga) on the feathers of a common buzzard (Buteo buteo). Louse infestation is suggestive of chronic disease and thus warrants a thorough general examination.

Physical examination

The use of suitable, humane **means of restraint** is essential for proficient and comprehensive physical examination (including palpation and auscultation) of avian patients. These techniques are described in Chapter 20 'Handling, restraint and anaesthesia'.

The bird's **nutritional status** should first be evaluated by gently palpating the breast area to assess the pectoral mass and the keel (carina sterni).

Body condition can be divided broadly into the following categories:

- overweight,
- optimal,
- underweight,
- markedly underweight and
- cachectic.

18.13 Budgerigar with characteristic pumice-like hyperkeratosis of the periocular skin due to infestation with Cnemidocoptes pilae. The round bore holes, also often seen in the cere and around the cloaca, legs and feet, are typical findings.

While optimal body condition is generally characterised by convex curvature of the breast muscles and a palpable but not prominent keel, it is important to be aware of potentially marked variation associated with breed, gender and the conditions under which the bird is kept. For example, the presence of fat deposits in the winter months may be a normal finding in wild raptors. Falconry birds, on the other hand, are maintained in a slightly 'underweight' condition by means of a precisely controlled diet and regular weighing. Requiring a great deal of experience and extensive knowledge of individual birds, this approach is used to achieve an optimal balance between the bird's physical condition and its desire to hunt.

In some cases, apparent **undernourishment** may be a manifestation of muscle wasting due to disuse in birds that are denied the opportunity to fly. Conversely, animals suffering **acute, rapidly progressive disease** may present with essentially normal body condition.

Auscultation should next be performed to avoid potentially misleading findings associated with stress. Sounds that may be encountered range from faint normal respiratory noises to stertor. The latter may, in cases of **sternostomosis** for example, be audible without the aid of a stethoscope when the bird is held to the examiner's ear. Mild respiratory noises induced by excitement are not pathological.

A detailed **visual appraisal** of the **feather coat** (see Chapter 17 'Common integument') and individual feathers should then be conducted. In the normal bird, the feather coat is smooth, shiny and lies against the body. Stripes in the feathers, oriented perpendicular to the shaft ('stress marks' or 'hunger traces') are indicative of longstanding disease present at the time that the affected feathers were forming. These are best appreciated with the aid of backlighting (see Chapter 17, Figure 17.42). The feathers should also be inspected for evidence of parasite infestation (Figure 18.12).

In addition, attention should be given to **cutaneous structures** including the claws, beak, cere, cutaneous papillae, comb, wattle and, where present, the uropygial gland.

Not infrequently, **haemorrhage** with extensive blood staining of the feathers and cage originates from damaged blood feathers or fractures of the claws. While this clinical presentation may appear dramatic to the owner, the injuries are rarely life-threatening.

Due to the relatively thin avian dermis, skin turgor is not a reliable indicator of **hydration status** in birds. A refill time of greater than 1–2 seconds in the ulnar vein indicates approximately 7 per cent dehydration. Extreme dehydration (10 per cent or more) is characterised by enophthalmus with a haematocrit of less than 10 per cent and total serum protein of less than 1 per cent. This constitutes a medical emergency. Evaluation of subcutaneous blood vessels through the paper-thin skin is relatively straightforward.

The **nostrils** and **nasal cavities** should then be inspected for abnormal appearance or soiling. Normal nasal mucosa appears pink, moist, smooth and glistening. Indicators of **upper airway disease** include:

- nasal discharge,
- sneezing, which may be accompanied by violent head movements,
- encrusted wing coverts and
- periocular swelling.

Gentle pressure on the nostrils may reveal a **nasal discharge**. Birds with respiratory disease clear discharges from their nose and trachea by 'sneezing'. This takes the form of a jerky expiration that may be accompanied by **violent head movements**. This action should not be equated with sneezing or coughing in mammals, as birds lack a muscular diaphragm. Crusting of the wing coverts with mucus results from the bird wiping its nose, and is suggestive of an upper respiratory infection.

Inflammation of the **infraorbital sinus** (e.g., due to mycoplasma infection) frequently manifests as swellings under or around the eye (Figure 18.14), usually accompanied by conjunctivitis. Palpation and paracentesis can be used for further investigation of suspected sinusitis.

In chickens, the **beak can be opened** by carefully pulling on the comb and wattle (normal finding: pink mucosa free of deposits). Gentle external pressure can be used to protrude the larynx into the oropharyngeal cavity, allowing inspection of the mucosa, choana, tongue and larynx with the aid of a light source.

In pigeons, the **oropharyngeal cavity** is opened by fixing the upper beak in place with the thumb and middle finger, while using the index finger to hold the lower beak open (Figure 18.16). Particular attention should be given to the identification of any **mucosal deposits in the oropharyngeal cavity.** These include:

18.14 Obvious swelling beneath the eye of a turkey (Meleagris gallopavo) due to infection of the infraorbital sinus with Mycoplasma gallisepticum.

18.15 Metaplastic enlargement of the lacrimal gland with eyelid displacement in an African grey parrot (Psittacus erithacus) due to hypovitaminosis A.

- tightly adherent yellow plugs suggestive of trichomoniasis,
- diphtheritic pseudomembranous lesions characteristic of avian pox and
- white deposits associated with candida infection.

Lesions typical of trichomonas infection ('yellow head') may also be found in raptors and psittacines.

Swellings seen under the tongue, or externally between the rami of the mandible, are usually associated with metaplasia of the sublingual salivary gland. These findings are suggestive of dietary hypovitaminosis A and consequent respiratory aspergillosis (which may be associated with dry respiratory noise).

The **crop** should be examined by palpation (normal finding: crop filled to variable degree, no palpable deposits on the crop wall) and by obtaining a crop swab. The latter is carried out by inserting a cotton bud moistened with physiological saline into the oral cavity via the left angle of the mouth and advancing it over the base of the tongue into the oesophagus (on right side of the throat) and into the crop. The cotton bud should be passed along the crop wall using a rotating movement to ensure the detection of parasites that may be present in the mucosal crypts.

Alternatively a **crop wash** can be performed. A curved irrigation cannula or crop tube is used to instil physiological saline into the crop and aspirate a sample for analysis (light microscopy, 40× to 100× magnification).

Microscopic examination must be performed while the crop sample is fresh and still at body temperature, as trichomonads begin to die off around 20 minutes after sample collection.

The **abdomen** should be examined for evidence of swelling, first by visual appraisal, then using palpation (Figure 18.17). Palpable structures (allowing for species variation) and potential pathological findings include:

- ventriculus and intestinal loops,
- pubic bone,
- eggs and
- swelling due to ascites or tumours.

An **increase in the distance between the pubic bones** indicates imminent oviposition. Neoplasia is an important consideration in budgerigars, in which long-term studies have revealed a tumour prevalence of up to 12 per cent. **Kidney tumours** may be palpable under the synsacrum. These frequently exert pressure upon the lumbosacral plexus, located between the kidney and the synsacrum, and may thus result in characteristic unilateral hind-limb lameness. In the case of very large tumours, dyspnoea may also be apparent.

Abdominal palpation is followed by inspection of the **cloaca** (Figure 18.18; normal findings: feathers around the vent are unsoiled, cloacal mucosa is pink, moist, smooth and glistening). The following features should be noted:

18.16 Appropriate technique for opening the beak of a pigeon, avoiding damage to the cere.

18.17 Abdominal palpation (e.g. for diagnosis of egg binding or intra-abdominal tumours) in a budgerigar (Melopsittacus undulatus).

- shape and symmetry,
- closure of the vent,
- distance between the pubic bones,
- evidence of injury and
- signs of inflammation.

Faecal contamination of the feathers surrounding the vent may indicate the presence of **diarrhoea**. Endoscopy (after the instillation of water into the cloaca) enables more detailed examination.

Examination of the limbs includes comparison of the left and right side (see technique described in Chapter 20). During palpation and passive manipulation of the limbs (Figure 18.1), the following should be noted:

- palpable bony landmarks,
- any joint swelling,
- abnormal movement not associated with joints, crepitus.

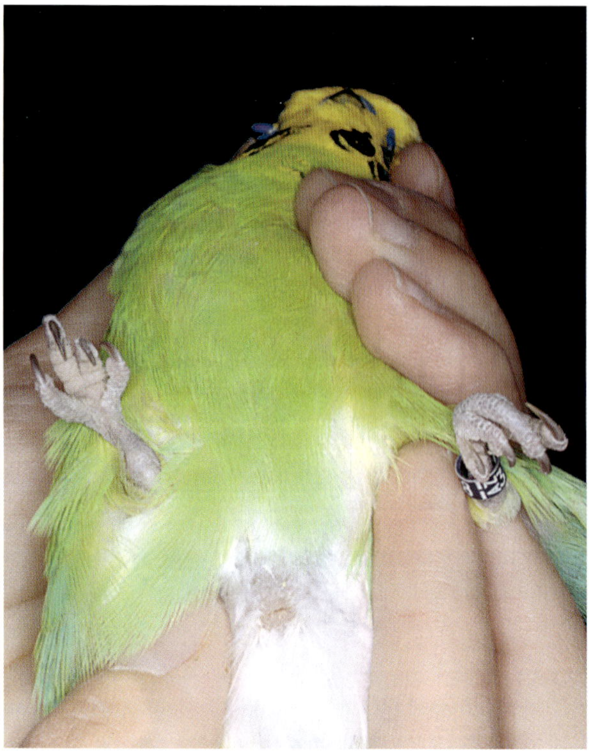

18.18 Visual inspection of the cloaca (e.g., for detection of injuries or parasite infestation) in a budgerigar (Melopsittacus undulatus).

Further investigation

Laboratory techniques

Laboratory assessment includes blood sampling (e.g., serology for salmonella, tuberculosis, herpesvirus and paramyxovirus) and examination of crop, nasal and cloacal swabs, as well as corneal impression smears, for the presence of viruses, bacteria, fungi and parasites.

19

Imaging techniques

R. Korbel, A. Probst and H.-G. Liebich

Imaging techniques used in avian medicine include modalities based on **radiation** (radiography, computed tomography), **sound waves** (ultrasound) and **magnetic fields** (magnetic resonance imaging). **Photography** should also be regarded as part of the practitioner's repertoire of diagnostic imaging techniques, since this is used on a daily basis for documenting normal and abnormal findings, and can be of considerable value in the investigation of disease processes.

Despite the widespread use of photography as a diagnostic aid, standardised protocols for this practice have not been developed. Some of the key aspects of photography in avian practice are therefore included in this chapter.

Until as recently as the 1920s and 30s, pictorial documentation of anatomical and clinical findings in general and specialised human and veterinary medicine consisted almost exclusively of hand-drawn diagrams (Figures 19.1 and 19.2).

A particularly striking example of this practice from the field of avian ophthalmology is *The Fundus Oculi in Birds* by C.A. Woods (Chicago, 1918), an atlas containing more than 200 colour illustrations of the posterior segment of the eye in a diverse range of avian species. Visualisation of the fundus through the dilated pupil of fresh cadavers was achieved using a combination of candlelight and mirrors.

In recent years, conventional photography based on silver halide film has been gradually replaced by digital techniques. Digital technology has undoubtedly been a great boon to the use of photography in routine avian practice. Nevertheless, it should also be considered that analogue photography may, in some circumstances, provide superior colour depth, contrast and resolution, compared with many commonly used digital cameras.

Diagnostic imaging techniques used in avian practice include:

- photography,
- radiography,
- sonography (diagnostic ultrasound, US),
- computed tomography (CT) and
- magnetic resonance imaging (MRI).

Photography

Photographic documentation of macroscopic anatomy and clinical findings (other than whole-body images) gen-

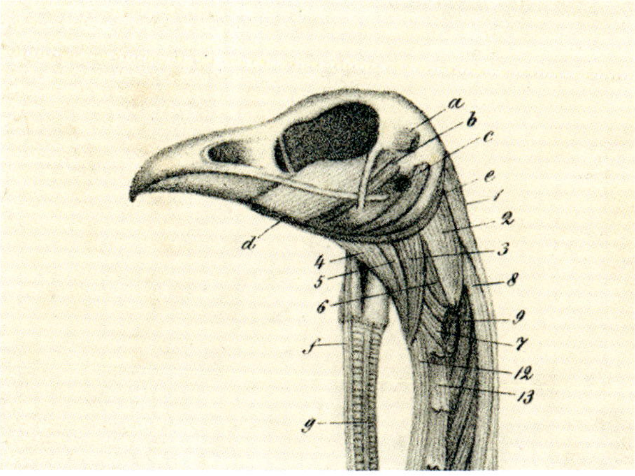

19.1 Illustration of the muscles of the jaw, hyobranchial apparatus, tongue, larynx and trachea of a chicken. From *Anatomie der Hausvögel* (*Anatomy of Domestic Birds*), Ernst Friedrich Gurlt, Berlin, 1849.

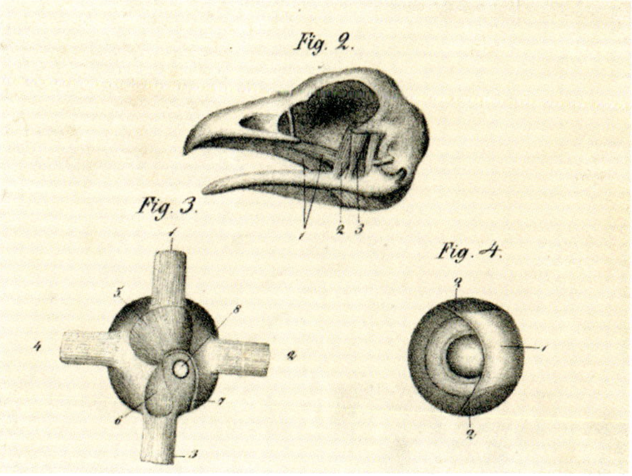

19.2 Illustration of the muscles of the jaw and the eyes of a chicken. From *Anatomie der Hausvögel* (*Anatomy of Domestic Birds*), Ernst Friedrich Gurlt, Berlin, 1849.

19.3 Contemporary photographic apparatus for hand-held or tripod-based photographic documentation of anatomical, clinical and pathological findings in birds: single lens reflex (SLR) camera with hybrid (digital-analogue) mechanism (interchangeable camera backs), 100-mm-macrolens and ring flash with four independently operable tubes and separate control unit.

erally requires magnification ratios ranging from 1:20 to 1:5. In particular fields, such as ophthalmology, ratios of up to 2:1 may be necessary.

The **basic equipment** required for this purpose is an SLR camera (ideally digital, resolution greater than 8 megapixels) (Figure 19.3).

Zoom lenses with a focal length of approximately 35–70mm (35mm format) are recommended to allow use of the camera at a practical distance. Macro lenses with a focal length of 100mm are useful for reducing shadowing. A ring flash with TTL ('through the lens') metering is essential for hand held photography and for elimination of shadows. For optimal light regulation and for maintaining depth of field, it is important that the ring flash has at least two – preferably four – flash tubes that can be operated independently (Figure 19.3).

Very large reproduction ratios can be achieved using close-up diopters, reverse-mounted lenses and bellows.

The auto-focus setting should ideally be disengaged, as this usually produces poor results, particularly in extreme close-ups. In most situations, focusing manually on the pertinent details within the visual field is preferable.

It is important to ensure that the deepest focus and focal range (depth of field) is achieved. This is accomplished using an aperture size (f-stop) of 11 or more.

Approximately two-thirds of the range of focus lies behind the focal plane, while one-third lies in front (under macro conditions, the ratio is closer to half-half). Depth of field decreases with increases in aperture size, focal length and magnification ratio. Mathematically, at a magnification ratio of 1:1 and f stop of 22 the depth of field is less than 1mm.

Thus when photographing anatomical specimens and avian patients, and generally when the camera is hand-held, it is advisable to predetermine the magnification ratio, pre-focus manually and then move the subject into the focal plane using the viewfinder.

The fast shutter speeds (<1/1000s) enabled by the use of an **appropriate flash** should eliminate blurring due to movement. For optimal results with shallow depths of field, particularly at high magnification ratios, the subject should be oriented, where possible, parallel to the focal plane (Figure 19.4).

Higher-quality, more diagnostically informative images can be produced by **placing the subject in front of a uniform, plain background**. Coloured, light absorbing non-reflective photographic cardboard is usually suitable for this purpose. To minimise shadowing, the distance between subject and background should be approximately 50cm.

For **photographing organs** (e.g., recording findings in *ex situ* organs) from above (vertical viewing angle), the tissue can be placed on a sheet of glass positioned horizontally, approximately 30cm above a coloured background. Maintaining an adequate distance between subject and background is also important when photographing from a horizontal perspective. Depending on the light source, a distance of approximately 30–50cm is optimal when using a 100mm macro lens. Shadow effects can be reduced through the use of a second light source, retro-illumination of the background and potentially also by digital image manipulation.

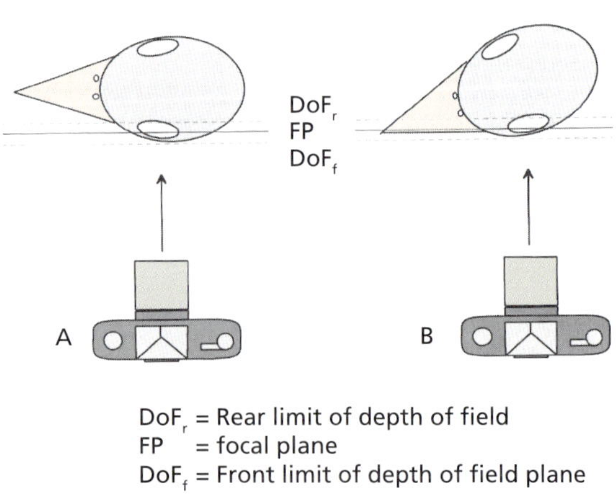

19.4 Optimisation of image quality by positioning of the patient with respect to a fixed focal plane and depth of field: in figure A, the beak lies outside the depth of field. By turning the head towards the focal plane (figure B) the beak comes into focus. Shadowing is avoided by placing the patient at an appropriate distance (at least 50cm) from the (preferably dark) background.

Radiography

In order to minimise the risk of stress and shock, radiographic examination of avian patients is ideally performed with the aid of gaseous anaesthesia (isoflurane or sevoflurane) using appropriate techniques for handling and restraint.

High-resolution (mammography) films without intensifying screens are useful for avian radiography. The following table serves as an approximate guide for current-exposure time product (mAs) values for birds of varying body weights at a focal spot-film distance of 70cm and a voltage of 50 kilovolts (50 kV) (noting that there can be considerable differences between x-ray machines):

- less than 50g body weight (e.g. budgerigars): 2.0 mAs,
- 100–250g body weight (e.g. cockatiels): 2.5 mAs and
- 250–1,000g body weight (e.g. mynas and macaws): 4.0 mAs.

The use of digital technology in avian radiography has found its way into routine practice as systems providing adequate resolution and contrast have become available at reasonable prices. Digital radiography systems providing highest resolution and contrast, such as the 'needle image plates'(NIP) based system (e.g., DX-M Digitizer system by AGFA Inc.) are highly recommended over conventional

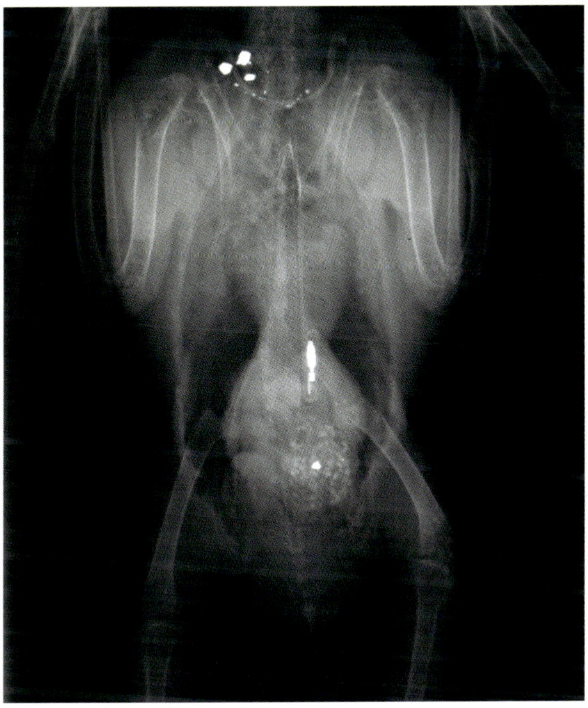

19.5 Radiograph of a bald eagle (Haliaeetus leucocephalus; ventrodorsal view): lead shot pellets are visible in the ventriculus.

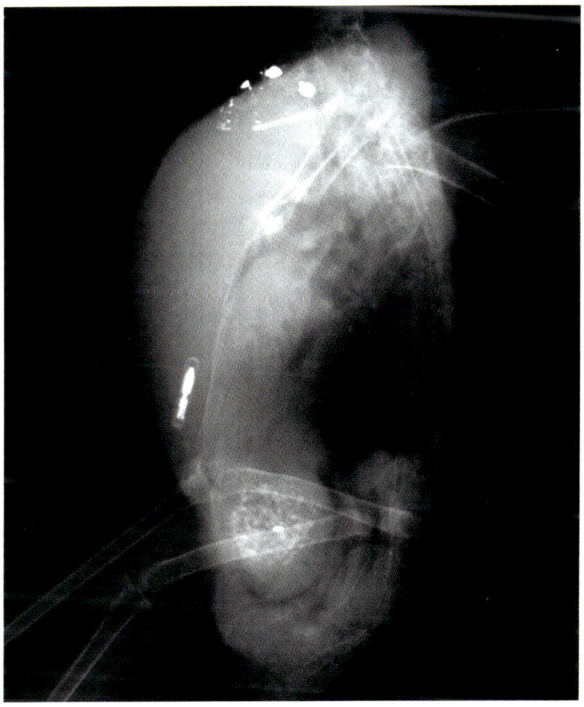

19.6 Radiograph of a bald eagle (Haliaeetus leucocephalus; lateral view): lead shot pellets are visible in the ventriculus.

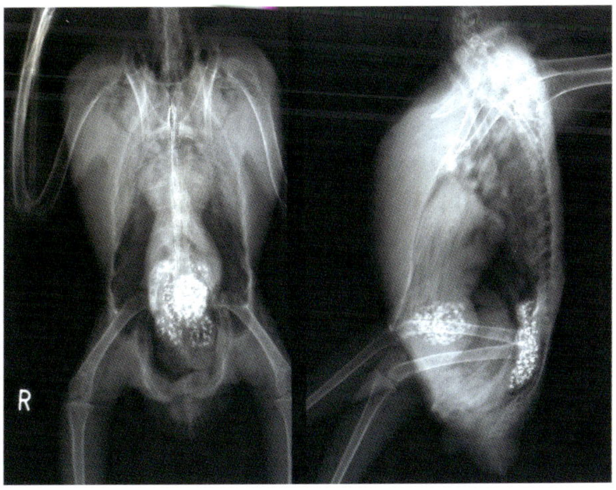

19.6a Digital radiograph (AGFA DX-M digitiser) of an African grey parrot (Psittacus erithacus) with renal gout.

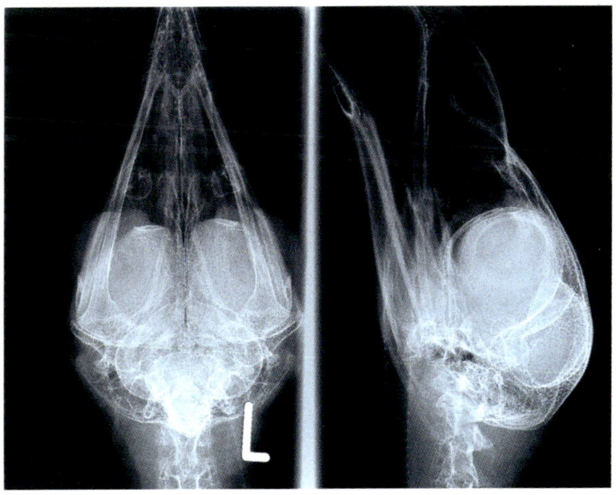

19.6b Digital radiograph (AGFA DX-M digitiser) of the skull of a golden eagle (Aquila chrysaetos).

'powder image plates' (PIP) due to the small body size of avian patients, where image resolution is crucial (Figures 19.6a and 19.6b).

Principles of radiographic image acquisition

For **routine radiographic examination** (identifying fractures, foreign bodies and organ disease), ventrodorsal and lateral views should be taken (Figures 19.5 to 19.12).

Restraint of an unanaesthetised or unsedated patient (e.g., using adhesive tape) should generally not be attempted due to the stress induced by immobilisation and the time required for obtaining radiographs (around two minutes).

When using isoflurane inhalation anaesthesia, patient positioning and image acquisition can be performed during the recovery period (approximately two minutes) following withdrawal of the anaesthetic gas. The duration of the recovery period varies between species and individuals, and is influenced by the general and nutritional condition of the bird.

Various devices have been developed in the US for **restraining birds without the need for handling by veterinary personnel**. This is necessary in jurisdictions where staff are not permitted to remain near the patient during irradiation. As well as reducing exposure of staff to radiation, this technique enables appropriate restraint and positioning of conscious patients, where concern exists about possible side-effects of anaesthesia.

In patients that are unlikely to tolerate either anaesthesia or restraint, imaging of the **unrestrained standing or recumbent bird** using a horizontally directed beam can be attempted. However, artefacts resulting from inadequate positioning cannot be ruled out using this approach.

To facilitate interpretation of radiographic images, it is important not only to ensure correct exposure (see above) but also to achieve **axial symmetry** in the positioning of the patient. This is particularly pertinent given the tremendous morphological diversity among the more than 10,000 avian species and 20,000 subspecies, as it enables comparison of the left and right sides of the body.

Depending on species, normal findings within the body cavity include a typical 'hour-glass'-like narrowing between the cardiac and hepatic silhouettes and location of the (grit-filled) ventriculus along an imaginary line drawn through the hip joints (e.g., in parrots). These serve as important reference points for identifying **organ displacement or enlargement.**

Positioning for the ventrodorsal view

PROCEDURE

The pelvic limbs of the dorsally recumbent patient are extended caudolaterally. In small birds that are difficult to handle with lead gloves, gauze tape can be tied to the feet and held by the gloved handler, who can thus keep their hands clear of the beam (Figure 19.7). Unless specialised views are indicated, the whole body is generally included in both the ventrodorsal and lateral views (Figures 19.11 and 19.12).

Important anatomical reference points for correct positioning of the bird in dorsal recumbency are the sternal carina and the vertebral column. Care must also be taken to ensure that the shoulder and hip joints are positioned symmetrically. In so doing, superimposition of the elbows and knees in long-limbed birds should be avoided.

Positioning for the lateral view

PROCEDURE

With the bird in lateral recumbency, both hindlimbs are extended caudoventrally (using gauze tape in small birds). The wings are fixed in a parallel position and drawn craniodorsally (i.e., opposite direction to the legs) over the back, applying the same amount of tension as to the hindlimbs (Figure 19.8). Due to the difficulty associated with handling birds with lead gloves, the pre-positioned bird is passed to the gloved handler by a second person, who then also takes the radiograph. Reference to alternative methods of restraint has been made above.

For the lateral view, anatomical considerations for ensuring correct positioning include alignment of both shoulder joints and hip joints.

POTENTIAL COMPLICATIONS

Sources of error in radiography include unduly prolonged procedure times resulting in stress following anaesthetic recovery, and use of excessive pressure or tension in the restraint and positioning of the patient. Particular care in patient handling is required when using lead gloves, as these interfere with the dexterity of the operator.

Contrast radiography

Contrast radiography can be employed to supplement plain radiography in examining the gastrointestinal tract. Barium sulphate is generally used, administered orally at 20–25ml/kg. When gastrointestinal perforation or atony is suspected, the use of iodine-based contrast materials is indicated to reduce the risk of peritonitis or settling out of barium sulphate, particularly in dehydrated patients. Timing of the radiographic sequence varies according to the structures under investigation and species-related variation in transit times. In psittacines, the following schedule serves as a useful guide:

- crop: immediately after administration,
- ventriculus and proventriculus: 20–30 minutes after administration,
- intestinal tract: approximately 1–1.5 hours after administration.

19.7 Positioning for radiography (ventral view). Even with small birds, lead gloves can be used (with appropriate care) for positioning the patient after induction of anaesthesia. Gauze tape is used as an aid for extending the legs.

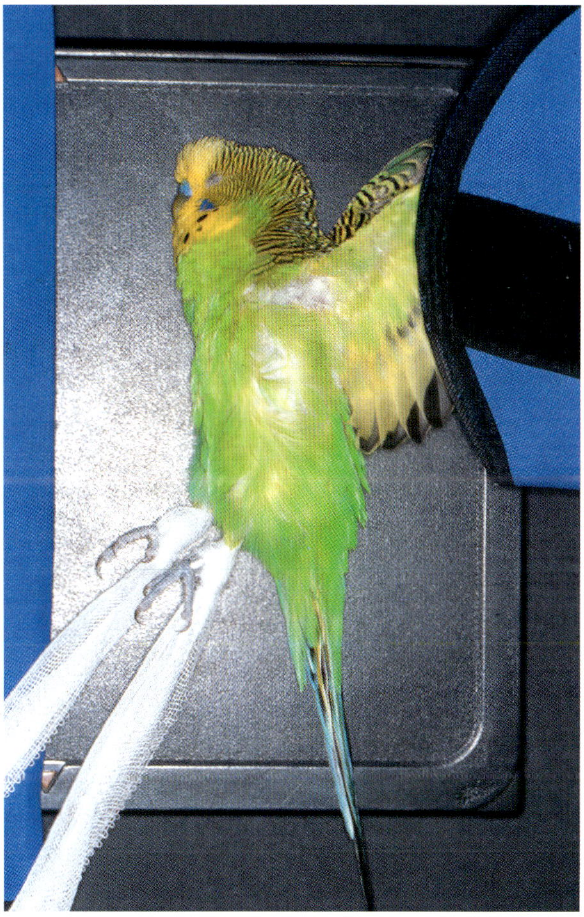

19.8 Positioning for the lateral radiographic view. The wings are extended and fixed over the dorsum, with simultaneous application of gentle tension to the legs (with the aid of gauze tape in small birds).

The passage of contrast material through the intestine may be considerably accelerated in cases of enteritis. In contrast, diseases that interfere with intestinal motility, such as proventricular dilatation disease or lead poisoning, may result in a marked increase in transit times. Contrast radiography plays an important role in the diagnosis of tumours (e.g., in budgerigars). Localisation of the tumour is based on typical displacement of contrast medium-filled intestinal loops.

Examples include:

- ventral displacement by kidney tumours,
- caudal displacement by tumours of the gonads or spleen (tumours involving these organs are difficult to differentiate from one another radiographically),
- dorsal displacement by liver tumours due to encroachment by neoplastic tissue between the abdominal wall and the intestine.

Sonography (ultrasound)

Sonography (usually B mode) can be used as an alternative or supplementary diagnostic imaging technique, particularly for examining the architecture of internal organs. Due to the limited surface area available for application of the ultrasound probe, microconvex and sector scanners with frequencies of 7.5–10MHz or greater are typically used for birds weighing up to 1,500g.

Transducers with a sector angle of 60 degrees are suitable for avian sonography. For specialised applications, such as sonography of the eye, higher resolution transducers may be indicated. Ultrasonographic examination can be conducted without anaesthesia. Conventional acoustic coupling gels can be used, although gel standoff pads may be required in small birds.

Patient preparation protocols are the same as for anaesthesia (see Chapter 20 'Handling, restraint and anaesthesia'), including a period of fasting.

For **sonographic examination of the abdomen**, the feathers are parted and the probe is placed between the xiphoid process of the sternum and the pubic bone

19.9 Plexiglass restraint board for radiographic examination of birds, with or without anaesthesia. Plexiglass plate (maximum 0.5cm thickness) is well suited to the construction of customised restraint apparatus.

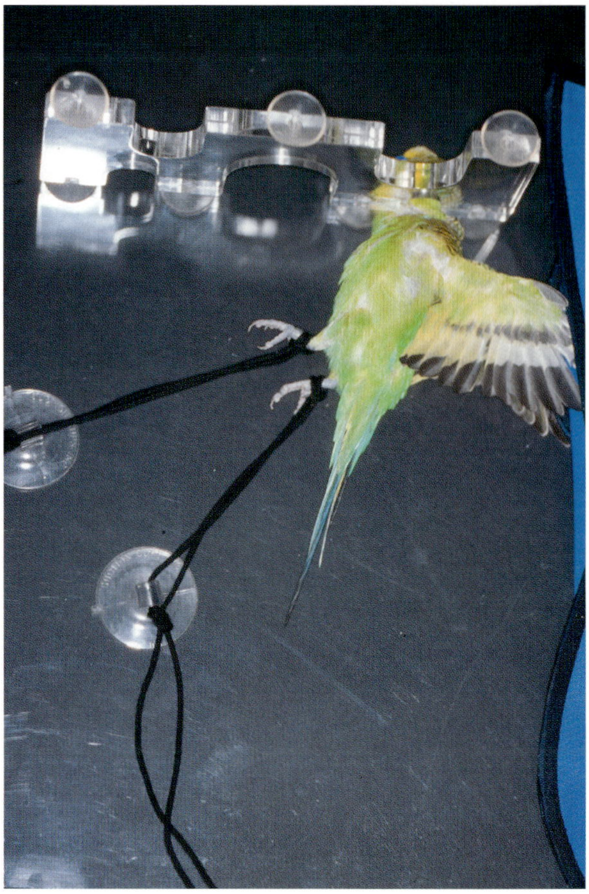

19.10 Use of plexiglass restraint board and additional aids for appropriate fixation of the head and limbs for radiographic examination of birds.

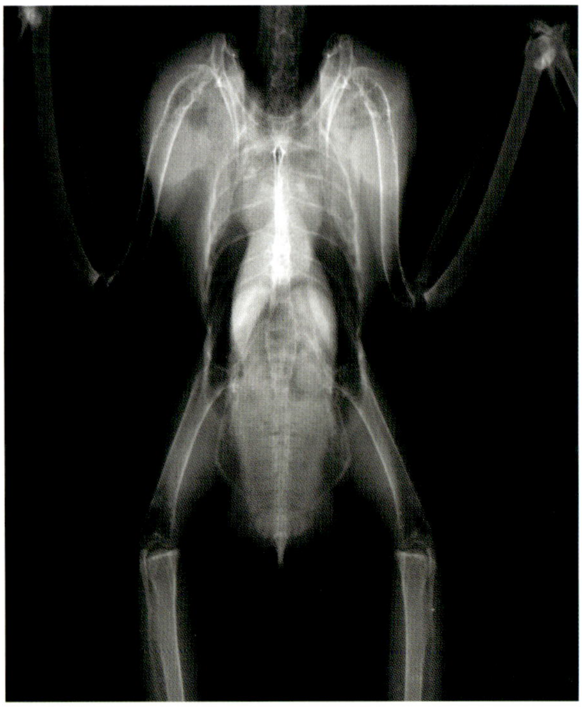

19.11 Radiograph of a normal common buzzard (Buteo buteo; ventrodorsal view).

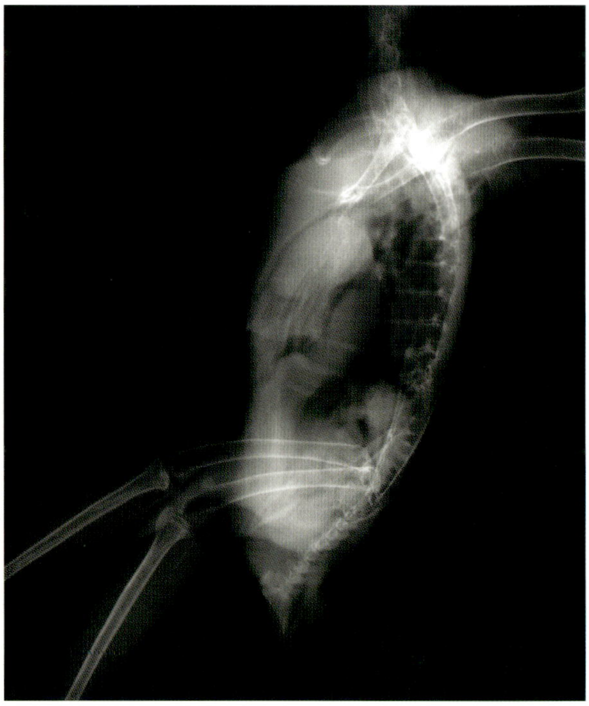

19.12 Radiograph of a normal common buzzard (Buteo buteo; lateral view).

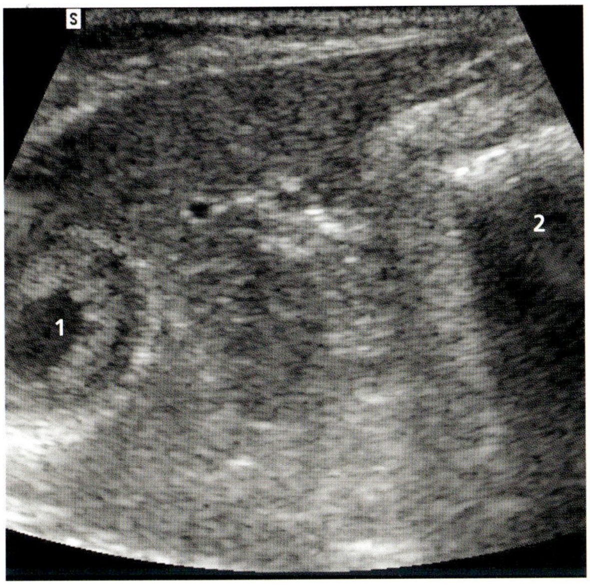

19.13 Ultrasound image (longitudinal section) of the liver of a chicken and adjacent (cranially) left heart chamber (1). The cuticle in the ventriculus produces an acoustic shadow (2). Courtesy of PD Dr S. Reese, Munich.

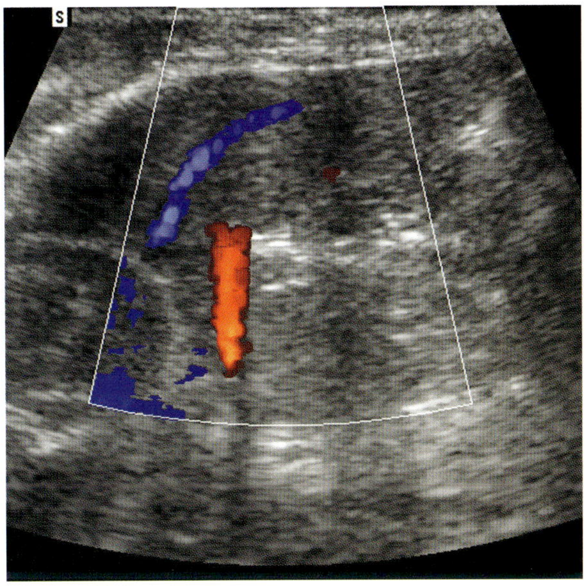

19.14 Colour doppler ultrasound image indicating blood flow in the major vessels of the liver. Blood flowing towards the heart in the right hepatic vein is represented in blue, while blood passing through the right hepatic portal vein (oriented in the opposite direction) is shown in red. Courtesy of PD Dr S. Reese, Munich.

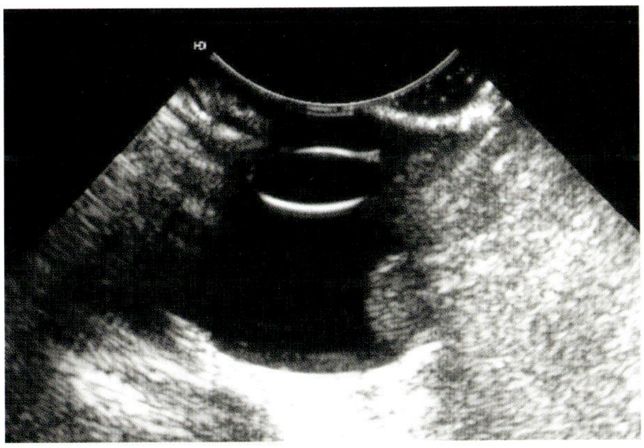

19.15 Sonogram of the left eye and pecten (pecten oculi) of a common buzzard (Buteo buteo).

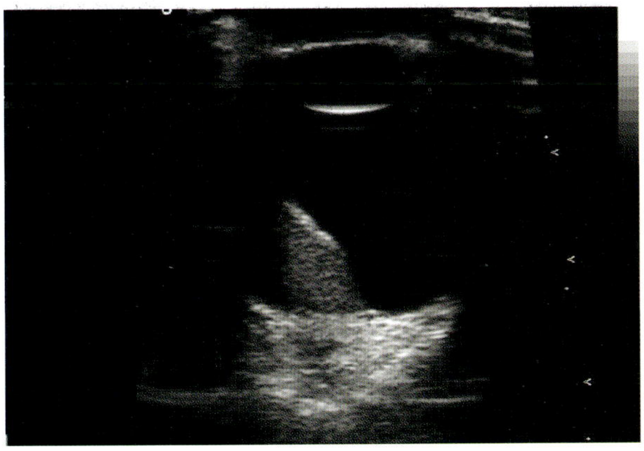

19.16 Sonogram of the left eye and pecten (pecten oculi) of a myna (Gracula religiosa).

(cranial and caudal approach). The air sacs reflect sound waves and must be avoided where possible. Depending on species, better visualisation of individual organs may be achieved by placing the probe along the left or right costal arch (lateral approach). Organs that may be examined sonographically include the liver and gallbladder, spleen, gastrointestinal tract, parts of the urogenital system and the heart (Figures 19.13 and 19.14).

The eye (Figures 19.15 and 19.16; see also Chapter 15 'The eye') can also be examined using B mode ultrasonography. As an alternative to placing the probe directly on the sensitive corneal epithelium, the procedure can be performed through the closed eyelid. This is particularly practical in owls, due to their larger upper eyelid.

The main indication for ophthalmic sonography is the presence of opacities in the anterior refractive media (e.g., keratitis, cataracts), as these prohibit direct examination of intraocular structures using techniques such as ophthalmoscopy (e.g., for diagnosis of vitreal haemorrhages and detachment of the pecten, commonly seen in trauma patients).

Doppler sonography as well as **3D and 4D-modes** have also been utilised in avian medicine. These techniques increase the diagnostic potential of ultrasound, permitting examination of organ vascularisation and three-dimensional evaluation of neoplastic lesions.

Computed tomography and magnetic resonance imaging

Computed tomography and **magnetic resonance imaging** (Figures 19.17 to 19.20) are the most advanced diagnostic imaging modalities available for identifying abnormalities of internal organs in live birds. In contrast to radiography, computed tomography enables early detection and localisation of focal and diffuse changes, and can be used to analyse tissue density. Moreover, internal structures can be represented in three dimensions. The combination of sectional imaging and 3-D reconstruction permits the assessment of spatial relationships between organs (e.g., the gonads, stomachs and nearby structures) and is of considerable value in localising foreign bodies.

Disadvantages of these techniques include the need to sedate and possibly restrain the patient (including the attendant risks of such interventions), and the relatively high cost of the procedure. Thus, they are generally only used when specifically indicated, or for particularly valuable birds (e.g., breeding and falconry birds).

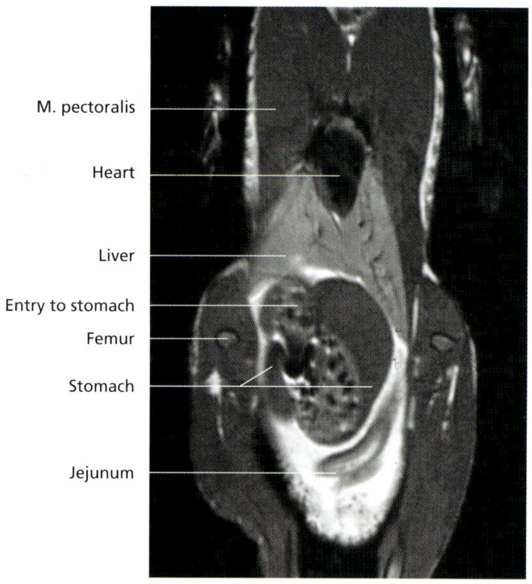

19.17 Magnetic resonance image (spin echo, T1-weighted) of a hen (dorsal plane) including heart, liver, stomach and intestine.

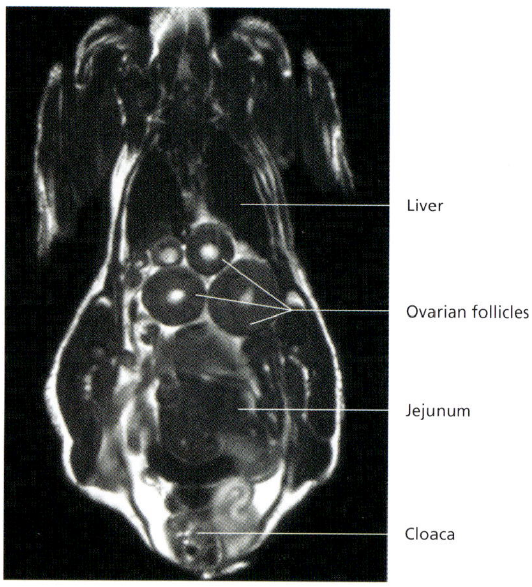

19.18 Magnetic resonance image (spin echo, T1-weighted) of a hen (dorsal plane) including liver, ovarian follicles and intestinal segments.

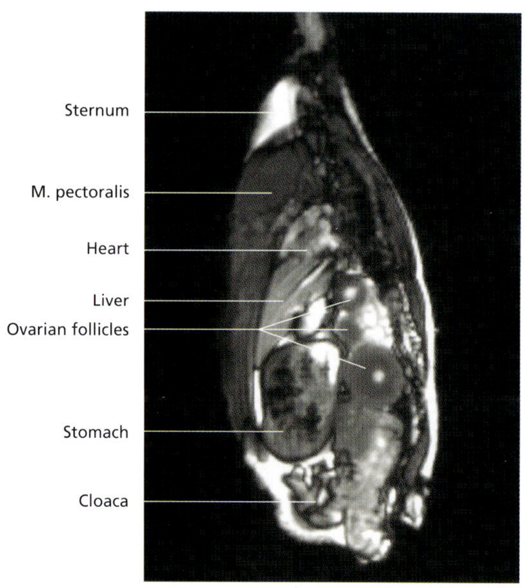

19.19 Magnetic resonance image (spin echo, T2-weighted) of a hen (sagittal plane) including heart, liver, ovarian follicles and segments of stomach and intestine.

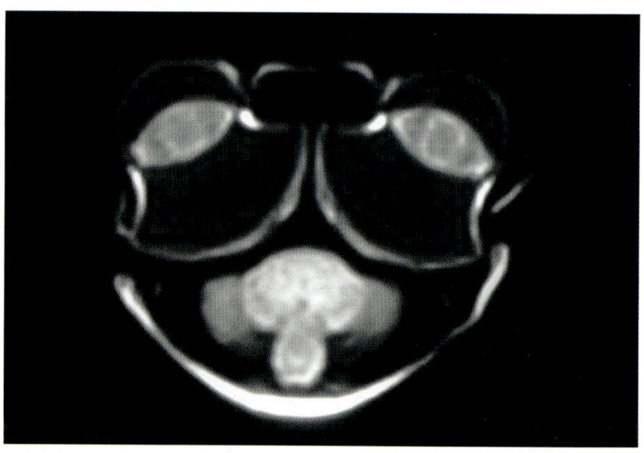

19.20 Magnetic resonance image (spin echo, T2-weighted, Repetition Time (TR): 2000 ms, Echo Time (TE): 100 ms, dorsal plane, 3mm slice thickness) of the head of a tawny owl (Strix aluco): lens architecture (with lens core and annular pad) and interorbital septum are shown. Courtesy of M. Konar and R. Korbel, Vienna.

Handling, restraint and anaesthesia

20

R. Korbel, S. Reese and H.-G. Liebich

In avian medicine, proficiency in routine handling and restraint of patients is essential for meeting the diagnostic and therapeutic needs of individual birds across a range of species (Figure 20.1). Handling errors can often be attributed to a failure to account for the special characteristics of avian anatomy and physiology.

Correct use of restraint is essential for thorough physical examination and for appropriate subsequent diagnostic and therapeutic intervention. Handling requirements vary according to the circumstance of the patient(s). In the management of flocks, particularly commercial poultry, practicality dictates that therapeutic agents are usually delivered via the food or water supply. In individual birds, on the other hand, administration of medication by injection is often preferable.

There remains a widespread tendency to believe that handling birds, particularly smaller patients, is likely to result in death from shock. Moreover, there is concern that the handler might injure the bird, or vice versa. Such reservations can lead to a situation where birds are examined by visual appraisal alone, and medication is delivered in the food or drinking water. However, this does not constitute a professional approach to case management.

20.1 Capture of a large psittacine: the head and neck are temporarily restrained by pressing the patient against the side of the cage. A layer of absorbent paper is used to create a hygienic barrier.

The techniques described below are suitable for handling and restraining birds in a clinical context. Particular consideration is given, from an anatomical and physiological perspective, to identifying and avoiding potential sources of error.

Theoretical principles

In addition to relevant animal welfare guidelines, factors governing the types and duration of restraint of avian patients include the general condition of the bird, and various particular anatomical and physiological characteristics of members of the class Aves.

Several of the following parameters are interdependent and their detrimental effects are additive. However, the use of handling procedures that are optimised for different species and individual birds can minimise the **risk of shock** and reduce the likelihood of **injury** to either the bird or the veterinarian.

Wrapping the patient in a cloth or towel can facilitate restraint and decrease the risk of injury. For clarity, these aids have been omitted from the images presented below.

Respiratory dynamics

Features that distinguish the respiratory apparatus of birds from that of mammals include the absence of a diaphragm, comparatively small (relative to bodyweight) lungs that are fixed to surrounding structures, and the presence of air sacs (see Chapter 7 'Respiratory system' and Figures 20.2 and 20.3).

During **inspiration**, the ribs move laterally and the sternum is displaced cranially and ventrally, largely due to the contraction of the m. intercostales externi and the mm. levatores costarum (Figure 20.2). The resulting increase in volume within the body cavity causes air to flow into the lungs and air sacs. Air flow through the parabronchi is unidirectional during both inspiration and expiration, making the avian respiratory system highly efficient (Figure 20.3).

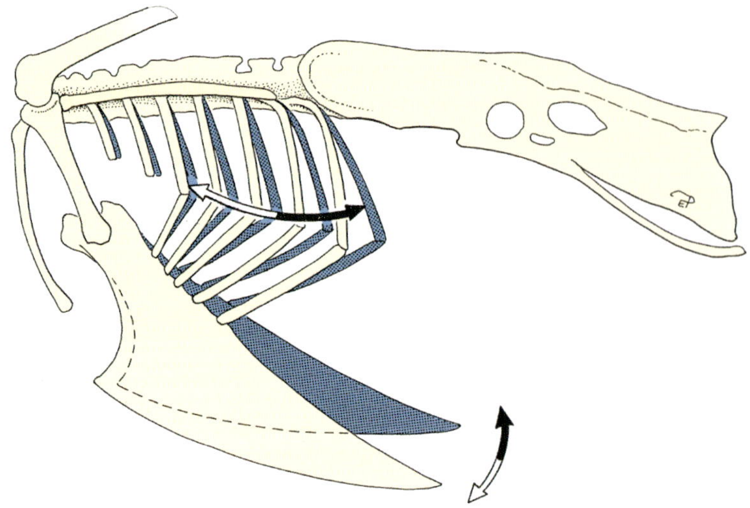

20.2 Respiratory dynamics in the bird. The arrows indicate the direction of movement of the bony components of the thorax in the chicken: **Inspiration** – expansion of the thoracic cavity through cranioventral displacement of the sternum and lateral movement of the ribs. **Expiration** – contraction of the thoracic cavity through caudodorsal displacement of the sternum and medial movement of the ribs.

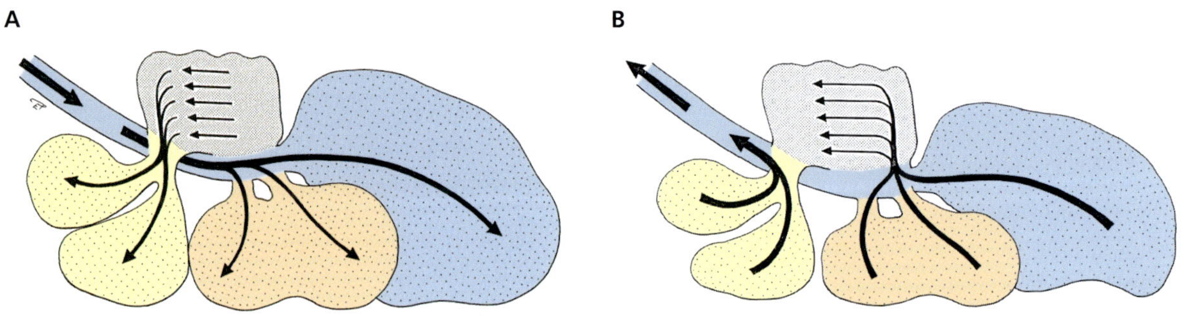

20.3 A) Schematic represenation of air flow in the avian respiratory system during **inspiration**: the caudal air sacs (caudal thoracic air sac and abdominal air sac) receive unused inspired air from the primary bronchi, while partially used air from the lungs flows into the cranial air sacs (cranial thoracic air sac and clavicular air sac). B) Schematic representation of air flow in the avian respiratory system during expiration: unused air from the caudal group of air sacs flows into the lung, while used air is expelled from the cranial group of air sacs via the primary bronchi and the trachea.

The mechanics of respiration are further aided by the weight of the viscera which, in the bird's normal upright posture, exert pressure in a cranioventral direction.

When the bird is turned on its back during clinical examination or surgery, these anatomical relationships are reversed. Breathing is compromised as the weight of the pectoral muscles and viscera must now be opposed during inspiration. When combined with the pressure placed on the sternum or abdomen during handling, this can result in inadequate ventilation and possible **asphyxiation**.

Compression of the breast and abdominal regions, effectively immobilising the bird in expiration, is the main cause of sudden death during physical examination and must therefore be expressly avoided.

Thermoregulation

In contrast to mammals, which predominantly lose heat through **evaporation**, birds experience a greater proportion of heat loss through the skin (particularly on the legs) through **convection** and **conduction**. Healthy birds can increase the rate of heat loss by spreading their contour feathers, thus opening up the feather coat, and by spreading their wings (see Chapter 17 'Common integument').

When birds are restrained, the flight response may be elicited, resulting in muscle contraction and increased heat generation. Moreover, a large proportion of the animal's body may be covered by the examiner's hand, particularly with small patients. This reduces the capacity for heat loss and may lead to hyperthermia.

Since the body temperature of birds is relatively high, organ damage due to hyperthermia can occur quite rapidly, compared with mammals. In most bird species the average core temperature is 40–43°C, though in some cases it is considerably lower (e.g., 36.1–36.9°C in the ostrich).

Hypothermia is an important consideration in the context of anaesthesia.

Reflex activity and heart rate

Occupying a relatively low position on the evolutionary scale, birds are among those animals that are **primarily driven by reflexive responses**. Dominant among these is the flight response, the expression of which is dependent upon the degree of domestication. The flight response induces a moderate degree of arousal. Inappropriate attempts at capture (e.g., bird is able to fly around the room or is chased around the cage) and prolonged periods of restraint further increase the level of excitement. This leads to a dramatic increase in heart rate and may result in circulatory collapse and death.

Birds with severe illness are at greater risk of shock, as are certain species such as canaries, grouse, pheasants, quail, gyrfalcons and various zoo birds. Owls, on the other hand, are relatively robust.

Skeleton

Much of the avian skeleton is pneumatised and the long bones are characterised by a thin cortex (see Chapter 1 'Introduction'). These features increase the risk of fracture when birds are handled inexpertly, particularly when the limbs are fixed in position away from the body.

Methods

Preparation

Capture of the bird establishes control over the patient and allows for temporary restraint. History-taking and visual appraisal (see Chapter 18 'Clinical examination') should precede capture, as the arousal associated with handling renders subsequent observations unreliable.

Before any handling procedures are carried out, particuarly if the patient is obviously seriously compromised, the owner should be advised of the risk of circulatory collapse or stress-moulting (significant feather shedding despite appropriate handling). This prevents subsequent misunderstandings or recriminations on the part of the owner.

Due to their emotional bond with the bird (particularly with tame birds), the owner should always be discouraged from providing **assistance**. Inexperienced owners also underestimate the risk of injury. Moreover, depending on local regulations, the attending veterinarian may be held responsible for any injuries sustained by the owner under these circumstances.

Exceptions can be made in the case of experienced individuals, such as breeders or falconers. Owners of tame birds should leave the room while the patient is restrained to avoid breaching the animal's trust. It is important to note that approaching the bird from above, or engaging its gaze from the front (bilaterally), can trigger fight or flight responses similar to those elicited by an oncoming predator.

As birds respond primarily to visual stimuli, **dimming the light** in the examination room facilitates capture and reduces the likelihood that escaped patients will sustain injury by flying into windows, noting that birds typically flee towards a light source. To further minimise the risk of injury, window glass should be frosted or, at a minimum, covered with blinds. Draping the ceiling has also proven useful in practice for retrieving birds that have escaped from their cage or transport container. The tendency of birds to fly towards the light should also be taken into account when capturing aviary birds under dim lighting conditions. For this reason, aviaries should ideally have an anteroom or vestibule with a separate door.

Toys that present potential risks (which are, in any case, of questionable merit on animal welfare grounds) should be removed from the cage before initiating capture. If the cage has a small door, it is advisable to detach the top of the cage and approach the bird via the floor opening.

Capture of the patient

Caged birds should be grasped using a calm but purposeful action, preferably from the rear so that the wings can be secured gently against the body (Figure 20.4).

Gloves should be used for handling birds with powerful beaks and claws (e.g., raptors) to minimise the risk of injury. For the purposes of hygiene, the bird should be grasped through a paper towel. The likelihood of injury can be reduced by grasping the bird immediately around the head and neck and holding it against the side or floor of the cage. For large birds, the other hand can then be used to stabilise the wings and feet, as restraining only the head and neck creates the risk of cervical vertebral fracture.

20.4 Restraint of a juvenile bald eagle (Haliaeetus leucocephalus) by grasping the hindlimbs and pressing the wings against the bird's body. A leather jacket, gloves and protective glasses are used to prevent injury.

Complications associated with capture usually occur when the process takes too long or when inappropriate technique is used. The former allows the patient to assume a defensive position (supine on the cage floor) or to fly around in the cage. This creates the **risk of shock**, as does excessively forceful handling.

Nervous birds, and those that have experienced frequent examination, may adopt a pre-emptive defensive stance, making capture more difficult. It is advisable to transport such birds in a suitably sized and constructed solid-walled container (e.g., cardboard or wooden box with air holes) from which the bird can be removed more easily. By providing a calming dark environment, these containers also reduce the risk of injury during transport.

Minimal restraint should be used with tame and juvenile patients. The head can usually be restrained successfully after scratching the bird's neck. Gloves should be avoided when handling these patients, as they can make the bird hand-shy and elicit abnormal behaviours (e.g., feather plucking). Use of a towel, which the bird can bite as a means of distraction, is often sufficient to **facilitate capture**. Tame birds should not be retrieved from the hand or shoulder of the owner as this risks injury to the owner.

Wild birds are usually strongly prone to exhibiting a **flight response.** To catch sick wild birds in the open, the animal should be covered with a towel through which it can then be grasped. Depending on species, care should be taken to avoid injury inflicted by the bird's claws (e.g., **birds that kill with their feet**, including owls, goshawks and eagles) or beak (e.g., **birds that kill by biting**, such as falcons).

Particular caution is needed with birds such as cormorants**,** grey herons and storks to avoid lightning fast, reflex-driven **stabbing** movements of their pointed beaks in the direction of the handler's eye. This reaction is elicited by reflection of light from the cornea of the handler, mimicking the mirroring of light from potential prey (e.g., the scale of a fish).

Chickens and water birds are handled differently, both in domestic and commercial situations. Chickens are **grasped by the feet**. Alternatively, in barns and open runs, a poultry hook is used to snare the bird just above the feet. Water birds are initially secured **by the neck** and temporarily restrained by holding down the head.

Restraint

The aim of restraint is to allow birds that have been provisionally secured during capture to be positioned optimally for the required interventions, while minimising the likelihood of injury to patient and handler (Figure 20.4). Appropriate restraint should allow for the patient to be relatively relaxed, even during prolonged periods of immobilisation. This is achieved in large part by **shielding the head** thus reducing stimulation of the dominant visual sense (see above).

A variety of techniques can be used, depending on the nature of the intended procedure. The following techniques are described below:

- thumb and finger head restraint,
- thumb and two-finger head restraint ('helmet grip'),
- two-finger hold ('ringer's hold'),
- technique for holding pigeons,
- technique for holding pigeons for examination of the pectoral and pelvic limbs,
- 'bunch of flowers' restraint,
- techniques for restraining poultry and water birds,
- techniques for use in surgery.

Thumb and finger restraint

METHOD

The head is secured from behind by placing the thumb and index finger on either side of the head near the mandibular joint. When using one hand (suitable for small birds up to the size of a budgerigar) the animal's body rests in the handler's palm, which can be used to hold the wings gently against the body. The hindlimbs are extended and secured by the ring and little fingers to prevent the bird from grasping instinctively at instruments used in examination and treatment (e.g., syringes). The other hand remains free to perform the required tasks (Figure 20.6). With larger birds, the second hand is used to hold the extended hindlimbs against the base of the tail, simultaneously securing the tips of the wings (two-handed technique).

APPLICATION

This means of restraint is suitable for palpation of the breast and abdomen, and for administration of intramuscular injections. Other applications include oral delivery of medications and, when the bird is held by an assistant, spreading and palpation of the wings.

POTENTIAL SOURCES OF ERROR

- placing pressure on the eyes with the thumb and/or index finger,
- compromising respiration by placing pressure on the sternum and/or abdomen.

Thumb and two-finger restraint (helmet grip)

METHOD

In this variation of the two-finger hold, the thumb and middle finger are placed on either side of the head near the mandibular joint, and the index finger is positioned against the occipital bone for additional support (Figure 20. 5). The wings and hindlimbs are held in the same way as for the thumb and finger hold.

APPLICATION

This technique is useful for restraining larger and stronger patients that may be able to twist their head out of the thumb and finger hold, in which the head is secured at only two points.

POTENTIAL SOURCES OF ERROR

- as for the thumb and finger hold.

Two-finger restraint ('ringer's hold')

METHOD

The index and middle fingers are placed on either side of the upper cervical region, securing the head and extending the neck (Figure 20.7).

APPLICATION

This technique enables the handler to retain the use of their thumb while using a one-handed grip (e.g., for removing rings, trimming claws or examining the wings).

POTENTIAL SOURCES OF ERROR

- excessive torsion of the neck,
- excessively tight restraint, resulting in inhibition of respiration.

Restraining pigeons

METHOD

The extended legs are held between the index and middle fingers and the wing tips are secured against the tail base by the thumb (Figures 20.8 and 20.9). Directing the patient's head towards the examiner creates an artificial obstacle that reduces visual stimuli and reduces the likelihood of attempts at flight. The other hand can also be used to support the breast; for example, during prolonged examinations. This technique can also be used for Galliformes and Phasianiformes. Pigeon fanciers generally condition their birds to this grip to prevent escape attempts and to minimise stress during handling.

APPLICATION

As well as being useful for general handling, this technique is suitable for examination of the head region, collection of crop swabs and visual inspection of the eyes and oral cavity.

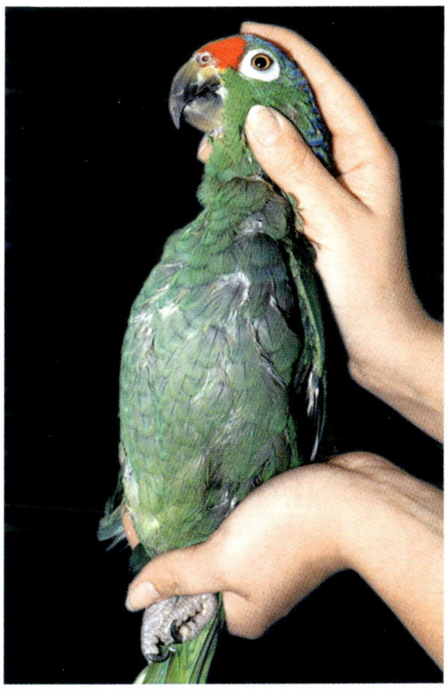

20.5 'Helmet grip': the head is held by placing the thumb and middle fingers over the mandibular joints, rotation of the head is prevented by placing the index finger against the back of the skull. Useful in relatively strong birds.

20.6 Thumb and finger hold: the thumb and finger are placed on either side of the head near the mandibular joint (note: avoid pressure on the eyes), the wings are fixed by the palm of the hand and the extended hindlimbs are secured by the ring and little fingers (note: avoid pressure on the chest and abdomen).

20.7 The 'ringer's hold' permits the handler to restrain the patient while also positioning the ring-bearing hindlimb between the thumb and ring finger.

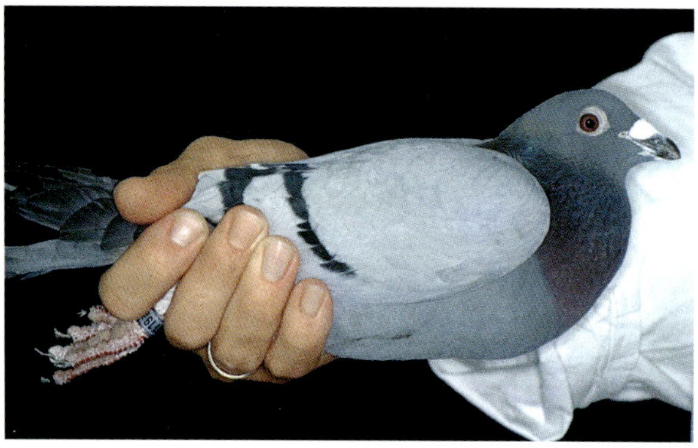

20.8 Restraining a pigeon by grasping the hindlegs between the index and middle fingers and securing the wings against the base of the tail with the thumb. Orienting the pigeon's head towards the handler serves to calm the bird by reducing visual stimuli.

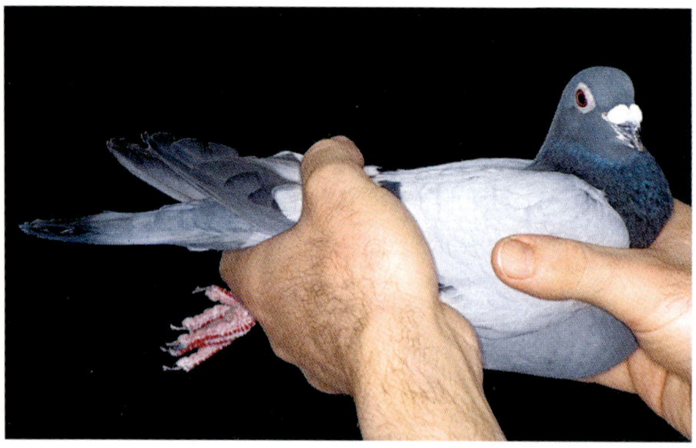

20.9 Use of the second hand to support the pigeon's breast region permits longer periods of restraint as well as visual inspection of the bird's head.

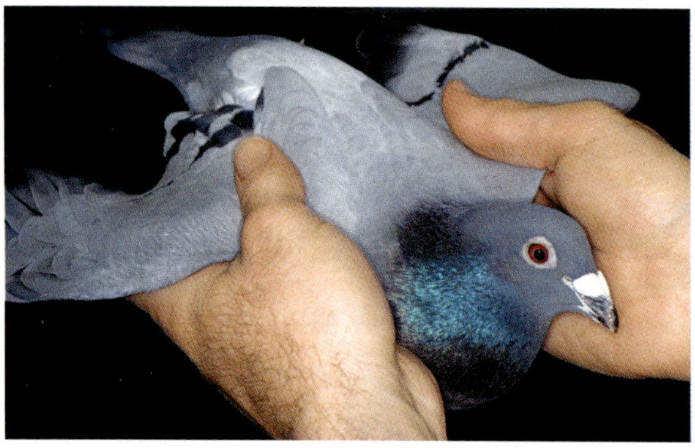

20.10 During routine examination of the wings, the forelimbs can be palpated simultaneously using the thumb and index finger. The remaining fingers are used to hold the bird on both sides of the breast region.

POTENTIAL SOURCES OF ERROR

- holding the body and the wings too far caudally and distally, allowing the bird to flap its wings,
- risk of pulling out tail feathers if the bird is instinctively grasped by the tail in response to an escape attempt.

Correct execution of this technique is of particular significance when dealing with pigeon fanciers, as the examiner's proficiency in bird handling is often interpreted as a general indication of their medical expertise (or lack thereof).

Holding pigeons for examination of the thoracic and pelvic limbs

The limbs are examined as a unit, generally commencing with the forelimbs then proceeding to the hindlimbs (Figures 20.10 and 20.11).

EXAMINATION OF THE THORACIC LIMBS

METHOD

With the examiner's hands in 'prayer position', the patient's breast region is fixed bilaterally using the middle, ring and little fingers. The patient's head is directed towards the examiner to reduce visual stimuli (see above).

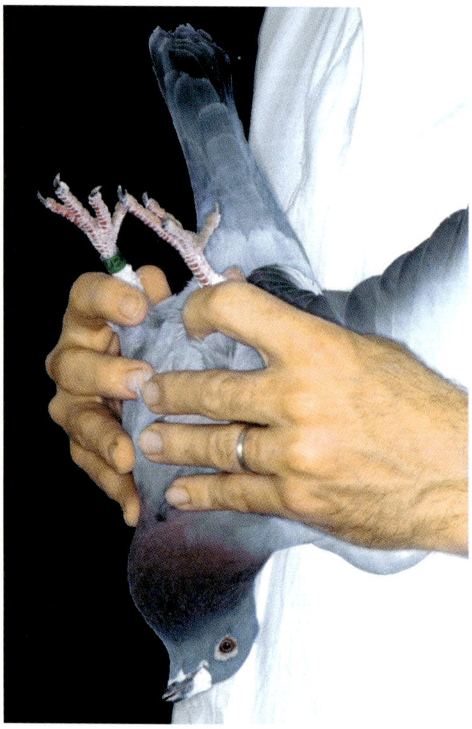

20.11 Routine examination of the hindlimbs of pigeons includes simultaneous palpation of both limbs with the thumb and index finger. The patient is restrained in an inverted position using the heels of the hands to secure the gently extended wings against the examiner's body.

The free thumb and index finger can then be used to palpate the wings simultaneously in a proximo-distal direction (Figure 20.10).

APPLICATION

Simultaneous (comparative) palpation of both wings is performed as part of routine physical examination, and when examination of the wings is specifically indicated.

POTENTIAL SOURCES OF ERROR

- inhibition of respiration by placing excessive pressure on the rib cage.

EXAMINATION OF THE PELVIC LIMBS

METHOD

The grip depicted in Figure 20.11 is achieved by transitioning smoothly from that shown in Figure 20.10. When examination of the wings is complete, the patient is rotated towards the examiner until its head is directed downwards. The heel of each hand is used to hold the bird against the examiner's chest, by pressing against the underside of the partially extended wings. Thumb and index finger are then used to examine both legs concurrently (Figure 20.11).

20.12 In the 'bunch of flowers' grip, the wings and extended hindlimbs are held together in a fist-like grasp, positioned around the caudal synsacrum.

APPLICATION

Palpation of the hindlimbs (usually simultaneously) is performed as part of routine physical examination, and when examination of the hindlimbs is specifically indicated.

POTENTIAL SOURCES OF ERROR

- elicitation of defensive responses (and subsequent excitement) by excessively prolonged examination and restraint.

'Bunch of flowers' grip

METHOD

In this simple grip, the patient is restrained in an upright position by simultaneously grasping the wing tips and caudally extended hindlimbs, incorporating the caudal synsacral region and the base of the tail (Figure 20.12).

APPLICATION

This technique enables the restraint and manipulation of the patient by a single person. It can be used for:

- restraining birds as large as pigeons without exposing the examiner to the risk of bites,
- palpation of the breast region,
- visual inspection of the head and
- administration of intramuscular injections.

POTENTIAL SOURCES OF ERROR

- holding the body and the wings too far caudally and distally, allowing the bird to flap its wings,
- risk of pulling out tail feathers if the bird is instinctively grasped by the tail in response to an escape attempt.

Restraint of chickens, raptors and waterfowl

Methods for restraining chickens and water birds are an extension of the technique used for capture (see 'Capture of the patient'). Chickens and wild raptors can be immobilized for short periods by holding the hindlimbs with the body hanging in an inverted position (Figure 20.13).

For the purposes of examination and transport of chickens and waterfowl over short distances, the preferred method is to hold the upper wings with the body freely suspended (Figure 20.14). Chickens can be temporarily restrained (e.g., when collecting several birds from a flock) by drawing the wings over the dorsum and interlocking them at the level of the carpometacarpal joint (Figure 20.15).

Specialised methods of restraint may be required for particular species, such as zoo birds. Penguins are

20.13 Chickens and raptors can be restrained or transported for short periods by grasping both hindlimbs and allowing the bird's body to hang freely.

20.14 Waterfowl can be restrained or transported for short periods by grasping the proximal forelimbs, with the body allowed to hang freely.

20.15 Chickens can be immobilised by interlocking the wings over the back.

restrained by holding the bird around the breast region (under the wings) with the hindlegs hanging freely, while restricting visual stimuli. This method reduces the risk of beak-related injury of the handler and stress-induced aspiration of regurgitated crop contents by the bird.

Fixation for surgery

The practice of tying down patients, commonly used when performing surgery on mammals, is rarely employed for birds of small to medium size. In addition to previously stated reasons, the use of sticking plaster or restraint boards in combination with heating pads is undesirable due to the need to quickly return patients to sternal recumbency after surgery (see 'Respiratory dynamics').

Restraint of surgical patients is merely used to position the bird in a manner best suited to the intended procedure. Exceptions include potentially problematic anaesthesia in water birds, where restraint may be required when the bird is insufficiently relaxed. Methods for routine surgical positioning are generally adaptations of the techniques described above.

Total immobilisation is important for certain procedures such as microsurgery (e.g., ophthalmic surgery). This can be achieved using air sac perfusion anaesthesia, which produces reversible perfusion-induced inhibition of respiration, ensuring complete immobility.

METHODS

For surgery involving the body cavity, the patient is placed in dorsal recumbency and the moderately extended wings are held gently against the operating surface using the palms of the hands. The thumbs and index fingers can be used to keep the hindlimbs clear of the operating field by retracting them caudolaterally.

For endoscopy/surgical sexing the patient is usually placed on their right side as an endoscopic approach from the right side is hindered by the stomach (Figure 20.16). An assistant holds the patient in position by gently abducting the left wing and securing the head using the 'helmet' grip. The hindlimbs are extended either cranioventrally or caudoventrally, depending on the species of the patient, and upon whether the operator prefers to access the body cavity from cranial or caudal to the thigh.

Detaching biting or grasping birds

The first priority is to remain calm, particularly as the specialised clutching mechanism of the claws of raptors prevents them from releasing their grip when aroused. As the bird begins to settle down, an assistant can carefully loosen the grip exerted by the beak or claws. Often this can also be achieved simply by blowing on the patient's head or releasing the bird from a restraining hold. Use of force to remove biting or grasping birds can result in fractures of the beak and toes. The bird must not be splashed with water due to the risk of aspiration. Ideally, the need for any such intervention is avoided by the careful use of appropriate methods of restraint.

Anaesthesia

In a broad sense, anaesthesia can be considered a form of restraint. As well as being used for surgical intervention, anaesthesia is used prophylactically for avoiding stress.

A distinction is made between anaesthesia and sedation, the latter potentially being suitable for procedures such as transport and capture.

Methods of **sedation** include:

- diazepam administered at 0.5–1.5mg/kg, intramuscular or oral (via food or drinking water).

It must be emphasised that the manner in which the patient is handled has a significant impact on the duration of anaesthetic induction and the efficacy of the anesthetic protocol. Calm, quiet capture and handling is very important in this regard, as is shielding of the patient from visual and auditory stimuli.

Notwithstanding its relatively greater cost, **inhalation anaesthesia** offers the advantages of being simpler to perform, better tolerated by the patient and associated with short recovery times (Figure 20.16).

Agents used for **inhalation anaesthesia** include:

- isoflurane: induction with 4–5 per cent isoflurane in oxygen (1.2–1.5l/min) using a semi-open system; maintenance at 0.7–4.0 per cent, depending on species.

Of the agents available for **injectable anaesthesia,** ketamine hydrochloride (21–40mg/kg i.m.) has several disadvantages. It produces inadequate analgesia and protracted recoveries that may be accompanied by excitement. This predisposes the patient to hypothermia, injury and aspiration of crop material. The recovering patient should therefore be wrapped in a towel and placed in sternal recumbency in an adequately humidified (approximately 70 per cent relative humidity) recovery box pre-heated to 28–30°C. Alternatively, a fully reversible injectable protocol (mixture of medetomidine 0.35mg/kg, midazolam 4.5mg/kg, fentanyl 6µg/kg administered i.m.; reversed within 3–4 minutes by the adminstration of a mixture of atipamezole 2mg/kg, sarmazenil 0.6mg/kg and naloxone 0.16µg/kg) can be considered, though this may not achieve a surgical plane of anaesthesia.

A further anaesthetic technique known as **air sac perfusion anaesthesia** makes use of the particular anatomical features of the avian respiratory tract (see Chapter 7 'Respiratory system'). This procedure involves retrograde perfusion of the lung-air sac system (oxygen 0.3L/min, isoflurane 0.5–2 per cent, sevoflurane 1–4 per cent), preferably via the left caudal thoracic air sac.

Air sac perfusion leads to a drop in partial CO_2 pressure ('washout') to below the threshold required for stimulation of respiration (approximately 40mm Hg in the domestic

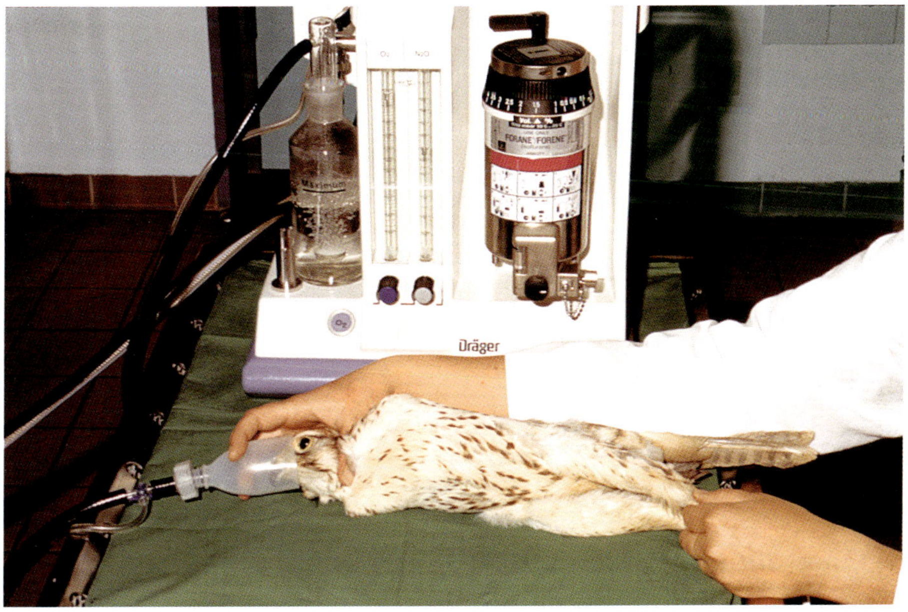

20.16 Appropriate positioning of a saker falcon (Falco cherrug) for endoscopy. Inhalation anaesthesia is performed using a mask, with an isoflurane-oxygen mixture in a semi-open system.

pigeon). The resulting apnoea persists for the duration of perfusion. Spontaneous breathing resumes around 2–4 minutes after perfusion ceases, when the CO_2 in the respiratory tract has returned to threshhold levels. Alternatively, respiration can be induced by the admixture of 1 per cent CO_2. Advantages of this well-tolerated and readily controllable technique include free surgical access to the head region, a reduction in intra-ocular pressure and complete immobilisation of the patient (beneficial for microsurgical techniques including ophthalmic surgery). It also provides a means of emergency ventilation of patients with obstructions of the upper airway.

Management of shock

Inappropriate handling and restraint commonly triggers a sequence of events that result in **shock**. In this situation, further interventions should be kept to a minimum, depending on the circumstances.

The uppermost priority is to place the bird in sternal recumbency in a pre-heated recovery box, with addition of oxygen as required, and to eliminate all visual and auditory stimuli. If shock is the result of a handling procedure (e.g., claw trimming) rather than disease, birds managed in this way generally recover quickly.

Birds requiring medical intervention should promptly be administered fluids pre-warmed to body temperature (lactated ringer's solution, 21–30ml/kg SC or IV). In cases of blood loss due to injury or intraoperative haemorrhage, rehydration should be carried out over several days to restore fluid to the intracellular compartment. Fluid administration can be performed in various ways (see Chapter 21 'Medication and blood collection techniques'). Blood should also be collected for determination of haematocrit and total plasma protein concentration.

Use of corticosteroids is problematic in birds due to their potent immunosuppressive effect. They should only be administered when specifically indicated and the concurrent use of antibiotics for prevention of secondary infection should be considered.

The relative advantages of intravenous fluid replacement (rapid efficacy) should be weighed on a case-by-case basis against the potential disadvantages of having to handle the patient (e.g., stress associated with compression of the jugular vein when using the 'ringer's hold'). In some situations, subcutaneous adminstration of fluids may be preferable (see Chapter 21 'Medication and blood collection techniques').

The following protocol applies to the **prevention** and **management** of shock:

Initial management:

- place patient in a pre-heated recovery box (approx. 30°C),
- dim the ambient light,
- desist from further non-essential handling.

Therapeutic intervention:

- weigh the patient,
- administer diazepam (0.5–1mg/kg),
- administer dexamethasone (5mg/kg), preferably IV in emergency situations,
- administer fluids (at body temperature, 21–25ml/kg):
  - 40 per cent glucose solution,
  - 40 per cent lactated ringer's solution,
  - 20 per cent amino acid solution;
- depending on circumstances, commence short term broad spectrum antiobiotic prophylaxis (e.g., enrofloxacin 10mg/kg; cefotaxime 100mg/kg).

Analgesia and pain management

It is important to stress that, while birds share with domestic mammals and humans a capacity for pain perception, they have relatively limited means of demonstrating that they are experiencing pain. As such, the use of analgesics should always be considered in cases where the presence of disease, or the need for manipulations or interventions, is likely to be painful. The guiding principle here is that pre-emptive adminsitration of analgesia is preferable to retrospective pain management (i.e., **prevention rather than treatment of pain**).

The response to analgesic agents, including duration of efficacy and side effects, can vary greatly between species. At excessive doses, auto-antagonism may be observed. Preferred analgesics in avian medicine include opioids and non-steroidal anti-inflammatory drugs (NSAIDs).

The main agents and their clinical characteristics are summarised below:

- NSAIDs: meloxicam (0.3–0.5mg/kg); peak effect 30–90 minutes post-administration; side-effects include pronounced hypothermia and respiratory depression.
- Opioids: butorphanol (1–3mg/kg) and buprenorphine (0.5mg/kg); more potent analgesia than NSAIDs; duration of effect up to seven hours post-administration; marked side effects with somnolence and protracted post-anaesthetic recovery in raptors.

21

Medication and blood collection techniques

R. Korbel and H. E. König

Theoretical principles

Administering medications

The use of appropriate techniques for medication and blood collection is of considerable importance in avian diagnostics and therapeutics. This is exemplified by the contrast between the oral and parenteral routes in the treatment of individual patients in companion bird medicine.

Delivery of medications per os has several limitations, including:

- a relatively slow rise in blood and tissue concentrations following enteral absorption,
- poor control over dose rates due to species variation in water intake and/or selective food intake, and
- potential impacts of the pH within the digestive tract or interactions with gastrointestinal flora.

Administration of drugs by injection results in a rapid and predictable onset of action. This facilitates prompt and targeted intervention in disease processes that, due to the high avian metabolic rate, can progress very quickly.

Thus, while the oral route is routinely used in commercial flocks, largely for economic reasons, it is inappropriate for treatment of acute disease in individual birds.

The most common indication for **venipuncture** in avian medicine is collection of blood. Micro-methods for determining laboratory values, performing rapid diagnostic tests and conducting gender determination have become increasingly established in recent years. This is particularly advantageous for avian practitioners as these techniques reduce the ratio between the amount of blood required for the sample and the maximum blood volume that can be safely collected from the patient (1 per cent of body weight).

Methods

Restraining the patient

Appropriate handling of the patient (see Chapter 20 'Handling, restraint and anaesthesia') by veterinary personnel is vital for avoiding complications.

Inadequate restraint can induce stress and shock, particularly as the resting avian heart rate is relatively high. Furthermore, birds are at risk of sustaining injury if they attempt to escape or defend themselves. Especially in small patients, this can rapidly lead to life-threatening situations. Use of the examiner's hand or a towel to shield the patient's head, thus limiting visual stimuli, is useful for calming the bird. Noise should also be kept to a minimum.

Preventing haemorrhage and achieving haemostasis

Haemorrhage can be avoided by tensing the skin prior to injection and releasing the skin before the needle is removed. The skin spreads over the injection site, creating a reliable seal. Minor bleeding can be controlled by the application of pressure. Small haematomas also exert pressure on the injection site.

Haemostasis is very important, particularly in small birds, as any blood lost to haemorrhage contributes to the relatively limited maximum volume that can safely be removed from the circulation. It is crucial, therefore, to reduce the risk of vessel laceration during blood collection by using appropriate restraint. When performing venipuncture, the hand in which the syringe is held must be in direct contact with the patient, as this allows the operator to react appropriately (hand moves with the bird) if the bird attempts to defend itself.

As well as causing undesirable blood loss, minor haemorrhages result in soiling of the feathers and objects within the bird's cage. This can be a source of anxiety for owners, who typically overestimate its significance.

Disinfection of the skin

Disinfection of the skin and feathers can only be achieved to a limited extent. Moistening of the water-repellent plumage with alcohol serves primarily to separate the feathers in order to visualise the vessels and other anatomical structures between the feather tracts. Visual access is important because the veins tend to 'roll' under the skin, due to a relative lack of subcutaneous tissue.

Excessive dabbing or rubbing of the skin should be avoided, as this can result in intracutaneous haemorrhages that obscure the structures underlying the paper-thin skin.

In species with a dense covering of down devoid of apteria (waterfowl, various zoo birds, etc.), the vessels cannot be seen through the skin. Instead, the location of the vessel can be identified by dampening the feathers and smoothing them against the body. Occlusion of the vein causes the feathers over the vessel to rise, exposing its subcutaneous course.

Needle and injectate specifications

Optimal needle gauge and length are determined by the size of the bird. The smaller the diameter of the needle, the greater the speed at which the injectate leaves the needle tip, increasing the risk of vessel perforation and haemorrhage. This applies particularly to viscous preparations that necessitate greater pressure on the plunger. It is advisable, therefore, to use the lowest possible gauge (largest diameter) needle that is practical for the size of the patient.

Due to the relatively high viscosity of avian blood (nucleated erythrocytes), relatively large bore needles are also indicated for blood collection. Generally, 22G to 25G needles are used, with 25G being suitable for most small birds. Siliconised or heparinised needles can be used to reduce the likelihood of coagulation within the needle.

With respect to injectable medications, **aqueous solutions** are preferable to oil-based products. In particular, the administration of cold formulations containing relatively large oil droplets is associated with the risk of embolism, local necrosis and even myocardial infarction.

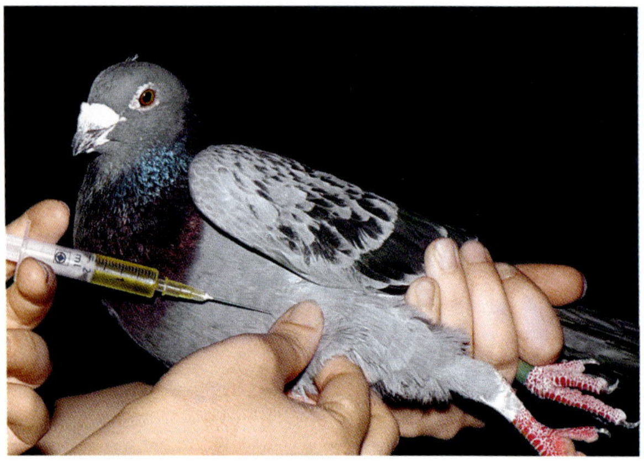

21.1 Subcutaneous injection into the pre-crural fold in a pigeon (Columba livia).

Wherever possible, it is advisable to use products with good tissue tolerance, as myopathy and necrosis are often observed in association with intramuscular injection. For the same reason, some products (e.g., calcium formulations) should be administered subcutaneously.

Methods of administration

Pharmacological agents and other substances can be administered by means of:

- plastic tube or crop needle,
- drops applied to the nares or conjunctiva,
- intramuscular injection,
- subcutaneous administration,
- intracutaneous or percutaneous administration,
- application of creams or powders,
- intraosseous (intramedullary) administration,
- intravenous and intracardiac injection and
- intrapulmonary injection.

Tube or crop needle

APPLICATION

Plastic tubing and crop needles are used for oral administration of medication and contrast material, performing crop washes and for medically indicated force feeding of young or inappetent birds.

RESTRAINT

The thumb and finger or helmet grips can be used, paying heed to the degree of extension of the neck (Figures 21.2 and 21.4). When handling birds such as raptors and larger psittacines, the use of a **mouth speculum** or **wooden gag** facilitates access to the oral cavity and prevents biting of the tube. The gag is inserted horizontally between the upper and lower beak. To ensure that it is not ejected by the bird's tongue, the thumb and index finger are used to apply upward pressure on angle of the beak and downward pressure on the upper beak (Figure 21.4).

The use of a **metal speculum** is associated with the risk of damage to the horn of the beak, especially when the area over which the speculum is in contact with the beak is small.

To access the oral cavity in pigeons (e.g., to administer tablets or obtain a crop swab), the upper beak is grasped from behind using the thumb and middle finger, and the index finger is used to depress the lower beak (see Chapter 18 'Clinical examination' and Figure 18.16).

METHOD

The material is delivered directly into the crop to **avoid the risk of aspiration** (Figures 21.2 and 21.4). Curved

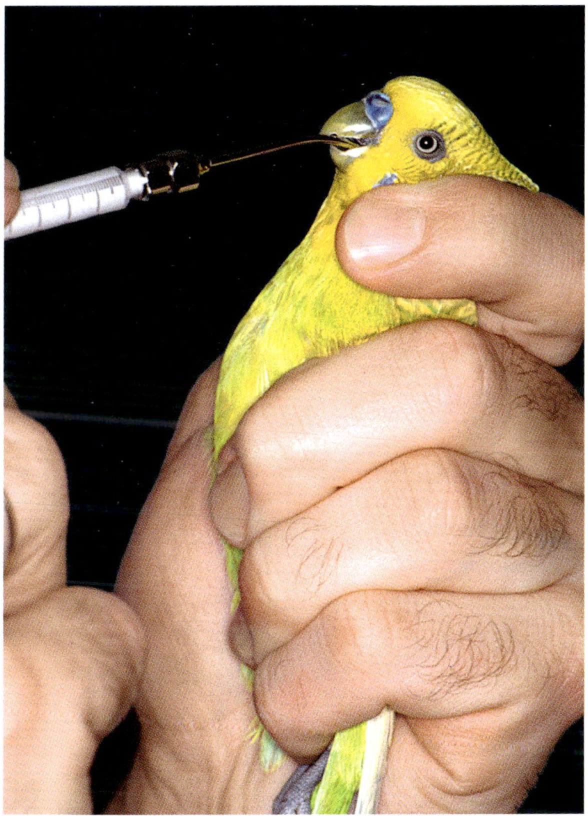

21.2 Oral administration of medication using a crop needle in a small bird: the crop needle is introduced at the angle of the beak on the left side, then directed towards the right side of the oral cavity and advanced across the tongue over the laryngeal opening. The patient is restrained using the thumb and finger hold.

21.3 After passing the needle over the laryngeal opening the syringe is tilted vertically so that it can be passed easily into the oesophagus, located on the right side of the neck, and advanced into the crop (the needle is inserted up to its hub).

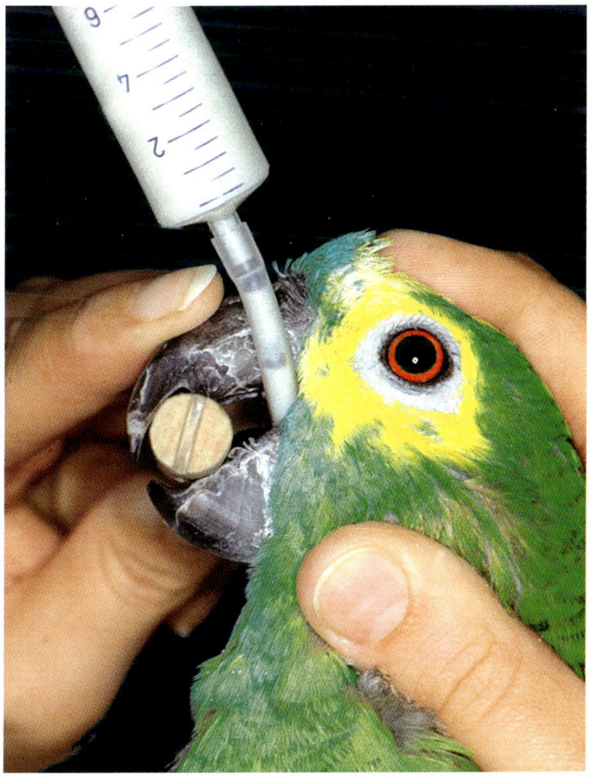

21.4 Oral dosing using a plastic tube in a larger patient: the beak is held open using a wooden gag, which is kept in place by placing pressure on the upper beak with the index finger and directing pressure towards the angle of the mouth with the thumb. The tube is introduced at the angle of the beak on the left side, then directed towards the oesophagus, located on the right side of the neck. Correct positioning of the tube is determined manually (respiratory sounds indicate incorrect placement).

crop needles are suitable for birds up to around 100g bodyweight, while a flexible plastic tube is used for larger patients. The **crop needle** is inserted horizontally at the angle of the beak on the left side. It is then tilted ventrally and guided manually into the crop (Figures 21.2 and 21.3). Similarly, **plastic tubing** is inserted at the left angle of the beak and directed diagonally over the laryngeal opening and base of the tongue towards the right side of the neck (Figure 21.4).

Careful restraint of the patient during this procedure is essential as serious injuries of the beak can have grave consequences.

For anatomical reasons, introduction of the needle or tube from the **left side** is particularly important. In birds with curved beaks, insertion of the needle from other angles may result in the tip of the needle being directed towards the left side of the body, i.e., opposite to that on which the avian oesophagus is located. Subsequent advancement of the needle may result in perforation of the pharynx. As a cautionary measure, material should (at least initially) be administered in small portions rather than as a complete bolus.

POTENTIAL SOURCES OF ERROR

- inadvertent introduction of the tube or needle into the trachea via the laryngeal opening; respiratory noise indicates incorrect positioning,
- insufficient advancement into the oesophagus; also associated with the risk of aspiration,
- perforation of the crop due to inappropriate manipulation of the needle / tube or introduction of the needle / tube into the right side of the mouth.

Application of drops to the nares or conjunctiva

APPLICATION

Topical application of drops is used to administer small volumes of liquid medication to small patients with minimal stress. Conjunctival instillation is also used as a means of vaccine administration in commercial operations.

METHOD

The drops are applied to the nares (Figure 21.5), from whence the liquid enters the oral cavity through the choana. It is then swallowed.

When applied to the conjunctiva, some of the dose is absorbed by the mucous membrane while the remainder reaches the oral cavity via the naso-lacrimal system.

POTENTIAL SOURCES OF ERROR

- incorrect handling can lead to excessive arousal, which may lead to aspiration.

Intramuscular injection

Muscles suitable for injection are the:

- m. supracoracoideus and
- m. iliotibialis lateralis.

M. SUPRACORACOIDEUS

APPLICATION

This is the standard site for administration of intramuscular injection.

21.5 Application of drops to the nares. The liquid passes through the choana into the oral cavity and is then swallowed.

RESTRAINT

The head is held between the thumb and index finger (either by the operator, for small birds, or by an assistant) using the thumb and finger or helmet grips (note: avoid placing pressure on the eyes). If the bird is **restrained by the operator** (birds up to around 100g body weight), the body of the patient rests in the operator's palm and the hindlimbs are secured by the ring and little fingers (note: avoid placing pressure on the abdomen) to prevent the bird from instinctively grasping the syringe (Figure 21.9).

METHOD

The sternal carina is located by palpation (Figures 21.6 to 21.8) or, in birds with little down (e.g., canaries), by blowing against the feathers. The needle is inserted into the m. supracoracoideus (formerly m. pectoralis profundus) at a slightly craniodorsal angle of entry (70–80 degrees), paramedian (and preferably close to) the sternal carina, at the level of **transition from the middle to the cranial third of the carina** (Figures 21.6 and 21.9).

If the injection is placed too superficially, **blood may ooze undetected** into the subcutis or into the septum between the m. pectoralis and m. supracoracoideus. This can lead to death hours after injection, without any obvious prior clinical signs. Injection too far caudal and lateral can result in **inadvertent puncture of the liver** (Figure 21.7), due to species variation in the extent of the medial sternal incisure (see Chapter 2 'Head and trunk'). To prevent haemorrhage, the middle finger of the restraining

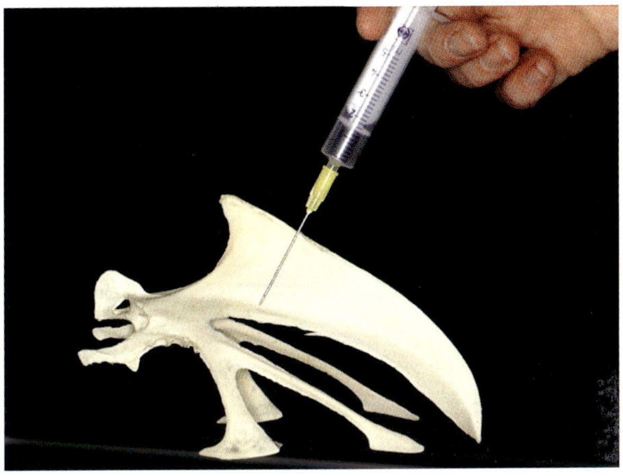

21.6 Correct positioning for injection into the m. supracoracoideus at the junction between the cranial and middle thirds of a line extending between the cranial and caudal sternal margins.

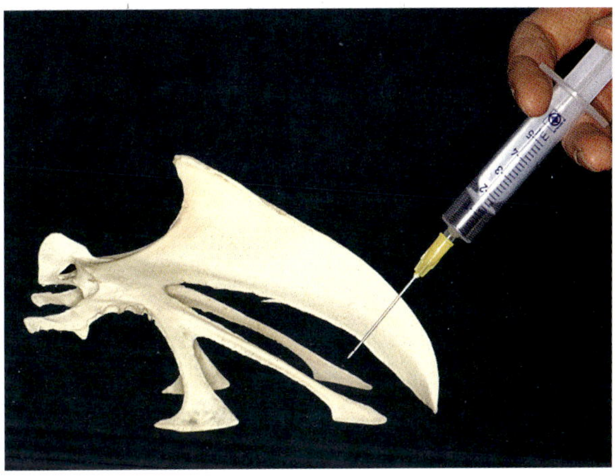

21.7 Incorrect positioning for injection into the m. supracoracoideus at the junction between the middle and caudal thirds of a line extending between the cranial and caudal sternal margins; results in penetration of the body cavity (liver) through the medial incisure.

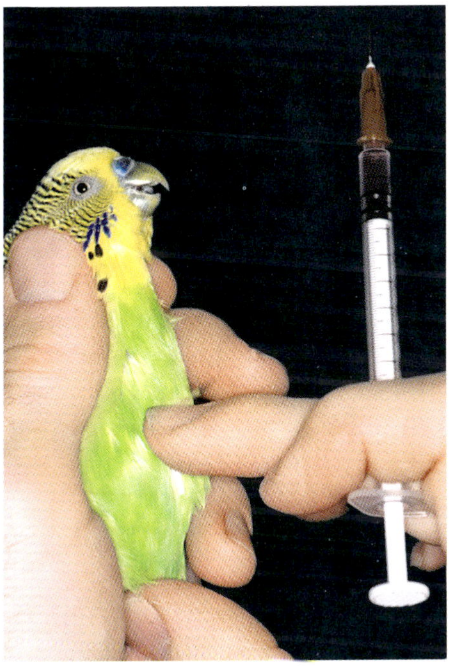

21.8 Technique for intramuscular injection into the m. supracoracoideus. The patient is restrained using the thumb and finger hold and the sternum is palpated to determine the point of entry (junction between the cranial and middle thirds of a line connecting the cranial and caudal sternal margins).

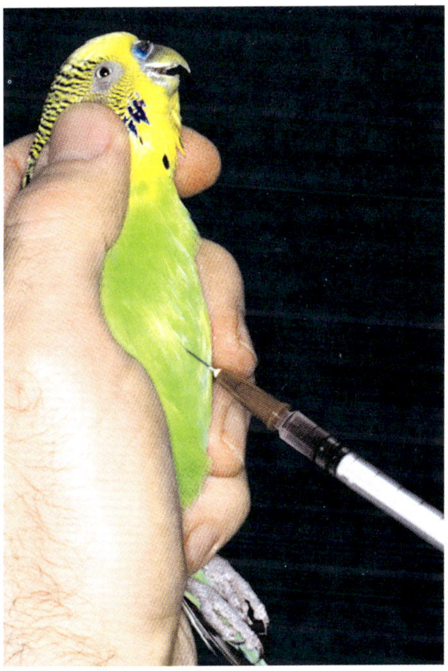

21.9 The injection is placed paramedian (left or right) to the sternal carina, at the junction between the cranial and middle thirds of the sternum. To prevent haematoma formation, the skin over the injection site is tensed using the middle finger of the restraining hand.

hand is used to tense the skin prior to injection, and to release the skin before the needle is removed (Figure 21.9).

M. ILIOTIBIALIS LATERALIS

APPLICATION

The m. iliotibialis lateralis can be used for intramuscular injection in juveniles, cachectic birds with poorly developed breast musculature and various ratites. It also serves as a site for intramuscular vaccination of day old chicks against Marek's disease.

RESTRAINT

Birds can be restrained for this procedure by securing both the head and relevant hindlimb using the index and middle fingers. The other (not injected) hindlimb is restrained by the ring and little fingers (Figure 21.12).

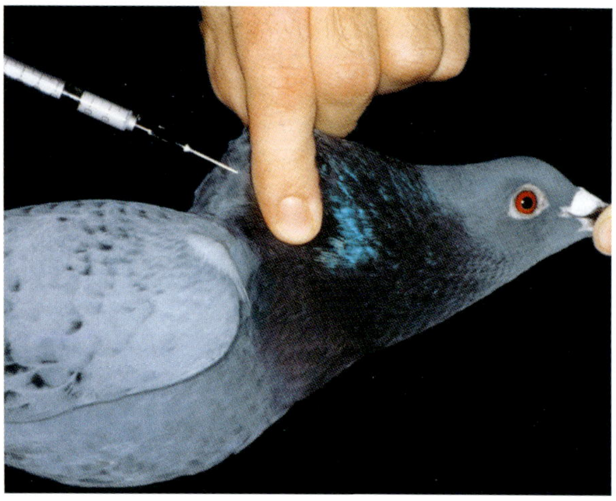

21.10 Subcutaneous injection into the caudal third of the neck of a domestic pigeon (Columba livia). The needle is directed cranially and the injectate is deposited in the subcutaneous connective tissue.

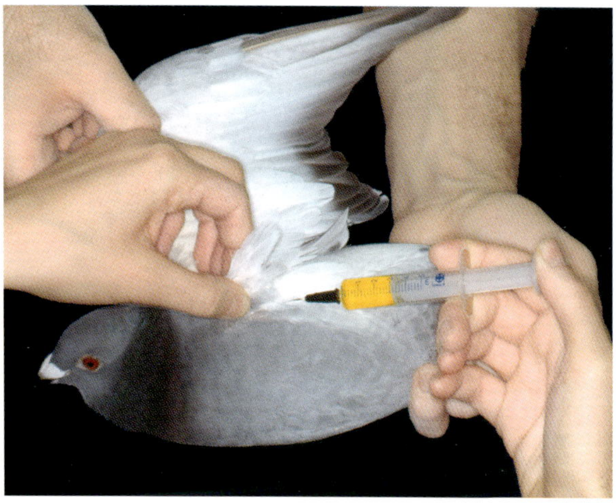

21.11 Subcutaneous injection into lateral flank of a domestic pigeon (Columba livia).

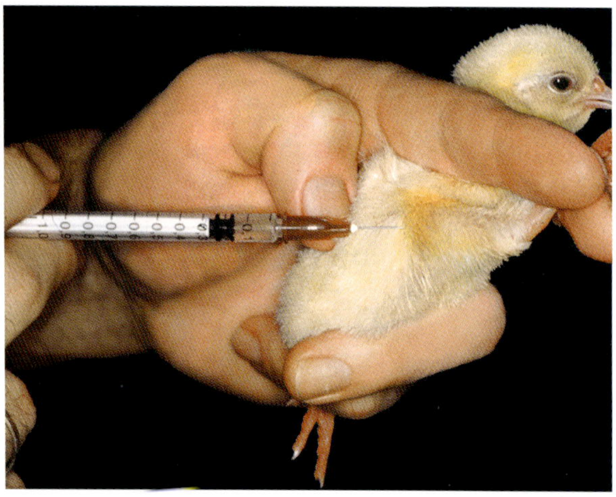

21.12 Injection into the m. iliotibialis lateralis. Appropriate restraint is effected using the 'ringer's hold'.

METHOD

The m. iliotibialis lateralis forms the lateral contour of the thigh. The injection is administered mid-thigh in a proximo-distal direction, parallel to the femur (Figure 21.12). This approach is associated with negligible risk to vessels and nerves.

Subcutaneous administration

Sites suitable for subcutaneous administration include the:

- nape of the neck (interscapular region),
- pre-crural fold and the
- lateral flank.

NAPE OF THE NECK

APPLICATION

This site is particularly suitable for **electrolyte and fluid replacement** and for vaccination of pigeons against paramyxovirus Type 1.

RESTRAINT

An assistant holds the hindlimbs between the index and middle fingers while also grasping the base of the tail (standard method for holding pigeons, see Figure 20.8). The neck is extended slightly and presented for injection (Figure 21.10).

METHOD

A fold of skin is raised in the caudal third of the neck using the thumb and index finger. The injection is placed in the **loose dorsomedial subcutaneous tissue**, after introducing the needle in a cranial direction at a relatively flat angle (Figure 21.10). Correct positioning of the needle (strictly dorsomedial) and formation of the injection bleb can be monitored with the hand used to raise the skin. If the procedure is performed appropriately, the incidence of complications is less than 1 per cent.

Due to the presence of a **venous plexus** (plexus venosus intracutaneus collaris) in the cranial neck and nape region (Figure 21.15), the risk of complications increases if the injection is placed too far cranial and lateral. Puncture of the venous plexus can result in persistent low-grade bleeding that may result in death some time (up to several hours) after injection.

Injection of **excessive volumes** of fluid can have similar consequences due to tearing of the connective tissue. Depending on the size of the bird, injected volumes should not exceed the following:

budgerigar size (30–40g)	1 ml
cockatiel size (around 100g)	2 ml
Amazon size (350–400g)	10 ml
buzzard or macaw size (900–1,100g)	20 ml

Injection in a caudal direction is associated with the risk of **penetrating the clavicular air sac**. Laterally oriented injections may damage the **jugular vein** and **vagus nerve**.

PRE-CRURAL FOLD

The patient is placed in lateral recumbency and the injection is placed into the pre-crural fold from the medial aspect, in a disto-proximal direction (Figure 21.1). This is the preferred site for subcutaneous fluid administration.

LATERAL FLANK

The patient is placed in a supine position with the head slightly elevated (beware of shock) and the injection is placed in a skin fold on the lateral flank, under the wing.

Due to the tautness of the subcutaneous tissue in this region, the volume of liquid that can be administered is relatively limited.

Intracutaneous and percutaneous administration

Techniques used for intracutaneous and percutaneous delivery include the:

- wing-web method,
- feather follicle method,
- spot-on-method and
- application of creams and powders.

WING-WEB METHOD

APPLICATION AND PATIENT RESTRAINT

This method is used for vaccination against **canary pox**. The hindlimbs and one wing are secured by an assistant. Firm yet careful restraint – sufficient to prevent attempts at self-defence and escape – is essential for efficient vaccination of species susceptible to canary pox (canaries, finches and woodland birds), as these birds are also prone to shock.

METHOD

After extension of one wing, a double-pronged wing-web applicator is used to deliver the dose into the skin of the propatagium (Figure 21.13). It is important to ensure that the **applicator needles do not come into contact with the feathers**, which might otherwise be drawn into the injection site or remove vaccine from the applicator.

Note: When vaccinating to control an outbreak, the prongs of the applicator should be sterilised between doses (e.g., using a gas flame) to prevent transmission of field virus strains.

FEATHER FOLLICLE METHOD

APPLICATION AND PATIENT RESTRAINT

This technique can be used for vaccinating against diseases such as **pigeon pox**. The patient is restrained using the standard pigeon hold, with the bird's head shielded by an assistant.

METHOD

Feathers on the lateral thigh are plucked along a length of approximately 1.5cm. A brush is then used to apply the vaccine to the feather follicles (Figure 21.14).

SPOT-ON-METHOD

APPLICATION

The spot-on method is used in treatment of **endoparasitic mites** (air sac mites) and for the administration of hormone and iodine formulations.

21.13 Intracutaneous injection ('wing-web method') of the propatagium for vaccination against canary pox.

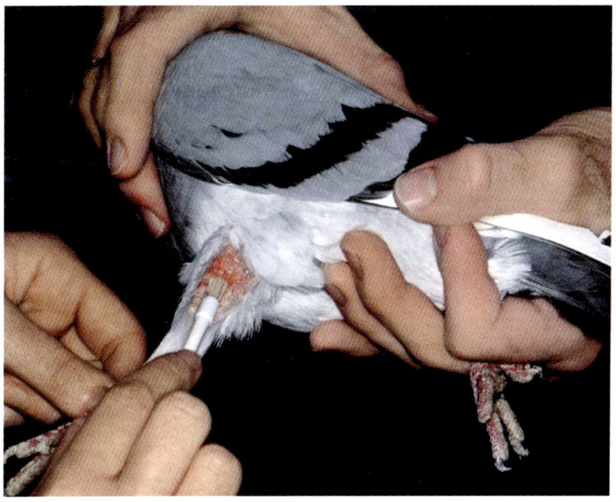

21.14 Intracutaneous application (feather follicle method) of pigeon pox vaccine to the follicles of freshly plucked feathers.

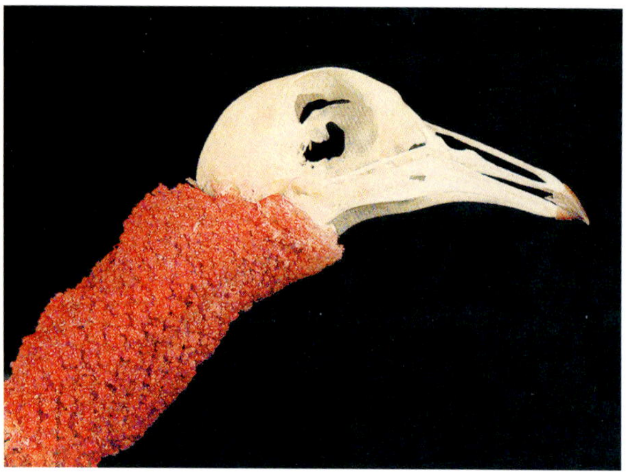

21.15 Plexus venosus intracutaneus collaris of a domestic pigeon. Courtesy of PD Dr S. Reese, Munich.

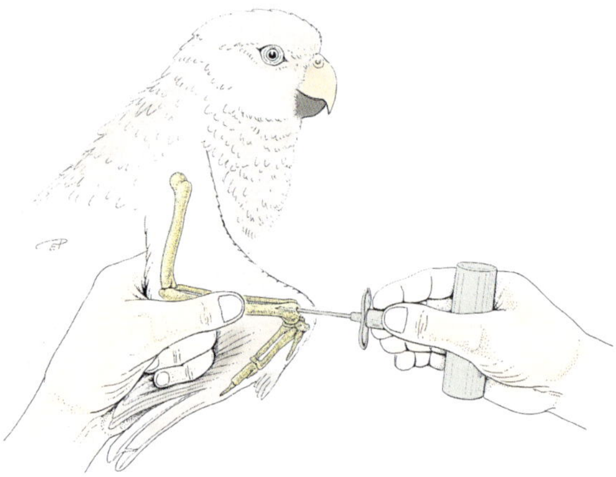

21.16 Technique for intraosseous administration of fluids into the medullary cavity of the ulna: the catheter is inserted into the distal ulna with the carpus fully flexed. Adequate flexion is important to avoid risk of injury to the joint.

METHOD

The product is dripped directly onto the skin, preferably onto an apterium between the scapulae. When using this method, bathing equipment must be removed from the patient's environment. Propylene glycol is the preferred carrier agent.

APPLICATION OF CREAMS AND POWDERS

Powder is applied after parting the feathers, against the direction of the feather coat. Access to bathing equipment should be prevented.

Intraosseous (intramedullary) administration

APPLICATION

The intraosseous route is used primarily for replacement of fluids and electrolytes in seriously ill patients (in lateral recumbency) and in emergency situations. Where possible, emergency patients should be treated in sternal recumbency, as their circulatory function is usually compromised.

METHOD

With the carpal or knee joint fully flexed, a catheter with a sharp stylet is introduced into the distal ulna or the proximal end of the tibiotarsus. When using the ulna, the catheter is inserted between the two palpable distal condyles, parallel to the median plane of the bone (Figure 21.16). On the tibia, the entry point lies just above the easily palpable cranial cnemial crest and just distal to the two femoral condyles. After introduction of the catheter the stylet is removed and the cannula is secured with a figure of eight suture. The catheter can be left in place up to 72 hours (patient in lateral recumbency) before removal.

When employing this technique it is essential to consider the following:

- pneumatised bones must not be used,
- hypertonic, alkaline solutions must not be used,
- catheters should be heparinised and flushed regularly,
- constant rate infusions at 10ml/h can be used for up to 48 hours.

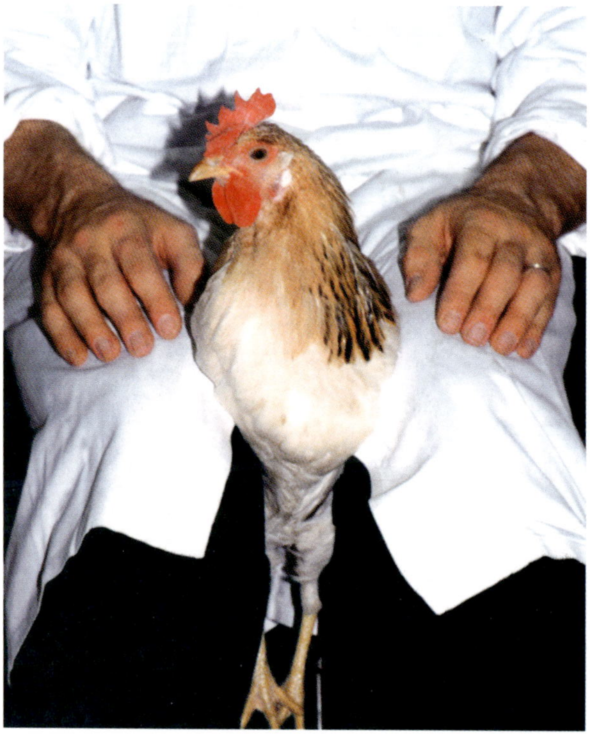

21.17 Appropriate restraint of a chicken for collection of blood from the right jugular vein. The operator applies gentle pressure with their knees.

Intravenous injection and blood collection

The following structures may be used for intravenous injection and blood collection:

- jugular vein,
- plantar superficial metatarsal vein,
- cutaneous thoracoabdominal vein,
- ulnar vein,
- heart.

JUGULAR VEIN

APPLICATION

Jugular venipuncture is considered the standard technique for blood collection in most birds (Figures 21.18 to 21.23). One exception is the pigeon, in which the deep position of the jugular makes it difficult to visualise.

The jugular vein is suitable for venipuncture in small birds (Figure 21.22) and can be used to collect relatively large volumes of blood, replace fluids and electrolytes and perform blood transfusions, with relatively little risk of haematoma formation.

RESTRAINT

The optimal method of restraint for jugular venipuncture varies according to the size of the bird.

Technique for solo operator, birds up to budgerigar size: The **'ringer's hold'** is used, exercising great care to ensure the patient is appropriately secured as, especially in small birds, defensive manoeuvres can quickly result in laceration of the vessel and potentially life-threatening haemorrhage.

Particularly for inexperienced operators, it is advisable to rest the restraining hand on a surface (Figure 21. 22).

Technique for solo operator, birds up to chicken size: The patient is positioned between the operator's knees

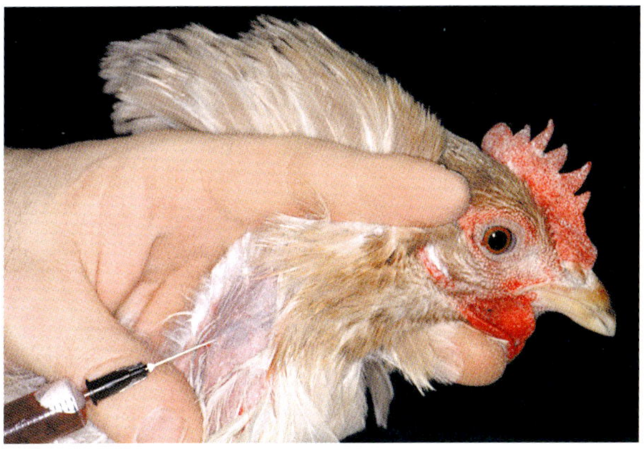

21.19 Collection of blood from the right jugular vein. For stability, the hub is rested on the thumb used to raise the vein.

21.20 Collection of blood from the right jugular vein in birds with a dense covering of down feathers, such as waterfowl (e.g. domestic goose). The feathers are smoothed against the body using an alcohol soaked gauze swab.

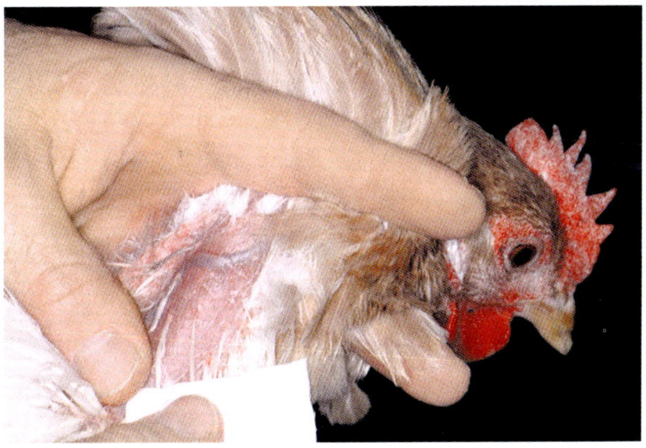

21.18 Visualisation of the right jugular vein in the lateral cervical apterium. With the bird restrained using the 'ringer's hold', the thumb is used to occlude the vein after parting the feathers with an alcohol-soaked gauze swab.

21.21 Venipuncture of the right jugular vein. Occluding the vein reveals its course by raising the overlying feathers.

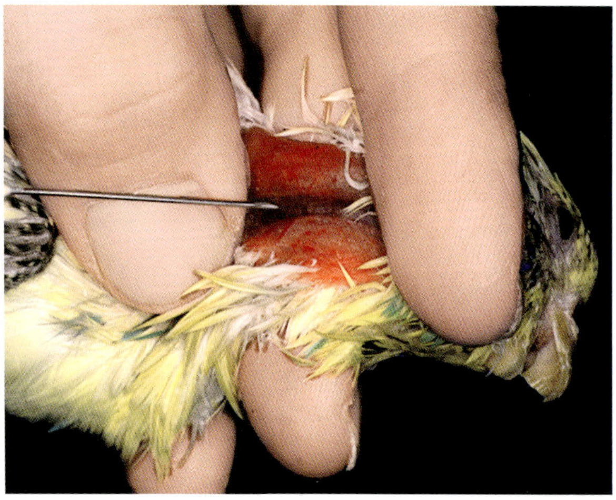

21.22 Collection of blood from the right jugular vein of a small bird (e.g. budgerigar) restrained by the operator using the 'ringer's hold'.

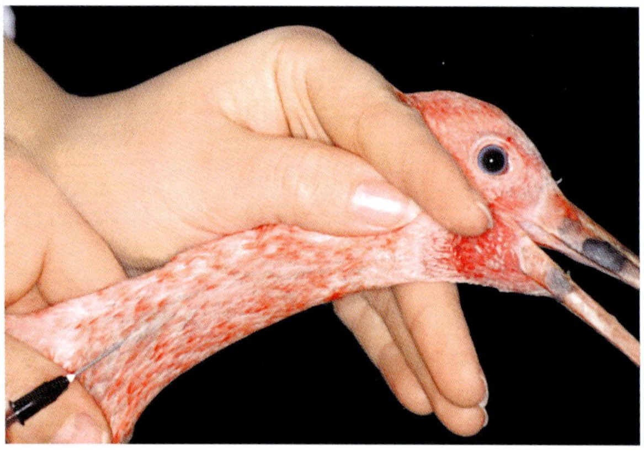

21.23 Collection of blood from the right jugular vein of birds with extensive down and lacking in apteria, such as the scarlet ibis (Eudocimus ruber): raising of feathers (that have been smoothed against the body with alcohol) by venous occlusion.

and thighs, and gentle pressure is used to secure the wings against the patient's body (Figure 21.17). It is important to ensure that the hindlimbs are hanging freely to prevent the animal from bracing itself against the edge of the operator's seat in a sudden escape attempt.

The head is restrained with the index and middle fingers by applying the 'ringer's hold' in the proximal cervical region. Ideally the left hand is used for restraint, as the right jugular vein is typically larger than the left (Figures 21.18 and 21.19).

Assisted technique: Potentially dangerous patients (e.g., raptors, parrots) and sensitive birds (e.g., grouse, some zoo species and larger birds) can be restrained by an assistant (e.g., using the 'helmet grip').

Reduction of visual stimuli by **shielding the patient's head** is useful for preventing excessive arousal, thus minimising the risk of stress and shock.

INJECTION AND BLOOD COLLECTION PROCEDURE

The feathers of the right, lower cervical region are moistened with alcohol and then parted. This area is especially well suited to venipuncture as it **lacks contour feathers** (in psittacines, finches, chickens and various other species), making the jugular vein relatively easy to visualise through the skin (Figures 21.18 and 21.23). For increased stability, the needle hub is rested on the thumb used to raise the vein (Figures 21.19, 21.21 to 21.23). Blood should be withdrawn relatively swiftly, but without haste to avoid collapsing the vein. Pressure is applied to the puncture site to prevent bleeding.

In waterfowl and many species of zoo birds, a lack of apteria and the presence of a dense covering of down prevents direct visualisation of the jugular vein. Alcohol is therefore used to smooth the feathers against the body. With the bird restrained using the 'ringer's hold', occlusion of the vein allows the vessel to be palpated easily (Figures 21.20, 21.21 and 21.23).

PLANTAR SUPERFICIAL METATARSAL VEIN

APPLICATION

This site is utilised for venipuncture in waterfowl, ratites as well as various zoo birds (e.g., penguins) and raptors.

RESTRAINT

For birds up to the size of a goose, an assistant turns the bird on its back and holds it in their arms, simultaneously securing the wings. The assistant uses one hand to **extend one of the hindlimbs** and occlude the vein at the distal tibiotarsus.

INJECTION AND BLOOD COLLECTION PROCEDURE

Venipuncture is performed distally on the **flexor aspect of the tibiotarsal joint**. The needle is directed **dorsomedially**, along the course of the vessel over the tarsometatarsus (Figure 21.24).

CUTANEOUS THORACOABDOMINAL VEIN

APPLICATION

The cutaneous thoracoabdominal vein is the preferred site for venipuncture in experimental protocols and for placement of indwelling catheters in chickens.

INJECTION AND BLOOD COLLECTION PROCEDURE

After visualising the diagonal course of the vessel across the chest wall, venipuncture is performed proximal to the convergence of its dorsal and ventral branches (Figure 21.25).

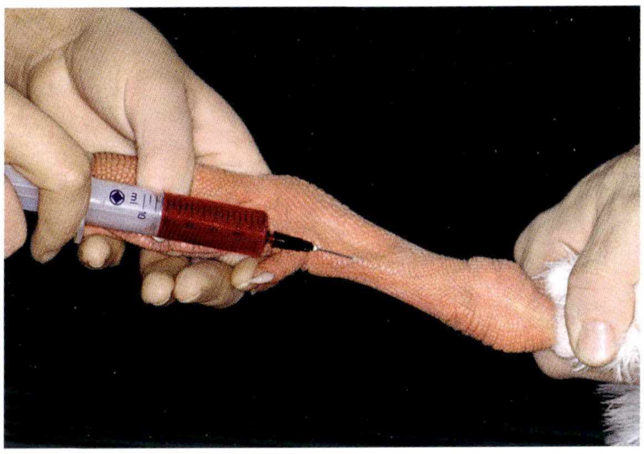

21.24 Collection of blood from the plantar superficial metatarsal vein in a domestic goose. Venipuncture (distoproximal) is performed after occlusion of the vein proximal to the intertarsal joint.

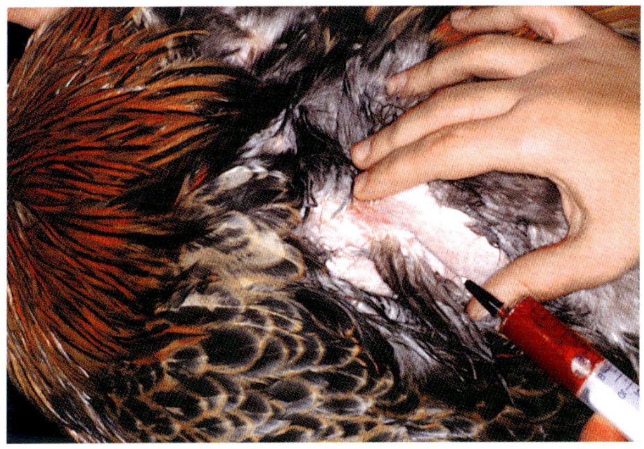

21.25 Collection of blood from the cutaneous thoracoabdominal vein. In chickens this vein is also suitable for placement of indwelling catheters.

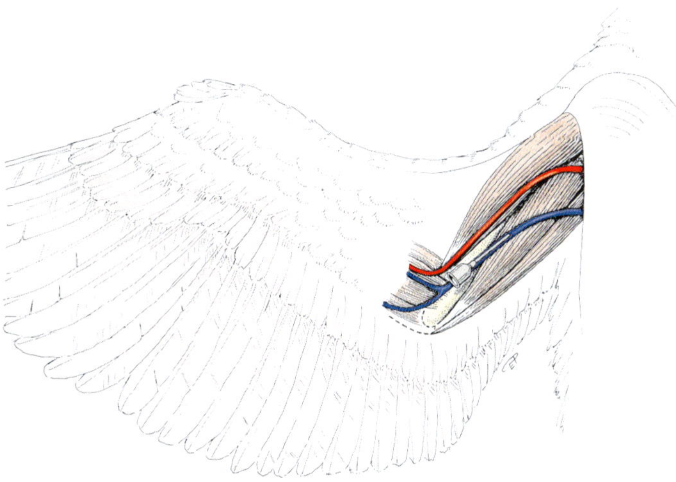

21.26 Topographic anatomy pertaining to collection of blood from the ulnar vein (schematic).

ULNAR VEIN

APPLICATION

Ulnar venipuncture (Figure 21.26) can be performed in most birds and is the method of choice if jugular venipuncture is not possible (e.g., in pigeons).

RESTRAINT

An assistant holds the patient in a modified **'bunch of flowers'** grip, with one of the wings extended (Figure 21.27). The head is covered during venipuncture.

INJECTION AND BLOOD COLLECTION PROCEDURE

After moistening the feathers with alcohol and parting the feather coat, the vein is occluded by holding the wing between the index and middle fingers (Figure 21.28). The oft-described technique of accessing the more distal tributaries of the ulnar vein (the deep and superficial ulnar veins) over the elbow joint is frequently associated with extensive **haematoma formation**, because the vein passes directly over bone and taut connective tissue. Although the vessel can readily be visualised here, venipuncture at this site is strongly discouraged. The preferred site is the **ventral brachial region** (Figures 21.26 and 21.30) where the vessel is underlain by the muscle belly of the m. humerotriceps. During venipuncture, the needle hub is rested on the thumb of the restraining hand (Figure 21.30). Prior to needle withdrawal, an assistant places pressure on the puncture site.

Euthanasia

Cardiac puncture

APPLICATION

Cardiac puncture is a humane form of euthanasia. As it results in relatively little bleeding, it offers the considerable advantage of being much less likely than other methods (intravenous or intrapulmonary routes) to result in the spread of pathogens through blood droplets or direct

21.27 Collection of blood from the ulnar vein: appropriate restraint of the patient by an assistant using the modified 'bunch of flowers' hold with extension of the wing.

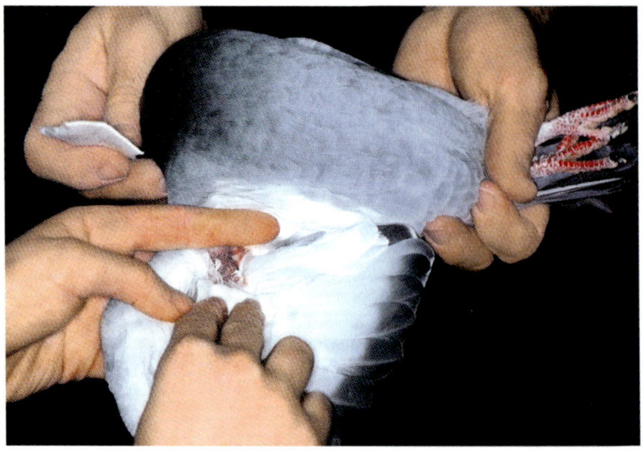

21.28 Visualisation of the vessels on the ventral surface of the brachium after parting of the feathers with alcohol and occlusion of the vein.

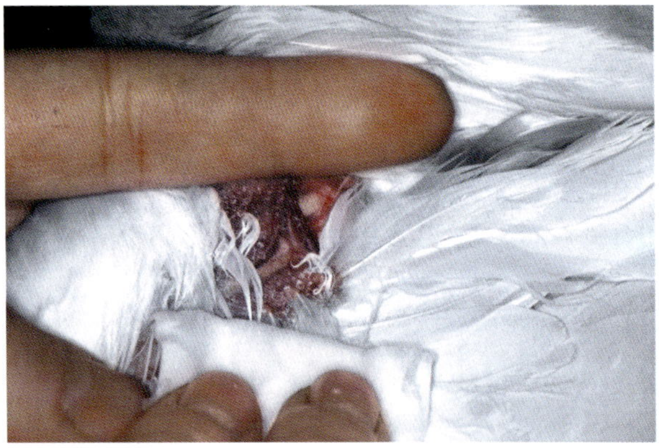

21.29 Parting of the feathers with alcohol and occlusion of the ulnar vein (between index and middle fingers) to reveal the convergence of the superficial and deep ulnar veins (distal) to form the ulnar vein (proximal) (detailed view of the topographical anatomy depicted in Figure 21.28). Note: Venipuncture over the elbow joint, where the vein is very easily visualised, frequently results in extensive haematoma formation and should therefore be rigorously avoided.

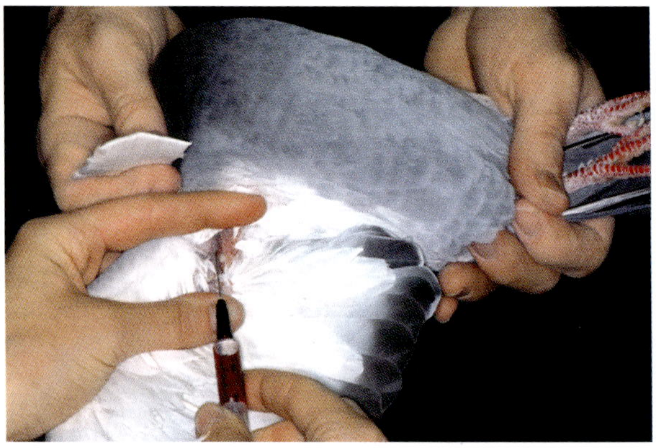

21.30 Collection of blood from the ulnar vein in the brachial region, proximal to the convergence of the superficial and deep ulnar veins over the m. humerotriceps. To avoid haematoma formation, mild pressure is applied prior to removal of the needle, with firmer compression of the venipuncture site after the needle is withdrawn.

contact. As such, it serves as a possible means of euthanasia in smaller commercial operations where disease outbreaks (e.g., avian influenza) necessitate culling, yet other methods (CO_2 gassing using small bags or boxes; CO_2 gassing of large tents or whole barns) are unavailable or impractical.

The use of cardiac puncture to collect blood for diagnostic purposes is rare, generally being limited to cases in which peripheral vessels are prohibitively small (e.g., day-old chicks). It is also employed when low blood pressure precludes the raising of a vein, yet the evaluation of blood parameters is imperative (e.g., rapid diagnosis or disease investigation in a flock).

This method is contraindicated in **patients that will undergo subsequent treatment**, as it can result in bleeding into the pericardial sac and potential death due to cardiac tamponade, particularly when repeated attempts are required.

RESTRAINT

The most appropriate method of restraint depends on the selected approach. For the lateral approach, the bird is placed in right lateral recumbency (see below) with its wings and hindlimbs secured. The hindlimbs are extended caudoventrally to permit palpation of the caudal margin of the sternum (Figures 21.31 and 21.35). For the cranial approach, preferable in small birds, the patient is placed in dorsal recumbency or, if an assistant is not available, fixed in a supine position using the 'ringer's hold' (Figure 21.34).

INJECTION AND BLOOD COLLECTION PROCEDURE

The required needle length depends on the approach (cranial or lateral) and the type of bird:

- adult chickens (Gallus gallus): 75–100mm,
- day-old chicks: 25–30mm,
- adult Muscovy or runner ducks: 75–100mm,
- adult domestic geese (Anser anser): 75–100mm and
- domestic turkeys (Meleagris gallopavo): 100–150mm.

Cardiac puncture using a lateral approach: This is the preferred approach for adult or larger birds. It is performed from the left side, as the liver extends into the puncture zone on the right side in several species.

The **entry point** is located at the **junction between the cranial and middle thirds** (cranial trisection point) of a line connecting the **shoulder joint** (art. humeri) and the midpoint of the **caudal sternal margin** (Figures 21.31 and 21.35).

21.31 Cardiac puncture (lateral approach). With the bird appropriately restrained in right lateral recumbency, the middle finger is used to locate the shoulder joint and the thumb is placed on the caudal sternal margin. The puncture site (index finger) is located at the cranial trisection point on a line connecting the thumb and middle finger.

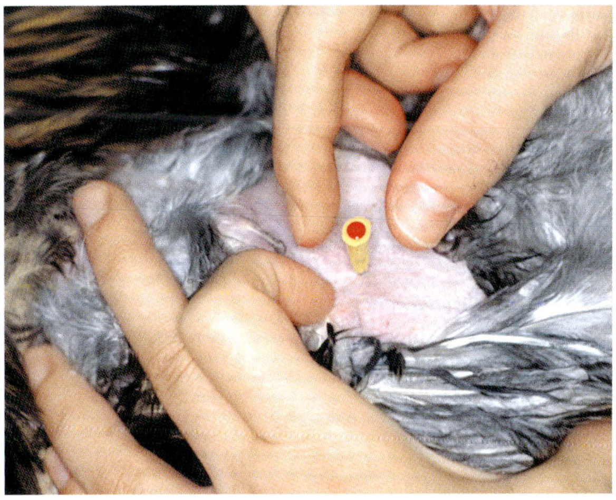

21.32 Puncture of the left ventricle, confirmed by blood rising in the needle hub. The point of convergence of the dorsal and ventral branches of the cutaneous thoracoabdominal vein (below the puncture site) serves as an additional anatomical landmark.

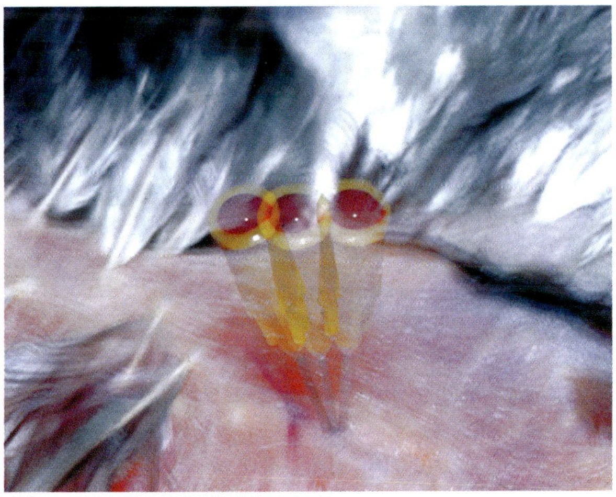

21.33 Correct placement of the needle for cardiac puncture confirmed by transmission of the cardiac rhythm to the needle and appearance of blood in the hub (stroboscopic flash with triple exposure, background is out of focus due to movement associated with respiration).

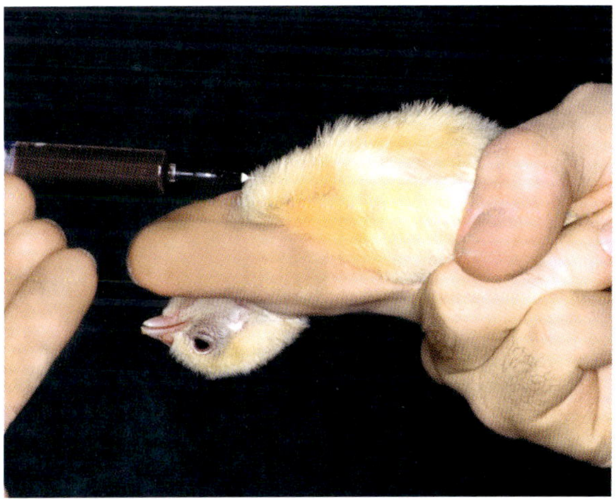

21.34 Collection of blood from the heart using a cranial approach (atrial puncture) with unassisted restraint of the small patient using the 'ringer's hold'.

As a further point of orientation, the dorsal and ventral branches of the **cutaneous thoracoabdominal vein** converge near the puncture site (Figures 21.32 and 21.35). Correct placement is confirmed by the appearance in the needle hub of blood (Figure 21.32) that pulses in synchrony with the cardiac rhythm (Figure 21.33).

Cardiac puncture using a cranial approach: The cranial approach is particularly well suited to smaller birds and day-old chicks. Using this approach, the needle passes through the cranial thoracic aperture between the two limbs of the furcula before entering the atria of the heart (Figure 21.34).

As the heart is relatively flattened within the body cavity, there is a risk of **damaging the great vessels** cranial to the heart, potentially resulting in cardiac tamponade. This interferes with blood collection and may lead to the death of the bird. In larger birds, particularly long needles are required for the cranial approach.

Intrapulmonary injection

APPLICATION

Pulmonary puncture and intrapulmonary injection under general anaesthesia is a rapid and practical means of **euthanasia**.

A significant disadvantage of this technique is the risk of pulmonary haemorrhage and expulsion of blood from the trachea. The use of irritant preparations (e.g., T 61) renders the lung unsuitable for subsequent pathological evaluation (see below).

METHOD

The needle is inserted in a dorsoventral direction into the third or fourth intercostal space, between the scapula and the thoracic vertebral column (i.e., paramedian). When localising this site, it should be noted that 2–6 thoracic vertebrae may be fused to form the notarium (a routine finding in pigeons and chickens).

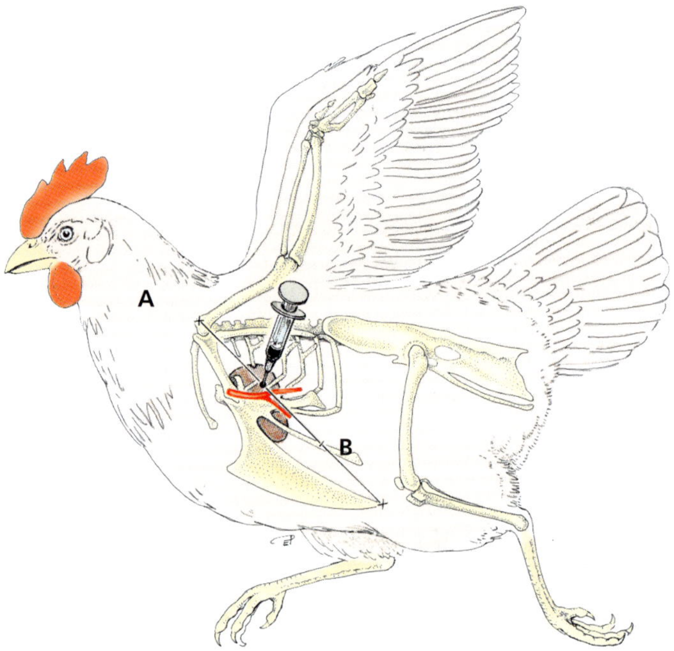

21.35 Cardiac puncture using a left lateral approach, with the bird restrained in right lateral recumbency. The middle finger is used to locate the shoulder joint (A) and the thumb is placed on the caudal sternal margin. The puncture site (marked using the index finger) is located at the cranial trisection point on a line connecting points A and B, just dorsal to the point at which the dorsal and ventral branches of the cutaneous thoracoabdominal vein converge.

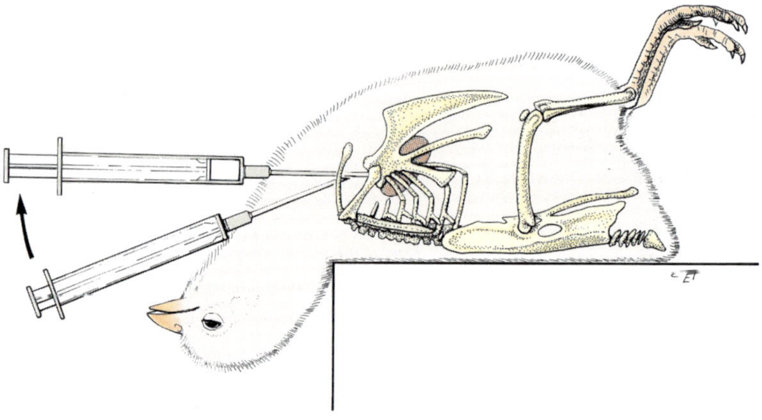

21.36 Cardiac puncture using the cranial approach in a day-old chick. The patient is placed in dorsal recumbency and the neck is hyperextended to permit insertion of the needle between the limbs of the furcula (near their junction) and through the cranial thoracic aperture into the heart. The needle is initially directed caudoventrally. Then, with light back-pressure applied to the syringe, the needle is advanced in a strictly caudal direction (parallel to the floor of the sternum) towards the base of the heart.

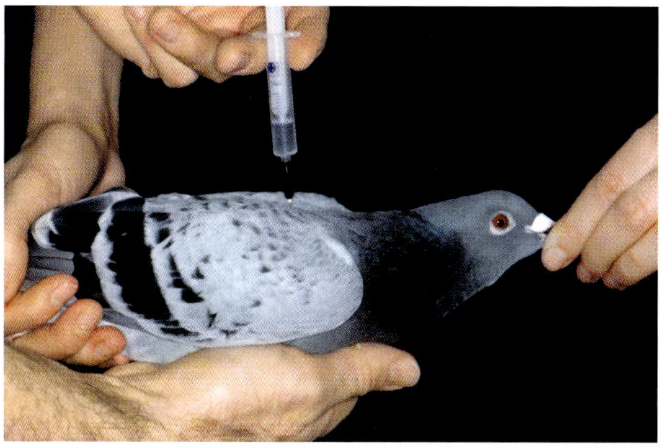

21.37 Use of intrapulmonary injection for euthanasia: the needle is inserted through the third or fourth intercostal space, between the notarium (os dorsale) and the scapula (lateral view).

DISADVANTAGES

Intrapulmonary administration of euthanasia solution results in widespread necrosis of the lungs and sometimes also the heart, making subsequent pathological assessment of these organs impossible. Furthermore, this procedure commonly results in blood vessel damage with **pulmonary haemorrhage** and egress of large volumes of blood from the oral cavity. It is important to be aware of this possibility, as owners often wish to be present when their bird is euthanised.

Humane euthanasia should therefore ideally be carried out by **intravenous administration** of euthanasia solution or, in small birds up to 100g, by means of chloroform inhalation. As well as being appropriate on welfare grounds, these procedures do not interfere with necropsy.

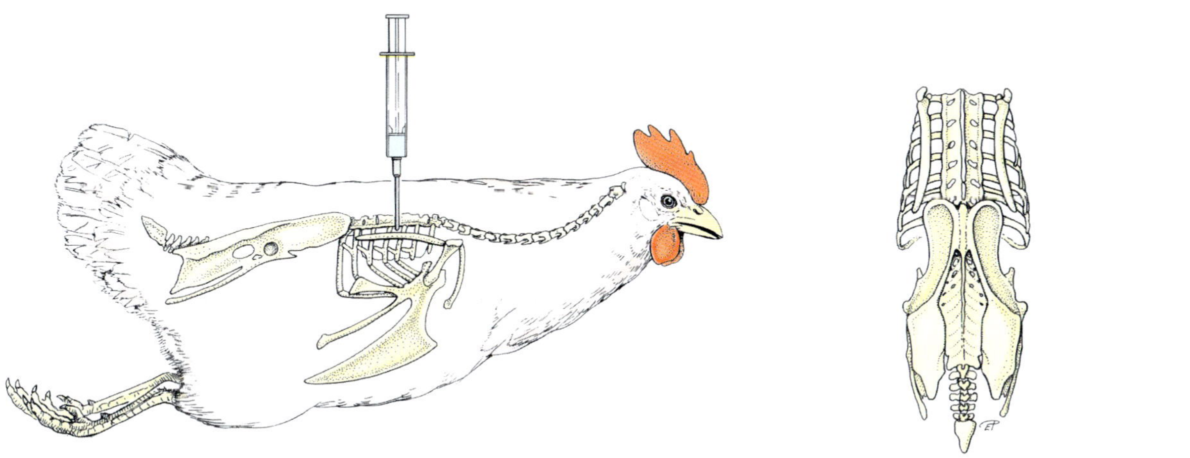

21.38 Use of intrapulmonary injection for euthanasia (schematic): the needle is inserted into the third or fourth intercostal space, between the fused vertebrae of the notarium (os dorsale) and the scapula (lateral view; insert shows dorsal view).

22

Endoscopy

R. Korbel and H.-G. Liebich

Indications

Due to the unique anatomical features of the avian respiratory system, birds are excellent candidates for endoscopy. The internal organs can be examined by introducing a rigid endoscope into the various air sacs (Figure 22.1), without the need for gaseous insufflation. The procedure is relatively straightforward, and may thus be considered along with radiography as an indispensable diagnostic aid in avian medicine. Endoscopy is also increasing in popularity as a minimally invasive technique for various clinical interventions.

Avian endoscopy was initially developed as a means of gender determination in **sexually monomorphic parrots**. This practice has since largely been replaced by non-invasive molecular genetic testing of blood or feather samples, using chromosomal analysis or repetitive DNA sequences. However, endoscopy also permits the direct examination of internal organs (including the gonads) and determination of reproductive status. As such, endoscopy offers the advantage of enabling simultaneous assessment of gender, functional reproductive capacity and the health of other organ systems. This is of considerable value in assessing a bird's fitness for breeding.

Specific indications for endoscopy include:

- reproductive disorders and internal organ dysfunction,
- organ biopsies,
- excretory dysfunction,
- abnormalities of the oesophagus and crop,
- dyspnoea,
- investigation of inconclusive radiographic findings,
- foreign bodies and endoparasitism,
- investigation of wound tracts.

Equipment

Relatively little equipment is required for endoscopy. **Rigid units** (**arthroscopes**) around 25cm in length and 1.9–4mm

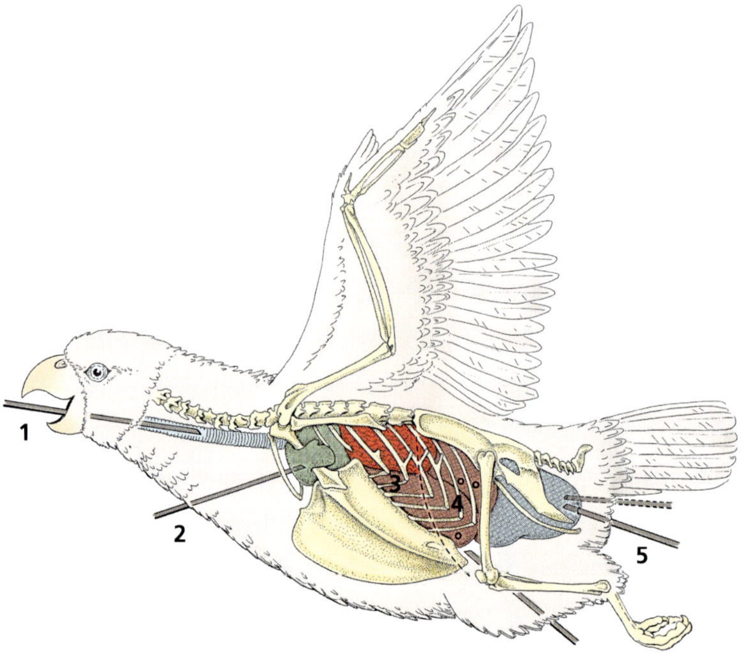

22.1 Endoscopic approaches to the trachea and oesophagus (plus proventriculus and ventriculus) (1), the unpaired clavicular air sac (2), the paired cranial (3) and caudal (4) thoracic air sac and abdominal air sac (5) in the bird (schematic). For routine sexing the standard approach is via the left caudal thoracic air sac, with the entry point located in the last intercostal space, behind the last rib or caudal to the thigh.

in diameter (ideally 2.7mm) are suitable for most applications in birds and can also be used for exotic species (reptiles, small mammals) (Figure 22.2). Flexible endoscopes are rarely used in avian practice.

Endoscopes with a **30-degree viewing angle** are useful as they provide a panoramic view when rotated about their optical axis. While 0-degree units are easier for beginners to use, they have a much more limited range of application. Relatively costly rod-lens (Hopkins) endoscopes are preferable to cheaper tubular systems, as the lower contrast provided by the latter devices (due to a greater number of air-glass interfaces) is inadequate for evaluation of small structures.

A **cool light source**, preferably a xenon vapour lamp, is required. Fitting of a still or video camera to the eyepiece of the instrument permits documentation of endoscopic findings (Figure 22.3).

Other essential pieces of equipment consist merely of a scalpel, curved forceps and absorbable 5–0 suture material. Where available, the use of an endoscope sheath is recommended, both for protection of the delicate endoscope and to facilitate the use of grasping and tissue forceps.

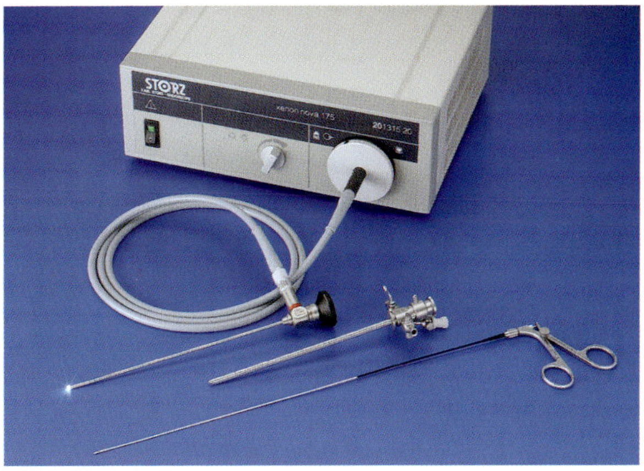

22.2 Basic equipment for endoscopy in birds and exotics: rigid 2.7mm endoscope with 30-degree viewing angle and protective endoscope sheath with instrument channels for flexible grasping and biopsy forceps.

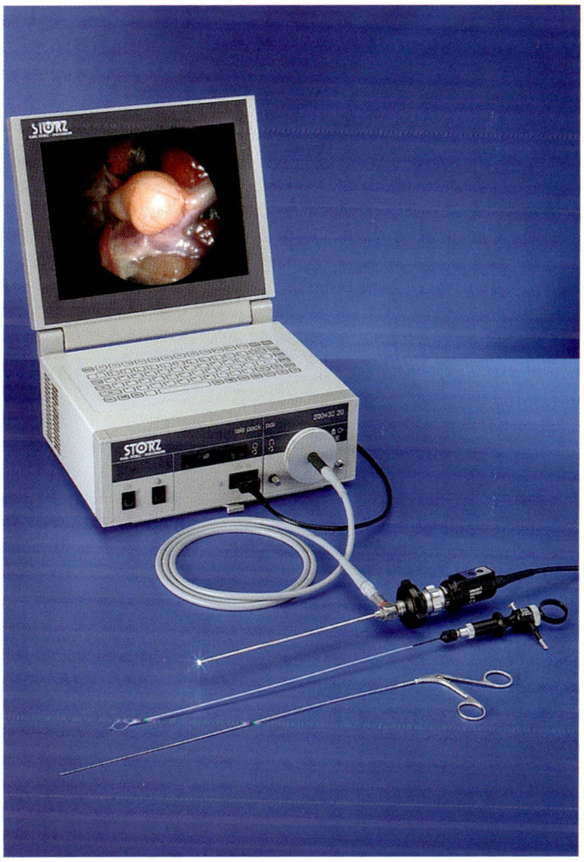

22.3 Video head and portable recording unit for live viewing of endoscopic findings using the camera head, or live display on a monitor with storage of still and video images.

Method

Endoscopy is an invasive and painful procedure and is therefore carried out under **general anaesthesia**, ideally **inhalation anaesthesia (isoflurane or sevoflurane) with endotracheal intubation**. Appropriate analgesia should also be provided (see Chapter 21 'Handling, restraint and anaesthesia').

Various approaches can be used, depending on the purpose of the procedure (Figure 22.1). Figures 22.6 to 22.13 provide an overview of the chronological sequence of an endoscopic examination, depicting various relevant normal findings.

The typical approach to the body cavity for **gender determination** is through the last intercostal space or caudal to the last rib. A less commonly used approach is made from caudal to the thigh (Figure 22.4). The body cavity is accessed on the **left side** as – with the exception of some raptors – only the left ovary is developed in the female (see Chapter 10 'Female genital organs'). The patient is placed in **right lateral recumbency** with the left wing extended away from the body and the left hindlimb retracted caudally. After being moistened sparingly with alcohol, the feathers are plucked from an area of approximately 1.5cm around the site of incision.

A 3–4 mm skin incision is made at the cranial border of the thigh (m. iliotibialis cranialis) at the level of the mid-femur. Curved forceps are inserted under the thigh muscles, after which the tip of the instrument is used to identify the last intercostal space or the area caudal to the last rib. Blunt puncture of the intercostal muscle or fascia is achieved using gentle pressure on the forceps (Figure 22.5).

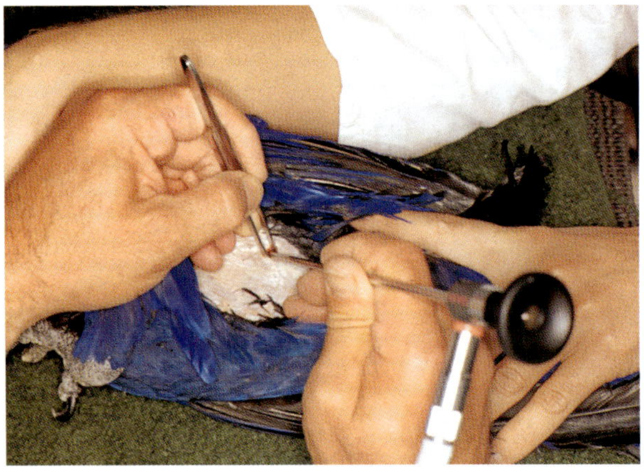

22.4 Entering the body cavity caudal to the muscles of the thigh reduces the risk of injury in birds in which the lungs extend more caudally, such as the hyacinth macaw (Anodorhynchus hyacinthinus).

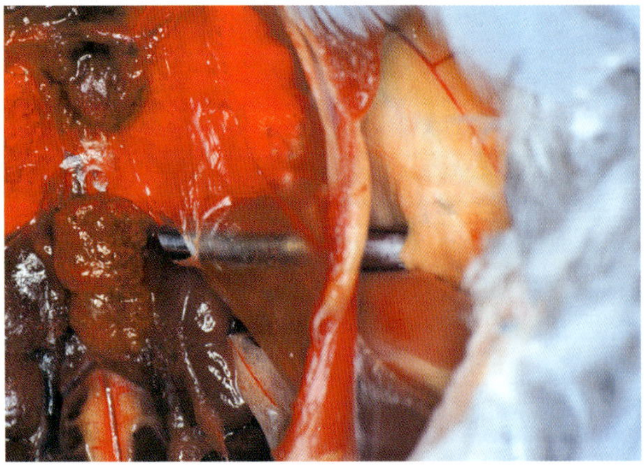

22.5 Endoscope *in situ* after penetration of the caudal thoracic air sac of a pigeon. The endoscope has been oriented towards the female gonads before puncturing the medial wall of the air sac.

A sonorous popping sound indicates **penetration of the caudal thoracic air sac**. The use of a pointed trocar for penetration of the body cavity is contraindicated due to the risk of organ damage.

The forceps are spread, and the endoscope is introduced into the body cavity through the resulting space. It is then directed craniodorsally through the caudal thoracic air sac towards the gonads (Figures 22.5 and 22.6). Blunt penetration of the medial wall of the air sac provides a clear view of the gonads and other internal organs (Figures 22.7 to 22.13).

The **gonads** are located **cranial to the kidney**, which serves as a useful anatomical reference point. It should be noted that there may be similarities in the endoscopic appearance of testes, mature ovarian follicles and intestinal loop flexures lying 'end on' in the immediate path of the endoscope. To avoid confusing these structures, it is important to be aware of their distinguishing anatomical features.

The **typically unpaired ovary** can be identified by its irregular surface or the presence of follicles of various size, including tertiary follicles with vascularized walls (Figure 22.10). Nearby structures amenable to endoscopic examination include the kidneys, liver, spleen, caudal lungs and air sacs (Figures 22.7 to 22.13).

The **male gonads appear as paired**, usually porcelain-coloured, spherical to bean-shaped structures that vary greatly in size according to the bird's reproductive status

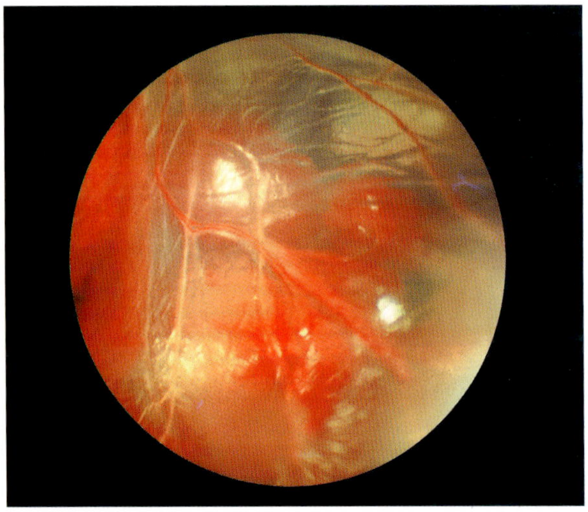

22.6 Endoscopic view upon introduction of the instrument into the left caudal thoracic air sac of a goshawk (Accipiter gentilis), revealing the relatively clear, semi-transparent medial wall of the air sac.

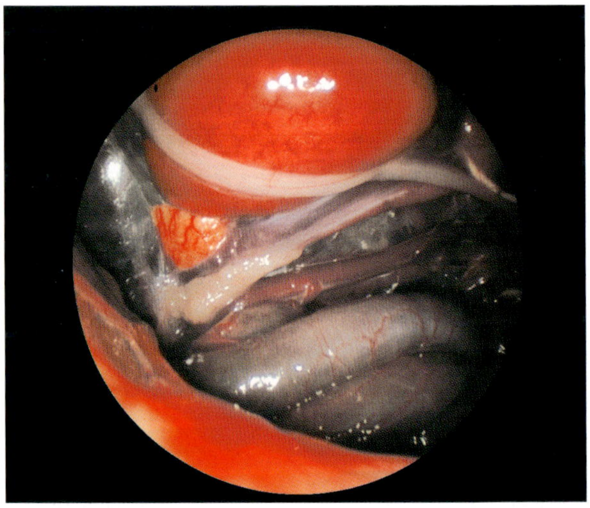

22.7 Cranial region of the body cavity of a goshawk (Accipiter gentilis) with cranial pole of the kidney (top) overlain by the ureter, triangular orange adrenal glands, juvenile female gonad, intestinal loops and caudal lung (bottom of image).

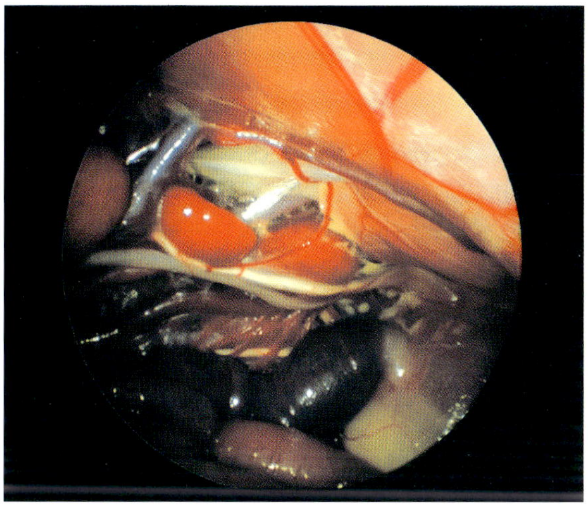

22.8 Caudal region of abdominal cavity of a goshawk (Accipiter gentilis) including the three renal divisions (cranial division partly obscured), the urate filled ureter, the abdominal aorta, parts of the bony pelvis and loops of jejunum and colon (bottom of image).

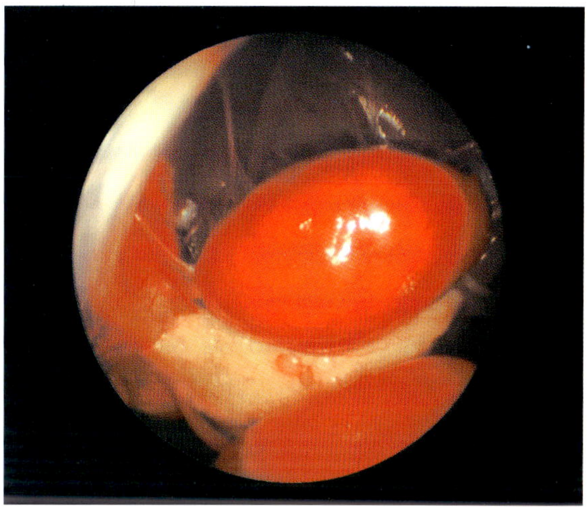

22.9 Juvenile, quiescent female gonad and cranial renal division in a crimson-rumped toucanet (Aulacorhynchus haematopygus).

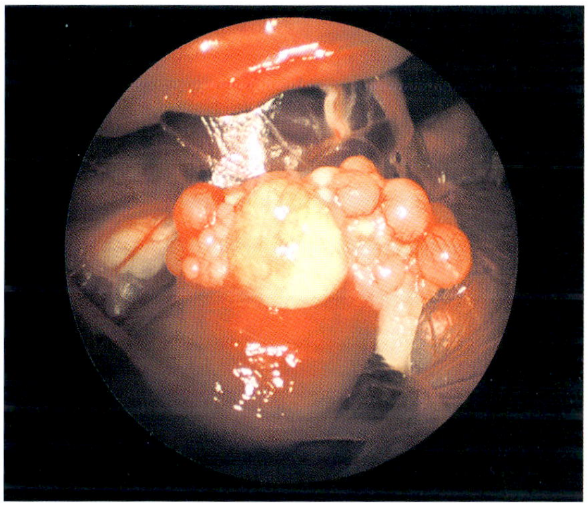

22.10 Active ovary with secondary and tertiary follicles in a goshawk (Accipiter gentilis).

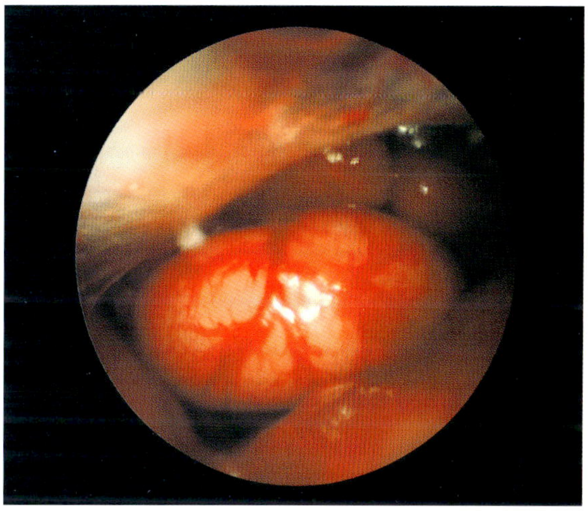

22.11 Left testicle of a domestic pigeon (Columba livia) with three characteristic parallel, arborising testicular veins.

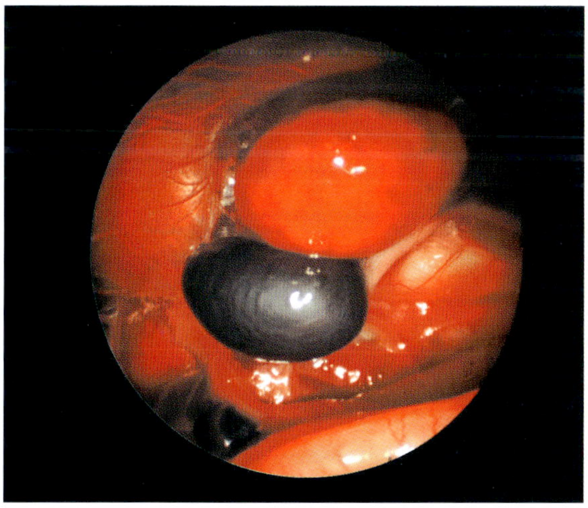

22.12 Cranial renal division and characteristic black-pigmented testicle of the crimson-rumped toucanet (Aulacorhynchus haematopygus). Black testes are also seen in cockatoos and various zoo birds.

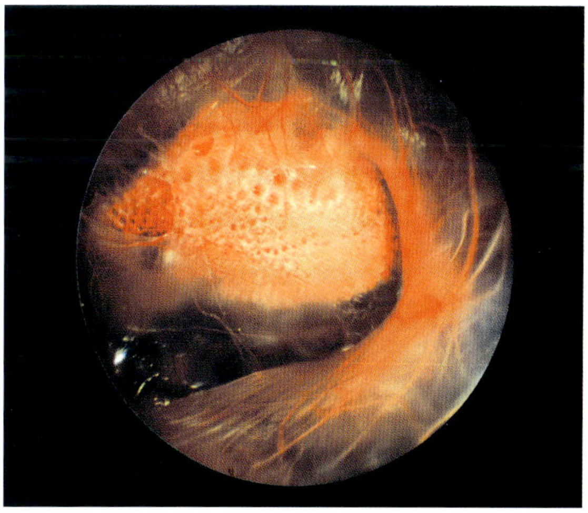

22.13 Caudal view of the lung with tertiary bronchi opening into the cranial thoracic air sac in a goshawk (Accipiter gentilis).

(Figure 22.11 and Chapter 9 'Male genital organs'). In some species (e.g., cockatoos), however, the testes appear black (Figure 22.12). The male gonads are further characterised by three arborising subcapsular vessels (external testicular veins, Figure 22.11) that allow them to be distinguished from ovarian follicles and loops of intestine.

As well as permitting superficial examination of the internal organs, endoscopy enables the identification of pathological changes such as renal gout, fungal granulomas on the air sacs and tuberculous granuloma formation on the liver or spleen. **Organ biopsies** can also be performed under visual guidance.

At the conclusion of the procedure, the skin incision is closed with one to two simple interrupted sutures. In the hands of a skilled operator, endoscopy is usually well tolerated by the patient in the post-operative period, as the air sac closes over rapidly.

Endoscopy can also be used to examine various other cavities and tracts including the:

- pharynx and nasal cavity,
- external acoustic meatus,
- cervical oesophagus and crop,
- proventriculus and ventriculus,
- cloaca and
- trachea to the level of the syrinx.

To ensure adequate spontaneous ventilation during endoscopic examination of the **respiratory tract**, including the trachea, a large bore trocar or catheter must be placed securely in the caudal thoracic air sac (so-called '**air sac perfusion**', see Chapter 20 'Handling, restraint and anaesthesia').

Any abnormalities identified during the endoscopic examination should be **documented** in the patient's record. This underscores the value of endoscopy in sexing of birds, compared with non-invasive methods that provide no additional information about the bird's suitability for breeding. The ability to show the findings directly to the owner in the form of video or photographs (Figure 22.3) greatly increases the acceptance of this technique.

Contraindications

Contraindications for endoscopy include:

- obesity with fat deposition in the body cavity (e.g., seasonally driven obesity in raptors in autumn),
- inadequate visibility leading to risk of iatrogenic organ trauma,
- moulting – increased likelihood of haematoma formation and bleeding into the body cavity with impairment of visibility.

Potential post-operative **complications** of endoscopy include subcutaneous emphysema. Possible causes of emphysema include operator inexperience resulting in excessive air sac damage. If emphysema is sufficiently extensive to impair movement associated with breathing, asphyxiation can result. Attempts at **repairing the air sac** are of little use as the tear is very difficult to locate. Instead, multiple punctures are made in the skin, and a drain is placed if required, to allow the air to escape freely from the subcutis until the air sac has healed of its own accord.

Excessive post-endoscopic haemorrhage is usually the result of vessel or organ damage due to inappropriate, excessively robust technique or lack of visibility. The risk of haemorrhage associated with controlled tissue trauma (e.g., hepatic or splenic biopsy) is much lower due to relatively faster clotting in birds, compared with mammals.

The vast morphological diversity among avian species is an important consideration when performing endoscopy in birds. Sound anatomical knowledge and a requisite degree of experience is essential for distinguishing between normal and abnormal findings, and avoiding iatrogenic injury, in the wide range of patients that may be encountered.

23

Surgical fracture management

R. Korbel, H.-G. Liebich and M. Meiners

General principles

Birds are **highly prone to fractures**. Sound knowledge of the topographical anatomy of the skeleton and surrounding structures is essential for appropriate orthopaedic fracture management. Surgical approaches in particular necessitate an appreciation of the fundamentals of topography, including the muscles, vessels and nerves overlying and adjacent to the bones (Figures 23.1 and 23.2). Species-related variation must also be taken into account.

Rather than providing a detailed account of individual techniques, this chapter deals primarily with general principles and specific aspects of surgical fracture management in birds, drawing on particular examples.

The unique avian way of life, especially the incorporation of flight, places birds at particularly high risk of fractures. Avian bones have a number of structural features that serve to counter this vulnerability. The pneumatisation of various bones (e.g., humerus, femur and sternum), for example, facilitates flight by contributing to the lightweight construction of the skeleton.

On the other hand, the **cortex** (compacta) of avian bones is **not as strong** or as pliable as that of mammals. The bones of birds are thus **more brittle**, and **comminuted fractures** are more common.

The incidence of fractures is influenced by the nature of the bird's environment and therefore varies for different patient cohorts. Nevertheless, while the incidence is relatively high in wild and falconry birds, fractures may be encountered in any type of bird, including aviary and cage birds, zoo birds and also pigeons, in which fractures

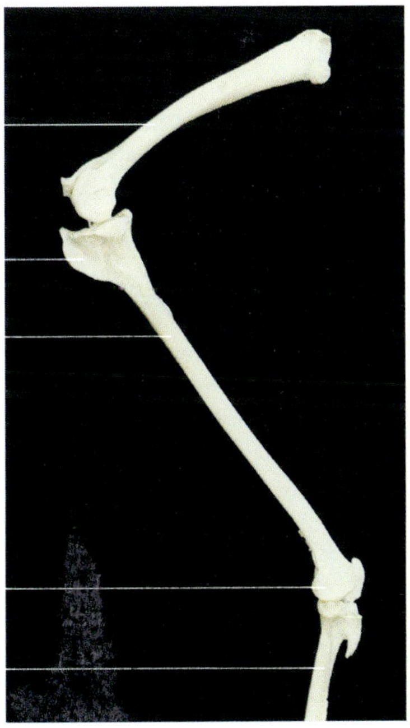

23.1 Anatomical relationships of the bones of the pelvic limb of a chicken (partial medial view).

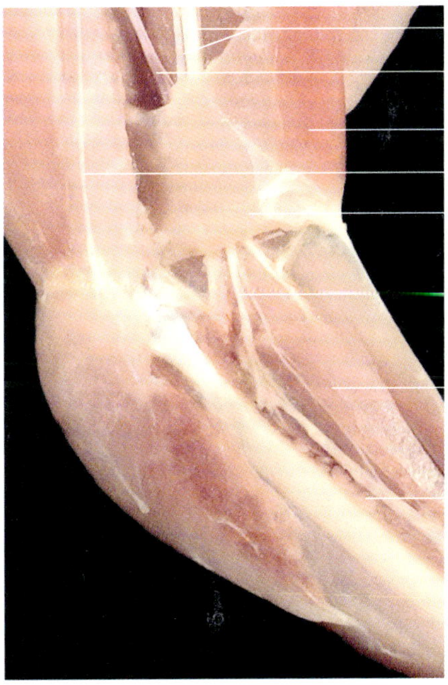

23.2 Topographical anatomy of the pelvic limb at the level of the knee joint and proximal tibiotarsus of a chicken (medial view).

are often a consequence of inappropriate housing or husbandry practices.

The most common cause of fractures is **trauma** resulting from collision with an obstacle or vehicle, from a gun-shot/hunting injury or from bite wounds (often complicated by infection with Pasteurella multocida).

Incorrect handling and restraint, including fixation of the limbs away from the body, increases the risk of fracture. Appropriate techniques are described in detail in Chapter 20 'Handling, restraint and anaesthesia'.

Various **metabolic disorders**, such as calcium and/or vitamin D_3 deficiency, can also play a role in fracture aetiology. In adult birds, these deficiencies may occur during times of increased demand, such as the laying period, manifesting clinically as **osteoporosis** with regression of the cortex. Insufficient bone mineralisation in growing birds can also compromise bone strength, potentially leading to **rickets**. The equivalent phenomenon in adult birds is referred to as **osteomalacia**.

Increased risk of spontaneous fracture may be observed in cases of **polyostotic hyperostosis** secondary to an oestrogen secreting ovarian or Sertoli cell tumour. The latter is commonly observed in budgerigars (Melopsittacus undulatus) (see also Chapter 9 'Male genital organs' and Chapter 17 'Common integument').

Clinical principles

The principal aim of avian fracture management is to restore the anatomical structure of the bone, so as to avoid any loss of function. This applies particularly to the forelimb and the pectoral girdle, as birds are able to compensate more readily for deformity of the hindlimbs than the wings. Even small deviations in the alignment of the bones or the length of the wings can interfere with flight. As the inability to fly severely curtails the bird's prospects of survival in the wild, it can constitute grounds for euthanasia.

The **prognosis for a full return to function** is variable and depends on the site of the fracture. Long-term telemetric monitoring of successfully rehabilitated fracture patients by the Raptor Centre, University of Minnesota, identified minimum estimated 12-month survival rates in the wild of just under 14 per cent.

Prognosis is also influenced by species-related anatomical variations. Forelimb fractures in swallows (Hirundo spp.), the common swift (Apus apus) and other birds with long, narrow wings are generally an indication for euthanasia, as the wing aerodynamics cannot be restored with sufficient precision to rehabilitate the bird for release into the wild.

In terms of equipment (Figures 23.6A and 23.6B), surgical technique and control of infection, the basic principles of fracture management are similar for birds and mammals. However, in addition to the aforementioned peculiarities of avian anatomy, a number of other factors must be taken into account when treating fractures in birds.

Fracture repair occurs primarily via **secondary bone healing** involving, in part, significant **callus formation**. This is particularly obvious radiographically (see Chapter 19 'Imaging techniques'), compared with mammals. An extensive callus can result in bridging of the gap between the radius and ulna, with formation of a synostosis. This impairs the 'drawing parallels' action of the antebrachial bones (longitudinal displacement of the ulna and radius with respect to one another), rendering the bird permanently unable to fly.

Primary bone healing occurs only when there is optimal fracture reduction, compression across the fracture site and absolute stability – the three fundamental prerequisites for uncomplicated bone healing. However, for various anatomical and functional reasons, it is very difficult to achieve these outcomes simultaneously in avian patients. Primary bone healing is therefore uncommon in birds.

Reference to a **standardised system for classifying fractures in birds** (Figures 23.3 to 23.5) is a useful basis for developing a surgical plan and formulating a prognosis. Selection of an appropriate orthopaedic intervention depends in part upon whether the fracture is **open** or **closed**. Where possible, closed fractures should not be exposed, due to the risk of complications such as infection and displacement of bone fragments. External coaptation may be sufficient for some patients. In the frequently encountered case of comminuted fractures (Figures 23.4 and 23.5), bone fragments should only be removed if they are infected, desiccated or necrotic. Otherwise, incorporation of bone splinters into the callus may aid in stabilising the fracture. Removal of bone fragments contributes to loss of bone mass, bone shortening and other associated consequences.

Fracture healing is relatively rapid, compared with mammals. In uncomplicated cases (e.g., metaphyseal fractures of the tibiotarsus in raptors) with optimal post-operative management (physiotherapy), full stability is restored in 3–4 weeks.

Commencing 3–4 days after surgery, **physiotherapy** should be performed 2–3 times per week under general anaesthesia. In addition to flexion and extension of the joints, attention should be given to stretching and relaxing the patagia (pro- and metapatagium). This overcomes the tendency for the joints to stiffen, and the **patagia** to undergo irreversible **contracture**, while the limbs are immobilised for the purposes of fracture healing.

Even in the presence of optimal bone healing, inadequate post-operative management can lead to partial or total loss of limb function. This can itself be an indication for euthanasia.

Birds are considerably less susceptible to **osteomyelitis** than mammals. Nevertheless, the administration of antibiotics (as well as analgesics) is indicated for open fractures or cases in which an external fixator is used. Antibiotic prophylaxis and measures for prevention of aspergillosis are particularly important in parrots and raptors.

To assist with **rehabilitation**, **bandages** are used as an **adjunct** to surgical fixation for immobilisation of the affected limb (Figures 23.30 to 23.33). It is also important to rest the bird for approximately two weeks, by placing it in an environment that limits its movement and is devoid of cage wire that could facilitate climbing. The use of dim lighting is also advisable.

Radiographs should be taken in the third week after surgery to assess the progress of fracture healing.

Rehabilitating birds to full flying capacity can be a time-consuming and labour-intensive process, due to atrophy of the flight muscles during the period of immobilisation.

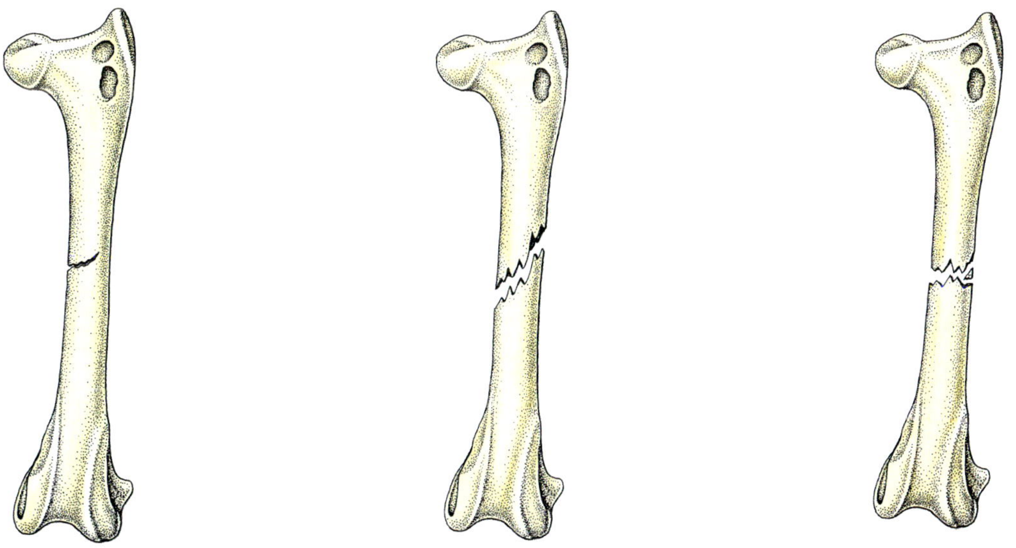

Fracture type 32 A 1 | Fracture type 32 A 2 | Fracture type 32 A 3

23.3 Classification of diaphyseal fractures of the femur in birds according to the system of Redig, Brown and Gordon (1996) and Ford (1997). Fractures are assigned an alphanumeric code according to their location and morphology. This system facilitates computer aided schematic fracture assessment for determination of prognosis and treatment. Types of fractures shown include simple or partial fractures of the femoral diaphysis (fracture type 32 A). Left: partial diaphyseal fracture (fracture type 32 A 1), middle: oblique diaphyseal fracture (fracture type 32 A 2), right: transverse diaphyseal fracture (fracture type 32 A 3).

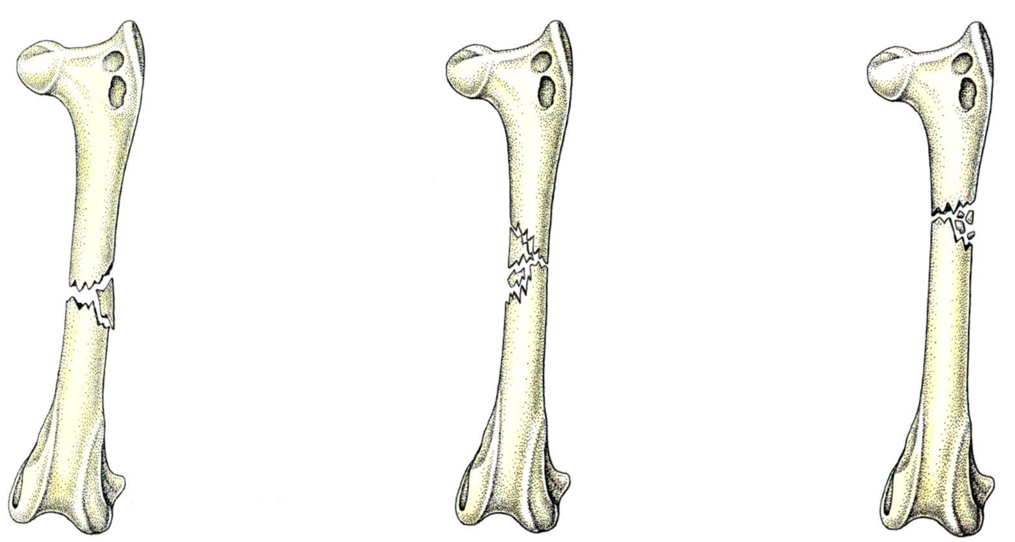

Fracture type 32 B 1 | Fracture type 32 B 2 | Fracture type 32 B 3

23.4 Classification of femoral fractures in birds (continued from Figure 23.3): femoral fracture with diaphyseal fragments (fracture type 32 B). Left: replaceable fragment (fracture type 32 B 1), middle: multiple replaceable fragments (fracture type 32 B 2), right: multiple non-replaceable fragments (fracture type 32 B 3).

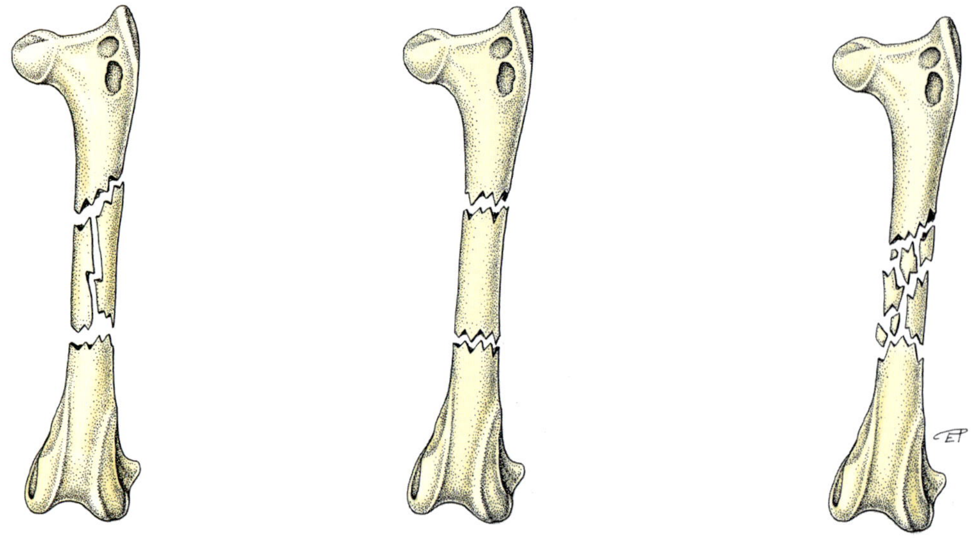

Fracture type 32 C 1 Fracture type 32 C 2 Fracture type 32 C 3

23.5 Classification of femoral fractures in birds (continued from Figures 23.3 and 23.4): comminuted diaphyseal fractures (fracture type 32 C). Left: reducible fragments (fracture type 32 C 1), middle: segmental fracture (fracture type 32 C 2), right: non-reducible fragments (fracture type 32 C 3).

Raptors should be housed in boxes with adjoining flying rooms and provided with training. Various training protocols can be used, including some derived from falconry (see below).

Techniques

Fractures of the long bones, including the humerus, radius and ulna, carpometacarpus, femur, tibiotarsus and tarsometatarsus, are commonly encountered in clinical practice. Surgical fixation of these fractures aims to return the patient to a physically normal and pain-free state. Achievement of this objective requires prompt and total stabilisation of the bone in correct longitudinal and axial alignment, as well as temporary resting of the limb while the fracture heals and limb function is restored.

Techniques used for fixation of long bone fractures in birds, in conjunction with bandaging as appropriate, include intramedullary pins, external fixators, interlocking nails and intramedullary stabilisation using xenotransplants. The use of cerclage wire is not appropriate.

In many cases, the method of choice is the **combination of an intramedullary pin and external fixator** (= intramedullary extraskeletal tie in fixator, hybrid fixator) (Figures 23.7 to 23.29). The advantages of this method include:

- effective stabilisation,
- shorter surgery times,
- straightforward positioning,
- precise rotational alignment,
- limited soft tissue trauma,
- faster healing,
- implant is completely removed,
- requires relatively inexpensive equipment,
- lightweight,
- greater patient comfort and mobility,
- good outcomes when combined with regular post-operative physiotherapy.

While producing excellent aesthetic results, surgical fixation may, depending on the technique used, also have certain drawbacks, including:

- slow healing,
- tendency for brittle avian bones to splinter during placement of cortical screws,
- potentially extensive soft tissue necessitated by exposure of a large section of bone,
- inadequate rotational alignment and
- the need for additional intervention for removal of pins and plates.

The equipment required for placement of a tie-in-fixator is shown in Figure 23.6. **Partially threaded positive (raised) profile miniature interface pins** (Fa. Imex, Texas, USA) should be used for transcortical pin placement. Negative profile pins tend to break at a weak point at the junction between the thread and the shaft. Pins of 0.9mm diameter (e.g., in budgerigars) or greater are used. The **intramedullary pin** (IM Pins, medullary nail) should fill approximately 50–60 per cent of the medullary cavity.

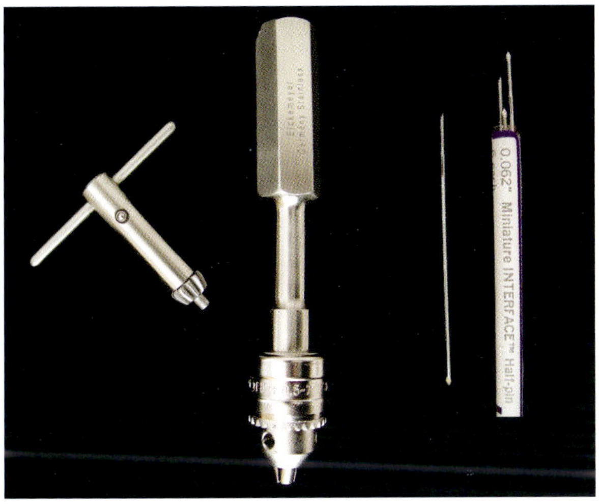

23.6A Equipment for treatment of long bone fractures using an external fixator and integrated intramedullary pin ('IM-ESF-tie-in-fixator'). From left to right: Hand drill with chuck key for pin placement, miniature-interface-pin (diameter 1.6mm; Imex®-Pin, USA) with partial positive profile thread. Pins with a negative profile are not recommended due to the tendency for formation of a weak point at the thread/ non-thread interface.

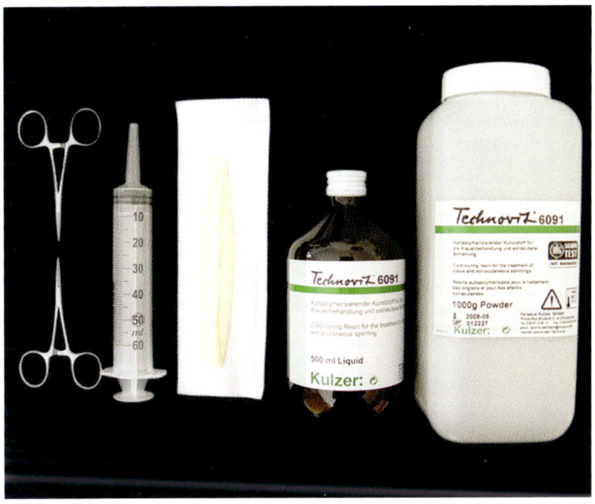

23.6B Materials for construction of a light polymethylmethacrylate (PMMA) connecting bar for joining transcortical and intramedullary pins. From left to right: baby mosquito® forceps, syringe with hub for inserting PMMA into a Penrose drain, two-component acrylic resin (Technovit®; liquid hardener and powdered resin).

The transcortical and intramedullary pins can be connected in various ways, ideally keeping the weight of the external apparatus to a minimum. A flexible connecting bar can be created at relatively low cost by filling a 6.25mm **Penrose drain** with **polymethylmethacrylate**.

More rigid alternatives include the use of a **FESSA type** tubular fixator or, in birds over 5kg, a **Kirschner-Ehmer** external fixation splint. To minimise weight in the latter system, carbon connecting rods are preferable to aluminium bars.

It is generally not necessary to cover the tie-in-fixator, although a loose elastic bandage can be applied to provide protection.

The key steps in orthopaedic management of avian fractures using a tie-in-fixator are depicted in Figures 23.7 to 23.9 (schematic), Figures 23.12 to 23.22 (demonstration) and 23.23 to 23.26 (in vivo), using the tibiotarsus as an example.

Fractures may be **managed initially** using only an external fixator, without an intramedullary pin. This limits movement at the fracture site while allowing treatment of soft tissue injuries without the risk of contracture. **Placement of the completed tie-in-fixator** can be undertaken up to three days later.

When using a tie-in-fixator for repair of femoral diaphyseal fractures, the intramedullary pin should penetrate the femur proximally on the trochanter, between the trochanteric fossa (concavity on medial aspect of trochanter) and the antitrochanteric articular surface (Figures 23.12 to 23.21). The transcortical external skeletal fixation (ESF)

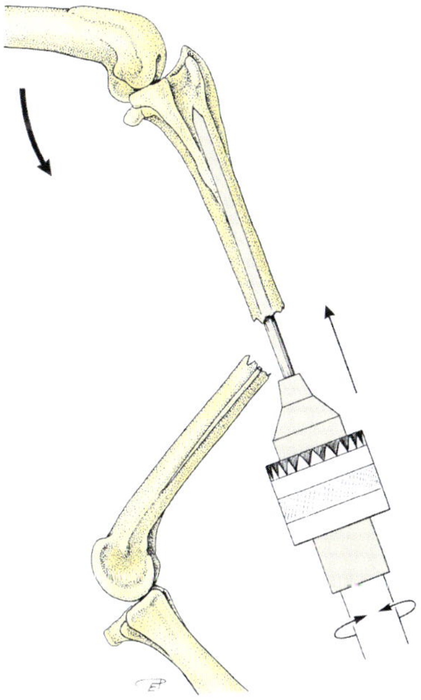

23.7 Principles and stages of fracture repair. Example – placement of a tie-in-fixator for management of an open fracture of the left tibiotarsus in a common buzzard (Buteo buteo): an intramedullary pin is inserted retrograde into the proximal fragment using a hand drill, prior to para-articular perforation of the cortex of the proximal tibiotarsus with the tip of the pin (after flexion of the knee joint); cf. also Figure 23.11.

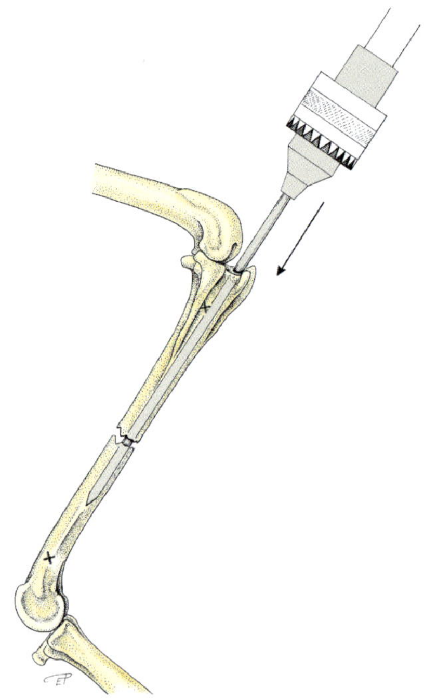

23.8 After transferring the chuck to the proximal protruding end of the intramedullary pin, the pin is drawn further proximally until its distal end extends approximately 0.5cm past the fracture site. The fracture is then reduced and the pin is advanced normograde, using a winding motion, into the spongy bone in the distal medullary cavity of the distal fragment (continued from Figure 23.7).

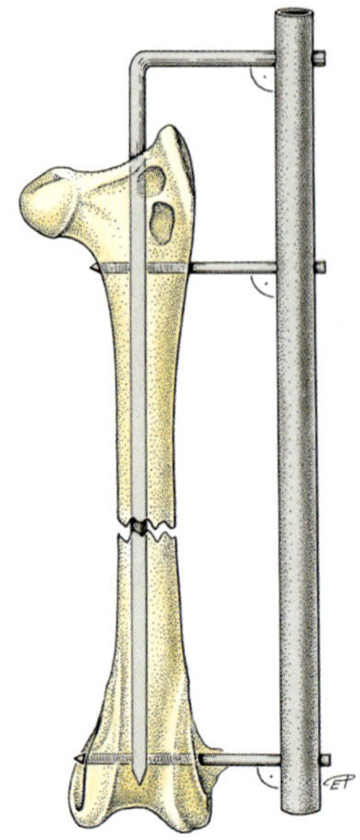

23.10 Repair of a femoral fracture in a common buzzard (Buteo buteo) using an external fixator and integrated intramedullary pin (tie-in-fixator) with a PMMA connecting bar (schematic).

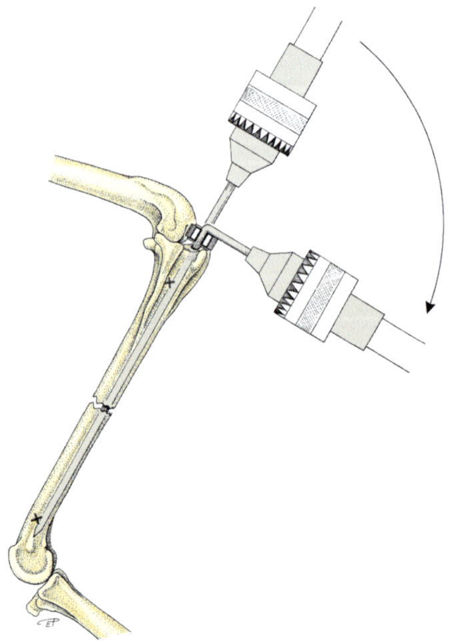

23.9 The hand drill is used to bend the protruding proximal section of the intramedullary pin through 90°, while grasping the pin near the joint with flat-nosed pliers (continued from Figure 23.8); cf. also Figure 23.16.

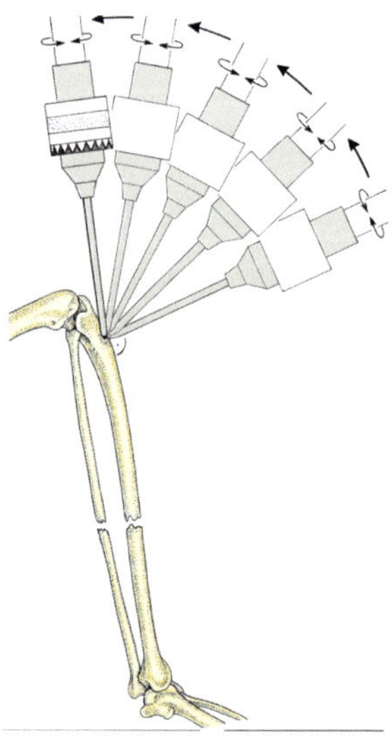

23.11 Schematic representation of the repair of a fracture of the left radius and ulna in a common buzzard (Buteo buteo). Advancement of an intramedullary pin into the medullary cavity of the ulna using a hand drill. To avoid iatrogenic fractures, the tip of the pin is initially applied to the cortex at right angles. While continually rotating the drill, the angle is gradually adjusted until the pin can be inserted in the desired proximo-distal direction after penetrating the cortex.

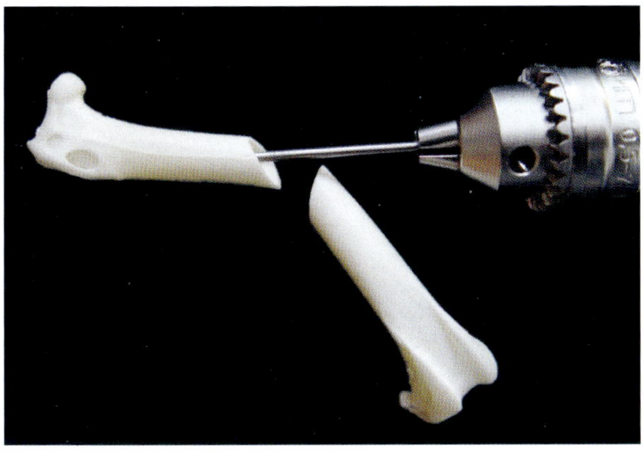

23.12 Anatomical principles and stages of fracture repair using an external fixator and integrated intramedullary pin (tie-in-fixator). Example – artificial fracture of the right femur of a common buzzard (Buteo buteo): retrograde insertion of an intramedullary pin into the proximal fracture segment using a hand drill. The femoral cortex is then penetrated, external to the hip joint.

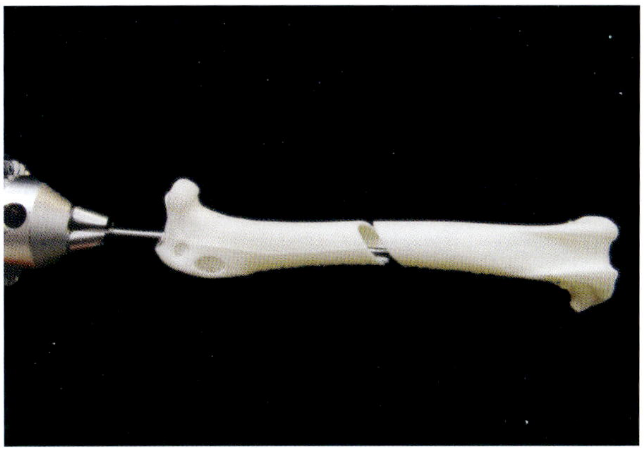

23.13 The chuck is transferred to the proximally protruding end of the intramedullary pin. Following alignment of the bone fragments, the pin is 'threaded' normograde into the distal fragment until it engages the spongy bone in the distal medullary cavity (continued from Figure 23.12).

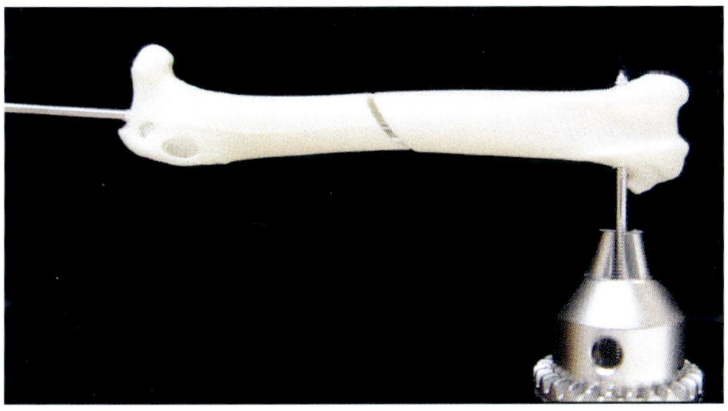

23.14 Placement of the distal transcortical pin: repeated circular movements of the drill, alternating between clockwise and anti-clockwise, are used to bed the tip of the transcortical pin in the cortex. The self-tapping pin is then advanced through the cortex of the lateral femoral condyle and the medullary cavity, past the intramedullary pin, and through the medial cortex. After 2.5–3 turns, the tip of the pin can be palpated, facilitating appropriate placement (continued from Figure 23.13).

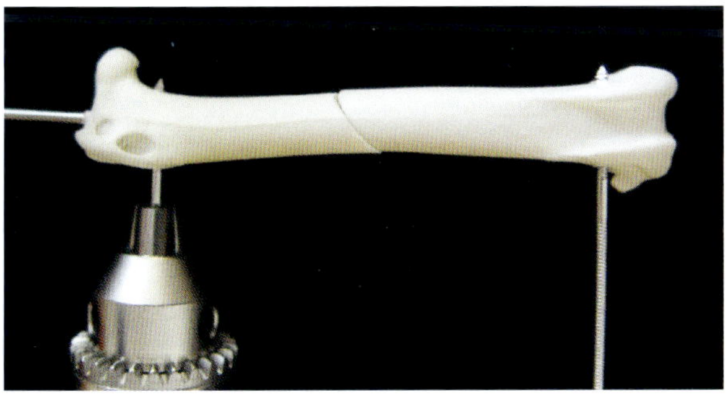

23.15 Placement of the proximal transcortical pin: the insertion point is on the lateral aspect of the trochanter (continued from Figure 23.14).

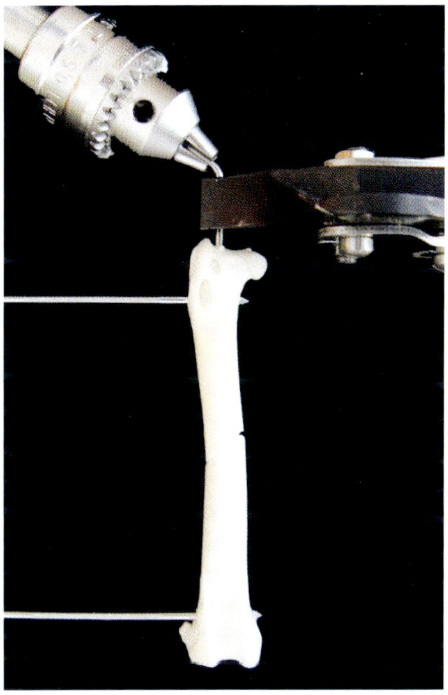

23.16 The proximally protruding end of the intramedullary pin is bent with the hand drill, using flat-nosed pliers to grip the pin near the joint (continued from Figure 23.15).

pins are inserted from the lateral side, perpendicular to the long axis of the bone. The distal insertion site is just distal to the lateral epicondyle, while the proximal pin is introduced on the lateral aspect of the trochanter (Figures 23.14 to 23.21).

Retrograde insertion of the intramedullary pin for fractures of the **tibiotarsus** is shown in Figures 23.7 to 23.9. The distal transcortical pin is inserted proximal to

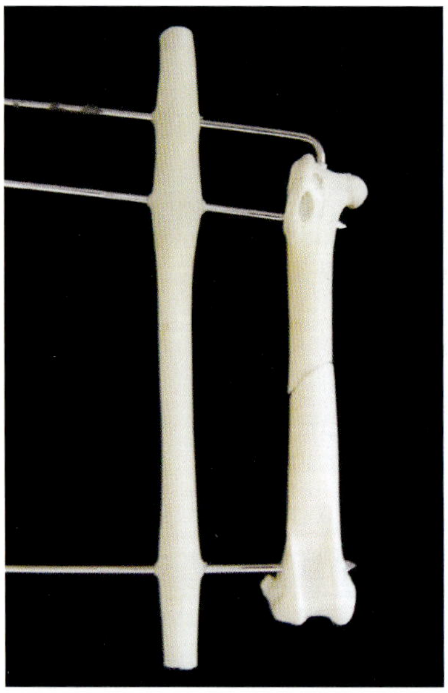

23.17 Construction of the connecting bar for integration of the intramedullary pin into the external fixator. A Penrose drain is mounted onto the transcortical pins and the bent section of the protruding intramedullary pin (continued from Figure 23.16).

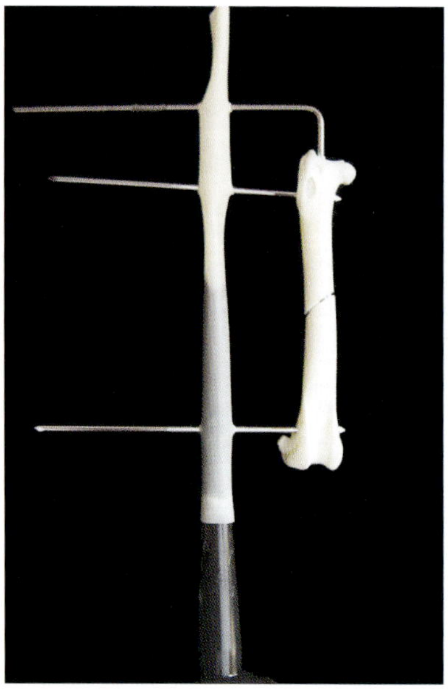

23.18 A syringe is used to fill the Penrose drain with rapidly hardening two-component acrylic resin (PMMA, Technovit®), ensuring correct alignment of the drain (continued from Figure 23.17).

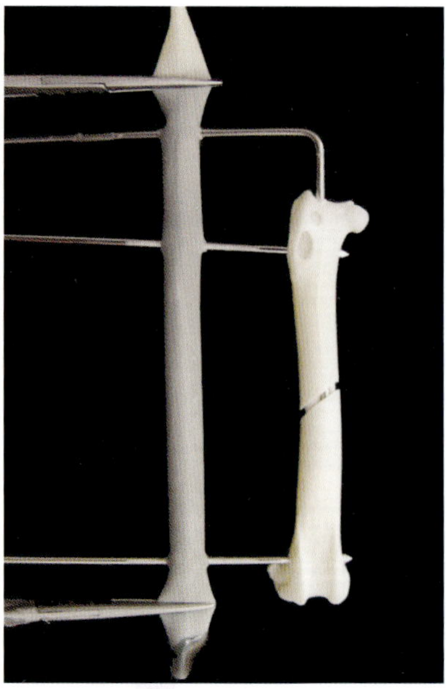

23.19 The ends of the drain are clamped (first the proximal end, opposite to the end at which the resin is introduced, then the distal end) until the resin has hardened (continued from Figure 23.18).

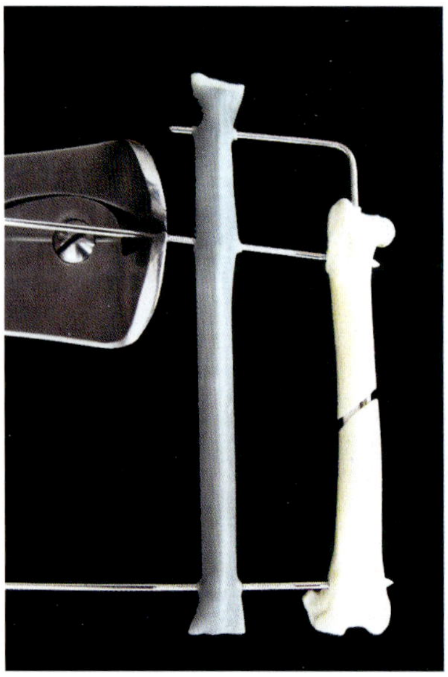

23.20 When the resin has hardened, the intramedullary and transcortical pins are trimmed to a suitable length using wire- or bolt cutters (continued from Figure 23.19).

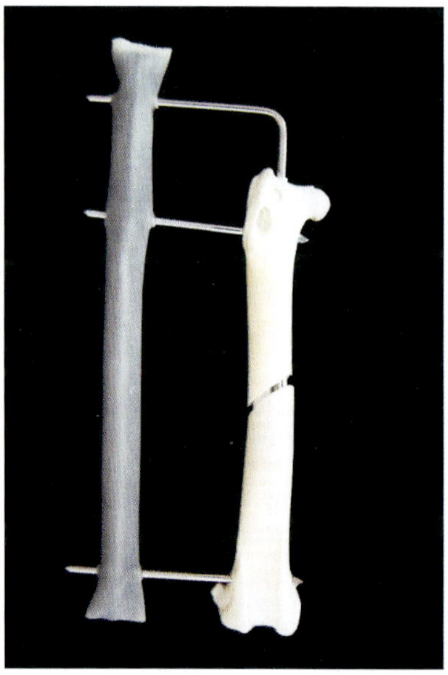

23.21 Completed fixation of a femoral fracture using a (unilateral) external fixator (Type 1) with integrated intramedullary pin (tie-in-fixator) (continued from Figure 23.20).

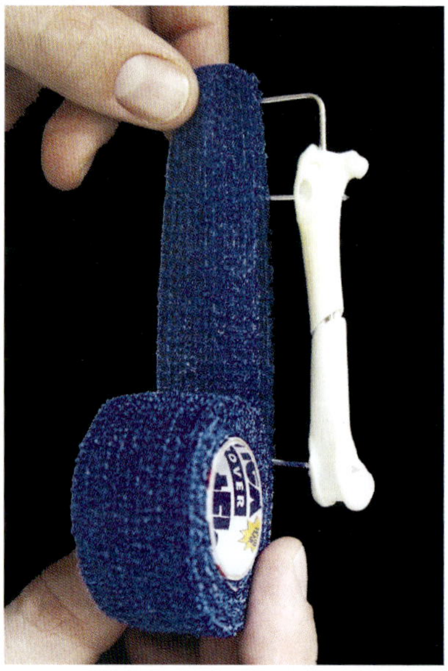

23.22 Application of an elastic bandage to prevent the external fixator from becoming entangled and to cover the transcutaneous drill holes after application of topical antibiotics (continued from Figure 23.21).

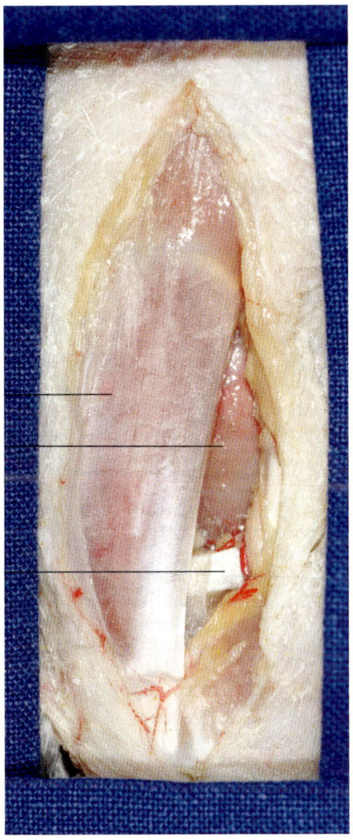

23.23 Use of an external fixator with integrated intrameduallary pin (tie-in-fixator) to repair a fracture of the left femur of a common buzzard (Buteo buteo): skin incision and appearance of superficial surgical field.

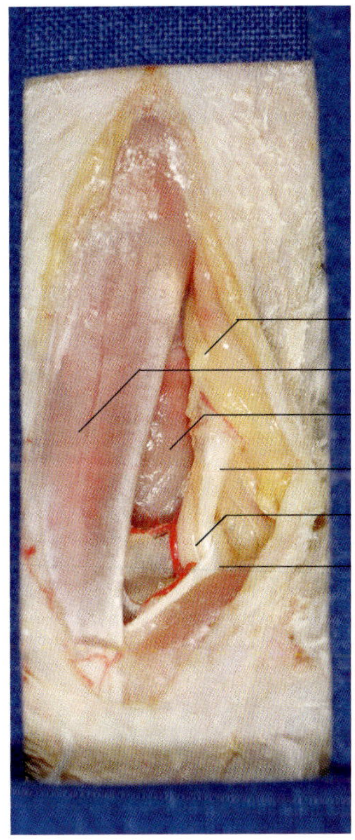

23.24 Tissues dissected to reveal the deep surgical field (continued from Figure 23.23).

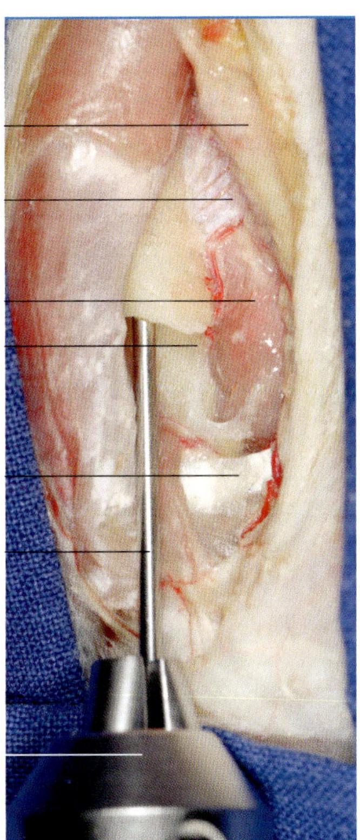

23.25 Retrograde insertion of an intramedullary pin (continued from Figure 23.24).

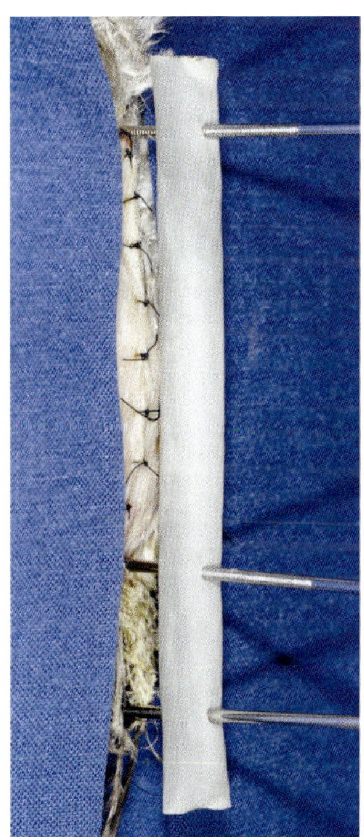

23.26 Connection of the transcortical pins (top and middle) and the bent end-portion of the intramedullary pin (bottom) using a polymethylmethacrylate connecting bar (continued from Figure 23.25).

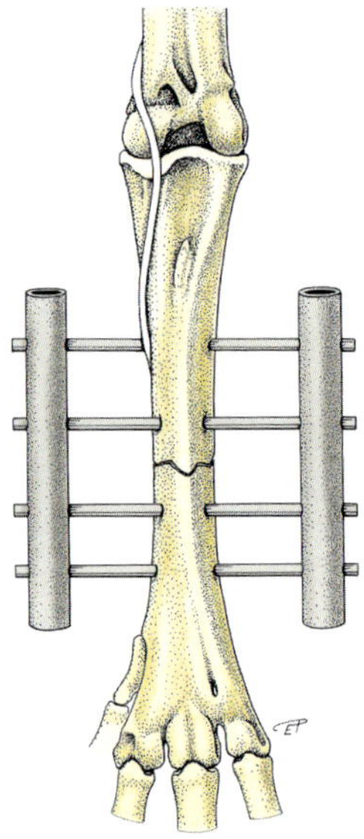

23.27 Fixation of a fracture of the right tarsometatarsus of a common buzzard (Buteo buteo) using a Type II external fixator (schematic).

23.28 Fixation of a fracture of the right tarsometatarsus of a common buzzard (Buteo buteo) using a Type II external fixator (anatomical specimen). Feathers should be removed from a wide area around the surgical site.

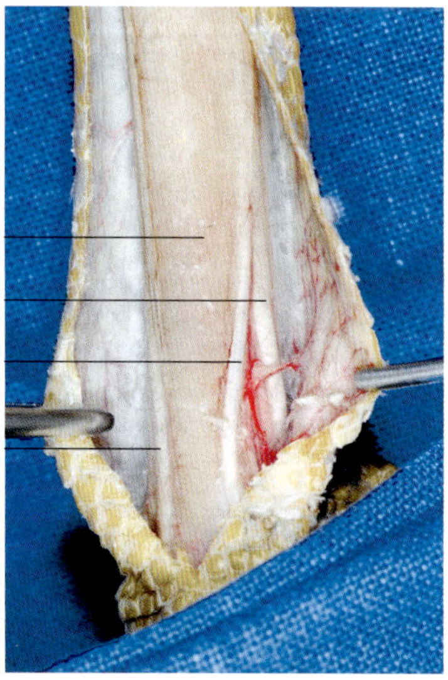

23.29 Use of a Type II external fixator to repair a fracture of the right tarsometatarsus in a common buzzard (Buteo buteo): surgical view following plantar skin incision.

the tuberculum retinacula m. fibularis (a protuberance immediately proximal to the lateral condyle), taking care to preserve the nearby extensor sulcus and supratendinal bridge. Proximally, the ESF pin is placed caudomedial to the head of the fibula.

When placing a tie-in-fixator in the pectoral limb, the wing is folded against the body while the connecting bar is attached. In this way, the body effectively acts as a guide, reducing the likelihood of misalignment or inappropriate positioning of the apparatus and subsequent compromise of flying ability. Note that when a bird is in dorsal recumbency in a fully relaxed state (under general anaesthesia), a wing in which the bones are correctly aligned will not lie flat on the underlying surface. Rather, it forms a downward facing concavity, with the wing tips elevated above the surface.

Removal of the **tie-in-fixator** is undertaken in stages so that the load on the healed bone is gradually increased. The first step in this process is to cut the section joining the intramedullary pin and the connecting bar.

The device shown in Figures 23.7 to 23.26, incorporating a single connecting rod, is referred to as a **'Type I'** (unilateral) tie-in-fixator. Others include **Type II** fixators (bilateral uniplanar) (Figures 23.27 to 23.29), used for treating fractures of the tarsometatarsus, and **Type III** (bilateral biplanar) configurations. **Cross-pin-tie-in-fixators** and **transarticular fixators** are also employed in avian fracture management.

Bandaging techniques

The main forms of external coaptation used in birds, other than splints, are the **figure-of-eight bandage** (Figure 23.30 and 23.31) and the **body wrap** (Figures 23.32 and 23.33). Both can be used either as the sole means of stabilisation and immobilisation of closed fractures, or as an adjunct to a tie-in-fixator. The figure-of-eight bandage is used in combination with the body wrap for fractures of the distal wing, while the body wrap is employed for fractures of the pectoral girdle and humerus. **Important anatomical landmarks** for application of a figure-of-eight bandage include the:

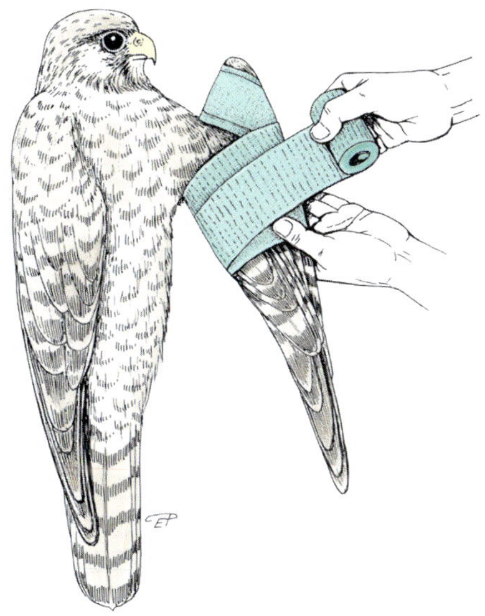

23.30 Application of a two-layer figure-of-eight bandage to facilitate stabilisation and immobilisation of the wing in management of fractures of the radius-ulna and distal limb (patient under general anaesthesia in dorsal recumbency): starting high in the axilla, the deep layer (gauze bandage) is passed dorsally over the wing onto the ventral surface of the same wing and back into the axilla. From the second pass onwards the bandage is wound in a figure of eight, incorporating the propatagium, advancing towards the tip of the wing in an overlapping pattern. Elastic bandage (e.g., Vetrap®) is the preferred material for the outer layer. Note: It is essential to ensure that the primary and secondary remiges are arranged parallel to one another at the commencement of bandaging. Tension and pressure on the patagium must be avoided (to prevent necrosis and contracture) and the bandage must always be wound away from the patient's body, with the bandage facing upwards as it is unrolled.

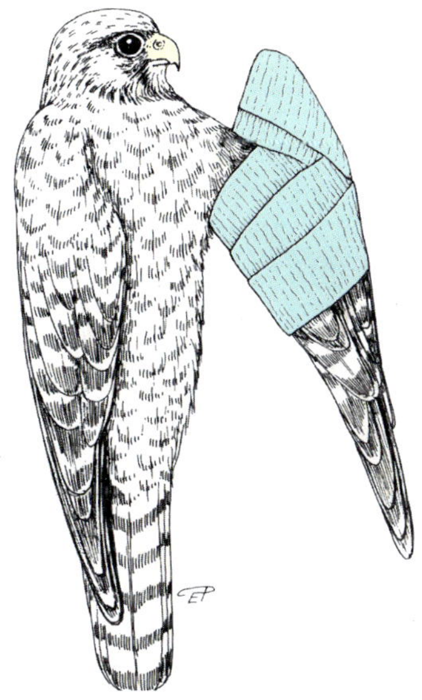

23.31 Completed figure-of-eight bandage as an aid to stabilisation and immobilisation of the wing in management of fractures of the radius-ulna and distal limb (continued from Figure 23.30).

23.32 Application of a body wrap to facilitate stabilisation and immobilisation of the wing in management of fractures of the pectoral girdle and humerus (patient under general anaesthesia in dorsal recumbency). For distal fractures, the body wrap can be combined with the figure-of-eight bandage. To allow the bird to maintain its balance, only one wing is included in the bandage. Commencing cranially in the axilla of the unaffected wing, the bandage is passed dorsally over the body. The bandage is then wound around the injured wing and the body in overlapping layers, securing the wing against the chest.

23.33 Completed body wrap combined with a figure-of-eight bandage. Note: Care must be taken to align the wing normally, with the tip directed only slightly towards the middle of tail. Particularly when using elastic wrap, it must be ensured that the bandage is not too tight to permit respiratory movements (checked by inserting a finger between the bandage and the sternum) and does not extend too far caudally (to avoid faecal contamination and obstruction of the cloaca) (continued from Figure 23.32).

- primary remiges,
- elbow joint (articulatio cubiti),
- innermost (proximal) secondary remiges,
- carpo-metacarpal joint (articulatio carpometacarpale) and
- pro- and metapatagium.

When applying the bandage it is important to ensure that:

- the proximal secondary remiges are included,
- compression of the patagium is avoided,
- hyperflexion of the carpometacarpal joint is avoided and the primary and secondary remiges are aligned in parallel,
- the inner gauze layer is completely covered.

In the application of a body wrap, care should be taken to:

- make allowance for respiratory movements,
- include only one wing (to allow for balance),
- allow sufficient room for movement in the pre-crural fold and
- keep the cloaca free of the bandage.

Foot bandaging (Figures 23.34 to 23.37) is used to **reduce pressure** on the vulnerable metatarsal pad in the management of diseases such as ulcerative pododermatitis (bumblefoot). These techniques are also engaged prophylactically in cases of unilateral limb fractures to avoid excessive pressure and injury in the healthy contralateral limb. It is important to note that both feet should always be bandaged. Shoe splints (Figures 23.34 and 23.35) consist

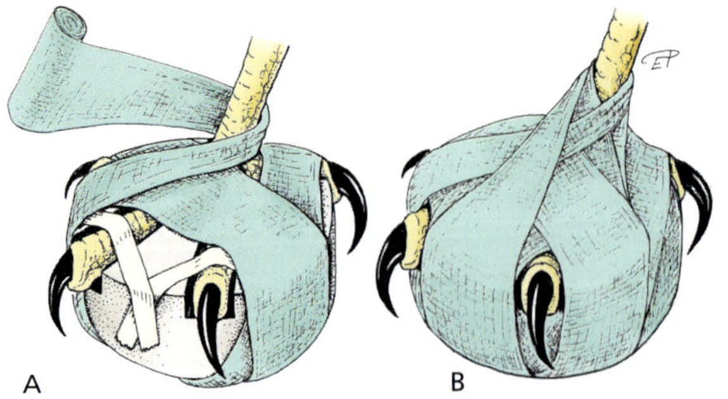

23.35 (A) Use of narrow adhesive strips for temporary positioning of the toes, and application of a gauze bandage in a figure-of-eight pattern to secure the splint to the foot. (B) Completed shoe splint. A second, water-resistant layer of bandage may be required. Shoe splints should always be applied to both feet (continued from Figure 23.34).

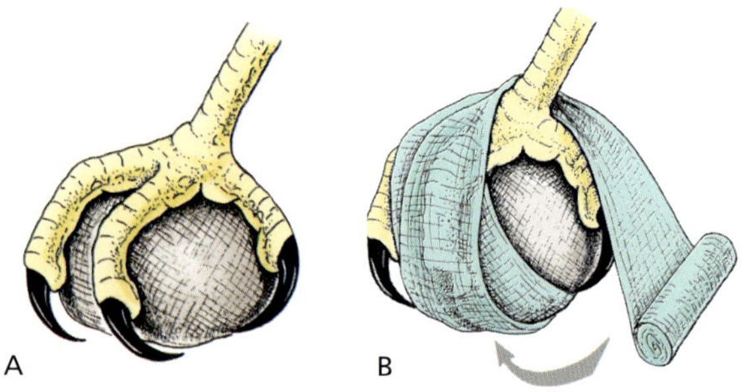

23.36 (A) Application of a two layer ball bandage for even distribution of pressure on the foot in the management of ulcerative pododermatitis ('bumblefoot'): a ball of gauze bandage is placed in the claw of a raptor. (B) The ball is secured by wrapping the claw in a figure-of-eight gauze bandage.

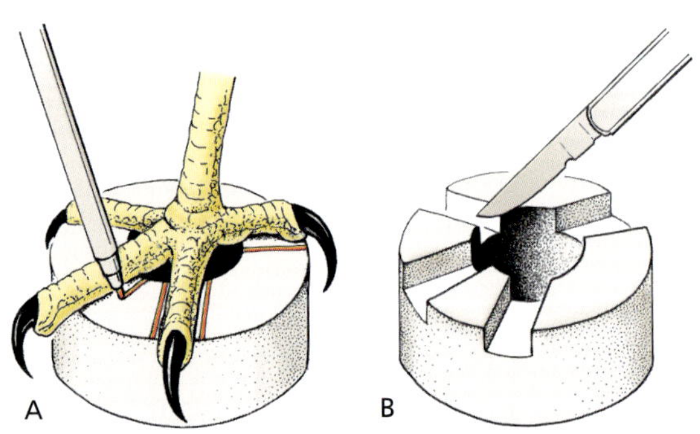

23.34 (A) Shoe splint for relief of pressure on the metatarsal pad in the management of ulcerative pododermatitis ('bumblefoot'): an outline of the foot is traced onto a polypropylene disc with a central opening. (B) Sections of polypropylene are excised to accommodate the digits. Note: the central opening is paramount.

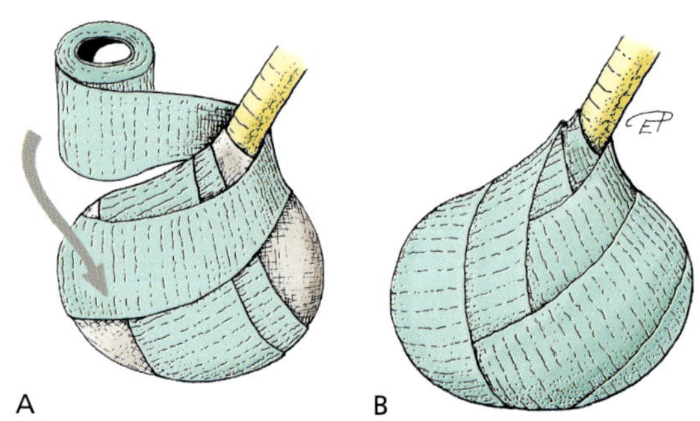

23.37 (A) Application of an outer water-resistant protective elastic bandage (e.g., Vetrap®). (B) Completed two-layer ball bandage. To avoid damage to the pad of the contralateral limb, the ball bandage should be applied to both feet (continued from Figure 23.36).

of a polypropylene disc with a central opening, thus relieving pressure on the pad. Ball bandages (Figures 23.36 and 23.37) are based on the principle of distributing pressure evenly across the whole foot.

Basic materials used for bandaging include absorbent gauze and elastic bandages such as Vetrap®. An additional layer may be incorporated to provide protection against moisture, faeces, litter and biting.

Frequently encountered **complications** associated with inadequate fracture fixation include:

- soft tissue damage resulting from incorrect placement of external fixators,
- contracture of the patagium due to lack of, or inadequate, physiotherapy,
- degenerative joint disease of the elbow due to lack of, or inadequate, physiotherapy,
- nerve damage (particularly tibial nerve) resulting from inadequate knowledge of topographical anatomy,
- ulcerative pododermatitis in the healthy contralateral limb due to lack of prophylactic foot bandaging.

24

Falconry and raptor medicine

R. Korbel and H.-G. Liebich

Throughout virtually all periods and realms of human society, raptors have been a subject of great interest. The enduring fascination with these birds stems from their distinctive appearance and unique way of life. Raptors are regarded as the embodiment of strength or as a symbol of humankind's dream of achieving flight, and often appear as representations of unbridled freedom on coats of arms and flags. As illustrated in Figure 24.1, the world of falconry has long been represented in imagery.

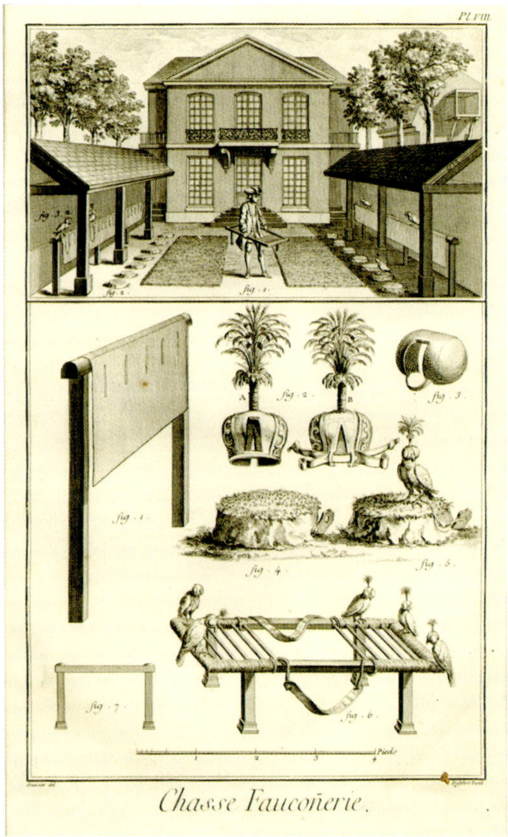

24.1 Nineteenth-century French etching of falconry mews showing equipment and training aids (Original: 'Dr Wentges Collection', Edith Haberland Wagner Trust, Munich).

24.2 Peregrine falcon (Falco peregrinus) with 'furniture', beating its wings ('bating').

Today, veterinary care is regularly sought for wild raptors and falconry birds. Familiarity with the specific nomenclature used for anatomical structures and clinical aspects of falconry – the so called language of falconry – is important when dealing with owners or falconers (Table 24.2). The following information builds upon the fundamentals and therapeutics described in previous chapters by explaining important terms used in falconry and wild raptor medicine, with brief reference to anatomical and clinical considerations in raptor husbandry. Particular attention is given to aspects of veterinary management that are common to trained and wild birds yet may, in some situations, require a divergent approach.

The act of hunting with a trained bird of prey is referred to as **'falconry'** or **'hawking'**. Raptors can be divided into two categories: those that kill with their sharp talons (Figure 24.5), using their beaks for feeding, and those that employ their hook-like beaks (Figure 24.8) to dispatch their quarry, using their feet to grasp and hold their prey. The beak of the latter group features a projection, the so-called **tomial tooth**, that enables these birds to use their beaks with a plier-like action.

An overview of the anatomical features of the different types of raptor is given in Table 24.1. All falcons use their beaks to kill, their feet serving purely to restrain

Table 24.1 Distinguishing characteristics of raptors that kill their prey with their feet versus their beaks.

Beak used to kill (feet used to grasp)	Feet used to kill
Falcons	Hawks (inc. Accipiters) Eagles Owls
Hooked beak with tomial tooth	Hooked cutting beak
Generally relatively weak feet and long toes, claws are relatively weak	Typically strong feet with powerful claws
Kill prey by biting (using the plier-like action afforded by the tomial tooth)	Kill prey by stabbing and crushing with their powerful feet
Feet are used to catch and immobilise prey/food	Beak is used solely for prehension

24.3 Falconer and peregrine falcon (Falco peregrinus), equipped for hawking.

24.4 Adult bald eagles (Haliaeetus leucocephalus) have a wingspan of more than 2m and a bodyweight of 4–6kg. Safe management of these birds requires sound knowledge of, and practical proficiency in, appropriate handling techniques.

24.5 Goshawks (Accipiter gentilis) belong to the group of raptors that kill with their feet (talons). These are characterized by a hooked cutting beak and oval shaped nostrils.

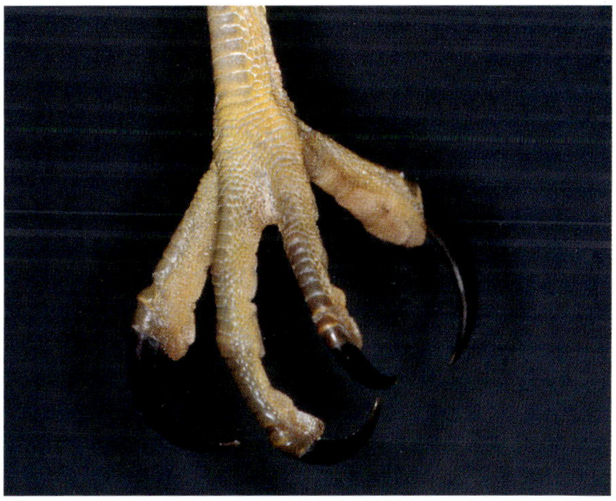

24.6 Foot of a goshawk (Accipiter gentilis) exhibiting typical features of raptors that kill with their talons. The first and second toes are particularly well developed.

24.7 The Eurasian eagle owl (Bubo bubo) is a typical example of a nocturnal raptor that kills with its feet, using its hooked cutting beak solely for prehension.

24.8 The peregrine falcon (Falco peregrinus) typifies raptors that kill with their hooked, plier-like beak. The round nostrils feature bony cone shaped projections ('baffles').

their prey. The first, backward-facing toe of all raptors, used in capturing prey, is referred to as the **'hallux'**. The second, medially directed toe serves to immobilise food or prey. Together, the first and second toes are referred to as the 'killing talons'. The remaining toes are used for grasping.

Falcons require a flat perch, referred to as a block. Curved, upholstered **bow perches** are used for raptors that prefer to grasp (e.g., hawks).

Diseases of the toes and their joints and tendons have serious consequences. Injury or lack of mobility of the second toe results in difficulty with prehension. Wild raptors lacking a functional hallux are unable to capture prey and thus cannot be rehabilitated. Similarly afflicted falconry birds can no longer be used for hawking.

24.9 Foot of a Eurasian eagle owl (Bubo bubo) (typical example of nocturnal raptors that kill with their feet). The toes are feathered as far as the distal interphalangeal joints.

24.10 Foot of a peregrine falcon (Falco peregrinus) (typical example of raptors that use their feet to grasp, rather than kill, their prey). The toes are long and prominent (characteristic of raptors that kill other birds) and feature stud-like digital pads

When handling raptors that **use their claws to kill**, factors over and above those identified in Chapter 20 'Handling, restraint and anaesthesia' need to be taken into account. In these birds, the tendinous grasping and binding mechanism (automatic digital flexor mechanism and digital tendon locking mechanism) and associated long digital flexor muscles function as a **locking gear system** (see Chapter 1 'Introduction' and Chapter 4 'Pelvic limb').

When the digital flexors are moderately contracted due to handling and arousal, careful opening of the tensed digits elicits a clicking noise akin to a ratchet. If a bird is handled inappropriately, this locking mechanism may result in reflex grasping of the handler's arm or hand, which the bird may be unable to disengage voluntarily.

Use of force to try to release the animal's grip risks damaging the joints and tendons and can lead to digital fractures. In these situations, the bird must be calmed by eliminating visual stimuli until the toes relax of their own accord.

The likelihood of sustaining bite injuries while restraining these birds is relatively low. In contrast, when handling raptors that kill prey with their beaks, the handler is at greater risk of bite wounds than injury inflicted by the claws.

When restraining large raptors close to the handler's body, it is important to be aware of the considerable force with which these birds can **beat their wings**. This applies in particular to wild birds that use their wings to assist them in capturing prey (e.g., golden eagles, to dislodge young goats from cliff faces).

To capture large raptors on the ground, a large towel should be wrapped around the bird. With the wings folded against the body, the hind legs can then be grasped through the towel and restrained. This simultaneously covers the animal's eyes and reduces excitement by removing visual stimuli.

More subtle measures can be employed for restraining 'manned' birds of prey (those that are familiar with their owner and trained to accept handling). **Hooding** and/or **shortening of the jesses** is often sufficient for basic examinations. These techniques make towels unnecessary, thus avoiding associated risks such as damage to the feathers.

The use of a **hood** (Figures 24.11, 24.13 and 24.14) to restrain falcons, and sometimes eagles, operates on the principle of calming the bird by eliminating visual stimuli. This technique is also used in the process of manning the bird (familiarising it with the falconer), and to prevent birds from flying from the fist prematurely at the sight of prey.

The principle of shielding the bird's head is widely used in the handling of highly visually oriented birds, including parrots and other species. It serves primarily to reduce the bird's overall level of arousal, thereby preventing stress and shock. In some cases, injections can be administered to calm birds without the need for additional restraint (e.g., hooded falconry birds perched on the handler's fist).

The hood is applied by grasping the **plume** with the thumb and index finger, drawing the hood along the chest to the lower beak, with the opening for the beak facing downwards, then rotating it onto the head.

The hood is closed (drawn) by pulling on the long hood straps (braces), and opened using the short knotted braces. If the bird is perched on the fist, the falconer can close and open the hood by pulling on the braces with their free hand and their teeth.

Styles of hood include Indian, Arab and Dutch varieties (Figures 24.13 and 24.14). They must be correctly sized for individual birds. The interior of the hood should be inspected as part of a physical examination. Traces of secretions are suggestive of an ill-fitting hood.

24.11 Lanner falcon with Dutch style hood (museum specimen). Hooding is used to calm manned falcons and eagles by removing visual stimuli.

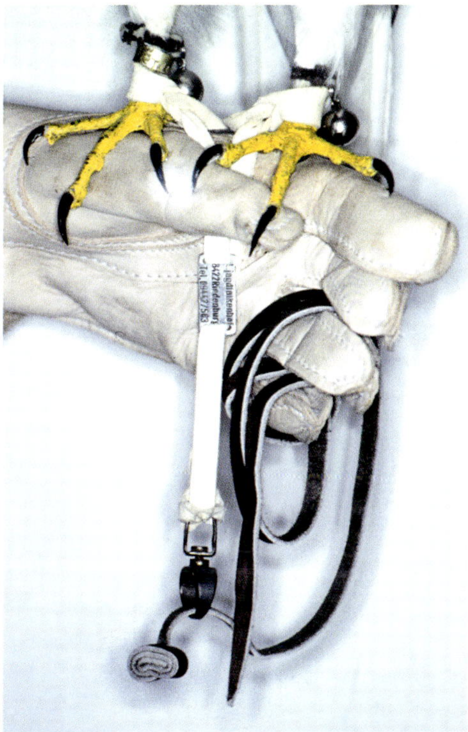

24.12 Correct restraint of a falcon using a gauntlet and 'traditional furniture'. Jesses (in this case bearing an identification tag) are tied around each leg with a special knot and are held between the thumb and index finger. The jesses are connected to the leash by means of a metal swivel. The bells are used for locating the bird.

24.13 Hood (Dutch plumed variety): front view showing opening for the beak and the plume consisting of feathers and other embellishments. The plume is used to manipulate the hood.

24.14 Hood (corresponds with Figure 24.13): rear view showing pleat for adjusting the width as well as short knotted braces for opening the hood (striking the braces) and long, plain braces for closing the hood (drawing the braces).

The equipment worn by a trained raptor, known as 'furniture' (Figure 24.12), consists of **jesses** (leather straps connected to the bird's legs, to which identification tags may be attached) and a metal joint, or **swivel**, by which the jesses are connected to the **leash**. **Bells** may also be attached to the feet to assist in locating birds that are obscured by vegetation (cover). In contemporary hawking, the function of bells is replaced by telemetry incorporating a mini-transmitter mounted on the bird (over the synsacrum/tail base) and a receiver.

A distinction is made between **traditional jesses**, consisting of strips of leather attached to the metatarsus by a special knot, and so-called **'Aylmeri' jesses**. The latter, preferred for animal welfare reasons, consist of leather anklets with grommets to which leather straps are attached.

When handling raptors it is important to avoid damage to the feathers. For this reason, less intrusive methods of restraint, such as holding the jesses, are preferred to the use of towels, which may cause quills to break if rubbed against the plumage against the direction of the feathers.

Feather damage may impair the bird's ability to fly and hunt. Even split-second errors in timing result in failure to capture prey, making the bird unfit for purpose.

In the management of wild raptors, where the integrity of the feather coat is a prerequisite for rehabilitation, it is particularly important to ensure that the remiges and rectrices are intact. Feathers that are broken or missing (e.g., due to hunting, handling or removal for orthopaedic intervention) should be replaced with an equivalent, previously moulted feather. By helping to restore their capacity for hunting, this technique contributes to the survival of birds released into the wild.

24.15 Arabian furniture.

24.16 Imping is used to repair feathers damaged during hunting or handling. Previously moulted, corresponding feathers from the same (or a similar) bird are mounted onto the feather stump to restore the animal's fitness for flight.

The procedure, known as **'imping'** (Figure 24.16), is performed by inserting a suitably lightweight and flexible timber or fibreglass dowelling rod (or other splinting material) into the stump of the broken feather, and mounting the replacement feather on the other end. The splinting material is thus effectively implanted in the follicle of the broken feather. Appropriately sized dowelling rods can be fashioned from cooking skewers or fibreglass trimmed from fishing rods. The dowel may be glued or sewn in place, the latter being more labour-intensive. Other birds of the same species can be used as a source of replacement feathers, which should be stored after moulting in their correct anatomical order (i.e., from primary to secondary remiges, and from outer to inner rectrices) (Figure 24.18). Repaired feathers can be identified radiographically by the presence of parallel spindle shaped shadows.

Feather defects manifesting as microscopic structural aberrations, scrolling and striations of the vanes (Figures 24.24 and 24.25) are occasionally encountered in wild raptors and falconry birds. These are the result of electrocution due to direct contact with powerlines or arcing between the wings and nearby high voltage lines. Despite little external evidence of damage, affected birds typically have massive internal organ damage and must be treated as emergency patients.

Lesions of this nature must be distinguished diagnostically from so called 'stress marks' or 'hunger traces' (see Chapter 17 'Common integument').

As part of the rehabilitation programme for orthopaedic patients, procedures used in the training of falconry birds, such as flying birds on a hand-held leash (creance), can also be employed with wild raptors (Figure 24.17).

It is important in this situation to open the jesses by removing the swivel before casting off. Otherwise the bird may become entangled in trees.

For (contrast) radiography (see Chapter 19 'Imaging techniques') and anaesthesia, the bird must be appropriately fasted. The last feed should occur more than 12 hours prior to the intended procedure. Care must also be taken to ensure that the pellet (casting) has been expelled before the procedure is carried out.

Due to the relatively high gastric pH of crepuscular raptors, the pellets of these species are characterised by the presence of small bony remnants (Figure 24.19). These are absent in the castings of diurnal birds of prey. Raptors that are unwell should be fed diets low in pellet forming material (skin, feathers and bones), as egestion of the pellet places additional strain on the bird.

The faecal and urinary excreta of raptors are referred to collectively as **mutes**.

To a limited extent, the **colour of the iris** can be used in determining the **age of raptors**. In sparrowhawks (Accipiter nisus), the iris is more or less pale yellow in young animals (Figure 24.20) and orange in older birds

24.17 Flying birds on a hand-held leash (creance) is a falconry technique that can be applied in a veterinary context for rehabilitation. In this case it is being used with a bald eagle (Haliaeetus leucocephalus) to build muscle and provide physiotherapy for the patagia and joints following orthopaedic management of a fracture of the radius and ulna.

24.18 Right wing of a juvenile goshawk with first to tenth 'beam' feathers (primary remiges), first to tenth 'flags' (secondary remiges), shoulder feathers, 'sarcel' (first primary), and alula (alular feathers).

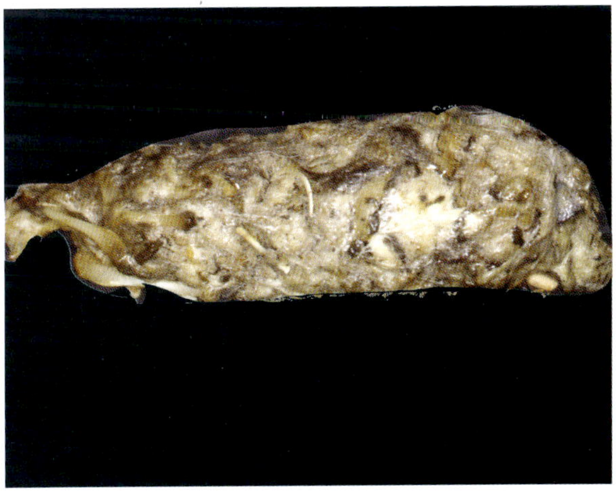

24.19 Typical casting of a crepuscular raptor (in this case great horned owl) containing small bony remnants of prey.

24.20 Juvenile Eurasian sparrowhawk (Accipiter nisus) with pale yellow iris.

24.21 Adult Eurasian sparrowhawk (Accipiter nisus, male) with orange iris.

24.22 Juvenile, ca. four-week-old great horned owl (Bubo virginianus) with cloudy lenses, typical of young altricial birds. The lens will gradually become clear from 4–6 weeks of age onwards. This phenomenon should not be confused with cataracts.

(Figure 24.21). Depending on the angle of incident light, the lens of **altricial raptors** is greyish-white up to 12 weeks post-hatching (Figure 24.22). With increasing age, the lens becomes clearer and assumes a bluish tinge. This is rarely observed in **precocious** species, typically diurnal raptors (Figure 24.23), and then only over a period of about two to three weeks. This phenomenon is commonly misdiagnosed as cataracts.

Male falcons are referred to as **tercels.** Etymologically, this term is derived from the word 'tertius', referring to the fact that males are up to one-third smaller than females.

An alternative explanation for this term comes from the observation that about one in three young birds in an eyrie is male. Other descriptive terms used in falconry include '**musket**' (male sparrowhawk), **'jerkin'** (male gyrfalcon), **'sakret'** (male saker falcon), **'lanneret'** (male lanner falcon) and '**haggard**' (wild bird trapped as an adult).

24.23 Juvenile, ca. 4 week old red-tailed hawk (Buteo jamaicensis). Apart from a slight bluish tinge, the lenses of precocious species are completely clear just a few days after hatching. The iris of this bird still exhibits the blue tone that transforms into yellow-brown in the adult.

24.24 Scanning electron micrograph showing a feather defect in the primary remiges of a peregrine falcon (Falco peregrinus), caused by electrocution.

24.25 Scanning electron micrograph of a feather defect caused by electrocution, detailed view of Figure 24.24.

On overview of falconry terms relevant to the veterinary management of wild raptors and falconry birds is provided in Table 24.2.

Table 24.2 Nomenclature used in falconry: selected terms of anatomical and clinical relevance in the veterinary care of raptors.

Falconry term	Anatomical/clinical significance
Alula	Alular feathers.
Beam feathers	Primary remiges (feathers associated with the manus).
Bells	Small round, usually paired bells attached to the tarsometatarsus with a strap; used for locating birds during hawking.
Bind, to (vb)	To grasp and hold prey during hunting.
Crissum	Undertail coverts.
Falconry, hawking	Hunting with trained birds of prey.
Falconry bird	Any manned, trained bird of prey used for hunting quarry; includes falcons, hawks, sparrowhawks, eagles and historically also ravens and shrikes.
Flags	Secondary remiges (the ten feathers attached to the ulna, between the primary remiges and the feathers of the shoulder).
Frayed feather	Feather with a worn or splayed vane; the result of chafing of, or damage to, the tips and terminal vanes of the remiges and rectrices due to contact with objects in the bird's environment.
Hallux	The toe that faces backwards, designated the first toe; the first and second (innermost forward facing toe) are referred to as the 'killing talons'.
Hood	Plumed head covering (the term 'rufter hood' describes hoods without plumes).
Imping	Repairing broken feathers using a replacement feather, a dowelling rod or other splinting material and glue.
Jesses	Leather straps attached to each tarsometatarsus; the conventional jess consists of a single piece of leather with three holes; the 'Aylmeri' jess combines the leather strip with anklets or bracelets; a hunting jess comprises a simple leather strap that is relatively easily displaced to avoid entanglement in cover.
Lure	Imitation quarry to which feathers or fur may be attached; used for training.
Manning	Taming raptors for handling and initial (creance) training for hawking.
Mantle, to (vb)	To extend the wings to cover/hide food.
Mute	Excreta (faecal and urinary) of raptors (the same term is used as a verb to describe the process of passing excreta).

Table 24.2 continued.

Falconry term	Anatomical/clinical significance
Pellet or castings	Indigestible components of the food (feathers, hair, bones) that are egested, or 'cast', after feeding (time between feeding and disgorgement of the pellet varies considerably).
Plume	Feathers adorning the top of a hood; used to manipulate the hood.
Sarcel	First (outermost) primary feather.
Seeling	Historical practice (now considered inhumane) of temporarily sewing the eyelids together to eliminate visual stimuli and acclimatise the bird to a hood.
Stress marks, hunger traces	Intermittent lines running through the vane perpendicular to the feather shaft; found particularly in birds that have experienced disease or malnutrition during feather formation, thus an important clinical indicator of chronic illness (e.g., parasitism or Salmonellosis); recurrent stress marks on one or two feathers indicate damage to those follicles.
Swivel	Metal joint for joining jesses to the leash.
Tercel	Male falcon (based on the fact that male falcons are around one-third smaller than females).
Train	Tail feathers (rectrices).

Glossary of terms

A

Alula	feathered first digit
Apteria (sing. apterium)	areas between contour feather tracts; usually contain down feathers or semi-plumes present

B

Barbae (sing. barba)	feather barb (ramus and barbule)
Barbulae (sing. barbula)	feather barbule

C

Calcar	spur; present in males of the order Galliformes
Cava hepatica	hepatic peritoneal cavity; formed by the transverse septum and the lateral ligaments of the liver
Cavum intestinale	intestinal peritoneal cavity; contains intestine, ovary, oviduct or testes
Cera	soft, thickened portion of the rhamphotheca surrounding the nostrils of some species
Chalazae (sing. chalaza)	twisted layer of albumen that suspends the yolk within the egg
Cloaca	terminal portion of the intestinal tract into which the urinary and genital tracts open
Collumella	sole avian auditory ossicle; equivalent to the stapes (stirrup) of mammals
Coprodeum	first (proximal) segment of the cloaca
Coracoideum	coracoid bone; part of the pectoral girdle
Cor lymphaticum	lymph heart; suction and pressure pump for circulation of lymph, located near the synsacrum

F

Filoplumae	filoplumes
Foramen pneumaticum	openings in bones for passage of intraosseous diverticulae of the air sacs
Furcula	'wishbone', paired clavicles

G

Glandula uropygialis	uropygial gland
Grit	small stones in the ventriculus (gizzard)

H

Hypotarsus	bony structure on the proximoplantar tarso-metatarsus; formed mostly from distal tarsal bones

I

Ingluvies	crop

J

Jugale	jugal bone; a bone of the facial portion of the skull; forms the jugal arch (arcus jugalis), together with the quadratojugal bone

L

Lagena	vesicular expansion of the cochlear duct at the tip of the cochlea
Latebra	white yolk; pendulous mass within the yellow yolk

M

Meckel's diverticulum	remnant of the yolk sac

N

Neopulmo	in phylogenetic terms, the 'new' (more recently developed) portion of the avian lung
Notarium	rigid bony unit formed by the fusion of cervical vertebrae 2–5

O

Operculum auris	skin fold that can close off the external acoustic meatus
Oropharynx	oral cavity and pharynx
Anulus ossicularis sclerae	scleral ring; ring of bones located at the boundary between the cornea and sclera

P

Paleopulmo	in phylogenetic terms, the 'older' portion of the avian lung
Parabronchus	tertiary bronchus; arises from secondary bronchial system
Patagium	skin folds of the wing
Pecten oculi	pecten, a derivate of the retina
Penna	feather
Pessulus	part of the syrinx
Phallus	copulatory organ
Pluma	down feather
Pneumocapillaris	air capillary; site of gas exchange in the avian lung
Podotheca	featherless thickened epidermis of the avian foot
Proctodeum	distal portion of the cloaca
Proventriculus	glandular stomach
Pterylae (sing. pteryla)	feather tracts; regions populated by contour feathers
Pulvinus	pad
Pyga	caudal part of the dorsal trunk
Pygostylus	fused last 4–6 caudal vertebrae

Q

Quadratojugale	quadratojugal bone; a bone of the facial portion of the skull; forms the jugal arch (arcus jugalis), together with the jugal bone
Quadratum	quadrate bone; a bone of the mandibular joint, equivalent to the incus (anvil) of mammals

R

Ramphotheca	keratinous covering of the beak
Rectrices (sing. rectrix)	flight feather of the tail
Remiges (sing. remex)	flight feather of the wing
Rostrum	beak

S

Septum horizontale	subdivides the coelomic cavity, forms ventral border of the lung
Septum obliquum	subdivides the coelomic cavity, extends from thoracic vertebrae to lateral sternal margin
Synsacrum	fused lumbar and caudal vertebrae
Syrinx	avian organ of phonation, also referred to as the caudal larynx (larynx caudalis)

T

Tarsometatarsus	product of fusion of the central and distal tarsal bones and the bones of the metatarsus
Tectrices (sing. tectrix)	covert feathers
Tela interdigitales	interdigital web
Tibiotarsus	product of fusion of the tibia and the proximal row of tarsal bones

U

Urodeum	middle portion of the cloaca

V

Ventus	vent; external opening of the cloaca
Vexillum	feather vane

Z

Zona flexoria craniofacialis	flexible region between the cranium and the upper beak
Zygapophysis	articular process of the vertebrae

Bibliography

Abdel-Magied, E.M. and King, A.S. (1978) The topographical anatomy and blood supply of the carotid body region of the domestic fowl. *Journal of Anatomy* **126**(3): 535–546.

Abourachid A. Kinematic parameters of terrestrial locomotion in cursorial (ratites), swimming (ducks), and striding birds (quail and guinea fowl). *Comp Biochem Physiol* 2001; Part A 131: 113–9.

Artmann A. Schnittbildanatomie der Kopf- und Halsregion der Gattung Ara unter besonderer Berücksichtigung des paranasalen Sinussystems. Diss med vet, Wien, 2000.

Baumel JJ. Aves nervous system. In: Sisson & Grossman's. *The Anatomy of the Domestic Animals*. 5th ed. Vol. 2. Philadelphia: Saunders 1975.

Baumel JJ. Nomina Anatomica Avium. 2nd ed. Cambridge and Massachusetts: *Nuttall Ornithological Club* 1993.

Berens v. Rautenfeld D, Budras K-D. The bursa cloacae of struthioniforms in comparison with the bursa of other birds. *J Morphology* 1982; 172: 123–38.

Berens von Rautenfeld, D., Wenzel-Hora, B., Budras, K.D. and Henschel, E. (1983) Morphologische Untersuchungen des Lymph-drainageweges nach Applikation von Intravitalfabstoffen und Röntgenkontrastmitteln. *Verhandlungen der Anatomischen Gesellschaft* **77**: 533–536.

Berens von Rautenfeld, D., Wenzel-Hora, B. and Hickel, E.-M. (1983) Lymphgefässsystem und Lymphographie beim Vogel, Teil 1: Anatomische Besonderheiten des Lymphgefässsystems. *Tierärztliche Praxis* **11**: 469–476.

Bischof, B. and Budras, K.D. (1994) The topography of the lymph heart in the domestic chicken (Gallus domesticus). *Lymphology* **26**(4): 177–185.

Breazile JE, Kuenzel WJ. Systema nervosum centrale. In: Baumel JJ. Nomina Anatomica Avium. 2nd ed. Cambridge and Massachusetts: Nuttall Ornithological Club 1993.

Budras K-D, Berens v. Rautenfeld D. Lymphbildung und Erektion des lymphatischen Kopulationsschwellkörpers beim Hahn (Gallus domesticus). Verh Anat Ges 1978; 72: 565–8.

Budras K-D, Berens v. Rautenfeld D. Zur funktionellen und topographischen Anatomie des Lymphherzens und Lymphgefäßsystems im Beckenbereich bei Wasser- und Laufvögeln. Anat Anz 1984; 156: 231–40.

Budras K-D, Hullinger RL, Berens v. Rautenfeld D. Lymph Heart Musculature in Birds. *J Morphology* 1987; 191: 77–87.

Butcher, G.D. and Miles, R.D. (2011) Factors causing poor pigmentation of brown shelled eggs. *CEVA eggs program online* **7**.

Chitty, J. and Lierz, M. (2008) *BSAVA Manual of Raptors, Pigeons and Passerine Birds*. BSAVA.

Coles, B. (2007) *Essentials of Avian Medicine and Surgery*, 3rd edn. Wiley-Blackwell.

Cotofan V, Cotofan O, König HE. Beitrag zum Studium des Thymus beim Hausgeflügel. Institutsarbeit. Inst Agr Jasi 1971; 2: 227–36.

Cotofan V, König HE, Cozariuc I. Das Koronarsystem des Hausgeflügels. Institutsarbeit. Inst Agr Jasi 1971; 2: 199–201.

Cura P, König HE. Die Gefäßversorgung der Halsregion einiger Hausgeflügelarten. Institutsarbeit. Inst Agr Jasi 1970; 2: 165–70.

Desserich, M., Ziswiler, V., Foelsch, D.W. (1984) Die sensorische Versorgung des Hühnerschnabels. *Revue suisse Zool* **90**: 799–807.

Doneley, B. (2011) *Avian Medicine and Surgery in Practice*. CRC Press.

Duncker HR. Vertebrate lungs: structure, topography and mechanics. A comparative perspective of the progressive integration of respiratory system, locomotor apparatus and ontogenetic development. Review. Respir Physiol Neurobiol 2004; 144: 111–24.

Dyce KM, Sack WD, Wensing CJG. Anatomie der Haustiere. Stuttgart: Enke 2002.

Eurell, J. and Frappier, B.L. (2006) *Dellmann's Textbook of Veterinary Histology*, 6th edn. Blackwell Publishing.

Evans HE. Anatomy of the Budgerigar. In: Petrak ML (ed). Diseases of cage and aviary birds. Philadelphia: Lea & Febiger 1982.

Ford, S. (1997) *Extern Project: Fracture Tracking System*. Washington State University, College of Veterinary Medicine.

Frank T, Probst A, König HE, Walter I. The syrinx of the male mallard (Anas platyrhynchos): Special anatomical features. Anat Histol Embryol 2006; 36: 121–6.

Frank T, Walter I, Probst A, König HE. Histological aspects of the syrinx of the male mallard (Anas platyrhynchos). Anat Histol Embryol 2006; 35: 396–401.

Franz V. Vergleichende Anatomie des Wirbeltierauges. In: Bolk, Göppert, Kallius, Lubosch. Handbuch der vergleichenden Anatomie der Wirbeltiere. Bd. II/2. Berlin, Wien: Urban & Schwarzenberg 1934.

Frewein J, Sinowatz F. Sehorgan. In: Nickel R, Schummer A, Seiferle E.

Gangl D. Die Muskeln der Hinterextremität des Strausses (Struthio camelus Linné 1758). Diss med vet, Wien, 2001.

Ghetie V, Chitescu St, Cotofan V, Hillebrand A. Atlas de Anatomie a Pasarilor Domestice. Bucuresti: Ed. Academiei Republici Socialiste Romania 1976.

Goller H. Versorgungsgebiete und Zentren der Gehirnnerven vom Huhn (Gallus domesticus). Berl Münch Tierärztl Wschr 1972; 85: 432–6.

Haimovici S, König HE, Popescu P. Beitrag zum Studium der Arterien des Halses der Vögel. Analele stiintifice Univ. A1. I Cuza Jasi 1972; 18: 309–15.

Hodges CD. The histology of the fowl. London: Academic Press 1974.

Hoppes, S.M., Tizard, I. and Shivaprasad, H.L. (2013) Avian bornavirus and proventricular dilatation disease diagnostics, pathology, prevalence, and control. *Veterinary Clinics of North America: Exotic Animal Practice* **16**: 339–355.

Huang, R., Zhi, Q., Izpisua-Belmonte, J.-C. and Patel, B.C.K. (1999) Origin and development of the avian tongue muscles. *Anatomy and Embryology* **200**: 137–152.

Hutchinson JR. Biomechanical modeling and sensitivy analysis of bipedal running ability. I. Extant taxa. J Morph 2004; 262: 421–40.

Jarvis, E.D., Güntürkün, O., Bruce, L., Csillag, A., Karten, H., Kuenzel, W., Medina, L., Paxinos, G., Perkel, D.J., Shimizu, T., Striedter, G., Wild, J.M., Ball, G.F., Dugas-Ford, J., Durand, S.E., Hough, G.E., Husband, S., Kubikova, L., Lee, D.W., Mello, C.V., Powers, A., Siang, C., Smulders, T.V., Wada, K., White, S.A., Yamamoto, K., Yu, J., Reiner, A. and Butler, A.B. (2005) Avian brains and a new understanding of vertebrate brain evolution. *Nature Reviews, Neuroscience* **6**: 151–159.

Kaleta, E.F. and Krautwald-Junghanns M.E. (2011) *Kompendium der Ziervogelkrankheiten*, 4th edn. Schlütersche Verlagsanstalt.

King AS, Mc Lelland J. Anatomie der Vögel. Stuttgart: Ulmer 1978.

King, A.S. and McLelland, J. (1980) *Form and Function in Birds*, Vol. 1. Academic Press.

King, A.S. and McLelland, J. (1981) *Form and Function in Birds*, Vol. 2. Academic Press.

King, A.S. and McLelland, J. (1985) *Form and Function in Birds*, Vol. 3. Academic Press.

King, A.S. and McLelland, J. (1989) *Form and Function in Birds*, Vol. 4. Academic Press.

Kolda J, Komarek V. Anatomie Domacich Ptaku. Praha: Ceskoslovenska Akademie Zemedelsych VED 1958.

Komarek V. Die männliche Kloake unserer Entenvögel. Anat Anz 1969; 124: 467–76.

Korbel R. Endoskopie bei Vögeln. In: Kraft W (Hrsg). Tierärztliche Endoskopie. Stuttgart, New York: Schattauer 1993; 177–90.

Korbel R. Augenkrankheiten bei Vögeln – Erkrankung des Auges. Luftsackperfusionsanästhesie und ophthalmologische Photographie. Habilitationsschrift München 1995.

Korbel R. Augenheilkunde bei Vögeln – Diagnostik und Therapie. Mit einer Einführung zu Augenkrankheiten bei Reptilien. Hannover: Schlütersche 2002.

Korbel, R. (2013) 2D and 3D ultrasonographical examination of the avian pecten. *Proceedings of the 34th Annual Conference and Expo – Association of Avian Veterinarians (AAV)*, Jacksonville/FL/USA: Session 10068.

Korbel, R. and Kösters, J. (1989) Epidemic deaths of wild birds after Aeromonas hydrophila infection. *Tierarztl Prax* **17**(3): 297–298.

Korbel, R. and Sigg, R. (2016) Corneal thickness in birds measured using Optical Coherence Tomography. *Proceedings of the 36th Annual Conference and Expo – Association of Avian Veterinarians (AAV)*, San Antonio/TX/USA: Session 246: Session 2105.

Korbel, R., Schulze, C., Velasco Gallego, M.L. and Rinder, M. (2013) Ocular coherence tomography (OCT) in the avian eye – clinical findings and evaluation protocols. *Proceedings of the 34th Annual Conference and Expo – Association of Avian Veterinarians (AAV)*, Jacksonville/FL/USA.

Krause R. Mikroskopische Anatomie der Wirbeltiere. Bd. II. Vögel und Reptilien. Berlin, Leipzig: de Gruyter 1921.

Lucas AM. Common integument. In: Sisson & Grossman's. The Anatomy of the Domestic Animals. 5th ed. Vol. 2. Philadelphia: Saunders 1975.

McLelland J. Avian Anatomy. London: Wolfe Publishing 1990.

McLelland J, King AS. The gross anatomy of the peritoneal coelomic cavities of Gallus domesticus. Anat Anz 1970; 127: 480–90.

McGeady, T.A., Quinn, P.J., FitzPatrick, E.S. and Ryan, M.T. (2006) *Veterinary Embryology*. Blackwell Publishing.

Nys, Y, Gautron, J, Garcia Ruiz, J.M. and Hincke, M.T. (2004) Avian eggshell mineralization: biochemical and functional characterization of matrix proteins. C. R. Palevol 3 (2004) 549–562.

Oppitz M. Die makroskopische Anatomie des Haushuhnes – Untersuchungen zur Publikationsgeschichte morphologischer Fragestellungen in der anatomischen Literatur. Diss med vet, Wien, 2006.

Peti-Peterdi, J. and Harris, R.C. (2010) Macula densa sensing and signaling mechanisms of renin release. *Journal of the American Society of Nephrology* **21**: 1093–1096.

Popescu A. Normal and pathological histology of bone development in birds. Bukarest: Publirom 1997.

Quinn TH, Baumel JJ. The digital tendon locking mechanism of the avian foot (Aves). Zoomorphology 1990; 109: 281–93.

Rajchard, J. (2009) Ultraviolet light perception by birds: a review. *Veterinaryni Medicina* **54**(8): 351–359.

Redig, P.T., Brown, P.A. and Gordon, J.J. (1996) Classification of fractures of avian long bones. *Proceedings of Association of Avian Veterinarians*, Tampa, 1996; Session # 2000.

Reese S, Dalamani G, Kaspers B. The avian lung-associated immune system. Review. Vet Res 2006; 37: 311–24.

Reiner, A., Perkel, D.J., Bruce, L.L., Butler, A.B., Csillag, A., Kuenzel, W., Medina, L., Paxinos, G., Shimizu, T., Striedter, G., Wild, M., Ball, G.F., Durand, S., Gütürkün, O., Lee, D.W., Mello, C.V., Powers, A., White, S.A., Hough, G., Kubikova, L., Smulders, T.V., Wada, K., Dugas-Ford, J., Husband, S., Yamamoto, K., Yu, J., Siang, C. and Jarvis, E.D. (2004) Revised nomenclature for avian telencephalon and some related brainstem nuclei. *Journal of Comparative Neurology* **473**, 377–414.

Rickart-Müller CR. Das Blutgefäßsystem der Niere des Haushuhnes (Gallus domesticus). Diss med vet, Gießen, 1968.

Romer A. Vergleichende Anatomie der Wirbeltiere. Hamburg, Berlin: Parey 1966.

Salomon F-V. Lehrbuch der Geflügelanatomie. Jena, Stuttgart: Gustav Fischer 1993.

Samuelson, D.A. (2007) *Textbook of Veterinary Histology*, Saunders Elsevier.

Schwarze E, Schröder L. Kompendium der Geflügelanatomie. 3. Aufl. Stuttgart, New York: Gustav Fischer 1979.

Sherwood, L., Klandorf, H. and Yancey, P.H. (2005) *Animal Physiology*, Brooks/Cole.

Skøtt, O. and Briggs, J.P. (1987) Direct demonstration of macula densa-mediated renin secretion. *Science*, **237**(4822): 1618–1620.

Stöger E. Zur Morphologie der Schilddrüse des Huhnes. Diss med vet, Wien, 1996.

Sweeney, S.J., Redig, P.T. and Tordoff, H.B. (1997) Morbidity, survival and productivity of rehabilitated peregrine falcons in the upper Midwestern US. *Journal of Raptor Research* **31**(4): 347–352.

Tadjalli, M., Ghazi, S.R. and Parto, P. (2009) Gross anatomy of the heart in Ostrich (Struthio camelus). *Iranian Journal of Veterinary Research* **10**(1): 21–27.

Tully, N.T., Dorrestein, G.M. and Jones, A.K. (2009) *Avian Medicine*, 2nd edn. Saunders Elsevier.

Vollmerhaus B. Körperhöhlen. In: Nickel R, Schummer A, Seiferle E. Lehrbuch der Anatomie der Haustiere. Bd. V.

Vollmerhaus B (Hrsg). Anatomie der Vögel. 3. Aufl. Berlin, Hamburg: Parey 2004.

Vollmerhaus B, Sinowatz F. Verdauungsapparat. In: Nickel R, Schummer A, Seiferle E. Lehrbuch der Anatomie der Haustiere. Bd. V. Vollmerhaus B (Hrsg). Anatomie der Vögel. 3. Aufl. Berlin, Hamburg: Parey 2004.

Wagner M. Die Osteologie der Hinterextremität und des Beckengürtels beim Afrikanischen Strauss (Struthio camelus Linné, 1758). Diss med vet, Wien, 2004.

Waibl H, Sinowatz F. Harn- und Geschlechtsapparat. In: Nickel R, Schummer A, Seiferle E. Lehrbuch der Anatomie der Haustiere. Bd. V. Vollmerhaus B (Hrsg). Anatomie der Vögel. 3. Aufl. Berlin, Hamburg: Parey 2004.

Walls GL. The vertebrate eye. Bulletin No. 19. Bloomfield, Hills, Michigan: Cranbook Institute of Science 1942.

Weissengruber GE, Forstenpointner G, Gangl D. Gut zu Fuß – funktionellanatomische Aspekte des bipeden Laufens beim Afrikanischen Strauß (Struthio camelus Linné, 1758). Wien Tierärztl Mschr 2003; 90: 67–78.

Wideman, J.R., Braun, E.J. and Anderson, G.L. (1981) Microanatomy of the renal cortex in the domestic fowl. *Journal of Morphology* **168**: 249–267.

Wlodaraski, K.H., Brodzikowska, A. and Kuzaka, B. (2014) *Folia Biologica* (Kraków) **62**(2).

Zengerling G. Struktur und Entwicklung des Immunsystems der Hühner-lunge. Diss med vet, München, 2006.

Index